Introduction to the Cellular and Molecular Biology of Cancer

Introduction to the Cellular and Molecular Biology of Cancer

Second Edition

Edited by

L. M. FRANKS and N. M. TEICH
Imperial Cancer Research Fund, London

Oxford   New York   Tokyo
OXFORD UNIVERSITY PRESS
1991

Oxford University Press, Walton Street, Oxford OX2 6DP
Oxford New York Toronto
Delhi Bombay Calcutta Madras Karachi
Petaling Jaya Singapore Hong Kong Tokyo
Nairobi Dar es Salaam Cape Town
Melbourne Auckland
and associated companies in
Berlin Ibadan

Oxford is a trade mark of Oxford University Press

Published in the United States
by Oxford University Press, New York

© The contributors listed on pp. xiii and xiv, 1991

All rights reserved. No part of this publication may be reproduced,
stored in a retrieval system, or transmitted, in any form or by any means,
electronic, mechanical, photocopying, recording, or otherwise, without
the prior permission of Oxford University Press

This book is sold subject to the condition that it shall not, by way
of trade or otherwise be lent, re-sold, hired out, or otherwise circulated
without the publisher's prior consent in any form of binding or cover
other than that in which it is published and without a similar condition
including this condition being imposed on the subsequent purchaser

British Library Cataloguing in Publication Data
Introduction to the cellular and molecular biology of
cancer.— 2nd ed.
1. Man. Cancers
I. Franks, L. M. (Leonard Maurice) 1921– II. Teich, N. M.
(Natalie M.)
616.994
ISBN 0-19-854732-3 (hardback)
0-19-854734-X (paperback)

Library of Congress Cataloging in Publication Data
Introduction to the cellular and molecular biology of cancer/edited
by L. M. Franks and N. M. Teich. — 2nd ed.
Includes bibliographical references.
Includes index.
1. Cancer. 2. Carcinogenesis. 3. Cancer—Molecular aspects.
I. Franks, L. M. (Leonard Maurice) II. Teich, N. M. (Natalie M.)
[DNLM: 1. Cytology. 2. Molecular Biology. 3. Neoplasms. QZ 200 I6186]
RC261.I58   1991   616.99'4071—dc20   90-7748
ISBN 0-19-854732-3 (hardback)
0-19-854734-X (paperback)

Typeset by Cotswold Typesetting Ltd, Cheltenham
Printed in Great Britain by
Bookcraft (Bath) Ltd
Midsomer Norton, Avon

Preface to the second edition

The second edition of this book—prepared sooner than we had expected—has given us an opportunity to correct some of the faults and errors pointed out by our readers and reviewers, as well as allowing us to bring the book up to date in a number of areas in which there have been rapid developments. In particular the chapters on the genetic and chromosomal changes, growth factors, immunotherapy, and epidemiology have been expanded and more information on viral and chemical carcinogenesis added to the appropriate sections. We have also clarified and added new information to most of the other chapters.

At some stage all authors and editors of introductory textbooks are faced with the awful choice of deciding what to leave out. When does completeness conflict with comprehension? Is the omission of this and that piece of information really a mortal sin or could the distinguished reviewer who pointed it out just happen to have been told about it by a passing graduate student? In the end of course we did what all editors must do and made our own choice.

We hope that this second edition will continue to be of use to its readers as an introduction to cancer studies and as a source of further information either in key references or in specialized reviews such as *Cancer Surveys*.

We should still appreciate comments and suggestions for further improvement.

London   L.M.F.
January 1990   N.M.T.

Preface to the first edition

Cancer holds a strange place in modern mythology. Although it is a common disease and it is true to say that one person in five will die of cancer, it is equally true to say that four out of five die of some other disease. Heart disease, for example, a much more common cause of death, does not seem to carry with it the gloomy overtones, not always justifiable, of a diagnosis of cancer. This seems to stem largely from the fact that we had so little knowledge of the cause of a disease which seemed to appear almost at random and proceed inexorably. At the turn of the century, when the ICRF was founded (in 1902), the clinical behaviour and pathology of the more common tumours was known but little else. Over the years clinicians, laboratory scientists and epidemiologists established a firm database. The behaviour patterns of many tumours, and in some cases even the causal agents, were known but how these agents transformed normal cells and influenced tumour cell behaviour remained a mystery.

The development of molecular biology opened up a major new approach to the molecular analysis of normal and tumour cells. We can now ask and begin to answer questions particularly about the genetic control of cell growth and behaviour, that have a bearing on our understanding not only of the family of diseases that we know as cancer but of the whole process of life itself. It is this, as much as finding a cause and cure for the disease, that gives cancer research its importance.

The initiating event which ultimately led to the publication of this book was the realization that many graduate students and research fellows who came to work in our Institute, although highly specialized in their own fields, had relatively little knowledge of cancer and there were few suitable text books to which they could be referred. Consequently, regular introductory courses were organized for new staff members at which 'experts' were asked to give a general introduction to their particular field of study. The talks were designed to give a background for the non-expert, as for example, molecular biology for the morphologist or cell biology for the protein chemist. The courses proved to be very popular. This book follows a similar pattern and has many of the same contributors—hence the fact that most are, or have been, connected with the Imperial Cancer Research Fund.

After a general introduction describing the pathology and natural history of the disease, each section gives a more detailed, but nevertheless general, survey of its particular area. We have tried to present principles rather than a mass of information, but inevitably some chapters are more detailed than others. Each chapter gives a short list of recommended reading which provides a source for seekers of further knowledge.

The topics covered have been selected with some care. Although some, particularly those concerned with treatment, may not at first glance appear to be directly related to cell and molecular biology, we feel that a knowledge of the methods used must give a wider understanding of the practical problems which may ultimately prove to be solvable by the application of modern scientific technology. On the other hand, knowledge of inherent cell behaviour (e.g. radiosensitivity, cell cycling, development of drug resistance, etc.) is important for the design of novel therapeutic approaches that rely less on empirical considerations.

Despite differences in the levels of technical details presented in some chapters, we hope that all are comprehensible. We have provided a fairly comprehensive glossary so that if some terms are not explained adequately in the text, do try the glossary. Finally, the editors would appreciate any comments, suggestions or corrections should a second edition prove desirable.

London   L.M.F.
December 1985   N.M.T.

Acknowledgements

Many people, as well as the named authors, have contributed to this book. It would be impossible to name all except to say that we are particularly grateful to colleagues and friends who read many of the chapters and made helpful comments.

Our special thanks are due to Mrs Christine Sinclair who was responsible for the final organization and processing of the typescript, without complaint and in spite of unexpected problems.

Contents

Contributors

G. E. ADAMS — MRC Radiology Unit, Chilton, Didcot, Oxon. OX11 0RD

A. BALMAIN — The Beatson Institute, Garscube Estate, Switchback Road, Bearsden, Glasgow G61 1BD

P. C. L. BEVERLEY — ICRF Human Tumour Immunology Group, University College and Middlesex School of Medicine, 91 Riding House Street, London, W1P 8BT

S. R. BLOOM — Department of Medicine, Royal Postgraduate Medical School, Hammersmith Hospital, London, W12 0HS

W. F. BODMER — Imperial Cancer Research Fund, P.O. Box 123, London, WC2A 3PX

R. COX — MRC Radiology Unit, Chilton, Didcot, Oxon. OX11 0RD

I. S. FENTIMAN — ICRF, Clinical Oncology Unit, Guy's Hospital, London, SE1 9RT

D. FORMAN — ICRF, Cancer Epidemiology Unit, Gibson Building, Radcliffe Infirmary, Oxford OX2 6HE

L. M. FRANKS — Imperial Cancer Research Fund, P.O. Box 123, London, WC2A 3PX

M. F. GREAVES — The Leukaemia Research Fund Centre, Institute of Cancer Research, Fulham Rd., London, SW3 6JB

B. E. GRIFFIN — Department of Virology, Royal Postgraduate Medical School, Hammersmith Hospital, London, W12 0HS

I. R. HART — Imperial Cancer Research Fund, P.O. Box 123, London, WC2A 3PX

W. I. P. MAINWARING — Department of Biochemistry, The University of Leeds, Leeds, LS2 9JT

J. S. MALPAS — ICRF Medical Oncology Unit, St Bartholomew's Hospital, London, EC1 7BE

M. C. Pike

Department of Preventive Medicine, USC School of Medicine, 2025 Zonal Avenue, Los Angeles, California 90033, USA

J. M. Polak

Department of Histochemistry, Royal Postgraduate Medical School, Hammersmith Hospital, London, W12 0HS

D. Sheer

Imperial Cancer Research Fund, P.O. Box 123, London, WC2A 3PX

N. M. Teich

Imperial Cancer Research Fund, P.O. Box 123, London, WC2A 3PX

P. E. Thorpe

Imperial Cancer Research Fund, P.O. Box 123, London, WC2A 3PX

M. D. Waterfield

The Ludwig Institute, Courtauld Building, 91 Riding House Street, London, W1P 8BT

E. Wawrzynczak

Institute of Cancer Research, Haddow Laboratories, 15 Cotswold Road, Sutton, Surrey SM2 5NG

R. A. Weiss

Institute of Cancer Research, Fulham Rd., London, SW3 6JB

C. Wigley

Dept. of Anatomy, U.M.D.S. (Guy's Campus), London Bridge, London SE1 9RT

J. A. Wyke

The Beatson Institute, Garscube Estate, Switchback Road, Bearsden, Glasgow G61 1BD

1

What is cancer?
L. M. FRANKS

1.1 Introduction

Cancer has been known since human societies first learnt to record their
activities. It was well known to the ancient Egyptians and to succeeding

civilizations but, as we shall see, most cancers develop late in life so that until the expectation of life began to increase from the middle of the 19th century onwards, the number of people surviving into the 'cancer age' was relatively small. Now that the common diseases of childhood and infectious diseases, the major causes of death in the past, have been controlled by improvements in public health and medical care, the proportion of older people at risk has increased dramatically. Although diseases of the heart and blood vessels are still the main cause of death in our ageing population, cancer is a major problem. At least one in five will develop cancer and for this reason alone its control, or even better, prevention, are important. But cancer research has an even wider significance. Cancer is not confined to man and the higher mammals but may affect almost all multicellular organisms, plant as well as animal. Since it involves disturbances in cell growth and development, a knowledge of the processes underlying the disease will help us to understand the basic mechanisms concerned with life itself.

About 140 years ago a German microscopist, Johannes Mueller showed that cancers were made up of cells, a discovery which began the search for changes which would help to pinpoint the specific differences between normal and cancer cells. Although we know a great deal about the structure and behaviour of tumour cells, the main questions remain unanswered. The rapid advances in biological technology, particularly in cell and molecular biology, now allow us to try to answer questions which even 10 years ago could not be approached. But even the most advanced technology is of no value if it is not applied to the appropriate area. The cancer biologist must ask the right questions and to do this he must be aware of the biological background of the disease process he is studying. In this book we try to provide a brief background to the epidemiology and clinical aspects of the group of diseases we describe as 'cancer', and try to interpret changes in structure and behaviour of normal and tumour cells at the biological and molecular levels, against this background. We also try to indicate the areas in which new and exciting discoveries are being made. This introductory chapter provides a brief account of the general biology, cellular pathology and aetiology of cancer and some general definitions. The succeeding chapters deal with specific aspects in more detail.

Cancer is a disorder of cells and, although it usually appears as a tumour (a swelling) made up of a mass of cells, the visible tumour is the end result of a whole series of changes which may have taken many years to develop. In this Chapter I discuss what is known about the changes which take place during the process of tumour development and consider tumour diagnosis and nomenclature. To understand this, we need to know a little about the structure of normal cells and tissues and the mechanisms which control their growth.

1.2 Normal cells and tissues

The tissues of the body can be divided into four main groups; the general supporting tissues collectively known as mesenchyme; the tissue specific cells—epithelium; the 'defence' cells—the reticuloendothelial system; and the nervous system. The mesenchyme consists of connective tissue, fibroblasts which make collagen fibres and associated proteins, bone, cartilage, muscle, blood vessels, and lymphatics. The epithelial cells are the specific cells of the different organs, e.g. skin, intestine, liver, glands, etc. The reticuloendothelial system consists of a wide group of cells, mostly derived from precursor cells in the bone marrow which give rise to all the red and white blood cells; in addition some of the cells (lymphocytes and macrophages) are distributed throughout the body either as free cells or as fixed constituents of other organs, e.g. in the liver, or as separate organs such as the spleen and lymph nodes. Lymph nodes are specialized nodules of lymphoid cells which are distributed throughout the body and act as filters for cells, bacteria, and other foreign matter. The nervous system is made up of the central nervous system (the brain and spinal cord and their coverings), and the peripheral nervous system of nerves leading from these central structures. Thus each tissue has its own specific cells, usually several different types, which maintain the structure and function of the individual tissue. Bone, for example, has one group of cells responsible for bone formation and a second group responsible for bone resorption when the need arises, as in the repair of fractures. The intestinal tract has many different epithelial cell types responsible for the different functions of the bowel, and so on.

The specific cells are grouped in organs which have a standard pattern (Fig. 1.1). There is a layer of epithelium, the tissue specific cells, separated from the supporting mesenchyme by a basement membrane. The supporting tissues (or stroma) are made up of connective tissue (collagen fibres) and fibroblasts (which make collagen), which may be supported on a layer of muscle and/or bone depending on the organ. Blood vessels, lymphatic vessels and nerves pass through the connective tissue and provide nutrients and nervous control among other things for the specific tissue cells. In some instances, e.g. the skin and intestinal tract, the epithelium which may be one or more cells thick depending on the tissue, covers surfaces. In others it may form a system of tubes (e.g. in the lung or kidney), or solid cords (e.g. liver), but the basic pattern remains the same. Different organs differ in structure only in the nature of the specific cells and the arrangement and distribution of the supporting mesenchyme.

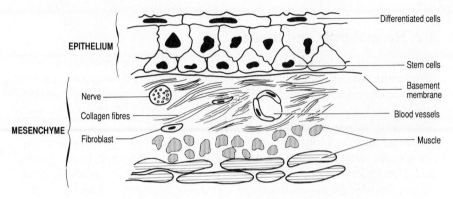

Fig. 1.1 A typical tissue showing epithelial and mesenchymal components.

1.3 Control of growth in normal cells

The mechanism of growth control is one of the most important and least understood areas in biology. In normal development and growth, there is a very precise mechanism which allows individual organs to reach a specific size which, for all practical purposes, is never exceeded. If a tissue is injured, the surviving cells in most organs begin to grow and replace the damaged cells. When this has been completed, the process stops, i.e. the normal growth controlling mechanism persists throughout life. Although most cells in the embryo can proliferate (increase in number), not all adult cells do retain this ability. In most organs there are special reserve or stem cells which are capable of growing in response to a stimulus, e.g. an injury, and developing into the organ specific cells. The more highly differentiated (developed) a cell is, e.g. muscle or nerve, the more likely it is to have lost its capacity to grow. In some organs, particularly the brain, the most highly differentiated cells, the nerve cells, can only proliferate in the embryo, although the special supporting cells in the brain continue to be able to grow. A consequence of this, as we shall see later, is that tumour, of nerve cells are only found in the very young and tumours of the brain in adults are almost invariably derived from the supporting cells.

In other tissues there is a rapid turnover of cells, particularly in the small intestine and the blood and immune system cells. A great deal of work has been done on the control of stem cell growth in the red and white cells (haemopoietic system), and the relationship of the factors involved to tumour development (see Chapter 12). Strangely enough, for reasons still unknown, rapid cell growth itself is not necessarily associated with an increased risk of tumour development; e.g. tumours of the small intestine are very rare.

From what little is known about growth control, it would appear that there are stimulatory and inhibitory factors which are normally in balance until a growth stimulus is required, either for repair or because extra work is required from a particular organ. These objectives may be achieved by hypertrophy, i.e. an increase in size of individual components, usually of structures which do not normally divide. An example is the increase in size of particular muscles in athletes. The alternative is the increase in number of the cells of the organ involved—hyperplasia. This may be in response to a physiological stimulus, e.g. some hormones, or in repair. These processes are all subject to normal growth control When the stimulus is removed the situation returns to the status quo.

We now know that there is a close relationship between growth factor production and tumour growth and that their production may be controlled by some genes (protooncogenes) which are genetically similar to some tumour-producing viruses. A number of different growth factors have now been identified and it is clear that they are widely distributed, that many do not act alone, and have many different functions. For example, one such factor TGF-β under different conditions may act as a growth stimulator, a growth inhibitor, or a regulator of gene activity in many different cell types. These findings have opened up whole new areas in cancer research. They are discussed in detail in Chapters 8–11.

1.4 The cell cycle

The method by which cells increase in number is similar in all somatic cells, and involves the growth of all cell components leading eventually to division of the cell into two new cells. Although the structural changes which take place have been known for many years, our knowledge of the molecular basis to the process is far from complete. Four stages are usually recognized: $G_1 \rightarrow S \rightarrow G_2 \rightarrow M$; G_1 is a gap or pause after stimulation where little seems to be happening although there is much biochemical activity; S is the phase of synthesis, particularly of DNA, to double the normal amount, although other components also increase; G_2 is a second gap period; and M is the stage of mitosis in which the nucleus breaks down to form chromosomes which in the normal cell separate into two identical groups, the nuclear membranes reform about each group, and the whole cell then divides into two identical cells. Cells may be blocked at particular stages in the cycle by drugs (see Chapter 17) or by physiological agents, or they may move out of the division cycle into a resting phase known as G_0. Other more complicated patterns of the cell cycle have been described but the one given here is sufficient for the purposes of this book.

One of the more interesting recent findings in this field was the discovery that there are two major control points in the cell cycle: one towards the end of G_1, known as the restriction point, and the other at the initiation of mitosis. G_2 cells deprived of growth factors before the restriction point will leave the cycle and enter G_0.

From work on yeast a gene, *cdc* 2/28, concerned with the production of a specific protein that is involved with both the passage through the restriction point and the initiation of mitosis, has been identified. The protein—a threonine or serine kinase (an enzyme that adds phosphate groups to amino acid residues)—does not act directly but is thought to activate other key proteins that in G_1 initiate S and in G_2 initiate mitosis. Some human cells have now been found to contain a similar gene. This work illustrates the complexity of the process of growth control but also shows how studies on simple organisms may help to explain changes in more complex individuals.

1.5 Tumour growth or neoplasia

It is not possible to define a tumour cell in absolute terms. Tumours are usually recognized by the fact that the cells have shown abnormal growth, so that a reasonably acceptable definition is that tumour cells differ from normal cells in that they are no longer responsive to normal growth controlling mechanisms. Since there are almost certainly many different factors involved, the altered cells may still respond to some but not to others. A further complication is that some tumour cells, especially soon after the cells have been transformed from the normal, may not be growing at all. In the present state of knowledge any definition must be 'operational'.

Given these qualifications we can classify tumours into three main groups

1. *Benign tumours* may arise in any tissue, grow locally, and cause damage by local pressure or obstruction. However, the common feature is that they do not spread to distant sites.

2. *In-situ tumours* usually develop in epithelium and are usually but not invariably, small. The cells have the morphological appearances of cancer cells (see p. 16) but remain in the epithelial layer. They do not invade the basement membrane and supporting mesenchyme. Some authorities recognize a stage of dysplasia-epithelial irregularity but not absolutely identifiable as cancer *in situ* which may sometimes precede

cancer *in situ*. Theoretically, cancers *in situ* may arise also in mesen-chymal, reticuloendothelial or nervous tissue but they have not been recognized.

3. *Cancers* are fully developed (malignant) tumours with a specific capacity to invade and destroy the underlying mesenchyme—local invasion. The tumour cells need nutrients that are provided in normal tissues through the blood stream. Some tumour cells produce a sub-stance, tumour angiogenesis factor (TAF), that stimulates the growth of blood vessels into the tumour, thus allowing continuous growth to occur. The new vessels are not very well formed and are easily damaged so that the invading tumour cells may penetrate these and lymphatic vessels. Tumour fragments may be carried in these vessels to local lymph nodes or to distant organs where they may produce secondary tumours (metastases). Cancers may arise in any tissue. Although there may be a progression from benign to malignant, this is far from invariable. Many benign tumours never become malignant.

Some of these problems of definition may be more easily understood if we consider the whole process of tumour induction and development (carcinogenesis)—see Chapters 5–11 for a further discussion.

1.6 The process of carcinogenesis

Carcinogenesis is a multistage process (Fig. 1.2). The application of a cancer producing agent (carcinogen) does not lead to the immediate production of a tumour. There are a series of changes after the initiation step induced by the carcinogen. The subsequent stages—tumour pro-motion—may be produced by the carcinogen or by other substances (promoting agents) which do not themselves produce tumours. Initiation, which is the primary and essential step in the process, is very rapid but once the initial change has taken place the initiated cells may persist for a considerable time, perhaps the lifespan of the individual. The most likely site for the primary event is in the genetic material (DNA), although there are other possibilities. The carcinogen is thought to damage or destroy specific genes probably in the stem cell population of the tissue involved (see Cairns 1975 for review).

Initiated cells remain latent until acted upon by promoting agents. Many of these 'transformed' cells may not grow at all or grow very slowly. It is at this stage that the influence of growth appears. Promoting agents are not carcinogenic in themselves but they do induce initiated cells to divide. Many agents will induce cell division, but only promoters will induce tumour development, so that although cell growth is necessary for tumour development there must also be other factors involved. The

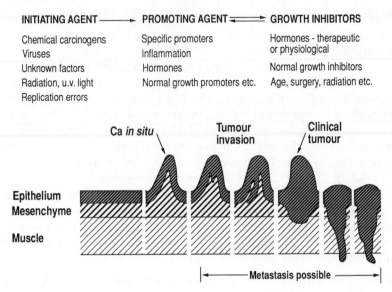

Fig. 1.2 Factors influencing tumour development showing progression from normal to invasive tumour.

suggestion is that promoting agents may interfere with the process of differentiation which normally takes place when cells move from the dividing stem cell population into functioning and usually non-dividing cells. Even though these growth promoting stimuli are acting on the cells, they may still be sensitive to the normal growth inhibiting factors in the body so that the final outcome depends on the balance between the factors and the extent of the changes in the initiated cells. This explains why preneoplastic or even apparently fully transformed tumours can be found but do not appear to be growing, and sometimes even regress.

The whole sequence of events in the process of tumour formation is almost certainly a consequence of gene changes although gene expression may be influenced by the host. We are now beginning to understand some of these changes although there are still many problems unsolved. The discovery that oncogenes of tumour producing viruses are related to genes (protooncogenes) (see Chapter 9) in normal, as well as some tumour, cells has led to intensive research into the relationship of these genes to normal and tumour growth and development. These genes have been localized to specific chromosomes and some to sites of chromosome abnormalities (see Chapter 10) in tumours. Much speculation now centres on the question whether the initiation, progression, and maintenance of some tumours depends on over-expression through gene amplification (an increase in the number of copies of a particular gene), or inappropriate expression (i.e. the wrong time) of normal genes or whether mutations in a critical region of a gene are necessary. A possible hypothesis is that a mutation may be necessary for the initiation event but that some or all of the later stages may depend on over- or inappropriate expression. These possibilities are discussed in the sections on carcinogenesis (Chapters 5–11).

One theory, for which there is now increasing evidence, was proposed by Knudson (described in more detail in Chapters 4 and 10). He suggested that at least two independent mutations are needed before tumours can develop. In cases of inherited (familial) cases of tumour predisposition, the first mutation is present in the germ cells (sperm or ovum) and is therefore inherited by every cell. Only one further mutation is required in these cases. In the more common non-familial cases two mutations (which may include gene deletions) in the same cell are required and the chances of this happening are consequently much smaller. It now seems certain that the changes must occur at the same site in each of the pair of homologous chromosomes and in some cases the exact chromosome has been identified (see Chapter 10). An attractive hypothesis for some tumours is that the deleted or altered genes normally produce a product that suppresses the expression of transforming growth factors (see Chapter 11) by another pair of genes. The term anti-oncogene has been used to describe DNA sequences (genes) that act as dominant suppressors of malignancy and the identification of such genes and their relationship to the genes identified by Knudson and others in familial tumours is a field in which there is now much activity (see Chapters 4, 9, and 10).

Another major and unexplained area is concerned with the time scale of carcinogenesis. The latent period between initiation and the appearance of tumours is one of the least understood aspects of tumour development. In man after exposure to industrial carcinogens, it may take over 20 years before tumours develop. Even in animals given massive doses of carcinogens, it may take up to a quarter or more of the total lifespan before tumours appear. Yet another unexplained fact is that only a very small number of cells 'initiated' by a carcinogen will eventually produce tumours—perhaps only one or two from many millions of treated cells.

1.7 Factors influencing the development of cancers

Many different factors are involved in the development of tumours. A cancer producing agent or carcinogen, and presumably promoting agents, must be present. Carcinogens may be chemical or physical, e.g. radiation or ultraviolet light which causes skin cancer in Caucasians exposed to tropical sunlight but rarely in coloured races. Identified chemical carcinogens include hydrocarbon carcinogens present in coal tar and a series of chemicals used in the rubber industry. Several specific promoting agents have now been identified and work on the mechanism of action of the different types is in progress (see Chapter 6). Animal experiments suggest that viruses may be associated with the initiation of some cancers—mainly the leukaemic group; viruses also seem to be associated with some types of human cancer. The role of oncogenes has already been mentioned. It is also becoming obvious that there is often an interaction between chemical carcinogens and viruses in tumour induction. A good example of this is seen in the association between hepatitis B virus and environmental chemicals in the development of liver cancer (see Chapters 6 and 8) and there is suggestive evidence in other tumours, particularly in cancer of the cervix (see Knox and Woodman 1988 for review). In other cases such as with cigarette smoking, no single agent has been isolated but cigarette smoke is a very complex mixture of chemicals many of which may contribute to the carcinogenic effect of smoking. We know that cigarette smoking leads to the development of lung cancer and that the more an individual smokes the greater his (or her) chance of developing lung cancer but all cigarette smokers do not develop lung cancer. There is considerable individual variation in response. We know from animal experiments and epidemiological studies (see Chapters 3, 4, and 6) that there is a genetic (DNA associated) basis for this. Analysis of these changes are now being done at a cellular and molecular level (see Chapters 9 and 10). Some genetically homogeneous inbred strains of mice are particularly susceptible to tumour induction by particular viruses or chemicals; some species are more susceptible than others to particular chemicals. In human populations some families and some races are more prone to develop certain cancers. This may be due to genetic or environmental factors. The chance of a particular individual developing cancer depends on the balance between the various factors concerned. For example, exposure to a massive dose of a carcinogen may override an inherent genetic resistance, or genetic susceptibility may be so high that the development of specific tumours is invariable. With some tumours, particularly lung cancer and some industrial cancers, exposure to the carcinogen alone is sufficient to almost override other factors, but for the so-called spontaneous tumours, i.e. those which develop without a so far recognizable cause, we have little idea of the relative importance of the various factors. Some of these factors are considered in more detail in Chapters 3–6.

Another factor which influences the type of cancer which develops is age. One of the few definite facts we have about cancer is that there is an age associated, organ specific tumour incidence. Most cancers in man and experimental animals can be divided into three main groups depending on their age incidence

1. Embryonic, e.g. neuroblastoma (tumours of embryonic nerve cells), embryonal tumours of kidney (Wilm's tumours), retinoblastomas etc.
2. Those predominantly in the young, e.g. some leukaemias, tumours of the bone, testis, etc.
3. Those with an increasing incidence with age, e.g. tumours of prostate, colon, bladder, skin, salivary gland, mammary gland etc.

The incidence of human tumours is considered in more detail in Chapter 3. There are at least three possible explanations for this last group of age associated tumours which includes the most frequently occurring human tumours.

1. There is a continuous exposure throughout life to low levels of a cumulative carcinogen.
2. With age there are humoral changes induced, i.e. in the cellular environment, by alterations in the immune or hormonal systems which allow or encourage neoplastic change to take place.
3. There are age associated changes in some cells which increase their susceptibility to neoplastic transformation.

We still do not know which of these explanations is correct or whether more than one process is involved.

1.8 The natural history of cancer or tumour progression

A series of changes takes place after a tissue cell is 'initiated' but the rate at which this occurs depends on changes in the cell and on changes in the host. Most chemical and physical cancer inducing agents are very highly reactive and when they react with DNA in the affected cell they usually damage many other sites as well as the relatively few which are thought to control neoplastic transformation. Thus the same agent may produce tumours in a given organ which differ greatly from each other, depending on the specific genes which have been altered or lost. At one extreme, if only the 'transforming' sites have been altered, the resulting tumour cells will still retain much of the normal differentiated structure and function of the cell from which they have arisen. In the skin, for example, it will still resemble a skin cell (Fig. 1.3) and may still produce normal skin products and be responsive to normal growth controlling factors. If the genes responsible for normal structure are more severely damaged, the resulting tumour cells have fewer normal properties. At the other extreme, the cells may have lost almost all the normal properties of the cell from which they have arisen. The loss of normal characteristics is known as dedifferentiation or anaplasia. The pathologist can grade tumours by making an approximate assessment of the degree of structural dedifferentiation by examining sections of tumours under the microscope. As a rule, there is an approximate correlation between the tumour grade and growth rate. The most differentiated tumours (low grade, i.e. Grade I) tend to be more slow growing, and the most anaplastic (high Grade III or IV) the more rapidly growing. Unfortunately this relationship is not absolute but it does give a useful guide to tumour behaviour. Human breast cancers have been graded in this way and it has been shown that about 80 per cent of patients with well differentiated Grade I breast cancers will be alive and well at five years (and much longer), but only 20 per cent of patients with Grade IV tumours will survive for this time. It is of course equally obvious from these figures that although 80 per cent of patients with Grade I cancers survive, 20 per cent with the same structural type of tumour do not; hence tumour growth is influenced by factors other than tumour structure, particularly the reaction by the patient's own defence mechanisms. Unfortunately we have few ideas about the nature of these mechanisms. In hormone responsive tissues such as the breast, the tumour cells may still retain some of the normal responsiveness to hormones (see Chapters 13 and 14). The pathologist's assessment of tumour grade is based only on alterations in structure and these are not invariably related to changes in function. Some cells may have lost their specific structural characters but still retain differentiated biochemical characters, and others may still appear structurally differentiated but have lost many normal functional attributes.

Another practical problem in the assessment of tumours is that tumours are not homogeneous (see Chapter 2 for fuller discussion) and some may contain areas with more than one tumour grade. Note that in the tumour shown in Figs 1.3 and 1.4, there is a progression from benign to malignant resembling that illustrated in Fig. 1.2, but the progression is in space rather than necessarily in time. It used to be thought that tumours arose from a single altered cell, i.e. were clonal in origin, but there is now some evidence to suggest that this may not be invariably true. But even if it were true, there is no doubt that by the time a tumour is detectable clinically, whether it has arisen from one or many cells, it has been present for a long time and the cells have had to go through a large number of cell divisions so that variation and selection of different cell populations have occurred. A tumour about 0.5 cm in diameter, which is just detectable, may contain over 500 million cells. The developed tumour usually consists of a mixed population of cells, which may differ in structure, function, growth potential, resistance to drugs or X-rays, and ability to invade and metastasize. Many of these characters may not

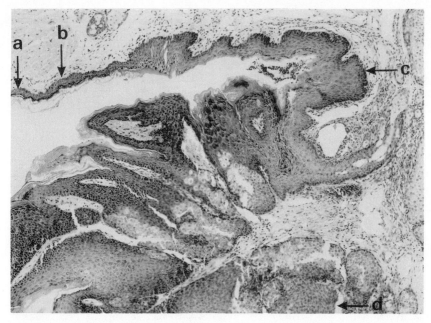

Fig. 1.3 Section (stained with haemotoxylin and eosin) of the edge of a squamous carcinoma of skin, with normal skin (a) on the left and increasing dysplasia (b) and (c) leading into the main mass of the tumour (d) below. (× 50)

be stable and may be influenced by the host response or by treatment. An obvious example is the destruction of X-ray sensitive cells by X-ray treatment. If the tumour also contains X-ray resistant cells, the cancer cells which are left after treatment will be X-ray resistant. Any individual character may vary independently.

Tumour progression is the development by a tumour of changes in one or more characters in its constituent cells. Although progression is usually towards greater malignancy, this is not invariably so. There are a number of cases—unfortunately small—in which rapidly growing tumours have ceased to grow or even disappeared completely. Although we do not yet have any explanation for this, it does show that there are natural mechanisms still to be discovered which will eventually allow us to control tumour growth.

1.9 The diagnosis of tumours

There are no absolute methods for diagnosing and assessing the degree of malignancy of tumours. Although many laboratories are trying to

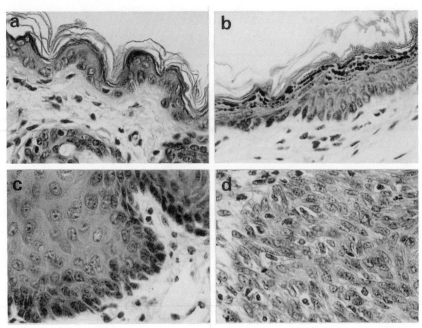

Fig. 1.4 Shows the areas marked in Fig. 1.3 but at a higher magnification. (a) Normal skin (compare with Fig. 1.1) showing mesenchyme below, covered by normal epithelium with basal cells and more differentiated superficial cells, covered by layers of keratin formed from the superficial cells. (× 360) (b) Dysplastic skin. There is an increase in the basal cells, which are more irregular than the normal, and there is a disturbance in the formation of keratin, which is clumped into an irregular dark mass in the surface layer instead of the normal flattened sheets seen in Fig. 1.4a, i.e. differentiation is disturbed. (× 360) (c) Cell overgrowth. The cells themselves are abnormal: they vary in shape and size, the nuclei are much larger than normal, and some are more deeply stained. The usually distinct separation between epithelium and stroma is not seen, suggesting that invasion may be taking place. The cells are still recognizable as skin cells. This would be diagnosed as a moderately well differentiated squamous carcinoma. (× 360) (d) The centre of tumour is made up of a mass of irregular spindle shaped cells with no recognizable skin features. This would be diagnosed as an anaplastic (undifferentiated) carcinoma. (× 360)

establish methods for doing this, none are entirely satisfactory. Microscopic examination of tissue is still the most reliable method for routine use. The function of the pathologist is to decide whether the structure of the cells in the tissue is sufficiently removed from the normal to allow a diagnosis of neoplasia to be made and, if so, whether the tumour is likely to be benign or malignant, its probable cell of origin, its degree of malignancy, and its extent of spread. For practical purposes the two techniques used are tumour grading and tumour staging. Tumour grading attempts to measure the degree of dedifferentiation in tumours and is based on histological and cytological criteria (Figs 1.3 and 1.4). Histological differentiation is concerned with alterations in the structure of the tissue, i.e. the relationship of cells to each other and to their underlying stroma. Cytological grading is based on the application of similar criteria to the structure of the specific tumour cells. Tumour staging assesses the extent of spread of tumours. Many investigators in the laboratory and the clinic have tried to find absolute markers of malignancy but, as I have already indicated, carcinogenesis is a multistage process. We should like to have markers for each stage of the process but, unfortunately, apart from histology, which as we have seen is not entirely satisfactory, we do not have such markers. This is a major deficiency in studying cancer. Many workers are now trying to identify tumour specific or tumour associated proteins, either by direct measurement or by developing specific antibodies to these proteins. This seems to be a promising approach not only in diagnosis (see Chapters 12 and 15), but also as a vehicle for carrying drugs or other agents which destroy cancer cells to their specific targets (see Chapter 18).

Thus at present there are few true diagnostic tests for malignancy except for genetic and chromosome associated changes discussed in Chapters 4 and 10. The production of abnormal hormones or normal hormones in abnormal amounts may in some instances be a useful guide to the presence of tumours (see Chapters 13 and 16) but are not true markers of malignancy. Experimental methods used for tumour identification are discussed on p. 27.

For the moment the most commonly used method of diagnosis depends on histology. Many millions of words have been written on tumour diagnosis and the World Health Organization has so far published 21 volumes on the structure and classification of tumours by an international panel of tumour pathologists. The following brief survey will only give a guide. I have chosen some of the examples, not because they are common, but because they illustrate some points more clearly than the more common tumours.

1.9.1 *Benign tumours*

Benign tumours usually resemble their tissue of origin but every tissue component need not be involved and the cells may or may not be in their normal relationship. Benign tumours arise in most tissues, increase in size but do not invade. They are usually separated from the surrounding normal tissue by a capsule of connective tissue. Cytologically the specific tumour cells do not differ substantially from the structure of the normal organ cells. Benign tumours of bone or cartilage may produce nodules of bone or cartilage indistinguishable from normal tissues. In epithelial tissues, groups of cells may also form local benign tumours made up of all tissue components. The covering or lining tissues of skin, intestinal tract, bladder, etc. may produce wart-like outgrowths containing all the tissue components, but closely packed to form a solid nodule. The common wart is a local outgrowth of all skin components. In other situations only one constituent cell may give rise to a benign tumour. The pituitary, for example, is a small gland at the base of the brain which produces many different hormones, each produced by a different type of cell, arranged in solid cords. Benign tumours of one cell type may develop and these tumours may then produce an excess of the particular hormone normally produced by that cell. Other benign tumours of the pituitary may contain more than one cell type, or produce more than one hormone, and some may be derived from non-hormone producing cells. The cells in all these tumours are arranged in solid cords as in the normal gland. The benign tumours do not invade the surrounding tissue but, if they increase in size, they may press on and damage the remaining normal cells or overlying nervous tissue or press on the optic nerves which pass nearby and lead to blindness. So that although the tumours themselves are benign they may cause serious disturbances by local pressure, or they may continue to produce excessive amounts of their normal product, which may in itself cause severe symptoms. Benign tumours of any other hormone producing gland such as the thyroid, adrenal, etc. may have similar effects. Alternatively benign tumours may damage the remaining normal cells and cause a loss of normal function.

Benign epithelial tumours arise in many other organs. There is a different pattern of tumour growth in organs with a tubular structure. Both the kidney and breast, for example, are made up of tubular structures with the epithelial tubules lined by several different epithelial cell types and surrounded by connective tissue. Benign tumours in these organs are made up of tubules usually with one, or less commonly two, different epithelial cell types, together with a variable amount of con-nective tissue.

1.9.2 *Malignant tumours*

Malignant tumours show two characteristic features—cellular abnor-malities (sometimes slight) and invasion of surrounding tissues. When both are present diagnosis is easy. The standard cellular criteria include a local increase in cell number, loss of the normal regular arrangement of cells, variation in cell shape and size, increase in nuclear size and density of staining (both of which reflect an increase in total DNA), an increase in mitotic activity (increased cell division), and the presence of abnormal mitoses and chromosomes (see Chapter 10). The diagnosis of carcinomas (malignant epithelial tumours) *in situ* depends on the recognition of these cellular changes in an area of epithelium, usually on a surface, the cervix of the uterus or skin, but it may occur in the bladder or other organs. The changes only involve the epithelium and there is no invasion of underlying tissues, i.e. the neoplastic cells remain where they began— *in situ*. The only definite evidence of malignancy is invasion of underlying tissues. In most cases this is easily recognized as the tumour cells destroy and replace the normal tissues. Sometimes tumour cells may be found invading blood or lymphatic vessels; they may then be carried to other parts of the body in blood or lymph and develop into secondary tumours (metastases) in these distant sites. This type of spread is characteristic of malignant tumours and is the major problem in treatment since a tumour that remains localized to its site of origin can usually be removed surgically or destroyed by radiation. The problem of metastasis is discussed in Chapter 2. Malignant tumours have no well-defined capsule and the tumour cells grow in a much more disorganized form than is found in benign tumours. The same criteria apply to all malignant tumours, whatever their tissue of origin.

1.10 The names of tumours (nomenclature) and the need for tissue diagnosis

Although the precise naming of tumours may seem to be an academic exercise, it is of great practical importance in deciding on treatment of each individual patient (see Chapters 16 and 17). Obviously it is important for each pathologist and surgeon to use the same name for the same type of tumour. Even after many years of effort by international organizations, there is still some confusion about names although fortunately more or less agreed versions are now coming into general use. But a more important point is that a knowledge of the type of tumour cell and the extent of spread are essential in planning treatment. Some tumours are known to be sensitive to drugs, hormones, or X-rays but others are resistant. Knowing the extent of spread will help to define the area for treatment by radiation or surgery or even whether surgery is possible. For these reasons, the surgeon will usually remove a piece, or the whole tumour if it is readily accessible, for examination by a pathologist. The tissue removed (a biopsy) is preserved by a chemical fixative and thin sections are prepared for examination under an optical or electron microscope (Chapter 14).

Although the names given to tumours seem to be confusing, there is a simple logical basis to tumour nomenclature. The terms tumour, growth, or neoplasm can be used to describe a malignant tumour. Tumours are

Table 1.1 Nomenclature of common tumours

Tissue	Basic cell type	Benign tumour	Malignant tumour
Skin	Squamous epithelium Basal cell Pigment cell	Papilloma Melanoma (naevus)	Squamous carcinoma Basal cell carcinoma[1] Malignant melanoma
Alimentary tract Lips, mouth, tongue, oesophagus	Squamous epithelium	Papilloma	Squamous carcinoma
Stomach Small bowel (rare) Large bowel	Columnar epithelium	Papillary adenoma	Carcinoma
Nasopharynx, larynx, lungs[2]	Bronchial (respiratory) epithelium	Adenoma (rare)	Carcinoma
Urinary system Bladder	Urothelium (transitional epithelium)	Papilloma	Carcinoma
Solid epithelial organs Liver, kidney, prostate, thyroid, pancreas, pituitary, etc.	Specific epithelium	Adenoma	Carcinoma
Gonads Ovary	Surface epithelium Germ cells	Serous cystadenoma Mucinous cystadenoma Teratoma	Serous cystadenocarcinoma Mucinous cystadenocarcinoma Teratocarcinoma Choriocarcinoma
Testis	Germ cells	Teratoma	Seminoma Embryonal carcinoma Choriocarcinoma Malignant teratoma (rare)

Tissue	Cell type	Benign	Malignant
Mesenchyme			
Fibrous tissue	Fibrocytes	Fibroma	Fibrosarcoma
Fat	Adipocytes	Lipoma	Liposarcoma
Bone	Osteocytes	Osteoma	Osteosarcoma
Cartilage	Chondrocytes	Chondroma	Chondrosarcoma
Smooth muscle[3]	Smooth muscle cells	Leiomyoma	Leiomyosarcoma
Striated muscle[4]	Muscle cells	Rhabdomyoma	Rhabdomyosarcoma
Blood vessels	Endothelium	Haemangioma	Haemangiosarcoma
Lymph vessels	Endothelium	Lymphangioma	Lymphangiosarcoma
Nervous system			
Nerve cells[5]			Neuroblastoma[5]
			Retinoblastoma
Supporting cells	Astrocytes		Astrocytoma[6]
	Oligodendrocytes		Oligodendrocytoma[6]
Covering cells (Central nervous system)	Meningeal cells	Meningioma	
Covering cells (Peripheral nervous system)	Perineurium	Neurofibroma	Neurofibrosarcoma
	Endoneurium		
Reticuloendothelial system[7]			
White blood cells[8]	Myeloid cells		Myeloid leukaemia
	Monocytes		Monocytic leukaemia
	Granulocytes		Granulocytic leukaemia
	Lymphocytes		Lymphatic leukaemia
Red blood cells	Erythrocytes		Erythroleukaemia
Lymph nodes	Lymphocytes		Non-Hodgkin's lymphoma
	Fixed reticuloendothelial cells		Hodgkin's disease
Embryonic type tissues	Mixed tissues	Teratoma	Teratocarcinoma

1 Invades locally; does not metastasize.
2 Lung tumours usually arise from lining epithelium of bronchi.
3 Muscle of intestine, bladder, blood vessels, etc.
4 Muscles under voluntary control, e.g. limb muscles; tumours very rare.
5 Nerve cell tumours in the very young only.
6 No absolute distinction between benign and malignant tumours possible; do not metastasize.
7 See p. 21.
8 See Chapter 12.

described by a generic name which specifies the general tissue of origin, i.e. mesenchyme, epithelium, or reticuloendothelial, and whether the tumour is benign or malignant. This generic name is qualified by the specific tissue of origin, e.g. kidney, breast, and this too may be qualified by further terms describing the cell of origin (if identifiable) and the pattern of growth. Some examples will make this clearer; a list is given in Table 1.1.

1.10.1 *Tumours of epithelium*

Benign tumours. Benign tumours of epithelium are usually described by their growth pattern and their tissue of origin. Benign tumours of skin may be papillary (a warty outgrowth or papilloma) or solid. A benign skin tumour derived from squamous (flattened) epithelium could be described as a squamous cell papilloma of skin. Benign tumours of glandular tissues are called adenomas and may be solid or papillary, e.g. solid or papillary adenoma of thyroid.

Malignant tumours. The generic name for malignant tumours of epithelium is carcinoma, e.g. carcinoma of skin. The common skin carcinomas may arise from the differentiated squamous cells or from the less differentiated basal cells, so that skin carcinomas may be described as squamous cell carcinomas or basal cell carcinomas. They may grow as flat (sessile) plaques or as warty (papillary) outgrowths. So that a tumour may for example be described as a papillary squamous cell carcinoma of skin. Its grade and the extent of invasion may also be given. The final pathologist's report may read 'moderately well differentiated (Grade II) squamous carcinoma of skin. The structure is mainly papillary but there is invasion of the underlying connective tissue; muscle is not involved'. This report tells the oncologist that the tumour is made up of squamous cells which are known to be sensitive to X-irradiation and that the extent of spread is limited, i.e. that it could easily be removed by local surgery. The final decision on treatment would then depend on the exact position of the tumour and, among other factors, whether surgery or irradiation would be easier or leave less scarring.

Malignant tumours of glandular tissues are also carcinomas but are sometimes described as adenocarcinomas, e.g. adenocarcinoma of breast, implying that the tumour has a glandular structure. As with the skin tumours, the cell type can be described (e.g. columnar cell or cuboidal) and if the cell of origin is known, this too can be added (e.g. ductal cuboidal cell adenocarcinoma of breast). The gross pattern of growth (sessile or papillary), and extent of spread can also be defined. Adenocarcinomas have a wider range of cellular patterns than tumours of covering epithelium. The cells may be arranged as large or small tubules or solid cords (trabeculae) or masses and this pattern will also be described. In some cases the tumour grade can be assessed.

Most tumours still retain some of the structural features of the cells from which they have arisen and, as we have seen, this allows the pathologist to make a rough assessment of the degree of malignancy by the extent to which the tumours have departed from the normal (grading); it also may allow the source of a secondary tumour to be established. But there are still problems. Some tumours may be so dedifferentiated that they no longer retain any structure which indicates the tissue of origin. In others some cells may develop in an abnormal way. A common event is that tumour cells from a glandular organ such as the breast, which are normally columnar in structure, may develop into squamous cells resembling those in skin tumours. This process is known as metaplasia and although confusing to the pathologist, does not as a rule influence the degree of malignancy. A final point is that in many tumours the structure is not homogeneous and more than one cell type, growth pattern, or grade of tumour may be present. All these features will be indicated in the pathologist's report.

1.10.2 Tumours of mesenchyme

Benign tumours. Benign tumours of mesenchyme are described by the cellular tissue from which they arise (see Table 1.1), although confusion may be induced if Latin or Greek roots are introduced. Benign tumours of fibrous tissue are fibromas, benign tumours of bone may be described as osteomas, and benign tumours of blood vessels as angiomas, but as can be seen from the Table the principles are simple.

Malignant tumours. The generic name for malignant tumours of mesenchyme is sarcoma and, as with carcinomas, this is qualified by the cell of origin and growth patterns. Thus a malignant tumour of bone cells is called a bone sarcoma or osteosarcoma but this can be qualified to describe behaviour. A tumour made of cells forming bone could be described as an osteogenic sarcoma and one with bone destroying cells described as an osteolytic sarcoma. Tumours derived from blood vessels are angiosarcomas, and so on. The extent of spread of sarcomas can also be defined and in principle sarcomas may also be graded in the same way as carcinomas, depending on the degree of dedifferentiation, but in practice this is rarely done since most of the sarcomas are in fact very rapidly growing.

1.10.3 Tumours of the reticuloendothelial system

This is a very complicated field. Benign tumours of the reticuloendothelial system do occur but, since tumours in this system vary considerably in their degree of malignancy and since the tumours may affect the whole system which is widely distributed throughout the body, it is difficult if not impossible to distinguish between a malignant tumour which has spread and a benign tumour which has originated in several different sites, i.e. has a multicentric origin.

Malignant tumours may arise from any of the cells of the reticulo-endothelial system. These cells develop from a population of multi-potential stem cells in the bone marrow and give rise to families of differentiated cells with widely differing structures and functions (see Chapter 12). They include circulating red and white cells, including T and B lymphocytes and other cells of the immune system (see Chapter 12), as well as fixed cells of the system in the spleen, lymph nodes, and other organs. The tumours that arise from these cells may retain some or all (or none!) of the functions of the parent cells.

The tumours can be conveniently divided into two main groups, those arising from blood forming cells—leukaemia, and those forming solid tumours, the lymphomas, but the distinction is not clear cut. In the first group, the tumour cells develop from precursor cells in the bone marrow and pass into the blood stream in the same way as normal blood cells so that the blood is filled with abnormal cells (leukaemic). Any of the stem cells of the bone marrow may give rise to leukaemias which may show any degree of differentiation, so that there may be undifferentiated stem cell leukaemias, or leukaemias with cells which retain some differ-entiated characters of normal white blood cells—myeloid (granulocytic or monocytic), lymphoid, or very rarely from red blood cell (erythroid) precursors. Although the striking feature of the leukaemias is that most of the tumour cells are in the blood stream, the cells may also penetrate normal tissues and form metastatic deposits in almost any organ. The biology of the leukaemias is described in Chapter 12. The solid tumours form a very mixed group. Their identification depends on the cell from which they are derived and the functions which they still express, particu-larly whether they are of T cell, B cell, or other origin. This distinction can now be made more precisely by using molecular markers (see Chapter 12).

One group of tumours, Hodgkin's lymphoma, identified by its clinical presentation and histological structure, is usually separated off from the non-Hodgkin's lymphomas, for which there is now a more or less agreed classification (the Keil Classification). The reader is advised not to dabble in this area without expert guidance but pick your expert carefully (see Habeshaw & Lauder 1988 for review).

1.10.4 *Tumours of the nervous system*

Benign and malignant tumours arise in the nervous system but, most remarkably, malignant tumours hardly ever spread outside the brain or spinal cord. Tumours of the nerve cells proper—neurons—only appear in the embryo or very shortly after birth. These are called neuroblastomas or, if they arise from the specialized nerve cell layer in the eye (the retina), retinoblastomas. Almost all other tumours in the brain and spinal cord arise from the supporting cells, e.g. astrocytes which give rise to astrocytomas, etc. or from the coverings of the brain (the meninges) which give rise to meningiomas. More details are given in Table 1.1. As with tumours of other sites, tumours of the nervous system can be graded by assessing the degree of differentiation.

1.10.5 *Tumours of mixed tissues*

Very rarely tumours which contain a whole range of different tissues may be found. These tumours, known as teratomas, are thought to arise from primitive cells of embryonic type and are usually found in the testis or ovary, but may occur elsewhere. They are sometimes benign but very often malignant change occurs in one component tissue.

1.11 Tumour staging and the spread of tumours— metastasis

Tumour metastasis is the major practical problem and a common cause of death in clinical cancer. Tumours invade the surrounding tissues and may grow out of the organ in which they arise and involve surrounding tissues (Fig. 1.2). During this local invasion, tumour cells may penetrate the lymphatics and be carried to the regional lymph nodes where they are arrested. Some are destroyed but others may grow and produce new tumours. If tumour cells get into blood vessels, they may be carried to any organ in the body. Again many are destroyed but others grow into secondary tumours. There are many unexplained problems. Carcinomas often involve lymph nodes but sarcomas rarely do. Some tumours give rise to secondary deposits more frequently in particular organs than others. Metastasis in the lungs, liver, and bone are common since these organs have many small blood vessels in which tumour cells in the blood become trapped, yet other organs like muscle and spleen, which also have many small blood vessels, are rarely the site of tumour deposits. Some of these problems are discussed in more detail in Chapter 2.

Tumour staging is used to give an assessment of the extent of spread of tumours. One of the more commonly used systems is that established by the International Union Against Cancer. This TNM system is based on an assessment of the primary tumour (T), the regional lymph nodes (N), and the presence or absence of metastases (M). Each of these categories is qualified by a number which indicates the precise extent of involvement according to clearly defined criteria.

1.12　How tumours present—some effects of tumours on the body

Tumours are usually diagnosed if they produce some effects (see Chapter 16). Tumours of the skin or of organs which can be easily examined such as breast, often present as a lump. Many cells in tumours die and these dead cells release enzymes which damage the overlying tissues so that a non-healing ulcer may form. Blood vessels at the base of the ulcer are damaged so that bleeding occurs. In the bowel or the urinary system, blood may be present in the stools, or in the urine, so that bleeding is a common presenting symptom in these organs. Many of the effects produced by tumours are due to the position of the tumour which may press on or destroy surrounding tissues or affect nerves and cause pain. Tumours in the bowel, for example, may cause obstruction either because the tumour mass grows into the cavity of the bowel, or by growing into the wall and destroying the muscle which normally moves the contents down the intestine. Tumours of the brain may present with headache caused by increased pressure inside the skull; tumours involving the bile ducts leading from the liver may cause jaundice, and so on. The physical effects obviously depend on the exact site of the tumour. Some tumours, particularly those which arise from hormone producing organs, may cause effects either by producing an excess of the hormones which the normal organ produces or they may cause a hormone deficiency by damaging the remaining normal gland cells. Less commonly they may produce abnormal hormones or hormones may be produced in tumours of organs which do not normally produce these substances (ectopic expression), e.g. some lung tumours may produce hormones normally produced by the pituitary gland. Anaemia due to bleeding from the tumour or due to some toxic effects on the bone marrow is a not uncommon presenting symptom. As well as these effects, many tumours may cause general wasting and loss of appetite (tumour cachexia) sometimes even though the primary tumour is still fairly small. The cause is unknown but it is thought to be due to some toxic product of the tumour.

But as well as these harmful effects, some tumours may stimulate the defence systems of the body so that they react against the tumours. Unfortunately we know very little about the way in which this occurs but it seems very likely that some of the unexplained differences in the growth and development of tumours in different individuals may be due in part to this host defence reaction. This is an important area in which research is still in its early stages, but promising results are beginning to appear (Chapter 15). For many years it has been known that the body sometimes produces substances that destroy cancer cells. Two of the substances, tumour necrosis factor (TNF) derived from macrophages and lymphotoxin (LT) derived from lymphocytes, have now been identified and their genes isolated. Using modern gene technology, it is possible to produce these substances in large enough quantities for clinical trials as a treatment. We now know that TNF, like other growth factors, is one of a family of cell regulatory proteins called cytokines and has many different functions. It seems that local, transient production of the factor may benefit the host but that sustained production may be harmful (see Balkwill and Fiers, 1989 for review of this group and similar factors now known as biological response modifiers).

1.13 How does cancer kill?

As we have seen, many cancers develop in older peole and a substantial number of patients do not die as a consequence of the disease but of some unrelated condition such as heart disease, incidental infections, or even as a result of an accident. Tumour related events may cause death directly or indirectly depending on the site of the tumour and the extent of spread. A common cause of death is due to involvement of vital organs, either by direct local invasion or from distant metastases, for example in the brain, lung, or liver. Rarely, death may be due to haemorrhage; more often anaemia and unexplained wasting may lead to decreased resistance to infection so that terminal bronchopneumonia or infection of the urinary tract (pyelonephritis) is common. In many cases it is not possible to establish the immediate cause of death.

1.14 Experimental methods in cancer research

Much of our knowledge of the development and growth of tumours is derived from a close study of cancers in patients by clinicians and pathologists. This has allowed us to define many of the problems to which we should like to find answers and, although the application of new techniques in cell and molecular biology to human tumours continues to provide us with valuable information, other methods have to be used to study changes which cannot be easily observed in man. These include, for example, observations on cell behaviour in the very early stages of carcinogenesis, the direct effects of carcinogens on the genome, the direct effects of drugs on tumours, and so on. Although cancer is a disorder of cells, it is influenced by changes in the environment in the host, so that for experimental analysis, we need methods which allow us to study the changes which occur in isolated cells as well as in the whole animal. We also need standardized methods for producing tumours and for some purposes we need to transplant tumours into a new host to study the effects of a different environment on growth and behaviour.

1.14.1 *Tumour induction and transplantation*

The induction of tumours by giving or applying carcinogenic agents to animals is an essential experimental tool not only for studying the process of carcinogenesis (see Chapters 6–8) but also for screening drugs or chemicals before use in man or for industrial processes. For some purposes, one needs samples of the same tumour for testing. The usual way to do this is to transplant the tumour into another animal. In ordinary populations there are large differences between individuals so that transplanted tumours (or normal tissues) are recognized as foreign by the new host and destroyed by the immune system. To avoid this, scientists have developed many 'pure line' (inbred) strains of mice. To do this, mice have been selected and inbred for many generations so that each individual in the colony is essentially genetically identical with any other (syngeneic). Tumours and normal tissues in these animals can be transplanted easily. A further refinement is that tumours can now be specially prepared and stored frozen in ampoules in liquid nitrogen at −193°C until needed for retransplantation into mice. Inbred strains particularly prone to develop a particular type of cancer or with a particular sensitivity to carcinogens have also been developed, so that the genetic basis for some tumours can be studied.

For human tumours or tumours from species in which no inbred strains are available, transplants can be made into animals in which the immune system has been impaired by treatment or in which there is a congenital defect in the immune system. One group of mice which have a congenital defect of this type also shows loss of hair. These 'nude' mice (*nu*) are used to maintain transplants of human and other tumours. Tumour or normal tissue grafts between individuals of the same species are allografts, between genetically identical individuals (e.g. identical twins or inbred strains) are isografts, and between foreign species (e.g. human tumour in nude mice) are xenografts. Unfortunately, not all tumours can be transplanted for reasons so far unknown.

1.14.2 *Tissue culture techniques*

For direct observation of tumour and normal cells isolated from their normal environment, tissue culture techniques are used. These methods allow studies on the direct effects of agents on living cells and the separation of different cell types from a mixed cell population, as well as the characterization of cell products. The most commonly used technique is cell culture in which fragments of tissue, tumour, or separated cells are put into sterile glass or plastic containers in a fluid nutrient medium and maintained at body temperature in an incubator, in an atmosphere of air and carbon dioxide (usually 5 per cent) similar to that *in vivo*. If the cultures are successful, cells grow out from the explants and fill the container. They can then be removed and transplanted to other containers or treated for storage in liquid nitrogen. These populations are mixed but single cells can be isolated and large numbers of genetically identical daughter cells (a clone) can be grown up from it and used for a more detailed study. Many tumour cell lines have been established and stored although only a small population of tumours will give rise to cell lines which can be maintained indefinitely. So far normal cells can only be maintained for relatively short periods, but have been used to study the induction of neoplastic transformation under closely controlled conditions. Cell culture can also be used to study the effects of drugs and the effects of cells on each other (cell interactions). A modification of the technique allows pieces of tumour or normal tissue to be maintained in an organized form ('organ cultures'). This method is particularly useful for looking at the effects which involve more than one cell type, e.g. some hormone effects. Cell culture has proved to be an essential basic technique for the development of modern cell and molecular biology, as will be seen later.

1.15 Experimental methods for tumour identification

Although the tissue diagnosis of most established tumours is usually straightforward using standard histological methods (see p. 13), there are still no absolute markers for malignant or premalignant cells, or methods for assessing the malignant potential (i.e. capacity for growth and spread) of tumours.

1.15.1 *Histochemical methods*

Refinements in microscopic techniques are sometimes of value. Histochemistry, in which chemical reactions are carried out on histological sections, is a technique that is still to be fully exploited. The precise cellular localization of enzyme reactions or tissue products can be established and by using sensitive microfluorimeters or microspectrophotometers, the amount of the end product can be measured. Tumour specific and tumour associated proteins can be identified by immunohistochemistry (see Chapters 15 and 18) in which a specific antibody to a particular protein can be applied to cells or to tissue sections and the bound antibody visualized by a second reaction specific to the antibody itself. These methods have been of great value in identifying the cell types in tumours, particularly of the lymphoid system, and in identifying structural components in cells, such as the cytokeratins found in epithelial cells, vimentin and desmin found mainly in mesenchymal cells, and specific proteins found mainly in cells of the nervous system. A technique similar in principle, in-situ hybridization, is also available for nucleic acids, using labelled probes specific for segments of nucleic acids or in viruses (see Chapter 14). Histochemical techniques and in-situ hybridization can be applied to light or electron microscope sections and the great virtues of the methods are that they allow the identification of small numbers of positive cells in a large mass of tissue and also allow the precise intracellular localization to be established.

1.15.2 *Other methods*

Many other techniques for tumour identification are still being explored. Oncogene products are being studied as potential tumour markers. Proton magnetic resonance is being used to study membrane changes in tumour cells and nuclear magnetic resonance is being used for tumour detection in the clinic. A more detailed assessment of these techniques can be found in *New approaches to tumour identification* (Britton 1987).

1.15.3 *Methods for genetic analysis*

Methods for family and population genetics are considered in Chapters 3 and 4 but a number of special techniques are needed for the analysis of genetic changes in cells. The development of these methods is responsible for the rapid increase in our knowledge of the structure of the mammalian genome. The ultimate aim is to define the chemical structure, order, and spacing of genes along the double helix that makes up the DNA which contains the whole genome (see Chapter 5) and to identify alterations that are associated with abnormal function. Low resolution physical maps can be made by determining the appearance of individual chromosomes (karyotyping and banding) and specific genes can be localized to particular areas on individual chromosomes (see Chapter 10). Higher resolution maps can be determined by molecular analysis (see Chapters 5 and 9).

DNA function can be studied by transferring DNA (transfer or transfection) from one cell to another and observing any changes in function. The whole genome can be transferred by fusing two cells together (whole cell fusion). Smaller amounts can be transferred by fragmenting the nucleus of the donor cell with chemicals and fusing the 'micro' cells that are produced (micro cell fusion). Individual chromosomes can also be separated and transferred to other cells (chromosome mediated gene transfer), as can pieces of separated DNA (DNA mediated gene transfer). The methods used in gene mapping are reviewed by Bentley *et al.* (1988). The whole field of somatic (non-germline) cell genetics and cancer growth and metastasis is considered in Franks (1988).

The influence of the cytoplasm on cell functions is studied by a reciprocal technique, i.e. cytoplasm from which the nucleus has been removed is fused with another whole cell (cytoplast fusions).

1.16 Conclusions

It must be obvious from this survey that there are many gaps that remain in our understanding of the phenomena that are concerned with the initiation and development of tumours. The succeeding chapters consider the present state of our knowledge and try to identify areas in which current research suggests that further studies may be profitable.

References and further reading

Alberts, B. *et al.* (ed.) (1989). *Molecular biology of the cell* (2nd edn). Garland Publishing, New York.

Ausubel, F. M. *et al.* (ed.) (1987). *Current protocols in molecular biology.* Wiley, New York. (An up-to-date manual of techniques.)

Balkwill, F. and Fiers, W. (ed.) (1989). The use of cytokines in cancer therapy. *Cancer surveys,* Vol. 8, No. 4. Oxford University Press.

Bentley, K. L., Ferguson-Smith, A. C., and Ruddle, F. H. (1988). A review of genomic physical mapping. In Somatic cell genetics and cancer. *Cancer surveys,* Vol. 7, No. 2 (ed. L. M. Franks), pp. 267–94. Oxford University Press.

Britton, K. E. (ed.) (1987). New approaches to tumour identification. *Cancer surveys,* Vol. 6, No. 2. Oxford University Press.

Cairns, J. (1975). The cancer problem. *Scientific American* **233**, 64–78.

Darnell, J., Lodish, H. F., and Baltimore, D. (ed.) (1986). *Molecular cell biology.* Scientific American Books, New York.

DeVita, T., Hellman, S., and Rosenberg, S. A. (ed.) (1985). *Cancer principles & practice of oncology* (2nd edn). J. B. Lippincott, Philadelphia. (A good general text on clinical aspects.)

Franks, L. M. (ed.) (1988). Somatic cell genetics and cancer. *Cancer surveys,* Vol. 7, No. 2. Oxford University Press.

Freshney, R. I. (ed.) (1987). *Culture of animal cells* (2nd edn). Alan R. Liss, New York. (A manual of basic techniques.)

Habeshaw, J. A. and Lauder, I. (ed.) (1988). *Malignant lymphomas.* Churchill Livingstone.

Knox, G. and Woodman, C. (ed.) (1988). Prospects for primary and secondary prevention of cervix cancer. *Cancer surveys,* Vol. 7, No. 3. Oxford University Press.

Oxford Text Book of Pathology (1989). (Good reference texts on general pathology.)

Watson, J. D., Hopkins, N. H., Roberts, J. W., Argetsinger Steitz, J., and Weiner, A. M. (ed.) (1987). *Molecular biology of the gene* (4th edn), Vols I and II. Benjamin/Cummings, Melno Park.

Willis, R. A. (ed.) (1967). *Pathology of tumours* (4th edn). Butterworth, London. (Over 20 years old but still the best general text on tumour pathology.)

Specialized reviews

Franks, L. M. (ed.) *Cancer surveys*, Cold Spring Harbor Laboratory Press. Published quarterly. Each issue provides an up-to-date review on a specific topic and covers clinical, experimental, and epidemiological aspects.

Klein, G. and Weinhouse, S. (ed.) *Advances in cancer research*. Academic Press. Published annually. Detailed reviews on individual topics.

DeVita, V. T., Hellman, S., and Rosenberg, S. A. (ed.) *Important advances in oncology*. J. B. Lippincott/Harper & Row Gower Medical Publishing. Published annually. Shorter reviews covering basic research and clinical aspects.

2

The spread of tumours
I. R. HART

2.1 Introduction

2.1.1 *Significance of tumour spread*

Metastasis is 'the transfer of disease from one organ or part to another not directly connected with it. It may be due to the transfer of pathogenic organisms or to the transfer of cells as in malignant tumours'. This transfer of cells is one of the fundamental problems of clinical oncology. Surgical removal, often combined with irradiation, frequently is success-ful in the treatment of primary tumours but widespread dissemination often defeats this mode of treatment. Cancer spread is responsible for a large proportion of cancer deaths and the relentless and seemingly intractable movement from primary site to distant organs is a major factor in people's fear of neoplastic disease.

2.1.2 *Pathogenesis of the process*

Tumour dissemination is a complex process where the eventual outcome depends on the result of a number of interactions between the tumour cells and host cells. There are five major steps involved in metastasis, though it should be realized that the process is a dynamic one which may pass from one step to another without interruption and a number of the steps will be operating concurrently. Following tumour development and growth there must be

(1) invasion and infiltration of surrounding normal host tissue with penetration of small lymphatic or vascular channels;
(2) release of neoplastic cells, either as single cells or small clumps, into the circulation;
(3) survival in the circulation;
(4) arrest in the capillary beds of distant organs; and
(5) penetration of the lymphatic or blood vessel walls followed by growth of the disseminated tumour cells (Fig. 2.1).

If all these steps are completed, the result will be formation of a secondary tumour in a distant organ.

2.1.3 *Tumour progression/evolution*

As mentioned in Chapter 1, tumours do not come into being with all their characteristics already developed. This view of cancer as static populations of cells has been replaced by one in which they are seen as much more dynamic entities where there is gradual acquisition of new characters as the tumour develops. This process has been termed tumour progression and, while there are exceptions, the general trend is for tumours to go from bad to worse. Thus with tumour progression there is a movement towards a more aggressive behavioural pattern and the ability to invade and metastasize may not be manifested until relatively late in the course of neoplastic development. Some of the possible mechanisms involved in this process of evolution and progression will be discussed later in the chapter but, for the present, it should be remembered that the fully malignant tumour cell, i.e. one that is able to invade and metastasize, may differ considerably in its character from a cell in the early stages of the transformation process.

2.2 Mechanisms of tumour invasion

Two of the five steps outlined in the pathogenesis of cancer spread depend on the ability of tumour cells to invade or infiltrate into areas of normal tissue. There is no evidence to show that the mechanisms used to gain access to the circulation are any different from the mechanisms used by the cells in moving out from the vessels in which they have become arrested so that both steps will be considered together here.

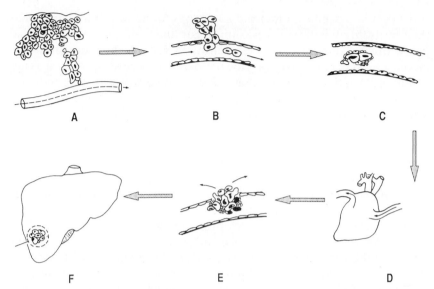

Fig. 2.1 The spread of malignant tumours. A. Primary tumour invades and spreads into adjacent normal tissue, eventually coming into contact with small blood vessels or lymphatics. B. These small vessels are penetrated by tumour cells which are released into the circulation. C. In the circulation a number of interactions occur between the released tumour cells and circulating host cells such as platelets, lymphocytes, and monocytes. D. The passage of individual neoplastic cells or small emboli throughout the body is made possible by a number of junctions between the lymphatics and blood vessels; few tumour cells survive this passage. E. Those tumour cells that survive must arrest in distant organs, possibly in mixed clumps containing both neoplastic cells and platelets or lymphocytes, breach the integrity of the vessel wall and move out into the surrounding normal tissues. F. Growth of such extravasating tumour cells gives rise to secondary tumour deposits (here shown growing in the liver) and the process may be repeated.

In general the mechanisms of tumour invasion are poorly understood though the three most likely possibilities are that it occurs as a result of:

(1) mechanical pressure;
(2) release of lytic enzymes; and
(3) the increased motility of individual tumour cells.

Obviously these mechanisms are not mutually exclusive and it is possible that in a given tumour any combination of the three may be involved and that the relative importance of each may vary depending both on the tumour type and its anatomical location.

The rapid proliferation of neoplastic cells may build up pressure which forces sheets, or fingers, of tumour cells along lines of least mechanical resistance in a manner somewhat analogous to the way that plants force their roots through the soil. Invasion, according to this hypothesis, is thus just a direct consequence of uncontrolled growth; pressure from the growing mass occludes (blocks) local blood vessels leading to local tissue death and a reduction in mechanical resistance which further aids the process. While it is true that the gross appearance of many malignant tumours conforms with this picture with finger-like projections of tumour cells emanating from the main growth, it is also true that there are many observations that cannot be explained by this hypothesis. Some highly invasive tumours grow more slowly than their benign counterparts; histological examinations often reveal clumps of neoplastic cells which in serial sections reveal no connection with the main tumour, and cancer cells often invade and penetrate loose tissues where it would not seem to be possible to build up any pressure effect.

Areas of normal host tissue adjacent to areas of tumour invasion are often severely disrupted and show considerable amounts of lytic damage. Because many animal and human tumours have higher levels of proteases and collagenases than corresponding benign or normal tissue, the concept that malignant tumours produce and secrete lytic enzymes that degrade normal tissue has become firmly established in the literature. However, technical difficulties have made any correlation between malignant behaviour and increased proteolytic enzyme activity difficult to interpret. Direct sampling of tissue (biopsy) may damage tissue and this may give rise to elevated levels of enzyme activity. Furthermore, tumours are not composed solely of neoplastic cells but contain both stromal and infiltrating reticuloendothelial cells. Many of the infiltrating cells, such as polymorphonuclear leukocytes and monocytes, contain high levels of those enzymes most likely to be involved in tissue degradation; indeed their presence may well contribute to the eventual invasive behaviour of the tumour either by releasing their own lytic enzymes or by behaving as inadvertent 'guides' for infiltrating neoplastic cells. Since the number of these infiltrating cells varies from tumour to tumour, or even between different parts of the same tumour, their contribution to overall enzyme activity also varies considerably. Immunohistochemical staining for different collagenases and proteases has located many of these enzymes at the periphery of growing tumours, frequently at the sites of overt tissue damage and tumour cell invasion. However, the central portions of tumours often are necrotic and the peripheral staining observed may reflect the production of these enzymes by living cells rather than direct involvement in invasion. For these reasons, perhaps the most compelling evidence on the role of proteolytic enzymes in tumour invasion has come from experimental studies where it has proved possible to examine enzyme production by tumour cells grown in tissue culture and then to correlate this capacity with the subsequent invasive behaviour of the cells after transplantation into animals. Using this approach it was possible to establish positive correlations between invasion and high levels of the enzymes cathepsin B, type IV collagenase, and plasminogen activator. These correlations are not universal in as much as different researchers have found conflicting results depending on the tumour types used. With data derived from naturally occurring tumours in man, the circumstantial evidence linking proteolytic enzymes with a role in tumour invasion is strong but the exact nature of the enzymes involved is still not established.

Evidence for the role of tumour cell motility in invasion is also equivocal. The finding of individual tumour cells or small clumps of tumour cells separate from the main tumour mass is difficult to explain without invoking the concept of tumour cell motility. Cinematography has been used to show that tumour cells in the body, as in tissue culture, are capable of active movement and migration. To what extent this ability is used in tumour invasion is unknown but it seems highly likely that the neoplastic cells do move through normal tissues by active locomotion. If motility does play a role in invasion, the next step is to determine to what extent is such movement directional in nature. Tumour cells in tissue culture can move towards substances which attract them (chemotactic factors) and changes in the surfaces on which they move may also affect the direction of movement. It is tempting to speculate that such mechanisms operate *in vivo* but, because of the difficulties in assessing such responses in the whole animal, firm evidence is lacking. However, it is possible that this situation could be improved in the next few years because of some recent biochemical investigations. Liotta and his colleagues (1986) at the National Cancer Institute in the United States have described the isolation and purification of a cell-motility stimulating molecule which they have termed 'Autocrine Motility Factor' (AMF). This 54 kilodalton (kDa) protein, first isolated from a human melanoma cell line, stimulates random motility of transformed cells but does not stimulate the motility of their untransformed counterparts. Further studies have shown that AMF is produced by, and stimulates a response in, a variety of neoplastic cells and has even been detected in the urine of patients with transitional cell carcinoma of the bladder. Thus it is likely that the production of a motility factor by a tumour cell could play a major role in the local invasive behaviour of cancer. Certainly, with the availability of complimentary DNA (cDNA) probes for in-situ hybridizations and monoclonal antibodies for immunohistochemistry, there will be opportunities to examine material from a wide range of invasive tumours to determine the extent of any correlation between AMF production and infiltrative behaviour. An even more exciting prospect is the possibility that blocking the response to AMF, or inhibiting the production of AMF, could lead to an inhibition of the invasive process with resultant therapeutic gains. Liotta (see Hart 1988) has reported that just such an inhibitor has affected profoundly the number of spontaneous metastases arising from a murine sarcoma. The generality, or indeed the likely significance, of this observation remains to be verified but it illustrates the underlying rationale for many investigations into the mechanisms of cancer spread; an improved understanding of the pathogenesis of the process at the molecular level will lead to the design of new methods of treatment.

It should be pointed out that tissues vary considerably in their ability to withstand tumour invasion. Tumours rarely penetrate the walls of arteries, arterioles, or even the larger veins while they readily invade capillaries and lymphatics. Such resistance is in part due to greater mechanical strength of the larger vessels but there is also a suggestion that certain of the tissues resistant to invasion, such as cartilage or the elastic fibres surrounding the larger vessels, are resistant because they release anti-proteolytic factors that inhibit proteases.

Such a mechanism is thought to be involved in the well-known, though not universally accepted, observation that cirrhotic livers are more resistant to metastatic involvement than are the normal organs. In a recent paper (Barsky and Gopalakrishna 1988) it has been reported that myofibroblasts (the cells involved in the collagenous bands found in cirrhosis) secrete a metalloproteinase inhibitor which not only inhibits the activity of type I and type IV collagenases but which also prevents the invasive behaviour of human tumour cells in an in-vitro assay of tumour cell invasion. This, it is argued, is why cirrhotic livers exhibit a strong resistance to metastasis as evident in post-mortem analyses. The presence of these factors in tissues with a natural resistance to invasion provides further support for the idea that proteolytic enzymes may play a role in mediating tumour spread.

2.3 Dissemination of tumour cells via lymphatics and/or blood vessels

Once tumour cells enter the lumen of lymphatic or blood vessels, they either remain at the site of penetration and grow there with a consequent occlusion of the vessel, or they release cells which are carried away in the lymph or blood. The release of individual cells or small emboli (clumps) has led to the suggestion that the cells of malignant tumours are less strongly attached to each other than cells of benign tumours and are more readily detached from the primary mass.

It is a common clinical observation that carcinomas, which are epithelial in origin, generally spread in the lymphatic system as well as in the blood, while the sarcomas of mesenchymal origin appear to spread via the haematogenous route. But this may be an arbitrary division. There are connections between the lymphatics and the blood vessels, and radiolabelled circulating tumour cells have been shown to be capable of moving between these two systems, either through direct veno-lymphatic communications (anastomoses) or from the lymphatics into the thoracic duct which empties into the jugular vein and the venous circulation. The preferential involvement of the lymph nodes with metastatic carcinoma may be a reflection of organ specific growth rather than a tendency to infiltrate specifically one system only. Alternatively it may be that there is a greater concentration of lymphatics in epithelial structures and a greater increase in the likelihood that these vessels rather than blood vessels will be penetrated.

In this chapter, the spread of tumours by either the lymphatic or the haematogenous route will be considered as a common process. From clinical observations and from experimental studies it is known that the mere presence of neoplastic cells in the circulation does not constitute metastasis. The process is inefficient and most of the cells released into the circulation die without forming a metastatic deposit. The death of many of the released cancer cells may be attributable to the controlling influence of the host's immune response but much may simply be the result of non-specific factors such as turbulence. The environment in the circulation is generally thought of as being hostile to disseminating tumour cells, but some of the interactions to which cells are exposed may aid their survival. Aggregation, either with other tumour cells or with host cells such as lymphocytes and platelets, may result in the formation of larger emboli (particles in blood stream) which are more easily filtered out in distant capillary beds. Surrounding the tumour cells by aggregating blood cells may also provide a protective outer layer which prevents damage to the central tumour cells.

To leave the circulation, cells must be arrested and implant in the capillary bed of an organ. Generally tumour cells do not adhere to the walls of the large vessels where, presumably, blood flow is sufficiently vigorous to sweep away attaching cells. Even though tumour cells are deformable (not rigid) and can pass through capillaries of narrower bore than their resting diameter, it is in the capillary bed that the cells are generally arrested. This arrest may be due to passive filtering, as in the case of emboli rather than individual cells, or it may represent an active process. It can be shown that tumour cells, like platelets, do not adhere to intact endothelium but attach preferentially to exposed basement membrane. Tissue culture studies suggest that the tumour cells themselves might stimulate endothelial cell retraction and loss, though the shedding of endothelial cells from the wall is a normal physiological process and the basement membrane is frequently being exposed at various sites in the vascular system. Some proteins such as fibronectin and laminin are involved in the attachment of normal cells to each other and to basement membrane and may also play a part in determining both the specificity and the kinetics of tumour cell attachment. Endothelial damage leads to platelet adherence, and tumour cell/platelet clumps may attach passively to the areas of endothelial retraction; arrest of the circulating neoplastic cell could then occur in the absence of any active ability or proclivity to attach to basement membrane.

A family of cell surface receptors termed the integrins (involved in a variety of functions including cell adhesion, migration during embryo-genesis, thrombosis, and lymphocyte function) (see Juliano 1987) has been identified based on the ability of these receptor molecules to recognize glycoprotein ligands (see Glossary) bearing the amino acid sequence Arg-Gly-Asp known as the RGD motif. Many tumour cells express integrin-like receptors at their cell surface and considerable interest has been generated by the demonstration that Arg-Gly-Asp-containing synthetic peptides can block both invasion *in vitro* and the formation of lung tumour nodules in mice co-injected with tumour cells and the inhibitory peptides (Gehlsen *et al.* 1988). It is true that such peptides have a very short half-life in the circulation, that attachment sites other than the RGD site may be utilized by disseminating cancer cells, that the peptides have to be used at very high plasma levels, and that, at the time of presentation, many cancer patients will already have established micrometastases and thus have passed the stage at which such treatments might be beneficial. Nonetheless, as with abrogation of AMF activity discussed earlier in this chapter, these types of approaches highlight the value of an understanding of the mechanisms of tumour spread in the design of novel therapeutic agents.

2.4 Patterns of metastatic spread

Some tumours often metastasize to particular organs. Thus osteosar-comas normally give rise to pulmonary metastases whereas neuroblasto-mas most commonly spread to the liver; the reason for this organ selectivity is unknown but two hypotheses have been proposed. In the 'mechanistic theory' the eventual site of metastasis development is a consequence of the anatomical location of a primary tumour; the number of viable tumour cells delivered to the capillary bed in the first organ encountered is due to the pattern of blood flow. For example, the venous blood flow from the large bowel goes to the liver through the portal veins and perhaps as a consequence, the liver is the commonest site for secondary deposits from bowel tumours. While this undoubtedly is true for some tumours it does not explain all patterns of tumour spread. Muscle is well vascularized and the kidney receives up to 25 per cent of cardiac blood output yet both these organs are infrequently involved in metastasis formation. Almost one hundred years ago Paget suggested the 'seed and soil' hypothesis, wherein the provision of a fertile environment (the soil) in which compatible tumour cells (the seed) could grow was the determining factor in deciding metastatic sites. Failure of an organ to develop metastases was not a consequence of the failure of disseminating cells to reach that site but was because of the inability of the organ to provide a favourable environment for growth. This hypothesis, perfectly suited to the Victorian era, with its biblical overtones of seeds falling on stony ground, has received strong support in recent years both from experimental results and clinical studies on the use of shunts to relieve extensive exudation of fluid in the peritoneal cavity (ascites) in terminally ill patients with abdominal cancers. Many of these patients produce considerable volumes of fluid in the abdomen leading to marked distress and discomfort; relief can be obtained by withdrawal of this fluid (paracentesis), but the procedure often needs to be repeated almost daily. To provide such patients with relief from this unpleasant consequence of their tumours, it has proved possible to insert an artificial shunt from the abdomen into the jugular vein; fluid is thus continually returned to the venous circulation and the uncomfortable build up of ascites is avoided. Within the ascitic fluid are large numbers of viable tumour cells and returning them to the jugular veins means the first capillary bed encountered is that located in the lungs. In spite of this, many patients who survive with this treatment for a number of weeks or more show no evidence of pulmonary metastases, even though many millions of viable cells have been passed into their lungs where, according to the mechanistic theory of metastasis development, they should have been filtered out of the circulation. While these results are most compatible with the 'seed-soil' hypothesis, there is no information on the number of cells retained in the lung.

Considerable attention in recent years has focused on the possible role of autocrine/paracrine (see Chapter 11) control of tumour cell proliferation in the development of neoplasia. Certainly it is possible that the local production of paracrine factors by specific organs, either mitogens or inhibitors of cell proliferation, or the stimulation of autocrine growth factor production and release by the tumour cells themselves in response to organ-derived signals, could have profound effects on patterns of metastatic involvement. The likely complexity of this type of growth regulation is underscored by the fact that peptide growth factors may inhibit or stimulate proliferation of different cell types or may even have conflicting activities on a single cell type depending on the nature of other signal molecules present. Unravelling the factors involved in the metastatic patterns of specific tumours is likely to be a difficult task but with the availability of increasing numbers of recombinant-derived peptide growth factors considerable insight into this aspect of tumour biology should be gained in the next few years.

One adjunct to this direct regulation of tumour cell growth that could also be involved in determining metastatic patterns is via organ related control of angiogenesis. The growth of solid tumours beyond a few millimetres diameter in size, a size at which they may be considered clinically irrelevant, depends on the formation of new blood vessels to supply these foci—a process termed neovascularization. A number of angiogenic factors, including acidic and basic fibroblast growth factor (aFGF and bFGF), transforming growth factor (TGF) α and β, angiogenin, heparin, tumour necrosis factor α and interleukin 1α, have been described. Many of these substances, such as TGF-α and bFGF, are produced by a variety of tumour cells and it may be that by the release of such peptides cancers stimulate the ingrowth of new capillaries. Again it is possible to invoke the idea of organ specific regulation of angiogenic factor release to explain why secondary growth may, or may not, occur in a particular anatomical site, though evidence for such a hypothesis is lacking at the present time. Many of the angiogenic factors, such as the FGFs, are distributed so ubiquitously throughout the body that at first sight they appear to be unlikely candidates for organ-related activity. However, these factors appear to be maintained in a functionally inactive state, either sequestered within the cells of origin or stored in the extracellular matrix, and the activation signal, possibly provided by the tumour cells themselves, might be organ specific and so account for preferential growth. It is also possible a third mechanism may be responsible for determining site specific metastasis. It has been suggested that there are specific interactions between cell surface proteins of tumour cells and organ specific proteins on the endothelial cells lining the capillaries or the exposed basement membrane in the capillary beds of different organs. Metastasis in this instance would not be the result of the increased delivery of tumour cells to the organ but the result of enhanced retention by selective adhesion. Certainly lymphocyte recirculation, a process which might be considered analogous to many of the steps in the dissemination pathway, is dependent on just such a specific interaction between cell surface molecules on the peripatetic lymphocyte and the endothelial cells lining the vessels of the lymphoid system. It is possible that variations in specificity or expression of the various matrix receptors of the integrin family (see above) in response to signals from the cell's environment could modulate the attachment behaviour of metastatic cells. Again results are likely to come from experimental model systems and then their significance has to be assessed by the difficult process of extrapolating to clinical observations.

2.5 The role of the immune system in modulating metastasis

The immune system reacts against foreign proteins either by direct attack by cells of the system or by the production of soluble antibodies against the proteins. The influence of the immune system on tumour growth is complicated and is discussed in detail in Chapter 15. As far as metastasis is concerned it must be noted that the process occurs in the presence of a mass of antigenic material, i.e. the primary tumour, so that this phenomenon might be thought to be highly susceptible to immune modulation. Results from experimental systems have confirmed this idea and shown that the immune system can exert a profound influence on the eventual outcome of the metastatic process. Interestingly, the immune system does not always exert an inhibitory effect on tumour spread but acts as a 'double-edged' sword. Thus lymphocyte aggregation with circulatory tumour cells can increase the size of emboli and assist in their arrest and lodgement. Furthermore, lymphocytes are capable of inducing angiogenesis and may facilitate the provision of nutrients to a proliferating secondary tumour.

A subpopulation of lymphocytes, termed NK cells (see Chapter 15), has been implicated as being of great importance in regulating metastatic spread in experimental animals. Athymic nude mice have high levels of these cells and it has been suggested that the known infrequency of metastatic spread of allogeneic and xenogeneic tumours implanted into these mice might be attributable directly to the efficacy of the NK cell system. Suppression of NK cell activity by various techniques has led to enhanced metastasis of certain tumours though, to repeat a refrain that is becoming very familiar in this Chapter, there appear to be no simple generalizations that can be drawn from such experiments since in other experimental systems the effect of NK cell depletion on metastasis seems to be minimal. It is a reasonable assumption that those tumour cells which have managed to survive the circulation have successfully avoided, or subverted, immune surveillance mechanisms. The molecular basis of such avoidance strategies has received increasing attention in recent years. Cytotoxic T lymphocytes recognize foreign antigens only in association with major histocompatibility (MHC) class I cell surface antigens. Consequently one way in which metastatic cells could avoid a T cell immune response is by a selective deficiency in MHC class I antigen expression. Much work has been aimed at examining this question in murine transplantable tumours but there are also some limited immuno-histochemistry studies on human material which support this concept. However, suppression of MHC class I gene expression is not likely to be the sole determinant in the balance between tumour cells and host immunity. Partly this may be due to what seems to be a directly contrasting mode of recognition operating within another anti-tumour effector system. Recent findings appear to suggest that elevated levels of class I antigens exert a protective effect against NK cell mediated cytotoxicity; that is, NK cells may recognize and respond to the lack, as distinct from the presence, of class I antigens. So, depending on whether the NK cell or the T lymphocyte plays the dominant role in immune limitation of spread of a specific tumour, it is possible that a single molecular lesion, the lack of class I antigen, could have either an inhibitory or a stimulatory effect on metastasis.

Class I antigens are not the only cell surface molecules involved in determining immune cell–cell interactions. Thus a variety of surface proteins on both lymphocytes and the target cell, termed lymphocyte function-associated antigens or LFAs (see Chapter 15), which have been assigned membership to the superfamily of adhesion molecules discussed above, serve to stabilize the adhesion reactions necessary for conjugate formation which leads to cell lysis. It is conceivable therefore that aberrations in expression of such molecules, or the ligands for such molecules, by the tumour target cells will lead to alterations in cytotoxicity which could be reflected by increased survival of disseminating cells. Again the availability of molecular probes should go a long way to answering such questions in the next few years.

Mononuclear phagocytes also appear to play a role in determining metastatic spread. Correlations have been established between the macrophage content of a series of tumours of similar histological origin and an inability to metastasize. Differences in absolute numbers of macrophages found in metastasizing or non-metastasizing tumours may be matched by differences in the functional capacity of such infiltrating cells. Thus there are reports that macrophages isolated from non-metastasizing or regressing tumours are cytotoxic whereas macrophages from progressing/metastasizing tumours are non-cytotoxic or may even stimulate tumour growth. Additionally circulating monocytes are cytotoxic to various tumour cells *in vitro* and if these cells are capable of exerting such effects *in vivo* it is possible that these cells in conjunction with NK cells may make a significant contribution to the elimination of circulating tumour cells.

The exact role of humoral (antibody mediated) immunity in the modulation of metastasis is no more clear than that of the cell mediated arm. Once again it would seem evident that any tumour capable of evoking an humoral immune response would be more susceptible to such a response while circulating as individual cells or as small emboli. Antibodies directed against tumour antigens have been identified in many experimental tumour systems and against a few naturally occurring human tumours. In cases of human melanoma there is a strong suggestion that the presence of circulating antibody correlates with the absence of metastasis. Antibody plus complement (see Glossary) may lead to direct lysis of circulating cells or may facilitate removal of such cells by mononuclear phagocytes. Physical coating of the cells with antibody might interfere with certain of the steps in the metastatic sequence such as aggregation or adherence. Much more work remains to be done before any definite conclusions can be drawn.

2.6 Tumour cell heterogeneity

2.6.1 Differences between primary and secondary tumours

All cells in a single tumour are not identical but there is range of population of cells expressing many different characters (phenotypes). Cells in a tumour may show differences in structure, e.g. morphology, growth rate, karyotype (chromosome pattern) or behaviour, e.g. invasion and metastasis. This diversity is a consequence of tumour progression and some of the possible mechanisms involved in generating this diversity will be discussed below. The concept of heterogeneity is accepted by most pathologists, biologists, and clinical oncologists. What is far more contentious is the idea, developed from experiments with transplantable rodent tumours, that metastases are derived from pre-existing subpopulations of cells in the primary tumour. Metastasis is an inefficient process and the majority of tumour cells released into the circulation do not give rise to secondary tumours. Do those few cells that survive do so fortuitously in a completely random manner, or is metastasis a selective process which allows the emergence of a pre-existent subpopulation of cells? Such cells presumably would possess certain characteristics that differ from those of the majority of cells. If only a few cells are capable of metastasizing then therapy should be targeted against those cells; the vast mass of cells in the tumour would not be life threatening. Again evidence for and against this concept comes from experimental studies because of the difficulties of analysing such characteristics in human tumours. Some workers have been able to show that cells derived from metastases are more metastatic than cells from the primary tumour whereas others, working with different tumour systems, have been unable to demonstrate this phenomenon. It seems highly likely that the process is a combination of both elements. Were metastasis entirely selective then it could be demonstrated in animal systems by the repeated selection of metastatic cells until 100 per cent efficiency was obtained; the injection of 100 metastatic cells should then lead to an eventual tumour burden of 100 metastatic nodules. This has never been achieved, even after selection; metastasis remains inefficient probably because metastasis is largely a random event where the destruction and elimination of circulating tumour cells is haphazard, regardless of whether cells are capable of forming metastatic tumours or not. There is, however, a selective aspect of the process, which is why metastatic variants can be isolated from heterogeneous populations of cells both by selection techniques and cloning procedures. This could help explain why there are many examples in the literature of differences between primary tumours and their metastases in terms of enzyme levels, karyotypes, drug sensitivity, and cellular oncogenes. Alternatively, since tumours are known to be heterogeneous, it is possible that the selection for these differences is fortuitous and is not associated with the process of metastasis *per se*. Whether metastatic deposits in cancer patients are the result of the proliferation of selected subpopulations of cells would seem to be the most important question in the pathogenesis of cancer spread.

2.6.2 *Epigenetic and genetic mechanisms for generating phenotypic diversity*

Since tumours are so obviously heterogeneous for a wide variety of characteristics, what is the source of this diversity? Studies based on individual markers in leukaemias and lymphomas have shown that these cancers appear to be almost universally monoclonal in origin, i.e. descended from a single transformed cell; the situation in the solid tumours is less clear-cut and there is a certain amount of evidence to suggest that some carcinomas might be multicellular in origin. Even if such tumours are monoclonal in origin, by the time of presentation in the clinic there has been tumour progression and the generation of diversity. To explain this diversity Nowell (1976) suggested that the transition from normal to transformed cell carried with it the acquisition of inherent genetic instability. This genetic instability, he suggested, allowed transformed cells to mutate at a higher rate than normal cells so that new variants were being produced continuously. Many of these variants would be eliminated by metabolic or immunological mechanisms but certain of these variants would possess selective growth advantages and these clones would grow to dominate the tumour populations. Sequential selection over time would lead to the emergence of sublines which would be increasingly abnormal both genetically and biologically. Genetic alterations occurring in progressing tumours could range from point mutations to gross aberrations such as loss or gain of complete chromosomes. These topics are discussed in more detail in Chapter 10. Certainly the malignant solid human cancers, which are able to metastasize, commonly show a degree of aneuploidy (variation in chromosome number) and mitotic variation. The hypothesis that neoplastic cells have greater genetic instability and mutate at a faster rate than normal cells has been tested experimentally. The measurement of mutation rates at different sites (loci) in the cellular genetic material (genome) has shown that, in experimental tumours, transformed cells are more genetically unstable than their normal counterparts. A similar change appears to accompany the transition from low to high metastatic activity. This last observation may help explain how metastatic cells develop from the original tumour cell population but, as with many findings in cancer research, raises many new questions.

The mechanisms by which these alterations in gene activity may influence tumour growth and metastasis may include overexpression of normal gene products, gene amplification, or mutation (aberration in gene structure and function). Alternatively epigenetic (non-genetic mechanisms altering gene expression) factors may be responsible for changes in metastatic behaviour. These mechanisms are discussed in detail elsewhere (Chapters 9 and 10) but active research in this field is opening up exciting possibilities not only in the general area of tumour growth but possibly in understanding the process of metastasis.

2.7 Experimental models/approaches to metastasis

It is apparent that relatively little is known about the exact mechanisms of tumour spread. In part this is a reflection of the complexity of the process and the fact that tumours of all types may not use identical mechanisms. In part it also reflects the difficulties involved in studying a dynamic process by essentially static observations. There is a wealth of data gathered on naturally occurring tumours in man from histological, surgical, and autopsy procedures. Many of the mechanisms likely to be involved have been inferred from these observations rather than by direct demonstration through experimental analysis. Biochemical studies on material from primary and secondary tumours obtained at the same time frequently is difficult because of problems in obtaining the material, so that considerable reliance has to be placed on the use of transplantable tumours in experimental animals to study the process of metastasis.

There are many advantages, and not a few disadvantages, associated with these animal models. The advantages arise from the ability to standardize procedures and the ease with which tumour cells growing as implants or as cell lines in tissue culture can be manipulated. The major disadvantage of the models is that the majority of the tumours represented are of mesenchymal origin, since these cells grow more readily in tissue culture whereas, as pointed out in Chapter 1, the vast majority of human solid cancers are epithelial in origin. Information derived from studies on mesenchymal cells may not be applicable directly to epithelial cells and there is a great need for the development of more realistic models of tumour spread (see Chapter 6). It may be that the use of epithelial lines of human tumour cells injected into athymic nude mice will provide such models. Notwithstanding these reservations about currently available transplantable tumour systems, it is true that a considerable amount of information has been derived from experimental studies and much of this has come because of the awareness of the cellular heterogeneity existing in tumours. While there has been some controversy over whether or not metastatic subpopulations of cells do pre-exist in the parental tumour, there is no doubt that general acceptance of this concept has led to the development of some powerful experimental tools. From a single parental tumour one can isolate sublines or variants which exhibit different metastatic capacities. When these differences are relatively stable, comparisons can be made between the variants to determine factors responsible for these phenotypic differences. Comparisons between metastatic and non-metastatic tumours of different origin, a situation somewhat akin to comparing apples and oranges, are thus avoided. Matched, homologous variants of high- and low-metastatic activity would seem to be ideal models for use in attempts to obtain the metastatic phenotype by gene transfer. At the time of writing no putative 'metastasis' gene has been isolated and well-characterized using DNA transfection techniques (discussed in Chapter 9) though considerable effort has been directed at such an approach (not least by the author!). What has been achieved though is the acquisition of metastatic capability in a small range of cell types by transfection with isolated oncogenes (see Chapter 9) suggesting that, for certain cell types, growth autonomy may be the only requirement for metastatic development. However, this is likely to be a finding that is very much dependent on the nature of the recipient cells and may simply indicate that the lines chosen to represent normality in the assay are actually far from normal in reality. A recent finding which resulted from the experimental utilization of high and low metastatic lines, which could account for this lack of success in the dominant transfer of the metastatic phenotype, has been the demonstration that enhanced malignant potential may be associated with the loss of a novel suppressor gene (Steeg *et al.* 1988; see also

Chapter 4). This gene, *nm*23, codes for a protein which has sequence homology with a developmentally regulated protein in *Drosophila* (the abnormal wing disc or *awd* gene). Mutations in *awd* cause abnormal tissue morphology in *Drosophila* and lend credence to the idea that loss of genes like *nm*23/*awd* which normally regulate development, rather than the acquisition of expression, is the mechanism underlying the achievement of the metastatic state (Rosengard *et al.* 1989). Clinical support for this scenario has come not only from the demonstration of low levels of *nm*23 RNA in human breast carcinomas of more aggressive behaviour (Bevilacqua *et al.* 1989) but also from the finding that the allelic deletions involving chromosome 18q late in colorectal cancer progression involve a gene specifying a protein with similarity to neural cell adhesion molecules (Fearon *et al.* 1990). Thus acquisition of metastatic competence may involve the loss of suppressor genes in a manner that is analogous to the role of suppressor genes in tumorigenicity (see Chapter 4).

Many of the associations between malignant behaviour and the possession of specific properties described in this chapter have been established using experimental analyses, and in the future it is certain that more associations will continue to be identified so that we may develop a more complete understanding of the pathogenesis of cancer spread. Until now modern molecular biology has played little part in unravelling this process but, given the availability of appropriate model systems, this is a situation that is likely to change dramatically over the next few years.

References and further reading

Barsky, S. H. and Gopalakrishna, R. (1988). High metallo-proteinase inhibitor content of human cirrhosis and its possible conference of metastasis resistance. *Journal of the National Cancer Institute* **80**, 102–8.

Bevilacqua, G., Sobel, M. E., Liotta, L. A., and Steeg, P. S. (1989). Association of low nm23 RNA levels in human primary infiltrating ductal breast carcinomas with lymph node involvement and other histopathological indicators of high metastatic potential. *Cancer Research* **49**, 5185–90.

Fidler, I. J., Gersten, D. M., and Hart, I. R. (1978). The biology of cancer invasion and metastasis. *Advances in Cancer Research* **28**, 149–205.

Fearon, E. R. *et al.* (1990). Identification of a chromosome 18q gene that is altered in colorectal cancers. *Science* **247**, 49–56.

Gehlsen, K. R., Argraves, W. S., Piersbacher, M. D., and Ruoslahti, E. (1988). Inhibition of in vitro tumor cell invasion by Arg-Gly-Asp-containing synthetic peptides. *Journal of Cell Biology* **106**, 925–30.

Hart, I. (ed.) (1988). Tumour progression and metastasis. *Cancer surveys*, Vol. 7, No. 4. Oxford University Press.

Juliano, R. L. (1987). Membrane receptors for extracellular matrix macro-molecules: relationship to cell adhesion and tumour metastasis. *Biochimica et Biophysica Acta* **907**, 261–78.

Liotta, L. A. and Hart, I. R. (ed.) (1982). *Tumour invasion and metastasis*. Martinus Nijhoff, The Hague.

Liotta, L. A. *et al.* (1986). Tumor cell autocrine motility factor. *Proceedings of the National Academy of Sciences, USA* **83**, 3302–6.

Nowell, P. C. (1976). The clonal evolution of tumour cell populations. *Science* **194**, 23–8.

Prehn, R. T. (1976). Do tumors grow because of the immune response of the host? *Transplantation Reviews* **28**, 34–42.

Rosengard, A. M. *et al.* (1989). Reduced Nm23/Awd protein in tumour metastasis and aberrant Drosophila development. *Nature* **342**, 177–80.

Steeg, P. S. *et al.* (1988). Evidence for a novel gene associated with low tumor metastatic potential. *Journal of the National Cancer Institute* **80**, 200–4.

Willis, R. A. (1972). *The spread of tumours in the human body*. Butterworth, London.

3

Epidemiology of cancer

M. C. PIKE and D. FORMAN

3.1 Introduction

Cancer epidemiology is the study of the pattern of cancer in populations and is essentially statistical, its methods and conclusions being expressed in terms of probabilities, for example 'Japanese women have less than a quarter the breast cancer risk of US women', 'women who have a baby before age 20 have only half the chance of getting breast cancer of women without children', or 'men exposed to benzene have an increased risk of leukaemia'. Its fundamental aim is, of course, to identify preventable (avoidable) causes of cancer, but it also has a critical role to play in many other areas of cancer research, particularly in the evaluation of screening tests to detect cancer at an early stage.

The most basic task of cancer epidemiology is simply to describe the occurrence of human cancer, noting differences, for example, between males and females, between persons of different ages, between different socio-economic classes, between persons in different occupations, between different time periods, between different areas of a country, and between different countries. This descriptive epidemiology has been a most fruitful source of ideas as to the possible causes of various cancers. For example, the enormous rise of lung cancer in men, but not in women, between 1920 and 1945 suggested that some recently introduced habit of men, but not of women, must be responsible, and cigarette smoking became the prime candidate. More recently the finding of large differences in the occurrence of colon cancer between different countries has led to intense investigation of the possible role of various aspects of the normal diet (such as fat or fibre content) as factors in the aetiology of this cancer.

3.2　Descriptive epidemiology

3.2.1　*Incidence rates*

To describe the differences in occurrence of a particular cancer between different groups in a meaningful way the most useful concept is that of an incidence rate—the probability of an individual in a particular group being newly diagnosed as having the particular cancer within a year. In epidemiological studies these probabilities are usually expressed not as fractions or decimals but as 'per 100 000', so that, for example, an incidence rate expressed as '200 per 100 000' is identical to an incidence rate of 0.2 per cent or a probability of 0.002. The incidence rate, for example, of breast cancer in females in England and Wales in 1983 was 83.6 per 100 000 (estimated by noting that there were 25 477 800 females in England and Wales on 1 July 1983 and that 21 297 new breast cancer cases were diagnosed in females in England and Wales in that year: 21 297/25 477 800 = 0.000836 or 83.6 per 100 000). We would estimate the equivalent figure for 1981–3 as 83.9 per 100 000 calculated by dividing the 64 079 new breast cancer cases in the three years by the appropriate population of females. Incidence rates are commonly affected by the sex distribution of the people in the reference group, and in this chapter we shall always be referring to single sex reference groups, e.g. the incidence rate of lung cancer in UK females in 1980.

3.2.2　*Age specific incidence rates*

Incidence rates are almost invariably strongly affected by the age of the people in the reference group. For aetiologically meaningful comparisons, incidence rates must be worked out separately for groups of persons all of a similar age. The collection of such incidence rates over a span of age groups is referred to as an age specific incidence curve.

Table 3.1 and Fig. 3.1 show the age specific incidence figures for cancer of the colon in females in England and Wales in the period 1979–82. These particular age specific incidence rates display the age incidence pattern seen with most of the important cancers, that is, a rapid increase in incidence with age (a more than 100-fold increase between age 25 and age 70). It has been found most useful to plot both the incidence rate and age on logarithmic scales, because the 'log–log' plot so obtained is, for cancers at many different sites, very close to a straight line (Fig. 3.2). Deviations from the straight line pattern provide import-ant insights into the aetiology of a number of cancers, in particular the major hormone associated cancers of women (breast, endometrium, and ovary).

The age specific incidence curve for breast cancer in females in England and Wales in 1979–82 is shown in Fig. 3.3; there is a roughly linear relationship of the logarithm of incidence with the logarithm of age until about age 50, but then the rate of increase of incidence with increasing age clearly slows down. This incidence curve therefore strongly suggests that something occurs around age 50 in women that acts as a brake on the rate of further breast cancer incidence increases— the obvious candidate is menopause. The protective effect of menopause

Table 3.1 Age specific inci-dence of cancer of the colon in females in England and Wales (1979–82)[1]

Age	Rate per 100 000
0–4	0.1
5–9	0.0
10–14	0.1
15–19	0.1
20–24	0.3
25–29	0.7
30–34	1.5
35–39	4.2
40–44	7.1
45–49	13.1
50–54	22.5
55–59	36.8
60–64	54.3
65–69	78.8
70–74	109.5

[1] From Muir *et al.* (1987).

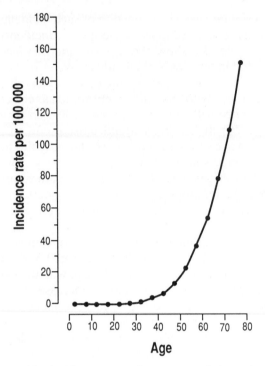

Fig. 3.1 Age specific incidence rates for cancer of the colon in females in England and Wales, 1979–82.

(either natural or resulting from removal of the ovaries) has now been definitely established (see p. 77), and this observation plays a central role in research into the causes and prevention of breast cancer.

The age specific incidence curves of endometrial and ovarian cancer have the same general shape as that of breast cancer. Early menopause also protects against both of these cancers, and provides key insights into the causes of both cancers. We discuss endometrial and breast cancer in more detail below.

The age specific incidence curve for cancer of the testis in males in England and Wales in 1979–82 is shown in Fig. 3.4; there is a peak in incidence around age 30 and then a steady fall. Such a pattern suggests that there is a limited group of susceptible men in the population and that they have almost all been diagnosed with the cancer by middle age. For this and other reasons it is sensible to postulate that the 'susceptible men' are those that have been exposed to some 'carcinogenic agent' either during development in the uterus or in the first few years of life. This hypothesis is now the subject of active research (see p. 83).

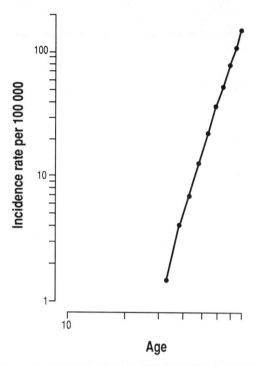

Fig. 3.2 Log–log plot of the age specific incidence rates for cancer of the colon in females in England and Wales, 1979–82.

If the incidence rates for a particular cancer are changing rapidly in time, then the log–log plot of incidence against age for a particular calendar period will not show the true relationship between incidence and age. If incidence rates are increasing, as they did for lung cancer when cigarette smoking became common, then the incidence at old age will be lower than the incidence that the population providing the incidence at young ages will have when they reach old age (see Table 3.5). The log–log plot will thus have too shallow a slope and may even reach a peak before old age and then decline. The opposite phenomenon will be observed if the incidence rates are decreasing. The correct interpretation of an age specific incidence curve, therefore, also requires an understanding of its changes over calendar time.

3.2.3 *Age standardized incidence rates*

Comparisons of age specific incidence curves between different groups for a particular cancer usually show that the pattern of change with age is very similar and it is only the level of incidence that varies. For example,

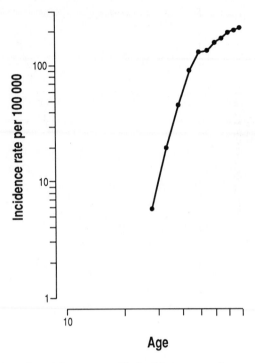

Fig. 3.3 Log–log plot of the age specific incidence rates for cancer of the breast in females in England and Wales, 1979–82.

Fig. 3.5 shows the age specific incidence of stomach cancer in men in Japan in 1973–7 and in the USA in 1969–71. For these situations, comparisons between the different groups can most simply be made by giving a single number representing the level of incidence for each group. The single number chosen is usually taken either as the sum of the incidence rates at each single year of age between, for example, 0 and 74 (the cumulative incidence method) or by using a weighted average of the age specific incidence rates (the age standardized incidence method). The cumulative incidence method gives a number that is the probability that a person gets the specific cancer within the age range considered (if he does not die from any other cause), while the age standardized method gives the incidence in a standard group whose age structure is represented by the weights used in calculating the weighted average; both methods give essentially the same comparative information.

    If the pattern of change of incidence with age is not similar in the groups being compared then single numbers cannot, of course, adequately represent the differences between the groups. Cumulative

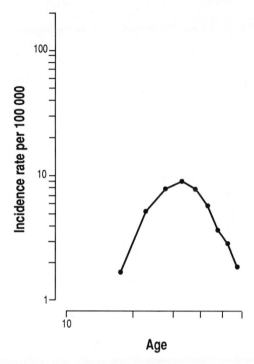

Fig. 3.4 Log–log plot of the age specific incidence rates for cancer of the testis in males in England and Wales, 1979–82.

incidence rates or age standardized rates will still provide valuable information on the occurrence of cancer in the different groups; they need to be supplemented with a description of the age specific patterns in the different groups.

Comparisons of cumulative or age standardized incidence rates between different groups is one of the most commonly used descriptive epidemiological methods. Between-country comparisons of age stan-dardized rates suggest that some 80 to 90 per cent of the current total cancer rate in the UK is caused by environmental and/or behavioural factors and may thus be preventable (discussed later).

3.2.4 Comparisons between populations

Table 3.2 shows the cumulative incidence rates (ages 0–74) of breast cancer in different populations around 1970 and around 1980; the rates vary over a more than sixfold range. Japan, China, India, and black Africa have the lowest rates, while the highest rates are observed in

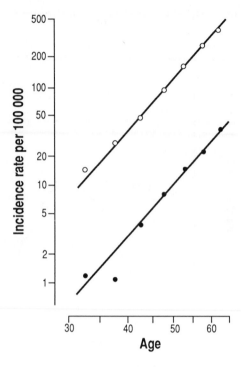

Fig. 3.5 Log–log plot of the age specific incidence rates for cancer of the stomach in males in Miyagi Prefecture, Japan, 1973–7 and in the USA (Third National Cancer Survey), 1969–71.

North American whites. It should also be noted that 16 out of the 19 populations shown in the table, for which rates are available at both points in time, show an increase during the decade considered; in many cases this occurred at a rate in excess of 1 per cent per year. Some part of this increase will undoubtedly be due to more complete cancer registration, but it would also seem that there has been a real increase in incidence.

Table 3.3 shows that Japanese, Chinese, and African immigrants to the USA have breast cancer rates intermediate between those of their country of origin and the USA white population. Blacks in Africa have rates less than one-fifth that of US whites, whereas US blacks have almost three-quarters that rate and at young ages their breast cancer rates are higher than that of US whites. These dramatic effects of migration on breast cancer rates show that environmental and/or behavioural factors are major determinants of breast cancer risk. The fact that a young Japanese American woman now has approximately the

Table 3.2 Cumulative incidence rates (ages 0–74) of breast cancer in different populations

Population	Cumulative incidence rate (%)	
	c. 1970[1]	c. 1980[2]
USA, San Francisco, white	8.9	9.9
Canada, British Columbia	8.8	7.5
Israel, Jews, Europe/USA born	6.5	7.8
USA, San Francisco, black	6.4	7.2
New Zealand, non-Maori	5.8	6.3
Sweden	5.8	6.8
New Zealand, Maori	5.7	6.8
USA, Hawaii, Japanese	5.3	5.5
England and Wales	5.3[3]	6.0
Brazil, Recife	5.1	5.0
Norway	4.9	5.8
USA, San Francisco, Chinese	4.8	4.9
Finland	3.7	4.9
Colombia, Cali	2.9	3.8
Israel, Jews, Africa/Asia born	2.9	5.6
India, Bombay	2.2	2.4
China, Shanghai	2.2[4]	2.1
Singapore, Chinese	2.2	2.9
Nigeria, Ibadan	1.7	N/A
Japan, Osaka	1.3	2.1
Senegal, Dakar	1.3[4]	N/A

[1] From Waterhouse *et al.* (1976).
[2] From Muir *et al.* (1987).
[3] Calculated from England and Wales Cancer Registration Statistics for 1971 (Office of Population Censuses and Surveys 1979).
[4] These rates are *c.* 1975 (Waterhouse *et al.* 1982).

same risk of breast cancer as a British woman means that the low rates of breast cancer in Asia are not simply due to a difference in genetic susceptibility.

Breast cancer is not unusual in showing large differences between incidence rates in different populations; for most cancer sites the pattern of disease in migrants comes to resemble that of their host country in a few generations as they adopt the local lifestyle. A similar effect to that for breast cancer is shown in Table 3.3 for colon cancer in men and in this case Japanese-American rates even exceed the rates in US whites.

Table 3.4 shows the cumulative incidence rates for the common cancer sites in the UK and the percentage reduction in these rates if the lowest observed rate in the world could be achieved in the UK—the reductions are over 70 per cent for most of the major sites, although some of the

Table 3.3 Cumulative incidence rates (ages 0–74) of female breast and male colon cancer in different populations[1] (*c. 1980*)

	Population	Cumulative incidence rate (%)
Female breast cancer	USA, white	8.84[2]
	USA, Japanese	5.03[3]
	Japan	2.21[4]
	USA, Chinese	4.99[3]
	China	2.03[5]
	USA, black	6.78[6]
	Africa, black	1.47[7]
Male colon cancer	USA, white	3.48[2]
	USA, Japanese	3.90[3]
	Japan	1.36[4]
	USA, Chinese	2.95[3]
	China	0.83[5]
	USA, black	3.23[6]
	Africa, black	0.20[7]

[1] From Muir *et al.* (1987).
[2] Weighted means of rates from 8 US Registries (Alameida, Bay Area, Los Angeles, Connecticut, Atlanta, New Orleans, Detroit, Hawaii).
[3] Weighted means of rates from 3 US Registries (Bay Area, Los Angeles, Hawaii).
[4] Weighted means of rates from 4 Japanese Registries (Hiroshima, Miyagi, Nagasaki, Osaka).
[5] Weighted means of rates from 2 Chinese Registries (Shanghai, Tianjin).
[6] Weighted means of rates from 7 US Registries (Alameida, Bay Area, Los Angeles, Connecticut, Atlanta, New Orleans, Detroit).
[7] Weighted means of rates from 2 African Registries [Dakar, Senegal (*c.* 1975) and Ibadan, Nigeria (*c.* 1970); Ibadan figure also includes rectal cancer with colon].

reductions in Table 3.4 may be inflated due to under-recording of cancers in the low incidence areas. There is clearly tremendous potential for preventing cancer if the environmental and/or behavioural factors responsible for the extreme variations in site specific cancer rates could be identified and people in the UK were prepared to adopt the low-risk behaviour patterns.

3.3 Identifying the causes

In this section we consider, in some detail, the nature of the research findings which underlie our current understanding of the causation of

Table 3.4 Comparison of England and Wales cumulative incidence rates (ages 0–74) for common cancer sites with the lowest incidence population (*c.* 1980)[1]

Primary site of cancer	Cumulative incidence rate (%)			
	England and Wales rate	Lowest incidence rate[2]	Lowest incidence population	Potential reduction
Males				
Lung	9.15	0.66	India, Madras	93%
Stomach	2.23	0.71	USA, Utah	68%
Prostate	2.17	0.10	China, Tianjin	95%
Bladder	1.99	0.21	India, Nagpur	89%
Colon	1.89	0.26	India, Bombay	86%
Rectum	1.60	0.35	India, Madras	78%
Pancreas	0.94	0.12	India, Madras	87%
Females				
Breast	5.96	1.87	Japan, Miyagi, Rural	69%
Lung	2.42	0.05	India, Nagpur	98%
Colon	1.65	0.27	India, Bombay	84%
Ovary	1.26	0.41	China, Tianjin	68%
Cervix	1.21	0.27	Israel, Non-Jews	78%
Endometrium	1.00	0.12	India, Nagpur	88%
Rectum	0.90	0.16	India, Nagpur	82%
Stomach	0.85	0.31	USA, Utah	64%

[1] Muir *et al.* (1987).
[2] Excluding cancer registry data based on small numbers (less than 1 million person-years).

five specific types of cancer (lung, stomach, endometrium, breast, and testis). We give a brief description of the known and strongly suspected risk factors for the complete range of cancer sites. The five selected sites discussed in this section have been selected for differing reasons and illustrate the principles involved. World-wide lung cancer rates have increased to epidemic proportions and we know precisely why. World-wide stomach cancer rates, in contrast, have been declining and as yet it is unclear why this has happened. Dietary practices are believed to be central to the development of stomach cancer and the difficulties involved in investigating diet and cancer associations are exemplified by research on stomach cancer. Breast, endometrial, and testicular cancer are all presumed to have a strong hormonal dependence. Endometrial cancer is particularly well characterized in this respect and almost every feature of its epidemiology is now understood. Breast and testicular cancer remain more enigmatic and require a more complex consideration of hormonal action.

3.3.1 Lung cancer

Table 3.4 shows that lung cancer is by far the most frequent cancer of UK men and it is also, after breast cancer, the most predominant cancer of UK women. This cancer occupies a unique position among the important cancers in that we have known for at least 35 years the cause of some 90 per cent of it, namely cigarette smoking. This is clearly avoidable. Present epidemiological research interest in lung cancer is mainly focused on identifying subgroups of persons at especially high risk of lung cancer if they smoke and in measuring the risk from passive smoking, i.e. from being exposed to other persons' cigarette smoke. The latter may be particularly important; if passive smoking does cause a significant amount of disease then more stringent control of smoking in public places is clearly called for.

Figure 3.6 shows the trends in male and female lung cancer mortality in England and Wales from 1916 to 1970. Some of the increase in the earlier period is due to improvements in diagnostic accuracy, but the size of the male risk and its continuing increase persuaded scientists and public health officials in the years immediately after the Second World War that the reason for the increase warranted urgent study. That

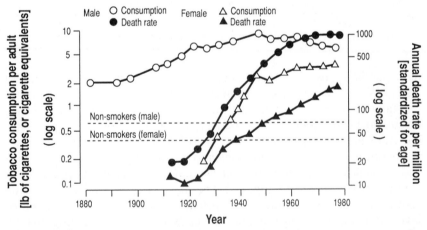

Fig. 3.6 Trends in male and female age standardized lung cancer mortality rates in England and Wales, 1916–70 and cigarette sales per adult. The dotted lines are lung cancer rates found in lifelong non-smokers in cohort studies conducted in the 1950s and 1960s—note that the observed national rates early this century were lower than these rates and these differences in all likelihood are a reasonable measure of the improvements in diagnostic accuracy that have been achieved over the years.

cigarette smoking could be the cause of the increase was suggested by a number of people and, unbeknown to scientists in the UK and USA, two epidemiological studies conducted in Germany in the late 1930s and early 1940s had, in fact, found a strong association between lung cancer and cigarette smoking. By the end of 1950 a further six case control studies had been published; these studies all compared the cigarette smoking history of lung cancer cases to that of similar men without the disease (controls) and all showed a clear relationship between increasing risk and increasing exposure.

Figure 3.6 also shows how, with women taking up smoking, their lung cancer rates have also increased and how well knowledge of the smoking habits of a population predicts its lung cancer rates some two to three decades later.

There was naturally a great deal of reluctance to accept the findings of the early case control studies—cigarette smoking was an almost universally accepted habit—and many further epidemiological studies were soon initiated, partly to answer criticisms that had been made of the case control studies, but also to investigate whether smoking was associated with cancers of other sites. A number of these epidemiological studies were cohort studies, i.e. studies in which large numbers of people with different smoking habits were identified (in the UK the smoking habits of a large number of doctors were recorded) and their subsequent mortality monitored. These cohort studies not only confirmed the lung cancer/cigarette smoking association but soon established that smoking was also causally associated with cancers at a number of other sites, e.g. tongue, mouth, pharynx, larynx, and bladder (Table 3.10).

Further studies of the relationship of cigarette smoking to lung cancer also showed that stopping smoking has an almost immediate effect on lung cancer risk. The absolute difference between the ex-smoker's lung cancer rate and the lung cancer rate of a non-smoker stays effectively constant with length of time after stopping, rather than continuing to increase as it does for the continuing smoker. These studies showed moreover that the age specific incidence rates of lung cancer of persons in late middle or old age depends on their lifelong cigarette smoking habits, so that age standardized lung cancer rates depend not only on current cigarette consumption in the population but on the cigarette consumption of young adults 40 or more years before. Furthermore, changes in population cigarette consumption may take decades to show up in reduced lung cancer rates in older people.

The tobacco industry responded to the ever mounting evidence of the carcinogenic effects of cigarette smoke by reducing the tar and nicotine content of cigarettes, first by introducing filters and later by modifying the tobacco. The tar content of the cigarettes is probably the relevant constituent as regards lung cancer, and the average tar yield per cigarette has steadily declined from approximately 32 mg/cigarette in 1960, to the current level of less than 14 mg/cigarette.

The average number of cigarettes smoked by men in Britain was approximately 10.5/day from 1950 to 1970, but has declined steadily since then and is now approximately 6.5/day. The average number of cigarettes smoked by women in Britain was less than 4/day up until 1960, rose to a maximum of 7/day in the mid 1970s, and then declined; it is now approximately 5/day.

The combined effect of the changed constituents of cigarettes and the average number of cigarettes smoked is shown in Table 3.5. There has been a very significant reduction in male lung cancer incidence; rates are now declining at all ages, except for those currently in their eighties. We can now look forward in this country to an overall two-thirds reduction in the male lung cancer rate even if no further reduction in tar content or cigarette smoking takes place. Female lung cancer rates are also showing a sharp decline at young ages, and although lung cancer rates of older women (over the age of 60 years) are still rising (due to the early adoption of smoking by the present cohort of older women), we can also confidently look forward to a substantial reduction in female lung cancer in the not too distant future.

Unfortunately in many other countries smoking rates are now increasing, especially in young people, in a manner similar to that which took

Table 3.5 Changes in lung cancer mortality rates in England and Wales, 1946 to 1985. (The highest rate for each age group is italicized and bold.)

Age	Death rate per 100 000					Reduction from peak rate to 1981–5
	1946–50	1956–60	1966–70	1976–80	1981–5	
Males						
30–34	*3.6*	3.5	2.5	1.6	1.2	67%
40–44	23.6	*25.1*	21.6	13.8	12.0	52%
50–54	95.4	*124.8*	116.0	99.6	77.0	38%
60–64	171.7	331.5	*369.5*	332.0	300.2	19%
70–74	140.0	387.8	621.0	*662.5*	629.8	5%
80–84	76.5	225.8	456.3	762.3	*833.6*	–
Females						
30–34	1.2	*1.4*	1.1	0.8	0.8	43%
40–44	4.8	6.0	*8.1*	6.2	6.0	26%
50–54	11.7	16.9	28.4	*35.9*	30.0	16%
60–64	22.1	32.8	51.5	84.6	*96.2*	–
70–74	31.6	44.5	73.1	110.5	*141.9*	–
80–84	28.0	45.1	65.7	106.2	*135.1*	–

place in the UK between 1900 and 1950. As these populations age there will be a large increase in their lung cancer rate. It has been estimated that world-wide as many as 3 million smoking-related lung cancer deaths will occur each year by 2025 compared with about 1 million now. For Europe the 2025 figure has been estimated at 670 000 (480 000 men, 190 000 women) up from the current figure of 260 000 (230 000 men, 30 000 women); the relatively larger increase for women reflects the changes in the number of young women currently taking up smoking. This is clearly a major public health challenge.

Two aspects of the relationship between cigarette smoking and cancer— identifying the genetic constitution of the smoker who is most at risk of lung cancer, and the carcinogenic effects of exposure to other people's cigarette smoke (so called 'passive' smoking)—are presently the focus of much interest. There are clearly large differences between the age at onset of lung cancer in different persons with the same smoking habits. These differences do not prove that different people have different susceptibilities to cigarette smoke induced lung cancer, it may all be simply a matter of chance; they do, however, strongly suggest that there may be different genetic susceptibilities involved. The host factor that has excited most interest is the gene complex coding for the cytochrome P450 enzymes. This controls the metabolic oxidative activation of various chemical carcinogens to an active cancer inducing form (see Chapter 6). The different levels of such enzymes, which are known to be genetically controlled in many animal species, could give rise to wide between-person variation in the generation of carcinogenic derivatives of the chemicals in cigarette smoke. Human studies have focused on two enzyme systems—aryl hydrocarbon hydroxylase (AHH) and debriso- quine 4-hydroxylase.

A number of polycyclic aromatic hydrocarbons (PAHs) that occur in cigarette smoke are potent carcinogens, and the activity of AHH has been closely linked to susceptibility to some PAH induced cancers in animals. The first few studies which compared the AHH activities of lung cancer cases with controls showed a clear difference between the cases and the controls, but subsequent studies have been most confusing and in many cases contradictory. The reasons for the confusion are probably methodological, such as the use of different assay and of different cell types to measure AHH activity. The use of cryopreserved tissue may enable one to circumvent some of these problems, but at present it cannot be stated whether or not AHH activity has any role in human lung cancer. Progress will probably require a different laboratory approach, such as identifying the genes controlling the system.

There has been much recent interest in the enzyme system that metabolizes the anti-hypertensive drug debrisoquine. It is known that

5–10 per cent of the population have a genetic constitution which makes them incapable of adequately metabolizing the drug and it is thought that such individuals may also be less able to metabolize cigarette smoke constituents into active carcinogens. Two studies have compared debrisoquine metabolism in lung cancer patients and controls, after carefully matching for smoking, and found significant differences between the groups. The authors concluded that either the gene system controlling this oxidation process is also directly involved in cigarette smoke carcinogenesis or that it is closely linked to a gene system which is. Unfortunately the methodology for carrying out debrisoquine metabolism studies is complex and time consuming. It had been hoped that the gene system could be studied directly using DNA screening tests, but this has not so far proved possible.

There is good evidence that parental, particularly maternal, smoking increases respiratory disease rates in young children, and evidence is slowly accumulating that pulmonary function of non-smokers is adversely affected by exposure to smoking by their spouses. Since cigarette smoke is clearly carcinogenic, it is reasonable to assume that involuntary (passive) smoking will also cause a certain amount of lung cancer; the epidemiological question of interest is not really whether or not involuntary smoking causes lung cancer, but how much lung cancer does it cause.

Epidemiological studies suggest that a never-smoker's risk of lung cancer may be increased by some 50 per cent from exposure to other persons' smoke. This estimate has been disputed by a number of researchers who suggest rather that the increased risk of lung cancer seen among never-smokers exposed to passive smoking is largely the effect of certain self-proclaimed never-smokers having at some time been actual smokers. A figure of a 10 per cent increase in risk is suggested by studies of cotinine (a measure of absorbed nicotine) in active and passive smokers. The relationship between the amount of nicotine absorbed and the amount of tar deposited in the lungs is not necessarily the same in passive and active smoking, however, and this could influence the postulated risk in either direction. At present we conclude that it is most reasonable to regard the 50 per cent figure as an upper limit of risk. Further carefully done epidemiological/biochemical studies should resolve some of these issues.

Although cigarette smoking is by far the major cause of lung cancer, exposure to a number of other substances has created a significant added risk of lung cancer for workers in certain industrial occupations. The most important of these has been exposures to asbestos. A substantial number of men previously exposed to high levels of asbestos, particularly in the shipbuilding and insulation industries, have very high lung cancer rates. This risk is now universally recognized and the asbestos levels permitted in industry have been very substantially reduced. Occupational exposure to PAHs from the combustion of fossil fuels has also been an important source of added lung cancer risk. High level PAH exposure has occurred particularly to men working in the fumes from coke ovens and in coal gas manufacturing. Other substantial risks to small groups of workers have resulted from exposure to some aspect of the manufacture or refining of chromates, nickel, and copper, and from exposure to radiation from the radioactive gas radon in the air of certain mines (see Chapters 6 and 7).

Radon is not confined to underground areas, but is present in the air in some houses and other buildings where it is not free to diffuse away. Recent advances in techniques for measuring exposure to radon have shown that levels are higher than was previously thought, and current estimates indicate that radon accounts for nearly half the average UK population exposure to ionizing radiation. In the UK the primary determinant of the indoor radon concentration is the underlying geology. In parts of Cornwall and some other areas of the UK the exposure of individuals may be more than 15 times the national average. In some other countries radon exposure from building materials and the radon content of water also influence heavily the indoor radon concentration. Extrapolation from studies of uranium and iron ore miners in North America and Sweden indicate that approximately 5 per cent of all lung cancers may be caused by exposure to indoor radon. If these estimates are correct, radon would be the second most important cause of lung cancer after smoking. Intensive investigation is under way to provide better quantitative information on these risks.

The general population has also been exposed in the past to significant amounts of PAHs in urban air, mainly from the uncontrolled burning of coal. Such general air pollution may have contributed to as much as 10 per cent of all lung cancer in heavily polluted cities but with the passing of various clean air legislation over the years, air pollution has been reduced dramatically and current levels are unlikely to be making more than a small contribution to lung cancer risk.

3.3.2 Stomach cancer

World-wide, stomach cancer is the second most important cancer after lung cancer, and it is the most important type in many countries where smoking has become popular relatively recently. The incidence rates for stomach cancer are declining in nearly all countries. This trend has gone furthest in the USA where stomach cancer has been declining since the 1920s and it is now only the sixth most important type. Even countries such as Japan and Chile, which have the highest rates in the world, are now showing decreases although the decline started much later than in the USA. The reduction in incidence cannot be attributed to public health policies or to planned interventions. Unlike the situation with lung cancer, no single major causative agent has been identified for stomach cancer and the reasons for its decline are largely unknown. It is of obvious importance to try to understand the reasons for the decline in order to accelerate its rate as, in most of the world, it is still a major contributor to overall cancer mortality.

Many factors have been investigated for a possible association with stomach cancer, but most attention has been paid to diet. This is understandable given that any carcinogens (or anti-carcinogens) in the diet will usually reside in the stomach for several hours before absorption in the small intestine. Diet is extremely difficult to investigate in epidemiological studies mainly because one is interested in obtaining a long-term diet history rather than just a record of what the person has consumed in the immediate past, and even a record of recent food and drink consumption tends to be very unreliable as individuals have great difficulty in recalling (and knowing) precisely what they eat and drink. The usual method of obtaining a diet history is for the cancer patient and matched control groups to indicate on a check list the frequency with which they consume a range of dietary items. At least 12 such dietary case control studies of stomach cancer have been carried out in the last few years.

From these studies four dietary factors stand out and, by and large results have been consistent in all studies. Increased consumption of vegetables and fresh fruit has an appreciable protective effect while consumption of salt and 'preserved foods' tends to increase the risk of stomach cancer. A summary of relevant results is given in Table 3.6. This shows that those in the highest intake group for fruit and vegetable consumption have relative risks of 0.5–0.7, which means they have a 30–50 per cent reduction in risk, compared to those in the lowest intake group. In contrast those with high intakes of salt and 'preserved foods' generally had relative risks in the range 1.5–2.5, indicating a 50–150 per cent increase in risk. The uniformity of these findings in studies from a variety of countries with diverse diets makes it likely that the results represent genuine causal associations.

The results in Table 3.6 beg the question of what mechanisms might be involved. Fruit and vegetables are known to contain a wide range of potentially protective nutrients and it is not possible to state with certainty which ones are critical. Indeed it may be precisely because there are several protective factors that plant products have a noticeable effect. A great deal of research has been carried out on the anti-oxidants contained in fruit and vegetables, particularly ascorbic acid and beta-

Table 3.6 Relative risks for selected dietary items in recent case control studies of stomach cancer

Country	Year published	Study size		Relative risk at highest intake level[1]				Type of preserved food
		Cases	Controls	Fruit	Vegetables	Salt	'Preserved foods'[2]	
China	1988	564	1131	0.6*	0.4*	1.4	1.4	Salted fish
China	1988	241	241	N/A	N/A	N/A	1.5*	Salted soya paste
Japan	1985	93	86	N/A	N/A	2.6*	2.0*	Pickled vegetables
Italy	1989	1016	1159	0.6*	0.6*	1.5*	2.4*	'Traditional' soup
Italy	1987	206	474	0.7	0.3*	1.5	1.6*	Ham
UK	1989	95	190	0.6	0.3*	6.2*	1.0	Smoked food
UK	1989	149	1934	0.4*	N/A	N/A	N/A	—
Greece	1985	110	110	0.8*	0.7*	N/A	N/A	—
Poland	1986	110	110	0.3*	0.6	N/A	N/A	—
USA, white	1985	194	195	0.5*	0.9	1.4	N/A	—
USA, black	1985	197	195	0.3*	0.5	1.8	2.3*	Home cured meat
Canada	1985	246	246	0.8*	0.8*	N/A	2.6*	Nitrite preserved food

* Relative risk statistically significant for test utilized in study.
[1] Figures show risk of developing stomach cancer for those consuming food at highest level of intake in comparison to those consuming at the lowest level of intake. The categorization of intake levels varies between studies but, most commonly, uses the range of intake in the control population divided into thirds.
[2] Relative risk presented for the item of preserved food showing the most extreme value in the study.

carotene. As many established carcinogens ultimately exert their effect via an oxidation pathway, potent anti-oxidants should act as anti-cancer agents. More specific evidence in support of such an association comes from studies in which serum levels of beta-carotene were measured in a defined healthy population and comparisons were made between those who subsequently developed stomach cancer and those who remained cancer free. Three such studies have reported results and all three showed a protective effect of high serum beta-carotene levels. In the one study which also measured serum ascorbic acid levels, high levels of ascorbic acid were also found to be protective.

There is much interest in determining whether taking very high doses of the vitamins ascorbic acid and beta-carotene reduces cancer risk and this is being evaluated in controlled trials (see Section 3.4.4). It is unlikely that vitamin supplementation alone will provide the optimal protective effect. Other dietary components are also likely to contribute, e.g. there is evidence from studies in China and Italy that chemicals contained in the allium family of vegetables (onions, garlic, etc.) are protective specifically for stomach cancer and this is not thought to be a vitamin associated effect.

High salt intake increases the risk of stomach cancer, probably by acting as a chronic irritant to the stomach mucosal lining. Over a long time period this would lead to impaired functioning of the mucosa which may result in direct exposure of the stomach epithelial cells to gastric juice. Any toxic or carcinogenic agents in the diet would therefore have a much more intimate contact with target cells. There is also animal evidence to suggest that increased salt intake causes transformed cells to proliferate more rapidly.

Included in the category of 'preserved foods' are a diversity of products and preservation methods. Clearly preservation *per se* is not a single factor and one needs to carefully investigate each of the individual food items. In some cases carcinogens have been identified, e.g. *N*-nitroso compounds in certain preserved meats, pickles, and fish, and in these instances, efforts can be made to reduce or eliminate them. In other cases carcinogens may be present in extremely small quantities and may only pose a problem in communities with a heavy dietary dependence on them. Some preserved foods, e.g. 'traditional' soup in Italy, may not be a problem *per se* but rather a particularly good indicator of an overall poor diet.

Generally a decrease in the dependency on preserved foodstuffs and salt, together with increased availability of fruit and vegetables, offers the best explanation for the world-wide decline in stomach cancer. It is unlikely that these four dietary factors are sufficient to explain the full extent of the decline. In the UK, for example, gastric cancer has declined markedly in the past 40–50 years and yet there is very little evidence to suggest that fruit and vegetable consumption has increased, and that salt and pickled food consumption has decreased by a commensurate amount even after allowing for a 20–30 year lag period between dietary change and change in cancer mortality. There have, of course, been other parallel changes in diet and dietary produce, e.g. the introduction of widespread domestic refrigeration and the consequent better food quality, may also play a part.

As well as diet, several other factors have been associated with risk of stomach cancer. Smoking increases the risk about twofold. There is also a strong gradient in risk associated with socio-economic status. Those in manual occupations tend to have a two to threefold increase in risk compared with those in non-manual occupations. Jobs which involve dust exposure, e.g. coal-mining, seem to have a particularly high risk. There are also genetic factors involved, as it has been known for some time that people with blood group A have an excess risk of certain types of stomach cancer.

Stomach cancer is frequently preceded by the development of chronic atrophic gastritis (CAG), i.e. inflammation of the stomach epithelium leading to cellular atrophy, and this has been regarded as a key aetiological event. Metaplastic and dysplastic transformation of epithelial cells occurs much more frequently in people with CAG and such changes frequently precede cancer. A consequence of CAG is loss of gastric acidity because of destruction of acid-secreting cells. All medical conditions that result in CAG and loss of acidity have an excess risk of stomach cancer. Patients with pernicious anaemia, for example, which is usually accompanied by CAG, have been shown to have a three- to fourfold increased risk of cancer.

Attention is now being paid to the causes of CAG. There are several agents that may act as irritants in the stomach and give rise to chronic inflammation. Salt has already been mentioned and coal and other dust particles, and perhaps regular use of certain drugs such as aspirin may act in the same way. Recently there has been a lot of interest in the role of the bacterium *Helicobacter pylori* as a cause of CAG. *Helicobacter pylori* is found exclusively in the stomach where it attaches to the luminal surface of the stomach epithelium beneath the surface mucous layer. There is now considerable evidence to show that *H. pylori* infection can rapidly cause inflammation and acute gastritis, although infection is frequently asymptomatic. Once established in the stomach, *H. pylori* infection seems to be resistant to natural host defence mechanisms, rarely resolves spontaneously, and can be eradicated only with aggressive antibiotic treatment. It is possible that the persistence of the organism will result in a continual inflammatory response and progression from acute to chronic gastritis and subsequently to CAG. It is not yet known whether *H. pylori* alone is sufficient for this progression or whether other cofactors are needed. Either way, the *H. pylori* associated gastritis seems to be a precursor of changes that predispose to stomach cancer. *Helicobacter pylori* infection is a relatively common occurrence in many populations and early evidence suggests that, geographically, there is some association between *H. pylori* prevalence and the rate of gastric cancer. Infection at a young age may be particularly important as this would expose the stomach to a chronic inflammatory stimulus for several decades.

Our understanding of what happens to the gastric epithelium after the establishment of CAG owes much to the work of Correa and his colleagues (Correa 1988). He has suggested that the increase in gastric juice pH which accompanies CAG, leads to colonization of the stomach by species of bacteria which can catalyse the formation of intragastric nitrite from nitrate. This nitrite may then nitrosate protein substrates and form carcinogenic *N*-nitroso compounds (NOCs). Such a sequence of events is biologically plausible and all of the individual steps have been demonstrated in studies in man. Also, anti-oxidants such as ascorbic acid are very effective at specifically inhibiting the formation of NOC in the stomach, which may partly explain the protective action of fruit and vegetables. What has yet to be established is whether endogenously synthesized NOCs make any significant contribution to the development of human stomach cancer and, if so, what factors control the rate of formation of these compounds.

Much epidemiological research has been focused on the role of dietary and water-borne nitrate in the synthesis of NOC. The complexity of assessing nitrate exposure has meant that most studies have tended to be correlational in design, i.e. overall population exposures to nitrate have been estimated and compared with relevant stomach cancer rates. This method of inquiry is inevitably crude, but if most of these studies gave consistently positive results it would be hard to avoid the conclusion that nitrate exposure was of significance. The results have been conflicting, a major problem being that in most countries where these studies have been carried out, nitrate intake is mainly derived from vegetable consumption. This means that high nitrate intake tends to be associated with a low cancer risk perhaps because it is also correlated with a high level of ascorbic acid intake. Some studies have looked more specifically at nitrate in drinking water but these too show no consistent pattern of association with stomach cancer. There is only one study that has identified individuals known to have been exposed to high levels of nitrate. They were then followed to see if they developed more stomach cancer than expected. This study involved a group of fertiliser manufacturing workers who inhale and ingest nitrate during processing work. Although this was a small study (there were only 12 cases of stomach cancer in the exposed group), there was no indication of any excess of stomach cancer. The lack of positive results does not mean that Correa's model is wrong but that nitrate exposure may not be the rate limiting factor in endogenous NOC formation. It may be that, once CAG is established, there is always sufficient nitrite present in gastric juice to cause nitrosation. In such circumstances what will then be of importance is the presence of nitrosatable protein substrates and modifying agents such as ascorbic acid, which may catalyse or inhibit the nitrosation process. Alternatively NOCs may be just one of several groups of carcinogens present in the stomach from which the epithelial cells are normally adequately protected. The role of CAG would be to diminish both physical and biochemical defence mechanisms by destroying the mucosal layer and cellular integrity.

In summary, there are several leads to understanding the aetiology of stomach cancer but there is no preventive advice that can be given for stomach cancer that would have as dramatic an impact as stopping smoking would have for lung cancer. It is highly likely that increasing fruit and vegetable consumption and decreasing salt intake will reduce risk and one immediate task for epidemiology is to attempt to quantify the level of risk reduction more precisely. It is debatable whether preserved food consumption in the UK makes any material contribution to risk, but in other countries it might. All these dietary modifications would have the additional benefit of reducing circulatory disease. The role of *H. pylori* in stomach cancer requires further investigation. It is apparent that it could be a major cause of CAG but its mode of transmission remains a mystery. Similarly the hypothesis relating nitrates and *N*-nitroso compounds to stomach cancer requires further research before it can be translated into preventive action.

3.3.3 *Endometrial cancer*

Although no carcinogen responsible for endometrial cancer has been identified, the biological basis of the large variation in endometrial cancer rates between different populations and between different groups of women within a population (e.g. between obese and slender women) is now virtually completely understood.

Epidemiological studies have shown that the risk of developing endometrial cancer decreases with an early age at menopause and high parity, increases markedly with increasing weight and with use of oestrogen replacement therapy (ERT) in perimenopausal and postmenopausal women, and decreases markedly with combination type oral contraceptive (COC) use (Table 3.7). These risk factors can all be explained in terms of the 'unopposed oestrogen hypothesis' for endometrial cancer.

To appreciate the unopposed oestrogen hypothesis it is essential to understand the production and serum concentrations of the two female hormones, oestrogens (a class of hormones) and progesterone (see also Chapter 13). During the first half of the menstrual cycle—the follicular phase before ovulation takes place—the ovary mainly produces oestrogens, the most important of which is oestradiol (E2). During the second half of the menstrual cycle after ovulation occurs—the luteal phase—the ovary produces both oestrogens and progesterone (Pg). The fluctuating serum concentrations of E2 and Pg during the normal cycle are shown in Fig. 3.7. In the postmenopausal period almost no Pg is produced, but low levels of oestrogens are produced from conversion (mainly in fat cells) of

Table 3.7 Effects of age at menopause, parity, weight, oestrogen replacement therapy, and combination type oral contraceptives on endometrial cancer risk[1]

Risk factor		Relative risk
Menopause (years)	≤48	1.0
	49–51	1.2
	≥52	1.7
Parity	0	1.00
	1	0.54
	2	0.22
	3	0.12
	≥ 4	0.06
Oestrogen replacement therapy (years)	0	1.0
	<1	1.4
	1	2.0
	5–9	6.4
	≥10	7.6
Current weight (lbs)	≤129	1.00
	130–149	1.45
	150–169	1.95
	170–189	9.60
	≥190	17.70
Combination type oral contraceptive use (years)	0	1.00
	<2	0.75
	2–3	0.79
	4–5	0.28
	≥ 6	0.14

[1] See Key and Pike (1988a).

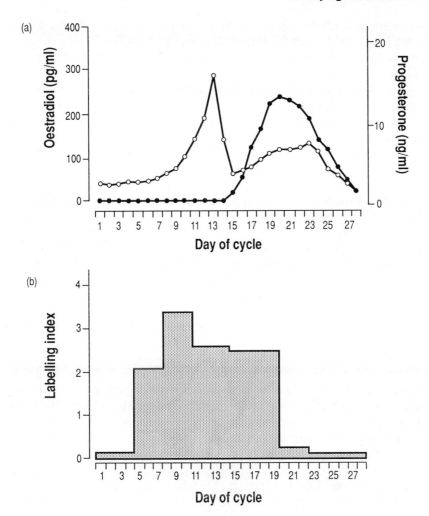

Fig. 3.7 (a) Serum concentrations of E2 (○) and Pg (●) by day of cycle (day 1 is first day of menses and 28-day cycle assumed with ovulation on day 14); and (b) endometrial mitotic rate by day of cycle.

adrenal androgens (male hormones). (These hormones are produced in both sexes.). The concentration of E2 in a slender postmenopausal woman is around 10 per cent (5 pg/ml) of the lowest level found during the normal menstrual cycle.

It is also essential to understand the nature of COCs and ERT. COCs, the commonly prescribed oral contraceptive, contain both a synthetic oestrogen (ethinyl-oestradiol) and one or other of a variety of synthetic progestogens (drugs which mimic the action of Pg). ERT, as usually prescribed until recently, simply contains a mixture of oestrogens (all 'natural', but some of equine origin, as the drug is commonly made from the urine of pregnant mares).

The unopposed oestrogen hypothesis maintains that endometrial cancer risk is increased by exposure to oestrogen which is not opposed by progesterone (or a synthetic progestogen), and this increased risk is caused by the increased mitotic activity (cell division) of the endometrium induced by such exposure because oestrogens stimulate mitosis in endometrial cells, only in the absence of progestogens. In women with normal menstrual cycles the mitotic rate rises rapidly from a very low level during menses to reach a near maximal level early on in the cycle, probably by day 5, and then stays roughly constant for 14 days after which it drops to a very low level with the post-ovulation increase in Pg (see Fig. 3.7). The peak mitotic rate is induced by the early relatively low follicular serum E2 concentration (50 pg/ml); later increases in E2 do not increase the mitotic rate further. The very low E2 concentrations of slender postmenopausal women induce very little endometrial cell division.

A smoothed age incidence curve of endometrial cancer is shown in Fig. 3.8. Incidence rises rapidly until about age 50 (average age at menopause) and thereafter at a much reduced rate. The effects of the various risk factors can be most clearly comprehended by noting their effect on this curve. Early menopause reduces risk by reducing the unopposed E2 concentration from the relatively high level during the follicular phase of every menstrual cycle to the very low level of the postmenopausal period; the reduced slope in the postmenopausal period starts at menopause (Fig. 3.9a). Obesity markedly increases the risk of endometrial cancer in premenopausal as well as in postmenopausal women. The mechanism by which obesity increases risk in premenopausal women is Pg deficiency. Obesity is associated with an increased frequency of anovulation (cycles in which no ovulation takes place). There is twice as much endometrial cell division during an anovulatory cycle as during an ovulatory cycle because there will be near maximal endometrial mitotic activity throughout the cycle rather than for only half of the cycle (the serum E2 level will be consistently above 50 pg/ml as even in an anovulatory cycle there is some ovarian follicle development). The further increase in risk with obesity in postmenopausal women is brought about because in such women obesity leads to increased peripheral production of oestrogens from adrenal androgens, and hence to an increased serum concentration of E2. Figure 3.9b illustrates the effect on the age incidence curve. In this figure we show the predicted curve for a woman whose obesity makes her anovular from age 35. Her endometrial mitotic rate for ages 35 to 50 is twice the normal

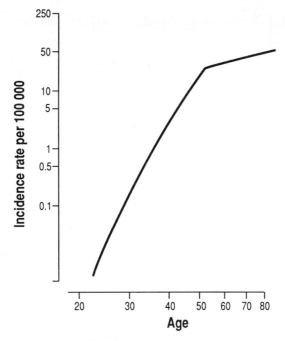

Fig. 3.8 Smoothed age incidence curve of endometrial cancer (from data for West Midlands Region of England, 1968–72).

premenopausal rate. After the menopause her mitotic rate is equal to the normal premenopausal rate (half the follicular rate because bioavailable E2 is approximately half the basal follicular level, but for double the time since there is no luteal phase).

The endometrial mitotic activity of a postmenopausal woman is greatly increased during ERT. The commonly taken dose of conjugated equine oestrogens (CEE, 'Premarin') taken as ERT is calculated to produce an endometrial mitotic rate of between 67 per cent and 100 per cent that occurring during the follicular phase of the menstrual cycle and the total mitotic activity over a 28-day period will be between 1.33 and two-times that of a premenopausal woman. The twofold figure is illustrated in Fig. 3.9c for five years of ERT use starting at menopause (taken as at age 50). Instead of the slope of the incidence curve decreasing at age 50, the slope will actually be steeper for the five years of ERT use. It is predicted that the increased risk will be lifelong, and that five years of such ERT use will increase lifelong risk some 3.5-fold.

Endometrial mitotic activity is near zero in women using COCs. Figure 3.9d illustrates what appears to happen with five years of COC use starting at age 28. The incidence curve will be nearly flat during the

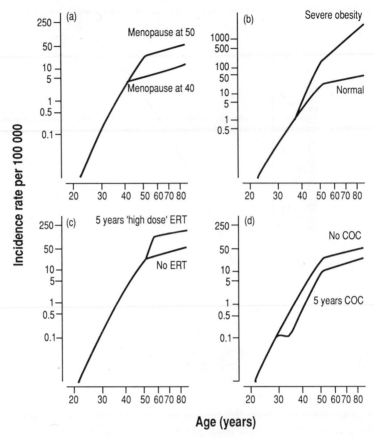

Fig. 3.9 Predicted effects of various 'hormonal' events on the age incidence curve of endometrial cancer (see text): (a) age at menopause of 40 compared to 50; (b) obesity affecting ovulation frequency from age 35 and oestrogen levels in the postmenopausal period; (c) 5 years of 'high-dose' ERT use from age 50 to 55; (d) 5 years of COC use from age 28 to 33.

time COCs are used, and will resume its upward trend when COC use is stopped. The protection against endometrial cancer is predicted to be lifelong, and it is predicted that five years of COC use will decrease lifelong risk by some 55 per cent.

Pregnancies reduce the risk of endometrial cancer because endometrial cells are not exposed to unopposed oestrogen during gestation. As can be seen in Table 3.7 there is a larger reduction in risk between parity 0 and parity 1 than with further increases in parity. This appears to be due to the increased frequency of anovular cycles in certain involuntarily nulliparous women.

3.3.4 *Breast cancer*

Hormones appear to hold the key to the understanding of human breast cancer, just as they do in certain animal species (see Chapter 13). Recent studies of the causes of the large variation in international breast cancer rates and of the effects on breast cancer of ERT use and of COC use have been particularly useful in advancing our understanding of the nature of the hormonal relationship.

Epidemiological research has established early menarche and late menopause as major risk factors for breast cancer (Table 3.8). A delay of three years in age at menarche has been found to reduce breast cancer risk by up to a half. Similarly the earlier a woman has her menopause the greater the reduction in her breast cancer risk: women with natural menopause before age 45 have only one-half the risk of women whose menopause occurs after age 55. The earlier a woman has her first birth the greater the reduction in her breast cancer risk; women with a first birth under age 20 have about one-half the risk of nulliparous women, but nulliparous women do not have as high a risk as women whose first birth is after age 35. This protective effect of first birth does not appear, however, until some years after the birth; before about age 32 parous women as a group are in fact more at risk of breast cancer than nulliparous women. The overall effect of obesity is to increase the risk of breast cancer, but this increased risk is restricted to older postmenopausal women; in premenopausal women obesity is associated with a

Table 3.8 Effects of age at menarche, age at first birth, and age at menopause on breast cancer risk[1]

Risk factor		Relative risk
Menarche (years)	⩽ 11	1.00
	12	0.90
	⩾ 13	0.50
First birth (years)	⩽ 19	0.83
	20–24	1.00
	25–29	1.30
	30–34	1.57
	⩾ 35	2.03
	Nulliparous	1.67
Menopause (years)	40–44	1.00
	45–49	1.27
	50–54	1.47
	55–59	2.03

[1] See Key and Pike (1988*b*).

decreased risk of breast cancer. ERT use in postmenopausal women causes a relatively small increase in risk, while studies of COC use have found either no effect or a relatively small increase in risk.

The success of the unopposed oestrogen hypothesis in explaining and predicting the epidemiology of endometrial cancer has stimulated the effort to develop an equivalent hypothesis for breast cancer.

There is some breast cell division during the pre-ovulatory follicular phase of the menstrual cycle, but the mitotic rate is approximately four-fold higher during the luteal phase. One interpretation of these data is that E2 alone induces some cell division, but that E2 and Pg together (post-ovulatory luteal phase) induce considerably more cell division— the 'oestrogen-plus-progestogen hypothesis'. An alternative interpretation is that breast cell division may only be induced by oestrogens with Pg having little or no effect. This 'oestrogen alone hypothesis' requires there to be a dose response relationship between breast cell division and E2 serum concentration in the range of E2 concentrations occurring during the normal menstrual cycle; there is no such relationship with endometrial cell division, where peak mitotic activity is induced by the low early follicular phase serum oestrogen levels. The breast cell division studies and the lack of a protective effect of COC use, show that an 'unopposed oestrogen hypothesis' for breast cancer is untenable. Current epidemiological evidence is compatible with both the oestrogen alone and the oestrogen plus progestogen hypotheses, although recent evidence tends to favour the latter.

A smoothed age incidence curve of breast cancer is shown in Fig. 3.10. As noted above, there is a steeply sloping line in the premenopausal period and a line with a much shallower slope in the postmenopausal period, with a gradual transition between the two lines during the peri-menopausal period. As with endometrial cancer, the effects of the various risk factors can be most clearly comprehended by noting their effect on the age incidence curve.

Early menarche increases breast cancer risk and, for a given age at menarche, the sooner regular cycles are established the greater the risk of breast cancer. Both E2 and Pg are increased at menarche and by regular cycling, and these observations are compatible with the oestrogen alone and the oestrogen plus progestogen hypotheses. Figure 3.11a illustrates the effect of a three year delay in menarche.

Menopause results in a drastic reduction in both E2 and Pg, and the protective effect of menopause is predicted by both hypotheses. Figure 3.11b compares menopause at age 50 with menopause at age 40. The latter curve has the reduced slope of the postmenopausal period starting at age 40 and the predicted breast cancer risk is decreased by almost 45 per cent.

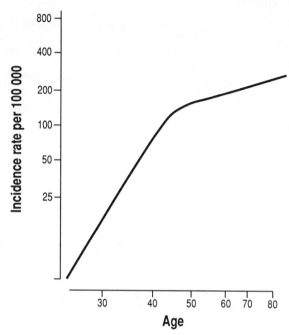

Fig. 3.10 Smoothed age incidence curve of breast cancer (from data for US white women, 1969–71: see Key and Pike 1988*b*).

The contradictory effects of obesity, i.e. an increased risk of breast cancer in postmenopausal women but a reduced risk in premenopausal women, are quite different from the effects of obesity on endometrial cancer, but are, in fact, predicted by both hypotheses. The increased anovulation associated with premenopausal obesity will decrease exposure to both E2 and Pg. This decreased E2 of premenopausal obesity is important as regards breast cell division (on the oestrogen alone hypothesis) since the breast cell division rate induced by the early follicular phase E2 level is much less than that induced (on this hypothesis) by the late follicular and luteal phase oestrogen levels. The decreased risk associated with premenopausal obesity is then gradually eliminated and an increased risk finally achieved by the increased oestrogen levels associated with postmenopausal obesity. Figure 3.11c illustrates the situation. This figure assumes that the breast cell mitotic rate of the obese woman is (i) decreased by 30 per cent in the premenopausal period starting at age 35 and (ii) doubled in the postmenopausal period: the breast cancer risk of such a severely obese woman is reduced by 20 per cent at age 45, is unaltered at age 65, and is increased by 25 per cent at age 80.

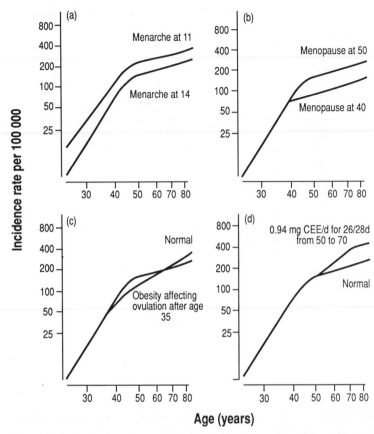

Fig. 3.11 Predicted effects of various 'hormonal' events on the age incidence curve for breast cancer (see text): (a) age at menarche of 11 compared to 14; (b) age at menopause of 40 compared to 50; (c) obesity affecting ovulation frequency from age 35 and oestrogen levels in the postmenopausal period; (d) 20 years of 'high-dose' ERT use from age 50 to 70.

Epidemiological studies of ERT show that an increase in exposure to oestrogen (unopposed by a progestogen) causes a small increase in breast cancer risk. We can estimate from these studies that 20 years of 'old style' standard ERT use causes an approximately 75 per cent increase in risk of breast cancer (Fig. 3.11d). Calculations show that the mitotic rate implied by this 1.75 relative risk is threefold higher than the normal postmenopausal rate and some three-sevenths the average pre-menopausal rate. This is in line with the known data on oestrogen levels achieved on ERT. The consistency of the oestrogen alone hypothesis with this relatively small relative risk depends on the dose response relationship between oestrogens and breast cell division. ERT produces effective oestrogen levels that are sufficient to induce substantial endometrial cell division, but, as we have noted, this does not require high oestrogen levels, and the effective oestrogen level achieved by ERT is certainly less than the effective oestrogen level achieved during the luteal phase of the menstrual cycle. The epidemiological results are thus completely in line with predictions.

COCs have either no effect on breast cancer risk or they slightly increase risk. COCs effectively block ovarian production of oestrogens and Pg, but consist themselves of an oestrogen and a progestogen both at fairly high levels; the epidemiological results (and mitotic rate results from women on COCs) suggest that the breast-effective hormone dose of oestrogen plus progestogen is similar in a normally cycling woman and in a woman on COCs. These epidemiological results are compatible with both the oestrogen alone and the oestrogen plus progestogen hypotheses. Evidence from a recently published Swedish study suggests that the addition of a progestogen to ERT substantially increases the breast cancer risk associated with ERT. If this result is confirmed this will essentially decide between the two hypotheses in favour of the oestrogen plus progestogen hypothesis, and should substantially alter current prescribing habits in which a progestogen is commonly added to ERT.

Detailed study has been made of the extent to which differences in the four major risk factors (ages at menarche, first birth, and menopause, and weight) between Japanese and American women could explain the large differences between their breast cancer rates. Age at first birth has not differed greatly between the two populations, and the distribution of age at menopause appears to vary little between the populations, but Japanese females born around 1900 had an average age at menarche some two years later and they weighed some 22 kg less at age 70 than the 1900 cohort of US white women. Although these differences steadily decreased over the years with the improved nutrition of the Japanese, they accounted for as much as two-thirds of the difference in the observed breast cancer rates in the two countries. They could not, however, account for all the differences—US white women still had 2.5 times the breast cancer risk of Japanese women even after allowing for these factors. To account for this 2.5-fold difference requires a 20 per cent lower breast cell mitotic rate in the premenopausal Japanese women. Such a reduction in premenopausal mitotic rate would result from an approximately 20 per cent reduction in premenopausal hormone levels (averaged over the menstrual cycle). A number of studies in the early 1970s showed lower premenopausal urinary oestrogens in Japanese compared to US women. Recent serum studies, which paid special attention to obtaining samples from Oriental women who were maintaining an old style lifestyle, show clearly lower E2 levels in the

Oriental women than in US or UK women. These results show that it is reasonable to conclude that the premenopausal E2 levels of the Japanese, who had the four- to six-fold lower breast cancer rates, could easily have been 20 per cent lower than the levels in Western whites. We therefore appear to have a complete hormonal explanation of the difference between Japanese and US breast cancer rates.

International breast cancer incidence and mortality rates are particularly highly correlated with *per capita* consumption of fat (Fig. 3.12). Epidemiological studies of individual breast cancer cases and controls have however provided little support for a role for fat consumption (other than as a major contributor to calories and hence to early menarche and increased postmenopausal weight); and a large prospective study found no relationship between serum cholesterol levels and breast cancer incidence. These studies can, however, be criticized on a number of grounds and further progress in evaluating dietary factors may be more easily achieved by relating premenopausal E2 levels to diet. A number of studies which have compared different racial groups found no evidence that total calories or per cent calories from fat had any effect on premenopausal E2 levels but two studies of US white omnivores and vegetarians, in contrast, show greatly reduced E2 in the vegetarians who consumed a much reduced percentage of calories from fat. Further careful work is clearly needed.

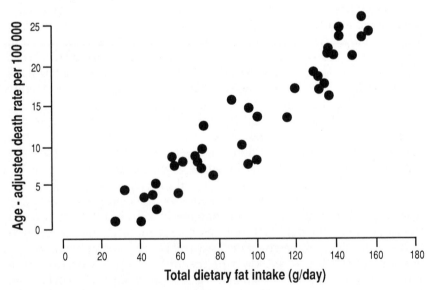

Fig. 3.12 Correlation between international breast cancer mortality rates and per capita consumption of fat.

On either the oestrogen alone or the oestrogen plus progestogen hypotheses the fact that COC use does not protect against breast cancer suggests, as we noted above, that the total hormone exposure of the breast of a women on COCs is approximately equivalent to her total hormone exposure during a normal menstrual cycle. Both hypotheses suggest that breast cancer risk associated with COC use would be reduced if the total hormone dose of COCs was reduced still further. This does not appear to be possible while maintaining the efficacy of COCs as good contraceptives. However, rather than rely on COC hormones to prevent ovulation one may instead use a luteinizing hormone releasing hormone (LHRH) agonist (LHRHA). LHRHAs will prevent ovulation and the amount of exogenous hormones required to be given to the woman will then be determined solely by the necessity to counteract the harmful effects of the lack of ovarian hormone production; it appears that only a low-dose oestrogen is needed, and calculations strongly suggest that such a contraceptive regimen will significantly reduce breast cancer risk when used by parous women (Pike *et al.* 1989). We are still so ignorant about the relevant events that occur in the breast between menarche and first birth that predicting the benefits or risks of using such drugs at relatively young ages is somewhat speculative. This contraceptive regimen is currently under study.

An alternative approach to the hormonal prevention of breast cancer based on the use of an anti-oestrogen in the postmenopausal period is under active study (Fentiman 1989).

3.3.5 *Cancer of the testis*

Table 3.9 shows that there is a very marked international variation in the incidence of testicular cancer. There has also been a large increase in the incidence rate of the tumour over time in many countries; in England and Wales the incidence rate has increased steadily since the turn of the century and the rate is currently doubling every 25 years. Several studies have attempted to identify specific causes of testis cancer but only urogenital abnormalities, in particular an undescended testis, have been firmly established as having an impact on risk. The distinctive age specific incidence pattern of testicular cancer (Fig. 3.4) must mean that some age related factor is playing a crucial role in causing the tumour. The relevant exposures may occur *in utero* or early in life and/or the germ cells may be susceptible to malignant transformation only at a critical period. What seems clear is that direct exposure to environmental agents cannot, by itself, explain the age-incidence pattern. Factors such as exposure to occupational chemicals can have, at most, only a limited aetiological role.

Table 3.9 Cumulative incidence rates (ages 0–74) of cancer of the testis in different populations[1] (*c.* 1980)

Population	Cumulative incidence rate (%)
Switzerland, Vaud, Urban	0.73
New Zealand, Maori	0.69
Denmark	0.62
Switzerland, Basel	0.60
Norway	0.45
New Zealand, non-Maori	0.41
USA, San Francisco, white	0.36
Israel, Jews, Europe/USA born	0.30
Sweden	0.26
England and Wales	0.26
Canada	0.23
Colombia, Cali	0.16
Finland	0.12
USA, San Francisco, Chinese	0.10
Japan, Osaka	0.10
Brazil, Fortaleza	0.09
USA, San Francisco, black	0.08
Singapore, Chinese	0.07
India, Bombay	0.07
USA, San Francisco, Japanese	0.05
Israel, non-Jews	0.05
China, Shanghai	0.05

[1] From Muir *et al.* (1987).

The one established major risk factor, namely a prior history of undescended testis (cryptorchidism), has been known for a long time. This condition is associated with an approximately fivefold increase in risk. Cryptorchidism is itself increasing and currently affects some 3.0 per cent of boys in the UK. It is also becoming apparent that cryptorchidism is just one of a range of congenital abnormalities that carry an increased risk of cancer of the testis. It seems that most congenital disorders of the urogenital system are associated with an increased risk of malignancy. These disorders range from gross inherited abnormalities of the genital tract to relatively minor defects, such as infantile inguinal hernia.

Uncorrected and surgically corrected cryptorchid testes appear to be almost invariably associated with the disruption of normal germ cell development. This testicular dysgenesis (loss of germinal epithelium and incomplete maturation of germ cells) starts at a very young age, progresses rapidly, and is often seen in association with carcinoma *in situ*

(CIS). It is noteworthy that in those situations where dysgenesis is observed in normally descended testes, as in infertile men, there also appears to be a high prevalence of CIS and subsequent malignant cancer. Although dysgenesis only rarely leads to cancer (some 1 per cent of cryptorchid males develop testicular cancer), dysgenesis may be the more immediate cause of the increased cancer risk, and this may occur in several other conditions in addition to cryptorchidism.

Further research on the nature and consequences of urogenital abnormalities is likely to reveal more about the aetiology of testicular cancer not only in men with these problems but possibly also in the large majority who develop the cancer in the absence of any gross urogenital abnormality. Henderson and co-workers have suggested, as a general model, that it is the persistence into adult life of 'primitive' germ cells in a 'developmentally arrested' form that eventually gives rise to a malignant condition (Henderson *et al.* 1988).

Several other factors have been suggested as being involved in cancer of the testis and there has been much interest in the roles of hormonal exposures and genetics. The circumstantial evidence for hormonal involvement is strong. Testicular development and function are both under extremely sensitive hormonal control and animal models have shown that *in utero* exogenous hormone exposures can result in testicular malformations, including cryptorchidism, in offspring. If exposure during the fetal period is important for humans, it could help explain why the incidence of the disease peaks some thirty years later.

A few American studies have shown that maternal use of hormonal preparations, including diethylstilboestrol (DES), during pregnancy will increase the risk of both undescended testis and testicular cancer in male offspring. Relatively few mothers have used such preparations. DES was used in a misguided belief that it could prevent miscarriage, but elevated endogenous maternal hormone levels might have a similar effect. Bioavailable oestrogen levels have been regarded as being particularly important. This is consistent with data from some studies that have associated nausea in pregnancy and high maternal weight with risk of testicular cancer. The first is symptomatic of excess bioavailable oestrogen; increased weight reduces the level of circulating sex-hormone binding globulin, which results in the same effect. The variables concerned are, however, both indirect indications of hormonal exposure, and these relevant findings have come from isolated studies only and have not always been confirmed in other investigations. A recent study found that bioavailable oestradiol levels were higher in first-trimester serum from mothers of cryptorchid boys than from mothers of controls; this offers more direct support for the hypothesis.

Post-natal hormonal exposure may also be relevant at least to the extent that testicular cancer does not occur until after puberty when the rise in pituitary hormones stimulating spermatogenesis and testosterone production occurs. One study found early age at puberty to be a risk factor and this could indicate an association between cancer risk and androgen levels. However, if anything, testosterone levels seem to be inversely related to testicular cancer. Despite lower cancer rates, serum testosterone levels are significantly higher in American black men than American white men, and a history of acne, which is associated with high levels of testosterone, has been associated with a decreased risk of testicular cancer. If adolescent hormones are important, the evidence points to pituitary hormones rather than testosterone.

The role of genetics is also an area of active research. As we noted above, there is a large geographic variation in the incidence of cancer of the testis, rates being extremely low in black and Asian populations. This is true also for migrant populations from these populations to countries where the disease is relatively common and therefore appears not to be just a result of different environments. The association between testis cancer and urogenital abnormalities in relatives adds weight to the likelihood of genetic involvement. There is also some evidence of association with particuar HLA (histocompatibility) haplotypes, although there is little consistency between the studies that have investigated this. An increasing number of families with multiple cases of cancer of the testis are now being reported, and it is of relevance to note firstly a higher proportion of bilaterally affected patients in these families and secondly a greater histological concordance between identical twins than between other siblings. Some of these families are now being used to investigate linkage with genetic markers. It is plausible that a genetic lesion may underlie the propensity to testicular dysgenesis and associated urogenital defects.

3.4 Risk factors

Table 3.10 summarizes the known and major suspected risk factors for each cancer site in order of the percentage of all cancer deaths caused by the tumour in the UK in 1986. The incidence and mortality figures give a rough guide to the relative importance of the site as a source of cancer; the relation between the incidence and mortality figures is a measure of the fatality rate associated with the particular cancer; and the male to female ratio for sites common to both sexes suggests the importance of exposure to sex specific risk factors. The established risk factors have been divided into major and minor categories in terms of their importance to the total cancer burden in England and Wales for the specific site. This division is somewhat arbitrary but a major risk factor, perhaps in combination with other factors, might account for more than 25 per cent of a particular cancer. Suspected risk factors have only been listed if they are likely to belong in the major category if proven. Table 3.10 also lists, in a separate column, the major established risk factors for cancers in countries other than the UK.

Several aspects of Table 3.10 are noteworthy. Over half of all cancer mortality is accounted for by cancers of the lung, large bowel, breast, and stomach. The other 30 specific sites in Table 3.10 together, therefore, contribute less to total cancer mortality than these four sites. Although, scientifically, there will be much to learn from understanding the aetiological factors for an uncommon cancer, preventing part or even all of the mortality from such a cancer will have only a small impact on total cancer mortality. Minor risk factors for a common cancer may have a much greater public health significance than major risk factors for rarer forms of cancer.

3.4.1 *Smoking and alcohol*

Lung cancer, by itself, accounts for over a quarter of total cancer mortality in the UK. The relationship of smoking with lung cancer has been discussed above. Table 3.10 shows that smoking is also a major established risk factor for cancers of the lip, mouth, tongue, pharynx, larynx, oesophagus, bladder, and pancreas. It is, in addition, a minor risk factor for cancers of the stomach, kidney, and cervix. Smoking is, therefore, a direct and avoidable cause of an enormous cancer burden. No other known single factor has anything like the same degree of importance for cancer in the developed world.

Smoking related cancers in the upper respiratory and digestive tracts, i.e. mouth, tongue, pharynx, larynx, and oesophagus, also share alcohol as an established major risk factor. For the larynx and oesophagus, it has been demonstrated that smoking and alcohol act synergistically, i.e. their effects are more than additive and are, in fact, close to multiplicative. The results of one large study of oesophageal cancer found that a non-drinking smoker of 20 cigarettes per day has a 1.7-fold increased risk, while a non-smoking drinker of 100 g of alcohol per day had a 7.2-fold increased risk, but a 20 cigarettes per day drinker of 100 g of alcohol per day had a 12.1-fold increased risk. Such synergism is not uncommon with cancer risk factors, and shows that control of one risk factor may have a larger effect than one might predict from studies in which people, or experimental animals, were only exposed to the single agent.

3.4.2 *Occupation*

Established carcinogens that occur mainly in occupational settings are

Table 3.10 Risk factors and certain basic information for the U.K. by specific cancer site[1]

Site (in order of decreasing contribution to mortality)	Incidence (%)	Mortality (%)	M:F	Risk factors			
				Major established	Minor established	Likely major but unproven	Major established but not in the U.K.
Lung	18.1	25.6	2–5:1 at different ages	Smoking	Radon Asbestos Polycyclic aromatic hydrocarbons Arsenic Nickel refining Chromates Bischloromethyl ether	Low intake of beta-carotene rich vegetables[2]	—
Large bowel	12.1	12.3	0.9:1	—	Polyposis coli	Low fibre intake Low vegetable intake High fat intake High meat intake	*Schistosoma japonicum*
Breast	10.7	9.8	0.01:1	Early menarche Late menopause Late first birth	Postmenopausal obesity	High fat intake Hormones (endogenous & exogenous)	—
Stomach	5.5	7.0	1.5:1	Low socio-economic status Achlorhydria Low intake of fresh fruit and vegetables	Blood group A Smoking	*Helicobacter pylori* High salt intake High intake of certain preserved foods	—
Prostate	4.7	4.8	M	Black racial group	—	Hormones (endogenous) High fat intake	—
Pancreas	2.8	4.3	1.0:1	Smoking	Diabetes	—	—
Bladder	4.7	3.3	2.1:1	Smoking	Aromatic amines Certain anti-cancer drugs	—	*Schistosoma haemotobium*

						Very hot drinks	Vitamin (A?) deficiency
Oesophagus	2.2	3.3	1.4:1	Smoking Alcohol	—	—	—
Ovary	2.3	2.8	F	Early menarche Late menopause Low parity Oral contraceptives (protective)	—	—	—
Leukaemia	2.1	2.5	1.2:1	—	Ionizing radiation[3] Benzene Certain genetic syndromes Certain anti-cancer drugs	—	Human T-cell leukaemia virus I (HTLV I)
Non-Hodgkin's lymphoma	2.1	2.2	1.1:1	—	Immuno-suppression	—	Epstein Barr virus
Brain and nervous system	1.4	1.7	1.3:1	—	—	—	—
Kidney	1.6	1.6	1.6:1	—	Smoking Aromatic amines Phenacetin	—	—
Cervix	2.0	1.4	F	Multiple sexual partners Low socio-economic status	High parity	Human papillomavirus (specific types only) Smoking Genital hygiene	—
Myelomatosis	1.1	1.4	1.1:1	Black racial group	Ionizing radiation[3]	—	—
Tongue, mouth, and pharynx (excluding nasopharynx)	1.1	0.9	1.6:1	Smoking Alcohol	Oral tobacco	—	Betel quid chewing Inverted smoking

Table 3.10—*continued*

Site (in order of decreasing contribution to mortality)	Incidence (%)	Mortality (%)	M:F	Risk factors			Major established but not in England and Wales
				Major established	Minor established	Likely major but unproven	
Liver	0.5	0.9	1.5:1	Cirrhosis	Hepatitis B virus Anabolic steroids Oral contraceptives Vinyl chloride monomer Thorotrast Immuno-suppression	—	Hepatitis B virus Aflatoxin Liver fluke (*O. viverrini, L. siniensis*)
Skin (melanoma)	1.1	0.8	0.8:1	Sun or other ultraviolet light Benign melanocytic naevi White racial group	Xeroderma pigmentosa	Sunburn	—
Endometrium	1.6	0.7	F	Early menarche Late menopause Low parity Exogenous oestrogens (in absence of progestogens) Combined oral contraceptives (protective) Obesity	—	—	—
Gall-bladder and bile ducts	0.6	0.7	0.6:1	Obesity Gall stones	High parity	—	—
Larynx	0.9	0.6	3.8:1	Smoking Alcohol	—	Human papilloviruses (specific types only)	—

Site			M:F ratio				
Connective tissue	0.4	0.4	1.1:1	—	Immunosuppression (Kaposi's sarcoma only)	—	—
Hodgkin's disease	0.6	0.3	1.5:1	—	Previous infectious mononucleosis	—	—
Skin (non-melanoma)	11.1	0.3	1.2:1	Sun or other ultraviolet light / White racial group	Arsenic / Polycyclic aromatic hydrocarbons / Xeroderma pigmentosa	—	Tropical ulcers
Pleura and peritoneum	0.3	0.3	5.8:1	Asbestos (some types)	—	—	—
Thyroid	0.4	0.2	0.5:1	—	Ionizing radiation[3]	—	—
Bone	0.2	0.2	1.2:1	—	Paget's disease	—	—
Nose and nasal sinuses	0.2	0.2	1.7:1	—	Nickel refining / Hardwood furniture, leather manufacture / Isopropyl alcohol production / Smoking	—	—
Testis	0.5	0.1	M	White racial group	Undescended testis	Fetal exposure to endogenous hormones	—
Nasopharynx	0.1	0.1	2.4:1	—	—	—	Salted fish / Epstein Barr virus / HLA type
Salivary gland	0.3	0.1	1.3:1	—	—	—	—
Penis	0.1	0.1	M	Early circumcision (protective)	—	Human papillomavirus (specific types only)	—
Lip	0.2	0.04	9:1	Smoking (esp. pipe) / Sun or other ultraviolet light	—	—	—
Choriocarcinoma	0.01	0.003	F	—	—	—	—

[1] Based on Doll and Peto (1987). The incidence figure is for UK cancer registries around 1984; the mortality for UK, 1986; both expressed as proportion of all cancers M:F ratio is for mortality.

[2] Low intake of such vegetables may protect against several types of cancer, although most evidence relates to cancer of the lung (see text).

[3] Ionizing radiation can be a cause of nearly all cancers (excepting possibly chronic lymphatic leukaemia and Hodgkin's disease).

nearly all categorized as minor risk factors in Table 3.10 (with the exception of asbestos in relation to cancer of the pleura and peritoneum). This does not mean that they are of minor importance for everyone; indeed for workers in the relevant industries they may be of great importance. For example, bladder cancer has been shown to be caused by occupational exposure to a number of chemicals used in the dye and rubber industries. These chemicals (2-naphthylamine, benzidine, 3,3'-dichlorobenzidine, and 4-amino-biphenyl) belong to a class of chemicals, the aromatic amines, which are now known to be animal carcinogens (see Chapter 6). 2-Naphthylamine is particularly carcinogenic; all 19 distillers in one factory developed bladder cancer, and it is no longer used in UK industry. It has been estimated that at most no more than 10 per cent of current bladder tumours are due to occupational exposure. There are several other occupational carcinogens that in the past would have constituted significant health hazards for specific groups of workers. At present, however, only a small proportion of the total cancer mortality experienced by the population as a whole can be attributed to occupational carcinogens.

3.4.3 *Infection*

Although there is an enormous scientific interest in virus-associated cancer (see Chapter 8), infectious agents appear to be of relatively little importance to carcinogenesis in England and Wales. This is very different from the situation in a number of developing countries where, in particular, the risk of liver cancer is enormously raised in persons chronically infected with the hepatitis B virus. For this reason liver cancer is a major cancer in many developing countries. There is also evidence that the Epstein-Barr virus may be aetiologically related to nasopharyngeal cancer and to Burkitt's lymphoma, but this interpretation of the evidence is controversial (see Chapter 8). Bladder cancer risk appears to be associated with chronic infection with the parasite *Schistosoma haematobium.* In England and Wales no infectious agents have been firmly related to cancer (except for a small number of hepatitis B associated liver cancers), but there is a great deal of active research into the role of certain subtypes of the human papilloma virus in cervical and penile cancers (see Chapter 8), and into the role of the bacterium *Helicobacter pylori* in stomach cancer (see p. 69). There is also continuing interest in the role of viral infections in certain types of leukaemia and lymphoma although it is unlikely such associations will be of importance to the forms of these diseases seen most commonly in the UK.

3.4.4 *Diet*

Dietary factors have been the subject of much epidemiological research in recent years and there is a general belief that modifications to the diet could have a major impact on cancer rates. Table 3.10 indicates that stomach cancer certainly has a major causal dietary component and that lung, large bowel, and breast cancers may have as well. Thus all four major sites could be affected by diet in the UK.

Diet could influence the risk of cancer in many ways. Obesity itself contributes directly to certain cancers, in particular to cancer of the endometrium (see p. 71), so that total calorie intake is important. The direct dietary consumption of known chemical carcinogens also has some role to play. Aflatoxin, a fungal contaminant of peanuts appears to be a major cause of liver cancer in certain tropical countries, while 'natural' chemicals contained in salted fish and bracken fern, which are consumed in parts of Asia, have been linked to cancers of the naso-pharynx and oesophagus respectively. It has not been possible to show that dietary consumption of carcinogens plays a significant role in any human cancer in England and Wales.

Most current research on diet and cancer is concerned with dietary items that would have an indirect link with cancer. A high level of fat intake, for example, has been repeatedly suggested to be a cause of breast (see p. 82) and large bowel cancer. There are strong international correlations between per capita fat consumption and both of these cancers (see, e.g. Fig. 3.12), and studies of Japanese migrants to Hawaii have found an increased fat intake and an increased risk of both cancers. These correlational studies are, however, the crudest form of epidemio-logical evidence, and studies of fat consumption of cancer patients and controls have provided relatively little support for such relationships. Moreover, prospective studies of serum cholesterol levels which have provided the most compelling evidence of a link between fat consump-tion and heart disease, have found no evidence whatsoever of a relation-ship with cancer. It appears, therefore, that if a high fat diet does increase the risk of either breast or large bowel cancer it will have to be some element of such a diet that does not raise serum cholesterol levels.

An exciting area of dietary research is the investigation of the potential protective effect against cancer of beta-carotene, a vitamin A compound. As mentioned in the section on stomach cancer, beta-carotene is one of several micro-nutrients that have anti-oxidant activity (others being ascorbic acid, tocopherol, other carotenoids, and selenium) and all might protect against cancer. Attention has been particularly directed towards beta-carotene because it accumulates in the body in direct proportion to intake (no homeostatic balance is maintained) and it has not been associated with any toxic side effects. Beta-carotene can, in addition, be converted to retinol which may have a different mode of protective action as an inducer of cell differentiation.

Nearly all epidemiological studies are consistent with a protective effect resulting from high beta-carotene intake, the most persuasive being those in which beta-carotene levels were measured in serum samples from cases and controls taken several years before the cases developed cancer. Currently, results are eagerly awaited from a number of intervention studies in which participants were randomly divided into two groups that received capsules of either beta-carotene or placebo.

Theoretically beta-carotene intake should reduce cancer risk at many sites. Much attention has been given to its putative role in preventing lung cancer. This is partly because subjects taking part in prospective studies, in which dietary information or serum samples are obtained before cancer develops, will tend to have more lung cancers than any other type. It should be emphasized that even if beta-carotene is genuinely protective against lung cancer, it is unlikely to be able to compensate for the detrimental effects of smoking.

3.4.5 *Other risk factors*

Several factors which are often associated with cancer are either not included in Table 3.10 or are categorized as minor factors. Industrial pollution, pesticide exposure, and psychological stress are not listed at all. A number of these omissions may reflect a failure of cancer epidemiology to investigate certain subjects (e.g. stress) adequately but these omissions also reflect the fact that although many chemicals are reported as positive in animal carcinogenicity or *in vitro* tests (see Chapter 6), they appear to make a negligible contribution to human cancer. In some cases this is because the doses employed in testing are orders of magnitude above those to which people are exposed. It is now also apparent that the human body is continually confronted with genotoxic (DNA damaging) agents not only from industrial chemicals but also from a wide range of products found naturally in many foods (Ames *et al.* 1987). It may, therefore, prove to be the case that many genotoxic agents will be of much less importance to human cancer than, say, host DNA repair mechanisms or factors which stimulate cells to proliferate. Clearly some genotoxins are of considerable importance, for example, u.v. light and the carcinogens present in tobacco smoke. Others may be of limited relevance to human cancer.

3.4.6 Summary

A comprehensive overview of our current understanding of the causes of all types of cancer has been made by Doll and Peto (1981, 1987) and their conclusions are summarized in Table 3.11. Tobacco (more specifically cigarette smoking) stands out in that it accounts for an estimated 30 per cent of all cancer mortality and it is the only factor for which all the associated deaths could undoubtedly be avoided by the implementation of known practical measures.

Dietary modifications could perhaps avoid a greater number of cancer deaths than changes in smoking (maybe 35 per cent), but our ignorance of the specific changes required means that there is a lot of uncertainty in the estimate of what is possible (10–70 per cent) and there are very few changes that could be immediately taken up that would be certain to reduce risk.

The contribution of occupational risk factors to the total population

Table 3.11 Proportions of cancer deaths in the USA[1,2] attributable to various factors

Factor or class of factors	Per cent of all cancer deaths		
	Best estimate that potentially could be avoided	Range of acceptable estimates	Known to be avoidable now
Diet	35	10–70	2
Tobacco	30	25–40	30
Reproductive factors	7	1–13	?[5]
Occupation	4	2–8	< 1
Alcohol	3	2–4	3
Geophysical factors	3[3]	2–4	< 1
Pollution	2	< 1–5	< 1
Medicines and medical procedures	1	0.5–3	< 1
Food additives	< 1	−5[4]–2	< 1
Industrial products	< 1	< 1–2	< 1
Infection	(10)	1–?	< 1

[1] After Doll and Peto (1981, 1987).
[2] Similar figures would apply to the UK.
[3] 'Only about 1%, not 3% could reasonably be described as "avoidable". Geophysical factors also cause a much greater proportion of non-fatal cancers (up to 30% of all cancers, depending on ethnic mix and latitude) because of the importance of u.v. light in causing the relatively non-fatal basal cell and squamous cell carcinomas of sunlight-exposed skin' (Doll and Peto 1981).
[4] 'Allowing for a possible protective effect of antioxidants and other preservatives' (Doll and Peto 1981).
[5] Manipulation of some reproductive factors (e.g. age at first birth) to avoid cancer is clearly possible but not necessarily practical or acceptable.

cancer burden is estimated as being no more than 8 per cent, and probably closer to 4 per cent. Occupational exposure to established carcinogens has been substantially reduced, e.g. asbestos, or entirely eliminated, e.g. 2-naphthylamine, in England and Wales, so that there are very few practical measures that could be implemented now which would guarantee a reduction in occupationally related cancer.

Reproductive factors, such as age at first birth and number of children, can account for some 13 per cent of cancer mortality. It is however not likely that women would manipulate such factors to prevent cancer, since these factors are so important to everyday living and having an early first birth, for example, may have other unwanted 'side-effects'. All hormone-related cancers may be controllable in the future by use of appropriate drugs in a manner similar to the way in which oral contraceptives reduce endometrial and ovarian cancer; this idea should not be thought of as interfering with nature any more than, for example, delaying first pregnancy.

Further details of the calculations that underlie Table 3.11 can be found in Doll and Peto (1981) while their more recent publication (1987) expands on our discussion of risk factors on a site by site basis.

Acknowledgement

We would like to thank Richard Doll, Richard Peto, and Sarah Darby for their contributions to specific sections of this chapter.

References and further reading

Ames, B. N., Magaw, R., and Gold, L. S. (1987). Ranking possible carcinogenic hazards. *Science* **239**, 271–80.

Correa, P. (1988). A human model of gastric carcinogenesis. *Cancer Research* **48**, 3554–60.

Doll, R. and Peto, R. (1981). *The causes of cancer: quantitative estimates of avoidable risks of cancer in the United States today.* Oxford University Press.

Doll, R. and Peto, R. (1987). Epidemiology of cancer. In *Oxford textbook of medicine* (ed. D. J. Weatherall, J. G. G. Ledingham, and D. A. Warrell). Oxford University Press.

Fentiman, I. S. (1989). The endocrine prevention of breast cancer. *British Journal of Cancer* **60**, 12–14.

Henderson, B. E., Ross, R., and Bernstein, L. (1988). Estrogens as a cause of human cancer. *Cancer Research* **48**, 246–53.

Key, T. J. A. and Pike, M. C. (1988a). The dose-effect relationship between 'unopposed' oestrogens and endometrial mitotic rate: Its central role in explaining and predicting endometrial cancer risk. *British Journal of Cancer* **57**, 205–12.

Key, T. J. A. and Pike, M. C. (1988*b*). The role of oestrogens and progestagens in the epidemiology and prevention of breast cancer. *European Journal of Cancer* **24**, 29–43.

Muir, C., Waterhouse, J., Mack, T., Powell, J., and Whelan, S. (1987). *Cancer incidence in five continents*, Vol. V. IARC Scientific Publication No. 88. International Agency for Research on Cancer, Lyon.

Office of Population Censuses and Surveys (1979). *Cancer registration statistics, England and Wales, 1971*. MBI-1. HMSO, London.

Pike, M. C. *et al.* (1989). LHRH agonists and the prevention of breast and ovarian cancer. *British Journal of Cancer* **60**, 142–8.

Schottenfeld, D. and Fraumeni, J. F. Jr. (1990). *Cancer epidemiology and prevention* (2nd edn). W. B. Saunders, Philadelphia.

Vessey, M. P. and Gray, M. (1985). *Cancer risks and prevention*. Oxford University Press.

Waterhouse, J., Muir, C., Correa, P., and Powell, J. (1976). *Cancer incidence in five continents*, Vol. III. IARC Scientific Publication No. 15. International Agency for Research on Cancer, Lyon.

Waterhouse, J., Muir, C., Shanmugaratnam, K., and Powell, J. (1982). *Cancer incidence in five continents*, Vol. IV. IARC Scientific Publication No. 42. International Agency for Research on Cancer, Lyon.

4

Inherited susceptibility to cancer
W. F. BODMER

4.1 Introduction: cellular genetic basis for cancer

Epidemiological studies, especially of variations in cancer incidence in different populations and their migrants, as discussed in Chapter 3, strongly suggest that at least 80 per cent of cancer incidence is attributable in the broadest sense to environmental factors. Studies on the incidence of cancer in the relatives of patients with the disease support the view that, overall, the contribution of inheritance to cancer susceptibility is not as large as it is in some of the other major chronic diseases, such as heart, mental, and autoimmune diseases. At the cellular level, however, in the individual in whom cancer develops, genetic changes in the cells destined to form a malignant tumour are the key to the progression from a normal to a malignant cell. In this Chapter I shall briefly review the evidence for the importance of genetic changes at the cellular level in the development of a cancer, and then discuss more extensively the types of situations where there is an inherited basis for cancer susceptibility and their potential interrelationships with changes at the cellular level.

Most of the fundamental ideas on the causation of cancer were suggested in the early years of this century, or earlier, while the idea that exposure to substances in the environment could be a cause of cancer goes back at least to Percival Pott in 1775. He pointed out that chimney sweeps tended to get cancer of the scrotum because they were continuously exposed to soot. Thus he was the first to identify clearly not only an environmental carcinogen, but also an occupational cancer.

The major fundamental ideas about the causes of cancer are:

(1) that genetic changes, or mutations, in somatic cells of the body are the main initiating events, and are responsible for tumour progression;

(2) the related idea that changes in the chromosomes, either in their number or organization, for example by deletion or translocation (see Glossary), are key events (see Chapter 10);

(3) that the immune system plays a role in combating cancer through recognition of novel antigens on cancer cells (see Chapter 15);

(4) that some cancers are caused by viruses (see Chapter 8);

(5) that cancers represent a form of dedifferentiation or, more generally, perturbation of the differentiated state, that is associated with a loss of growth control.

Modern developments in genetics, cell biology, and virology, especially in recent years at the molecular level, now make it possible to establish in detail the importance of these fundamental concepts for the development of many cancers. These five ideas can all be interrelated through the basic assumption that a cancer develops through a series of genetic changes progressing from the initiated cell, whose progeny eventually give rise to the cancer, to the ultimate malignant cell.

4.2 Evidence for a genetic basis

There are several major lines of evidence to support the view that cancer is a genetic disease at the somatic cell level. Cancers, as populations of cells, tend to breed true as far as their cell types are concerned. This is the fundamental basis for the histopathological diagnosis of a cancer as, for example, a particular sort of lymphoma or carcinoma. Staying true to type, however, says little more than that the cells of a cancer reflect, on the whole, the properties of the tissue from which they originated. Since the process of cellular differentiation, by which tissues acquire and maintain their particular characteristics is not, in general due to genetic changes in the sense of mutations, the evidence that a cancer tends to retain the characteristics of the tissue from which it is derived does little more than relate carcinogenesis to the process of differentiation. Normal differences between types of cells are based on differential gene expression. This emphasizes the possibility that some of the steps in tumour progression may not necessarily be mutations, but may involve aberrant control of gene expression by mechanisms, such as DNA methylation, which may be analogous to those which control cellular differentiation. The term 'epigenetic' is often used to describe this situation, and to contrast it with genetic changes in the sense of mutations, namely alterations in the DNA sequence itself.

Genetic markers can be used to show that the vast majority of cancers are clonal in origin, that is are derived at some point by cell division from a single cell. The evidence for this depends on the fact that in female cells, which contain two X chromosomes, only one of them is active or fully functional, thus ensuring that male (XY) and female (XX) cells have the same level of X chromosome gene activity. The female receives one of her X chromosomes from her mother and the other from her father. The random process by which inactivation of one of the X chromosomes occurs in somatic cells means that in roughly half the cells it is the maternal X chromosome that is active while in the other half it is the paternal X chromosome which is active. Thus, a female who is heterozygous for two forms, or alleles, of a gene on the X chromosome is a mosaic of cells, half of which on average express one allele and the other half the other allele. Glucose-6-phosphate dehydrogenase (G6PD) occurs in some populations, especially those of African origin, in two different forms, A and B, which can be distinguished by the technique of electrophoresis. A female who is heterozygous G6PDA/B is thus a mosaic of cells approximately half of which express G6PDA while the other half express G6PDB. A tumour in such an heterozygous female, if it is clonal, should not be a mosaic. All the cells of the tumour will either be G6PDA or G6PDB depending on which allele was active in the initiated cell from which the tumour was ultimately derived (see Fig. 4.1). Studies on tumours, mainly leukaemias in G6PDA/B heterozygotes, pioneered by Gartler, Fialkow, and others, have shown that at least the vast majority of leukaemias are clonal in origin by this criterion. There are similar data in the mouse, using another X chromosome enzyme marker, to show that the vast majority of experimentally induced liver tumours are clonal in origin.

This evidence, of course, only shows that at some point in the development of a tumour it becomes clonal. It does not rule out the possibility that epigenetic influences, such as for example persistent immune reaction or tissue repair, may create a multicellular environment which favours the initiation of a tumour. In such a situation the tumour could

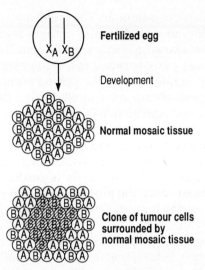

Fertilized egg

Development

Normal mosaic tissue

Clone of tumour cells
surrounded by
normal mosaic tissue

Fig. 4.1 Scheme to illustrate the use of the X-linked G6PD marker to show the clonal origin of tumours. The female heterozygote carries the allele A for G6PDA on one X-chromosome, and B for G6PDB on the other X chromosome. The two forms of the enzyme can be distinguished by their different electrophoretic mobilities on a gel. During development the fertilized egg divides to give rise to the somatic tissues. Each somatic cell has only one active X and since the mechanism of inactivation appears to be random, normal tissues will usually be a mosaic of two different sorts of cells with respect to G6PD activity, one expressing the A variant and the other expressing the B variant. If tumours arise from a single progenitor cell, then all the cells of the tumour will be derived by division from this progenitor and express the same G6PD type as the cell which initiated the tumour. Thus, tissue from a tumour will express only one of the two alleles (G6PDB in the Fig.), whereas samples of normal tissue will express a mixture of the two.

potentially be derived from any one of a number of cells in this altered cellular environment.

Another problem with this approach to establishing tumour clonality, especially for carcinomas, is that X chromosome clonality of a structure, such as a colonic crypt, may be established long before initiation of a tumour. In other words, even if a tumour was derived from several cells, if these all came from a localized 'patch' of epithelial tissue they would almost certainly already all be expressing the same X chromosome, either all paternal or all maternal, at the time of tumour initiation.

Chromosomal changes have been seen in tumours since they were first studied by Boveri and others in the early years of this century. However, as discussed in Chapter 10 it is only comparatively recently that quite specific chromosome changes have been identified that are usually characteristic of a particular tumour. The initial and classical example is the Philadelphia chromosome found in chronic myelocytic leukaemia (CML). This is now known to be a specific translocation, namely an exchange of particular parts between two chromosomes (in this case 9 and 22), and this is therefore a specific genetic event. The Philadelphia chromosome is almost always seen in all CML cells, and so is itself evidence for at least one genetic step in the development of this particular tumour. Added to this is the observation that this specific change has never been seen in a normal cell, and so the frequency with which it is produced must be exceedingly low. Thus, the fact that the Philadelphia chromosome is found in all CML cells is further evidence for the clonality of this tumour, since the probability of that particular chromosomal mutation occurring independently two or more times in the same tissue must be negligible.

The detailed genetic consequences of the Philadelphia chromosome translocation have now been elucidated most elegantly at the molecular level (see Chapters 9 and 10). The novel 'fused' gene formed by bringing together two different, normally unassociated genes from chromosomes 9 and 22 in the translocation now forms the basis for a remarkably sensitive test for residual leukaemic cells in the blood and bone marrow of CML patients after treatment. Many further examples of specific, clonal, chromosome changes in cancers are described in Chapter 10.

The majority of cancer causing agents, or carcinogens, are also mutagens, namely cause genetic mutations (see Chapter 6). One of the most widely used simple carcinogen screening tests is a test for mutagenicity using bacterial strains, the so called Ames test. Many agents act directly on DNA to cause genetic mutations and, in this case, there is often a good parallel between their mutagenic and their carcinogenic activity. There are, however, at least three important limitations to this approach for the detection of carcinogens. The first is that many carcinogens are turned into an active form in the body by various enzymes, in particular the P450 mixed function oxidases of the liver. In these cases, unless these enzymes are provided, usually as a liver extract, the test will be negative. The second limitation is that some agents can cause genetic changes, for example by interfering with chromosome organization, in a way that cannot be detected by bacterial mutagenesis assays. The third limitation is that the tumour promoters, which are chemicals that on their own cannot initiate a cancer, but which enormously increase the probability of a cancer developing once a cell is initiated, work by different mechanisms that cannot be detected by bacterial or other mutagenesis assays (see Chapter 6).

There are a number of rare inherited diseases that involve an inability to repair damaged DNA and so increase the mutation or chromosomal damage rates. These inherited syndromes are associated with marked increases in susceptibility to cancer (see p. 115).

A final piece of evidence for genetic changes in tumour cells comes from the exciting studies on oncogenes, mainly those identified in the oncogenic viruses. Using recombinant DNA techniques, specific genetic changes in the normal versions of the oncogenes have been demonstrated in a number of human and animal tumours, as discussed in several of the chapters in this book, especially Chapters 8, 9, and 10.

The general notion that changes in gene expression in somatic cells, mostly due to mutation (which in the broadest sense includes, for example, chromosome translocation), underlie the origin of cancer, unites the major fundamental ideas about its causes. Mutation, chromosome changes, the effects of viruses, novel determinants on the surface of tumour cells, and changes in the pattern of expression of differentiated gene products are all subsumed, in one way or another, under this general hypothesis, for which there is increasing direct experimental evidence.

Many lines of evidence suggest that tumour progression is a multistep process, presumably involving several successive genetic changes (see Chapter 6). Analysis of the increase in incidence of cancer as a function of age has been used by Doll, Armitage, Peto, and others to provide approximate estimates for the number of steps involved, which come to about four or five, at least for the carcinomas. This can, however, be no more than a very rough guess at the number of steps, and such estimates cannot clearly distinguish between genetic and epigenetic changes. One problem is that a single genetic change, for example in the control of a series of genes involved in a differentiation pathway, may lead to multiple changes in gene expression in one step. A possible example of this is intestinal metaplasia in the gastric epithelium. Intestinal metaplasia describes a focal region of the gastric epithelium which takes on, often almost completely, the phenotype of the intestine rather than the stomach. It is as if a switch has been thrown which changes the pattern of epithelial differentiation from that of the stomach to that of the intestine. This can be identified by a whole series of differences in gene expression associated with the intestinal phenotype. The significance of these lesions is that they appear to be the precursors to the majority of gastric carcinomas.

Each change in gene expression during the progression from the initial cell to the final malignant tumour must give rise to a selective advantage, in terms of enhanced growth rate or independence of growth from the effects of the immediate environment of the tumour. Otherwise the change would not be seen in all, or at least a substantial fraction, of the cells of the tumour. In this sense, tumour progression is an evolutionary process at the somatic level within the individual. Heterogeneity within a tumour can clearly arise both from different evolutionary sublines within any given tumour, and also from variations in expression associated with environmentally determined differences in the stage of residual differentiation of the cells within a tumour.

4.3 Genetic markers, DNA polymorphisms, and genetic linkage

The simplest family data providing evidence for inherited cancer susceptibility involve the patterns of inheritance expected for a clear-cut Mendelian trait due to a single recessive or dominant gene. A well-known example of a recessive inherited susceptibility is xeroderma pigmentosum (XP), an inherited DNA repair deficiency in which nearly all deficient individuals get skin cancer. In this case, affected individuals carry two defective copies of the relevant gene, one inherited from each of the parents, who will be unaffected carriers of the defective gene. The chance that an offspring of two such carriers is affected is the classical Mendelian ratio of one quarter. Retinoblastoma is an example of a dominantly inherited susceptibility, in this case to tumours of the eye. A single defective gene is enough to give rise to the susceptibility, so that an affected parent will on average produce 50 per cent affected offspring. There are many such examples of individually rare instances of clear-cut inherited cancers. Each of these cases may in itself be of fundamental interest, as will be discussed in the next section, though collectively their overall contribution to the incidence of cancer is likely to be very small.

To understand the basis of these inherited susceptibilities we need to identify and isolate the corresponding genes and establish their functions, and through that their relationship to cancer susceptibility. In the absence of a clear-cut clue as to the functional basis of an inherited susceptibility, an important first step in identifying the gene is to find out where it lies in the chromosomes.

The human genes are distributed amongst the 23 pairs of human chromosomes. Genes on different chromosomes are combined and passed on at random from parent to offspring. If a heterozygous A1/A2 individual is also heterozygous B1/B2 at a second locus on a different chromosome, then the combinations A1B1, A1B2, A2B1, and A2B2 are each passed on with a frequency of, on average, a quarter to each offspring. Suppose, on the other hand, that the A and B loci are on the same chromosome, so that, e.g. one of the pairs of homologous chromosomes in an individual carries A1B1 while the other carries A2B2. Then, if the genes are sufficiently close together, the combinations A1B1 and A2B2 will be passed on to the offspring more frequently than the 'recombinant' combinations A1B2 and A2B1. The frequency with which these recombinant, non-parental types are passed on is called the recombination fraction (r), and is an empirical measure of the distance between the A and B genes on the chromosome. The recombination fraction is thus less than half when genes are sufficiently close together on the same chromosome, and the genes are then said to be linked (see Table 4.1). Suppose for example the A1,A2 difference is the one that determines an inherited susceptibility to cancer, while the B1,B2 difference can be directly observed, as can for example the inherited HLA tissue types. Then the latter difference B1,B2 is said to be a marker difference, and it will be distorted in its distribution amongst individuals in a family with cancer due to the linkage between the A and the B genes. Thus the closer together the two genes are, the smaller is the recombination fraction r, and the greater will be the distortion in the distribution of the marker difference B1,B2 amongst individuals with cancer. For example, suppose A1 is a dominant susceptibility gene. Then, when linkage is close and the recombination fraction, r, very small, almost all the affected offspring of an A1B1/A2B2 individual will carry the B1 marker, whereas in the absence of linkage only one half would be expected to.

The first step in locating a susceptibility gene is to find a marker which is genetically linked, preferably quite closely, to the disease in question. This is done by studying the pattern of inheritance of different markers in families of susceptible individuals and looking for a marker that associates in the family with susceptibility, due to close genetic linkage. Often data from several families need to be combined, using special statistical techniques, to assess the significance of a genetic linkage to a marker.

In the past such studies have been done using blood groups and enzyme differences, but the number of such differences is limited. Now,

Table 4.1 Linkage and recombination

	Parent	Gametic (egg or sperm) combinations		
No linkage	A1 B1 A2 B2	A1B1, A2B2, A2B1, A1B2 passed on to offspring with equal frequencies of $\frac{1}{4}$		
		Parental	Recombinant	
Linkage	A1B1* A2B2	A1B1	A2B2	A1B2	A2B1
	frequencies	$\frac{1}{2}(1-r)$	$\frac{1}{2}(1-r)$	$\frac{1}{2}r$	$\frac{1}{2}r$

A1, A2, and B1, B2 are alleles at loci A and B respectively.

*The line separating A1B1 from A2B2 indicates that A1B1 are together on one chromosome and A2B2 on the other.

r is the 'recombination fraction'; the smaller r is the fewer recombinants are produced, and the closer together are the genes A and B. When, on the other hand, r is $\frac{1}{2}$, the result is as if there were no linkage.

however, using recombinant DNA techniques (see Chapter 5) a potentially unlimited source of genetic markers can be produced. Using a cloned DNA probe, differences in the sequence with which it is associated can be identified using restriction enzymes that cut the DNA at particular short sequences, together with the technique known as Southern blotting (see Fig. 4.2). The incidence of differences between individuals at the DNA level is such that there should be no difficulty in finding a set of such genetic differences, called restriction enzyme fragment polymorphisms or rflp, that covers all the chromosomes at a reasonable recombination interval. These differences and others that can be detected using the powerful PCR (polymerase chain reaction) technique can now be used systematically in cancer families to look for markers that are associated with, or closely linked to, genes giving rise to inherited susceptibility to cancer.

There are now many techniques by which a defined protein product or, more often, a specific human DNA sequence can be assigned to its

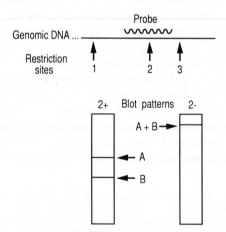

Fig. 4.2 DNA variation on a Southern blot. Restriction Fragment Length Polymorphism (RFLP): Genomic DNA cut by a given restriction enzyme is spread out on a gel by an electric current, according to its size (electrophoresis). The gel is then 'blotted' onto a suitable membrane (nylon or nitrocellulose paper) which is baked so that the cut pieces of DNA can be 'probed' with a radioactively labelled test sequence (Southern blot). The sequence is visualized by autoradiography which reveals the position where the probe has stuck to the blot by the radioactive decay from the label in the probe. Probe, here, refers to a DNA clone which overlaps segments A and B of the genome straddling position 2 where there is a difference for a restriction enzyme site. In 2 + individuals, the enzyme yields fragments A and B identified by the probe as on the gel on the left. In 2 − individuals there is no cut at position 2 and so the probe detects only the segment A + B which runs as a single band at a higher molecular weight as in the gel on the right.

position on a human chromosome. The key to this has been the development of techniques for the genetic analysis of cells in culture, somatic cell genetics, based on hybridizing human with other cells to form somatic cell hybrids (see Fig. 4.3). Human–mouse hybrid cells containing different combinations of human chromosomes can be used to associate a given human DNA sequence with a particular chromosome, or chromosome fragment, as a major step towards a localization. Other techniques now include the possibility of visualizing the position of a DNA sequence directly on a chromosome spread using in-situ hybridization coupled with autoradiography or the newer fluorescence based techniques as described in Chapter 10. [A more detailed description of the techniques for localizing human genes to their particular positions on the chromosomes can be found in various chapters in Franks (1988).] Once a gene has been localized, then this can now provide the basis for its identification and functional analysis.

4.4 Dominant inherited susceptibility and recessive genetic changes in tumours

4.4.1 Basic mechanisms

Mechanisms underlying inherited susceptibilities to cancer must be consistent with the general views described in Sections 4.1 and 4.2 as to its nature at the cellular level. There are, in principle, two basic types of mechanisms which can underlie an inherited susceptibility. The first could be through an influence on the particular cells from which a certain type of tumour is derived, and this can be thought of as essentially a tissue specific influence. The second may be through systemic effects, for example on the frequency with which mutations arise due to environmental effects, or on the efficiency with which potential carcinogens are metabolized. While the former mechanism should influence susceptibility to a particular form of cancer, the latter, which will be discussed later, in principle may give rise to inherited susceptibility to a wide range of cancers.

In 1971 Knudson pointed out that there should be some relationship between the genetic changes in a somatic cell that give rise to a cancer and changes in these same genes in the germ line, which will be passed on in the usual way from parent to offspring following Mendelian laws of inheritance. This follows from the fact that, when one of the particular genetic changes involved in initiation or progression of a tumour occurs in the germ line, all the somatic cells of an individual who has inherited this particular change will already carry one of the steps required for malignant change. This should increase the chance that a tumour will develop in such an individual, and so lead to an inherited susceptibility to

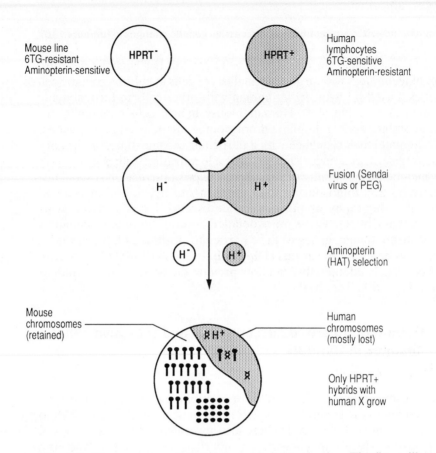

Fig. 4.3 Human–mouse cell fusion and hybrid selection. The figure illustrates the typical scheme for the production of human–mouse hybrids, using a drug resistance marker to select for the hybrid and either Sendai virus or polyethylene glycol (PEG) to fuse the cells. The mouse cells are a continuously growing cell line resistant to the drug 6TG (6-thioguanine) and as a result, sensitive to the drug aminopterin. Normal human lymphocytes are sensitive to 6TG, but resistant to aminopterin and cannot grow autonomously. After fusion of a mixed population of cells, only a small proportion will give rise to growing human–mouse hybrid cells and these must be selected out from amongst the background of mouse and human cells which have not fused, or have fused amongst themselves. Hybrid cells which result from fusion between the two different parent cell types will be resistant to aminopterin, like the human cell parent, but otherwise have the growth properties of mouse cell lines. Since normal human lymphocytes do not grow in culture, if the fused mixture of cells is suspended in the presence of the selective HAT medium containing aminopterin, only the human–mouse hybrid cells will be able to grow. The hybrid cells retain mouse chromosomes but lose most of the human chromosomes. HPRT is the enzyme which is deficient (−) in 6TG resistant aminopterin sensitive cells, but present (+) in normal (6TGS aminopterin R) cells. Since the HPRT gene is on the human X-chromosome, this chromosome must be retained in those hybrids which grow in the HAT-medium. Each hybrid will, apart from this, have its own particular combination of human chromosomes on the basis of which genetic segregation can be tested.

that particular cancer. Clearly, there could be such inherited changes which would involve, e.g. growth factors or their receptors, in a way that influenced the development of a variety of different tumours. In this sense, such a mechanism need not necessarily be tissue specific. Knudson emphasized that a corollary of these ideas was that identification of the gene involved in such an inherited susceptibility would also pinpoint a gene that might be involved in somatic changes in the same sort of tumour even when there was not an inherited susceptibility.

4.4.2 *Retinoblastoma*

Retinoblastoma, a rare tumour of the eye in children, was the first and still the best example of the application of Knudson's ideas. About 40 per cent of cases of retinoblastoma occur as a clear-cut, dominantly inherited Mendelian condition. The remainder are sporadic, in the sense that their first degree relatives (parents, siblings, or grandchildren) have a very low increased risk of getting retinoblastoma. The inherited, or familial, cases are most often bilateral, while the sporadic cases are often unilateral and in this sense less severe. Knudson argued that the familial cases of retinoblastoma were those in which one of the genetic changes essential for the development of the tumour was already inherited through the germ line. He also argued that the change in the *tumour* was recessive, so that the relevant gene on both homologous chromosomes had to be mutated or otherwise changed to become inactive. In the inherited form, one of the genes was already mutated and so only one further mutation, in the normal gene on the homologous chromosome, was needed for the development of the tumour. In the sporadic cases, on the other hand, independent mutational events in both homologous genes were required, and this had a very low chance of occurring. Hence the difference between the inherited and sporadic forms.

The inherited susceptibility gene on the basis of Knudson's model is dominant because, having inherited one abnormal gene, the probability of a second event occurring is sufficiently high that this gives the individual with the single abnormal gene an almost certain chance of developing a retinoblastoma. At the cellular level, however, the relevant genetic change is recessive since it is assumed that both copies of the gene must be inactivated in order to give rise to the tumour.

These ideas have now received dramatic support from chromosomal, genetic, and molecular studies. Thus, it was first observed that about 5 per cent of retinoblastoma patients had a small visible deletion of part of the long arm of chromosome 13. Then, using the fact that the esterase D enzyme variation had been mapped by somatic cell genetic studies to chromosome 13, it was shown by linkage of the esterase D marker to retinoblastoma in familial cases that these were all probably due to a mutation at the position on chromosome 13 which was deleted in that small minority of retinoblastoma patients.

These results were then extended to demonstrate in a simple and elegant way a role for the same gene in sporadic or non-familial cases of retinoblastoma by Cavanee *et al.* (1983). They used DNA probes which revealed RFLPs and which had been mapped to chromosome 13, again using somatic cell hybridization techniques. The argument was that, if Knudson's ideas were correct, such markers, when heterozygous in an individual, should often become homozygous in a tumour because a common mechanism for the second somatic change could be chromosome loss or somatic recombination (see Fig. 4.4). Thus, loss of the chromosome carrying the normal allele during tumour progression (which might sometimes be followed by reduplication of the remaining chromosome) would lead to a cell carrying only the defective retinoblastoma allele and at the same time only one of the RFLP marker alleles. Somatic recombination, in which there is exchange between homologous chromosomes in somatic cells, can be shown to lead to a similar effect. Thus, either of these processes would be revealed by the fact that in individuals who are heterozygous for the RFLP, as revealed for example by normal tissue surrounding a tumour, the tumour itself would express only one of the alleles detected by the chromosome 13 RFLP probe. In other words, while the individual is heterozygous for the marker, the tumour is homo- or hemizygous. Using this approach, Cavanee and his colleagues (1983) demonstrated such changes in a high proportion of both sporadic and familial retinoblastoma. This provided

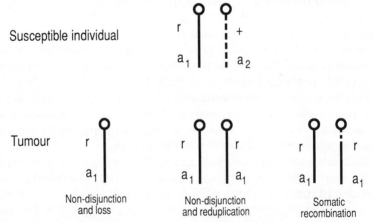

Fig. 4.4 Mechanisms for allele loss or homozygosity in retinoblastoma. r, + are the retinoblastoma and corresponding normal alleles a_1, a_2 heterozygous markers on chromosome 13. Tumour is homozygous a_1a_1, rr or hemizygous a_1r.

dramatic confirmation of Knudson's ideas on the recessive basis for some genetic changes in tumours.

Using this information the precise localization of the actual gene and its product have now been clearly identified. Clues as to its function have come from studies showing that the protein product, Rb, associates with the adenovirus E1A and certain other oncogene products (Whyte *et al.* 1988; see Chapter 9 for background on oncogenes). It has also been shown that, under some circumstances, the normal version of the retinoblastoma gene, when inserted into a transformed cell may reverse some of that cell's transformed properties (Lei *et al.* 1988). Most intriguingly it is known that the retinoblastoma gene product is present in most tissues of the body, even though the inherited defect gives rise mainly to retinoblastomas and also some osteosarcomas. Nevertheless, suggestive evidence has been obtained that the gene may also be defective in a significant proportion of breast and small cell lung carcinoma cases.

The elucidation of the inherited basis for retinoblastoma at the molecular level is a remarkable example of the power of modern techniques for molecular biology when combined with classical family studies. These approaches now make it possible to localize, identify, and analyse functionally any disease susceptibility gene and its product at the molecular level, particularly when there are both sporadic and familial forms of the disease.

4.4.3 *Familial adenomatous polyposis*

Colonic tumours are amongst the most common of the carcinomas. They are thought generally to arise as a rule from precancerous adenomas. Familial adenomatous polyposis (FAP, often also called adenomatous polyposis coli and the corresponding gene, APC) is a rare dominantly inherited susceptibility to colon cancer in which affected individuals develop from a few hundred to over a thousand adenomatous polyps in their large intestine. Since the polyps may be precancerous growths, amongst such a large number there is a high probability that at least one, if not more, will give rise to a carcinoma. Thus, almost invariably, an untreated individual with adenomatous polyposis will develop one or more colon carcinomas (Fig. 4.5). Early screening of individuals from affected families has been used to identify the polyps at a stage when prophylactic removal of the colon can prevent much of the risk of developing a carcinoma.

The clue to the localization of the gene for APC to chromosome 5 was a case report of a mentally retarded individual with multiple developmental abnormalities who had colorectal carcinoma with multiple polyps and died of complications characteristic of those seen in individuals with adenomatous polyposis. This individual was shown to have a small

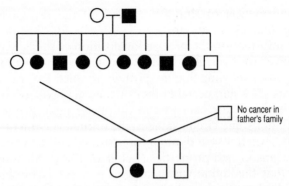

Fig. 4.5 Pedigree of the first clear-cut reported case of inherited polyposis coli (Lockhart-Mummery 1925). ●, affected; ○, unaffected. The disease is passed down through three generations from one affected parent. The pattern of inheritance is characteristic for a dominantly inherited trait.

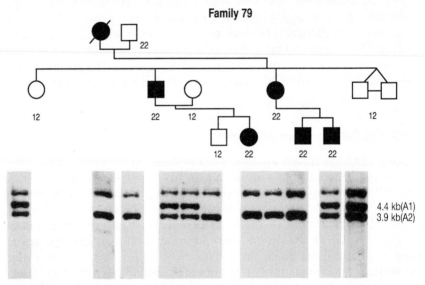

Fig. 4.6 Family 79 demonstrating segregation of FAP with a 3.9-kb fragment using the proble C11p11. The alleles are A1 at 4.4 kb and A2 at 3.9 kb. Symbols: □, male; ○, female; solid symbol, affected individuals.

deletion on the long arm of chromosome 5, suggesting that the polyposis gene might lie within this deleted region. A DNA probe which had been assigned to chromosome 5 revealed a polymorphism which was shown to be closely linked to APC (see Fig. 4.6; Bodmer *et al.* 1987). This C11P11 probe localizes to the region of the long arm of chromosome 5, 5q21-q22 which was missing in the one multiply abnormal sporadic case of FAP. Since that time a family with two brothers showing a similar deletion, both with multiple abnormalities and adenomatous polyposis, has been reported.

Following Knudson's arguments and the application of RFLP analysis to establishing the case for a recessive change in retinoblastomas, polymorphic DNA probes assigned to chromosome 5 were used to look for allele loss in sporadic colorectal carcinomas. Thus, using a highly polymorphic probe for chromosome 5, it has been shown that at least 20 to 40 per cent of sporadic colorectal carcinomas become homo- or hemizygous for chromosome 5 markers (see Fig. 4.7). This indicates that the APC gene becomes recessive in a relatively high proportion of colorectal carcinomas. This is the first clear example of such a recessive defect in one of the commonest of all cancers. As in the case of retinoblastoma, there are now many strategies which are being followed to clone the APC gene and establish its function.

4.4.4 *The general extent of recessive genetic changes in tumours*

There are many examples of individually rare instances of clear-cut inherited susceptibilities to particular cancers. Each of these may, as in

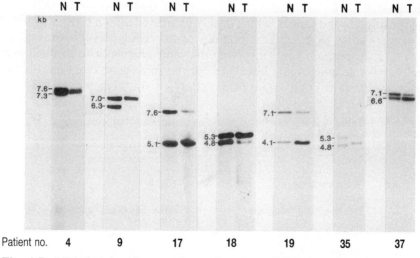

Fig. 4.7 Allele loss in primary colorectal tumours. DNA from matched normal (N) and tumour (T) pairs was tested with a highly polymorphic DNA probe known to map to the tip of the long arm of chromosome 5. Patient No. 9 is a typical clear example of allele loss (Band 6.3) from the tumour sample as compared to the normal tissue. In other cases the ratio of the intensity of one band in tumour tissue is much greater than another (for exmple, 5.1 versus 7.6 in patient No. 17's tumour). This is also clear evidence for allele loss as the tumour, when analysed, may often by contaminated with normal tissue.

the case of retinoblastoma and FAP, turn out to be of fundamental interest in providing clues to the genetic changes in non-familial cancers, even though the overall contribution to the incidence of cancer from these rare inherited susceptibilities is very small. In addition, the use of RFLPs to search for allele losses in tumours may reveal examples of recessive changes in tumours that are not associated with a familial inherited susceptibility. The case of multiple endocrine neoplasia type 2A (MEN 2A) is interesting in this respect, since allele loss has been shown for markers on chromosomes 1 and 22, while the gene for MEN 2A itself maps to chromosome 10. There is already a wide range of suggested genetic changes in tumours (see Table 4.2), a number of which have been found in common tumours such as of the colon, breast, bladder, and lung. It seems clear that these recessive changes form an important counterpart to the dominantly acting oncogenes found initially through the study of the oncogenic viruses (see Chapters 8 and 9). From Table 4.2 it appears other recessively acting genes important for more than one tumour are found on the short arm of chromosomes 3 and 17, now known to be the p53 oncogene.

The general scheme, based on Knudson's model, that fits the data on retinoblastoma, FAP, and other tumours where recessive genetic defects have been uncovered, is illustrated in Fig. 4.8. The individual who is heterozygous for a defective gene carries a dominantly inherited predisposition to the cancer, which is only expressed after a further somatic change leading to recessive expression at the cellular level in the cancer itself. The sporadic cases involving the same change arise through two successive genetic events knocking out the same gene function. Recessive mechanism could arise in a variety of ways, including negative

Table 4.2 Human malignancies associated with chromosome deletions or loss of RFLP alleles

Tumour	Chromosome region
Retinoblastoma	13q14
Wilms' tumour	11p13
Small cell lung carcinoma	3p13)24, 13p, 17p
Neuroblastoma	1p32)pter
Familial renal cell carcinoma	3p
Bladder carcinoma	11p
Uveal melanoma	2
Colorectal carcinoma	5q, 17p, 18
Breast carcinoma	13q, 17p
Bilateral acoustic neurofibromatosis	22q
Multiple endocrine neoplasia (Type 2A)	1, 22
Gastric carcinomas	13q

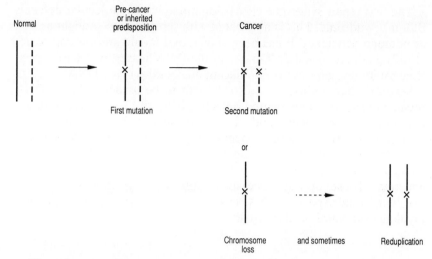

Fig. 4.8 Model for the two stages involved in recessive control of cancer.

control over the production of growth factors or their receptors. Recessive mechanisms could also arise from positive control of differentiation since blocking differentiation to achieve increased growth is likely to be one of the mechanisms leading to tumour progression.

4.5 Systemic inherited susceptibilities

4.5.1 *Repair deficiencies*

The best known example of a systemic effect leading to an inherited susceptibility to cancer is the recessively inherited xeroderma pigmentosum (XP), which is due to a deficiency in the ability to repair DNA. XP is characterized by a susceptibility to sunlight induced abnormalities of the skin, frequently followed by malignant skin cancer. Cells from individuals with the disease are unusually sensitive to the lethal effects of u.v. light and certain types of carcinogens, because they have defects in their ability to repair damaged DNA. It is now known that there are several different forms of the disease, probably due to different genetic mutations having similar effects. It seems likely that the reason why XP individuals predominantly get malignant skin cancers is that this tissue is most exposed to a mutagen, the u.v. light in sunlight. This exposure probably far exceeds that of internal organs from ingested mutagens, to which XP individuals would also be sensitive. There are, in fact, reports of patients with internal tumours.

The first repair syndrome clearly identified with a molecular defect is Bloom's syndrome. This is associated with sensitivity to sunlight as well as immune deficiency. It has been shown by Lindahl and his colleagues (1987) that there is a defect in one of the DNA ligases, which is responsible for joining gaps between adjacent nucleotides in the DNA.

Several other diseases are also thought to be associated with DNA repair defects, or other related defects, giving rise to sensitivity to u.v. light or other carcinogens, and all of these appear to be associated with a very significantly increased susceptibility to cancer. Notable amongst these disorders are ataxia telangiectasia, which is now also known to be a collection of diseases associated with different mutations, and Fanconi's anaemia. These are all recessively inherited syndromes, probably involving some aspect of DNA repair leading to increased chromosome breakage and multiple abnormalities.

4.5.2 *Carcinogen metabolism*

Inherited variations in the activity of enzymes that metabolize potential carcinogens are another source of inherited systemic susceptibilities to cancer. Amongst the best known such enzymes are the mono-oxygenases, or cytochrome P450 enzymes (Wolf 1986). These enzymes are known to metabolize many substances from inert into reactive, or carcinogenic, compounds. Thus, in the mouse there are differences between inbred strains in the level of the enzyme aryl hydrocarbon hydroxylase, which acts on certain hydrocarbons turning them into potent mutagens and carcinogens. These differences have been shown to be associated with differential effects of these hydrocarbons on the rate of tumour induction. Similar studies in man were initially promising and suggested a single gene difference in susceptibility to induction of lung tumours by cigarette smoking, but these observations have not been confirmed.

Inherited variations in various P450 enzyme activities are associated in man with differential responses to a wide variety of drugs. This is due to the fact that the activity of the drugs is modified by these enzymes. One example studied involves differences in the ability to metabolize the drug debrisoquine. About 10 per cent of the population metabolize this drug slowly and so have severe side effects when given the drug at therapeutic doses. This reaction appears to be associated with a recessively inherited difference in the relevant hydroxylating enzyme, the susceptible individuals being homozygous for a less active form of the enzyme. A controlled study of cigarette smokers with and without lung cancer has shown a striking sixfold lower frequency of the slow metabolizers of debrisoquine amongst lung cancer cases (Ayesh *et al.* 1984). If confirmed, this would be a most important example of an inherited systemic susceptibility to the carcinogenic effects of an environmental agent, cigarette smoke, due to differences in rates of carcinogen metabolism.

Most of the P450 enzymes have now been cloned. This means that it is now possible to study genetic variation in these enzymes at the DNA level and to test the association of slow and fast metabolism of debrisoquine with lung cancer by seeing whether detectable variation at the DNA level in the relevant enzyme, for example using RFLPs, associates with susceptibility to lung cancer amongst cigarette smokers. These and other variations, such as in the glutathione-S-transferases and certain acetylases, may provide important clues to inherited susceptibilities to cancer through metabolism of carcinogens.

4.5.3 *Immune response differences*

A third major example of inherited systemic effects concerns immune response differences. The major human histocompatibility (HLA) system controls two main sets of cell surface determinants which are involved in interactions between lymphocytes and other cells in the control of the immune response. The system is highly polymorphic, that is there are many differences between individuals with respect to the cell surface determinants; these differences make it necessary to match individuals for organ transplantation. HLA differences have also been shown to be associated with a variety of autoimmune or immune related disease, such as juvenile onset insulin dependent diabetes mellitus, rheumatoid arthritis, and ankylosing spondylitis, most probably through inherited differences in specific immune responses. A number of associations between HLA and different cancers have been suggested, most notably with nasopharyngeal carcinoma (NPC) and Kaposi's sarcoma, although in no case is the association as striking as that with the clear-cut autoimmune or immune related diseases. In the case of both NPC, associated with the Epstein Barr virus, and Kaposi's sarcoma, associated with the human immunodeficiency virus, an inherited immune response difference to the virus or virally induced cellular determinants is a plausible mechanism for an inherited difference in susceptibility to the cancer.

Inherited immune response differences associated with HLA variation as well as with variation in the other molecules of the immune system, notably antibodies and the T cell receptor, may be relevant in susceptibility to other virally induced human cancers. These include, for example, cervical and other cancers associated with the human papilloma viruses, the leukaemias associated with the various human T cell leukaemia viruses, and liver cancer associated with the hepatitis virus.

4.6 Types of family data and their interpretation

The simplest inherited susceptibilities are those, such as retinoblastoma, polyposis coli, the DNA repair deficiencies, and other inherited systemic susceptibilities, that follow a clear-cut Mendelian pattern of inheritance. In such cases, the nature of the genetic control is not in question, and the challenge is to identify the specific genes involved and interpret their functions at the molecular level. There are, however, many examples of inherited susceptibilities which are not so easy to interpret. For example, the associations of debrisoquine slow metabolizers or HLA variants and particular cancers were not identified through family studies, but by looking at the distribution of a particular genetic difference in patients with a given sort of cancer as compared to controls.

The classical approach to assessing a potential inherited contribution to a disease, in the absence of clear-cut Mendelian segregation, is to establish to what extent there is an increased incidence of the disease amongst the relatives of affected individuals. Often this involves specifi-cally the study of twins, contrasting identical with non-identical twins. If there is a major inherited component, then the disease incidence amongst co-identical twins of affected twins, the concordance, should be greater than that amongst co-non-identical twins. This is because the former share all their genes, while the latter on average share only half their genes, just as do any brothers or sisters. Studying twins brought up in the same household tends to average out the effects of environment. Twin studies tend to show a slightly increased concordance of cancer amongst identical as compared to non-identical twins, but the effect is marginal and the data very hard to obtain. More generally, studies on the incidence of particular forms of cancer amongst relatives of patients as compared to that in the general population have often indicated an approximately two- to fourfold increase in incidence amongst relatives, especially for breast and childhood cancers. It is, however, very difficult to interpret these relatively modest increases as necessarily due to genetic factors, since relatives also tend to share a common environment and this clearly could have a similar effect on incidence amongst relatives as do genetic factors. Thus, while such studies may suggest a limited genetic contribution to overall inherited susceptibility to certain cancers, they do not provide clear-cut answers and offer little or no prospect for further investigation.

Another approach to the problem of sorting out inherited susceptibilities, especially for the comparatively common cancers such as breast and colon cancer for which it is clear that the majority of cases do not show an obvious inherited component, is to ask whether there is, nevertheless, a subset of cases that tend to cluster in families. This would indicate a minority of cases associated with a clearly inherited susceptibility. Occasional very striking examples of clusters of cancers within a single family have often been described, and an example is shown in Fig. 4.9. The difficulty with this approach is in assessing whether the familial clustering is really significant. Obviously for a relatively common cancer, some cases will cluster in families simply by chance, and appropriate statistical methods must be used to distinguish these from familial clustering due to Mendelian segregation of a gene associated with an increased susceptibility. Statistical models for the expected distribution of different genetic types can be fitted to such families, but such models rarely provide a clear-cut answer to the interpretation of familial clusters.

The study of familial clustering as a basis for identifying inherited susceptibilities is also subject to another major difficulty. This is the fact, easily demonstrated, that even if a disease does not show any obvious familial clustering, it may nevertheless have a major genetic component. Suppose, for example, there exists a dominant gene which increases the chance of getting a particular form of cancer by a factor of 10, say from one in a thousand to one in a hundred. If the gene is rare, then in most families where it occurs it will only occur once in one of the parents. Then, for example, only in $(1/100)^2 = 1/10\ 000$ of such families with two offspring will both of them have the cancer, and the chance that a sib of an affected sib will be affected is only $1/200$. Even if the gene gives rise to a 10 per cent chance of getting the cancer, only 1 per cent of families with two children with the gene present once in either parent will have pairs of affected sibs. Nevertheless, even when the gene frequency is as low as 0.05 per cent, it could still be contributing as much as 50 per cent to the

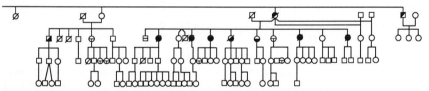

Fig. 4.9 Pedigree of a family with multiple cases of cancer (S. Cartwright and J. G. Bodmer, personal communication). Circles, females; ◓, carcinoma cervix; ◑, carcinoma liver; ●, carcinoma breast; ◒, carcinoma stomach; ⊕, cervical dysplasia; ⊕, ovarian cysts. Squares, males; ◪, carcinoma bronchus; ◣, skin basal cell carcinoma; ⊟, papilloma bladder. A diagonal slash through the symbol represents death of the individual.

total incidence of the particular form of cancer with which it is associated. Here, one would have a situation where a particular gene was responsible for 50 per cent of the incidence of one form of cancer, and yet because only 10 per cent of the people with the gene get the cancer, there would be very few examples of familial clustering. However, in those families where pairs of sibs are affected, most pairs will both carry the relevant susceptibility gene. And it is this that provides the clue to the objective study of such inherited susceptibilities. Because, if an inherited Mendelian marker difference can be found for a gene that is reasonably close to the one actually causing the inherited cancer susceptibility, then this marker will also tend to be associated with pairs of affected sibs. In other words, if one finds a genetic marker whose distribution amongst affected pairs of sibs is distorted as compared to Mendelian segregation, then this marker must be linked to a gene causing an inherited suceptibility.

This principle can be illustrated using as an example the association between HLA and Hodgkin's disease. A weak association between certain HLA markers and Hodgkin's disease was first observed in 1967. This was subsequently confirmed by many other studies, but its significance is only due to the fact that the association was studied so extensively. The maximum relative risk, a simple measure of the relative increase in the frequency of a particular HLA determinant amongst people with Hodgkin's disease as compared to controls, was only about 1.3–1.6 as compared, for example, to relative risks of close to 100 or more for the association of HLA type B27 with ankylosing spondylitis. Now, although the vast majority of cases of Hodgkin's disease are sporadic, a small proportion, perhaps up to 3 per cent of cases, occur in families with two or more affected individuals. Within such families, HLA typing can establish whether the affected pairs of sibs with Hodgkin's disease are HLA identical (namely have inherited the same HLA chromosome complement from each parent), share only one HLA chromosome but not the other, or have neither chromosome in common. The expected frequency of these three situations on the assumption of Mendelian segregation, and in the absence of any association within the families between HLA and Hodgkin's disease, is 1:2:1. Overall, amongst 32 sib pairs studied in this way, 16 were found to be HLA identical, 11 shared one HLA chromosome and 5 none, a highly significant departure from the expected Mendelian 1:2:1 or 8:16:8. The data thus clearly show an association between the HLA segregation in the families and Hodgkin's disease. This is exactly as expected if there is a gene in, or close to, the HLA region which confers susceptibility to Hodgkin's disease. Thus, in the case of Hodgkin's disease, the family data provide the most convincing evidence for an association with the HLA system. This approach can clearly be generalized to any situation where there is more than one member of a family with Hodgkin's disease. The question asked is whether the HLA distribution is distorted amongst the individuals with Hodgkin's disease, as compared to what could be expected from the normal pattern of Mendelian segregation.

The HLA system was originally chosen for study in the case of Hodgkin's disease because of the association between the mouse H2 system (the equivalent to HLA) and certain types of virally induced leukaemias. For most examples of familial clustering of cancer, there is no such clue as to which genetic marker should be studied. In this case all that one can advocate is a systematic search for a genetic marker which is distorted in its segregation amongst individuals with cancer in the family. While this may seem a haphazard approach the range of genetic markers available for such studies is now vastly increased using recombinant DNA techniques as described in Section 4.3, and so such systematic surveys are becoming a realistic possibility. Through them it should be possible to identify, for any significant familial clustering of an inherited cancer susceptibility, one or more genetic markers sufficiently close to the gene actually causing the susceptibility. Such a gene would be detected by a distortion in its expected Mendelian segregation amongst the individuals in the family who have cancer. In my view, this is now the only satisfactory way of establishing an inherited susceptibility, other than finding the gene which itself gives rise to that inherited susceptibility.

This is the general principle which underlies the analysis of the association between HLA and Hodgkin's disease discussed above. Clearly, the ability to detect a distortion in the marker distribution amongst individuals with cancer in families will be a function of how close the marker happens to be to the susceptibility gene. In practice, even a recombination fraction of 10 per cent would readily allow the detection of a segregation distortion. It can be calculated that approximately 250 markers regularly spaced at a 10 per cent recombination fraction interval are needed to cover the complete human chromosome set. This, therefore, in principle would be the maximum number of markers needed to be tested on a set of families in order to find one or more sufficiently close to the gene actually causing an inherited susceptibility for it to be identified.

4.7 Future prospects

Although inherited susceptibility to cancer may contribute no more than 20 per cent of overall cancer incidence, nevertheless this is both an important contribution in its own right and can help to provide major clues to the fundamental underlying causes of cancer, and to approaches for its prevention and treatment. Tissue specific inherited changes can provide clues to genetic changes taking place during tumour progression in non-inherited cancers. Inherited susceptibilities connected with systemic effects, such as deficiencies in DNA repair, carcinogen metabolism, and immune response, provide major clues to potentially controllable environmental factors which cause cancer. In all cases, the ultimate challenge is to identify the particular genetic differences and their functional basis.

The use of DNA polymorphisms to find markers linked to cancer susceptibility genes in principle provides an avenue to the eventual identification of the susceptiblity gene itself as in the case of retinoblastoma. The more closely the marker is associated within families with the cancer, the more likely it is to be near to the responsible gene and the easier it will be, therefore, eventually to isolate the gene itself. The technical problems are still formidable but the rate of progress in recombinant DNA technology is such that one must surely expect the problem of identifying cancer susceptibility genes and their functions to be solved within the foreseeable future. The analysis of retinoblastoma has been a model example and the search for the familial adenomatous polyposis gene is well under way.

In some cases an intelligent guess may provide a clue, once the chromosomal region within which a susceptibility gene lies has been identified. This proved to be the case in the identification of the involvement of the c-*myc* oncogene (lymphoma cells) and c-*abl* in chronic myelocytic leukaemias (see Chapters 9 and 10) and of the p53 gene as responsible for allele loss on chromosome 17p. As already emphasized, the genes for the P450 enzymes and other enzymes involved in carcinogen metabolism can now be studied using the DNA based techniques for their association with inherited cancer susceptibilities. Sooner or later also, all the genes for the DNA repair deficiency syndromes will be identified. Then, the question of whether there is an increased risk of cancer associated with an individual who carries just one copy of the defective DNA repair gene will become amenable to analysis using recombinant DNA techniques.

A genetic marker that is closely linked to an inherited susceptibility may have considerable practical value even if it does not immediately lead to the identification of the specific genetic function involved in the susceptibility. First of all, such a marker may help to recognize heterogeneity in the predisposition, since different subsets of susceptibles may show different patterns of linkage to different genetic markers. Second, within families, the linked marker defines a high risk group, the identification of which may be very valuable. For example, individuals identified as being at high risk may be treated prophylactically, as is now the case for polyposis coli. Such individuals may also be useful for studies of the physiology of the difference between high and low risk groups, which should help to identify the underlying functional basis for a particular inherited susceptibility. It may also be possible to do case control studies comparing high and low risk groups within families, in order to identify factors that may interact with a genetic predisposition.

Recombinant DNA technology leads to identification of the genetic steps, one at a time, that take a cell from the normal to the cancerous state. Already, for example, in the case of colorectal cancer, relevant genetic changes have been identified at specific regions on chromosomes 5, 17, 18, and perhaps 22, as well as the common change for one of the *ras* oncogenes and frequent changes in HLA expression on the tumours probably associated with escape from immune attack on the cancer. Many of these genetic changes can already be identified using archival material stored in the form of paraffin blocks from tumour biopsies. With the further development and refinement of DNA based techniques and approaches to growing out samples of cells from tumours and their surrounding tissues, it will become possible to identify all the genetic steps and their sequence during tumour progression, as well as their functional significance. That must be the ultimate classification of a tumour, through which one must hope to find new approaches either to prevention or early detection and to treatment.

Sooner or later we shall have essentially the whole DNA sequence of the human genes and some definition of all the basic functional units. When this situation is reached, having found a linked marker for a particular inherited susceptibility, it may be possible simply to look up the genes with relevant functions that are in its neighbourhood, and through that, focus on to the actual genetic difference responsible for the inherited susceptibility. There can be no doubt that the application of recombinant DNA techniques, coupled with epidemiological and genetic studies, will in due course unravel the genetic contribution to the initiation and progression of cancers both at the germ line and somatic cell levels.

References and further reading

Ayesh, R., Idle, J. R., Ritchie, J. C., Crothers, M. J., and Hetzel, M. R. (1984). Metabolic oxidation phenotypes as markers for susceptibility to lung cancer. *Nature* **312**, 169–70.
The paper on the association between debrisoquine metabolism and lung cancer due to cigarette smoking.
Bodmer, W. F. (ed.) (1982). Inheritance of susceptibility to cancer in man. *Cancer Surveys* **1**, 1–186.
Contains a range of articles covering many of the topics surveyed in this chapter.
Bodmer, W. F. *et al.* (1987). Localization of the gene for familial adenomatous polyposis is on chromosome 5. *Nature* **328**, 614–16.

Cavanee, W. K. *et al.* (1983). Expression of recessive alleles by chromosomal mechanisms in retinoblastoma. *Nature* **305**, 779–84.

Fialkow, P. J. (1972). Use of genetic markers to study cellular origin and development of tumors in human females. In: *Advances in Cancer Research* v. 15, pp. 191–226. Eds. George Klein and Sidney Weinhouse. Academic Press, USA.

Franks, L. M. (ed.) (1988). Somatic cell genetics and cancer. *Cancer Surveys* **7** (2).
A series of papers on genetic aspects of cancer as revealed by somatic cell genetic techniques.

Harnden, D., Morten, J., and Featherstone, T. (1984). Dominant susceptibility to cancer in man. *Advances in Cancer Research* **141**, 185–245.
A review of certain aspects of inherited susceptibility to cancer.

Huang, H-JS., Lee, J-K., Shew, J-Y. *et al.* (1988). Suppression of the neoplastic phenotype by replacement of the RB gene in human cancer cells. *Science* **242**, 1563–6.

Knudson, A. G. (1986) Genetics of human cancers: review. *Annual Reviews of Genetics* **20**, 231–51.
An up-to-date statement of Knudson's ideas.

Lockhart-Mummery, (1925). Cancer and heredity *Lancet* **1**, 427–9.

McKusick, V. (1988). *Mendelian genetics in man* (8th edn). Johns Hopkins University Press, Baltimore.
The standard reference catalogue for inherited human diseases, including cancer.

Mulvihill, J. J., Miller, R. W., and Fraumeni, J. F. Jr. (ed.) (1977). *The genetics of human cancer.* Raven Press, New York.
An earlier collection of papers on cancer genetics which is still a very useful survey.

Omenn, G. S. and Gelboin, H. V. (ed.) (1984). *Genetic variability in response to chemical exposures.* The Banbury Report 16. Cold Spring Harbor Laboratory, Cold Spring Harbor, New York.
A useful collection of papers on a whole variety of aspects of inherited differences in drug metabolism and their relationship to cancer incidence.

Solomon, E. *et al.* (1987). Chromosome 5 allele loss in colorectal carcinomas. *Nature* **328**, 616–19.
This paper and that by Bodmer *et al.* (1987) describe the localization of the familial adenomatous polyposis gene and the demonstration of its likely role as a recessive genetic change in a relatively high proportion of sporadic colorectal cancers.

Whyte, P. *et al.* (1988). *Nature* **334**, 124–9.
Describes the likely function of the retinoblastoma gene product and provides references to its initial identification.

Willis, A. E., Weksberg, R., Tomlinson, S. and Lindahl, T. (1987). Structural alterations of DNA ligase I in Bloom syndrome. *Proceedings of the National Academy of Science USA* **84**, 8016–20.

Wolf, C. R. (1986). Cytochrome P-450: Polymorphic multigene families involved in carcinogen activation. *Trends in Genetics* **2**, 209–14.
A review of the P450 enzyme systems.

5

Structure of DNA and its relationship to carcinogenesis
BEVERLY E. GRIFFIN

5.1 Introduction

Cells (and the intracellular substances secreted by them) make up the structural elements of the body. Within the nucleus of the cell resides its genetic information in the form of the polymeric material, deoxyribonucleic acid, or DNA. The integrity of this DNA is essential for the proper functioning of cells, their interactions, and, following naturally from this, the health of the whole organism. There are a few well-documented exceptions to DNA as the repository of genetic information. Some viruses carry their genetic information in ribonucleic acids, or RNA (see Chapter 8). Among these are viruses, designated *retroviruses*, that warrant serious consideration in any discussion about the genesis of cancer in avian and mammalian species. They code for an enzyme that converts their genomic RNA into DNA, which provides the origin of the term 'retro' or backward flow of information. Controversial suggestions have recently been put forward regarding the nature of the genetic agent in diseases such as scrapie in sheep and kuru in man, where no infectious DNA or RNA has been identified and chemical or enzymic reagents, which might be expected to destroy both, fail to abolish infectivity! It has been suggested by one school of scientists that the genetic information may indeed reside within protein molecules (so-called 'prions'), and models for protein replication have been put forward. However, at present, this suggestion remains speculative and unconvincing.

The introduction to this chapter in the previous edition divided the study of DNA into 'seven ages'. First was the age of the medical investigator, with the discovery by Miescher and colleagues in Germany over a century ago of a material in pus cells designated by them 'nuclein'. Their finding alerted the scientific world to the existence of a hitherto unknown, and possibly important, cellular component. The second age, that of the chemist, was necessary to provide the definition of the component parts of DNA (and RNA), their chemical nature, and how they are linked to make up the polymeric species. Following this comes the age of the geneticist and the discovery that DNAs contain the genetic information essential for the continuity of most organisms. In the fourth age the molecular biologist defined the mechanisms by which DNA could pass on its genetic information. From this group of scientists came the concept of the linear relationship whereby DNA specifies the structure of RNA which in turn specifies proteins (the so called 'central dogma' of molecular biology) and the concept of a triplet genetic code (see p. 136). The virologist provided many of the material and experimental designs for testing hypotheses proposed by molecular biologists and for identifying regulatory mechanisms that control gene expression, both quantitatively and qualitatively. The sixth age of DNA, that of the present, continues to belong to the biotechnologists and genetic engineers who turn academic exercises into practical reality, manipulating genes and their expression at will in the cause of medical or commercial progress and providing tools for probing the details of the biology of normal and abnormal cells. There are even firm plans for sequencing the human genome. At this junction it may be hoped that the history of DNA will not end like the famous Shakespearean diatribe on the 'seven ages of man' and terminate in 'second childishness and mere oblivion, sans teeth, sans eyes, sans taste, sans everything'. This seems highly unlikely, and one can predict optimistically that the 'seventh age of DNA' will complete the circle, returning to the cell and the cell biologist, who will draw on all the knowledge acquired over the last 100 or so years (as outlined briefly in the discussion that follows) to unravel the intricate interactions, balances, and counterbalances in the normal cell and contrast them with lesions that give rise to the malignant cell.

We now know that each specific character in an organism is coded by a gene, a unit of genetic information, which produces its effect by specifying the production of a particular protein. The genes consist of long strands of DNA arranged in a very specific order. The DNA exists in close association with a group of small basic proteins known as histones also arranged in an ordered manner with the protein molecules acting as wedges which have the correct shape to form the strands into coils. The structural unit is a nucleosome (Fig. 5.1) which is made up of a short

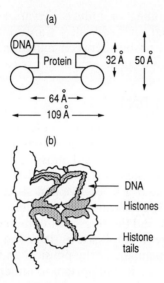

Fig. 5.1 Diagram showing substructure of a nucleosome. DNA strands in the nucleosome surround a core of histones and are continuous with linker strands of DNA above and below: (a) cross-section along axis; (b) from the side (from Richards *et al.* 1977).

length of DNA (about 200 nucleotide pairs—see p. 127) associated with a protein core of histones. The nucleosomes are attached to each other by a short piece of linker DNA (about 60 base pairs) like strings of beads which are themselves organized into coils or supercoils by associated non-histone proteins. Individual nucleosomes are too small to be genes (the average gene is thought to be about 1000 nucleotides), and act as packing devices. Recently it has proved possible to crystallize nucleosomes and thus precisely determine their three-dimensional structure. When genes are inactive, the DNA and protein molecules are closely packed. When the genes are active, that is, being transcribed, the protein DNA complex opens up in a different, more accessible structure to allow the process of gene expression to take place. Alterations in cell behaviour may be brought about by changes in the structure of the DNA— i.e., mutation—or by perturbations in the mechanisms which control gene expression. Although we now have a great deal of information on changes in DNA structure and their relationships to neoplastic development (to be discussed later in this chapter), our knowledge of control of gene expression is much more limited. There is convincing evidence that methylation of DNA, particularly at CpG base pairs (see p. 130) is one method by which control is exerted, in that such sites are often found to be transcriptionally inactive. Recent evidence suggests that methylation may enhance DNA coiling or induce curvatures in coils, provoking physical distortions that markedly alter transcription. Research in this area and into the relationship between mutation and gene expression should ultimately lead us to an understanding of cancer and of its control.

5.2 Components of DNA

DNA is composed of three relatively simple chemical species, namely, heterocyclic (nitrogenous) bases of which four are normally used, a five carbon atom sugar (deoxyribose), and phosphoric acid. The bases themselves are of two types, one a six membered ring species designated a 'pyrimidine', and the other a fused five and six membered ring species designated a 'purine'. By convention, pyrimidines as classes are abbreviated as Y, purines as R, and the rings are numbered as shown in Fig. 5.2. In spite of their apparent simplicity, purines and pyrimidines have the capacity for determining many of the physical and biological properties of individual DNAs. Thus, it is important to understand their chemistry. The structures of the two pyrimidine residues, cytosine (C) and thymine (T) that exist in DNA are shown in Fig. 5.2. Theoretically, both these species can exist in a number of tautomeric forms (see p. 130), although the isomer shown in the figure (Fig. 5.3) is that normally found in DNA. Nonetheless, it is relevant to consider the other forms, as illustrated for cytosine, since pyrimidines trapped in one of these alternatives (as for examples by alkylation) can lead to mutations in DNA. The various species result from a simple form of keto–enol (or imino–amino) tautomerism, that is, the interconversion between a double bonded oxygen atom ($=O$) and its singly bonded hydroxyl ($-OH$) counterpart which occurs by shift of a pair of electrons and hydrogen moiety or, alternatively, by comparable mechanism, between an $=NH$ and $-NH_2$ moiety.

Similarly, in the case of the two purines that make up DNA, guanine (G) and adenine (A), various tautomeric forms exist, but for native DNA, those shown in Fig. 5.2 persist. That is, for both pyrimidines and purines in DNA under normal circumstances the exocyclic oxygen atoms exist in the keto ($=O$) form, whereas the exocyclic nitrogens exist in an amino ($-NH_2$) form. This tautomeric preference in DNA can be altered, for example, by radical changes in pH or chemical modification, events frequently accompanied by important changes both in the physical and biological properties of the DNA.

Three of these heterocyclic bases, C, G, and A, also make up the building blocks of RNA. The fourth residue in DNA, thymine (T), is replaced in RNA, for reasons that are not wholly understood, by a similar molecule designated uracil (U) that lacks a methyl group at position 5. Similarly,

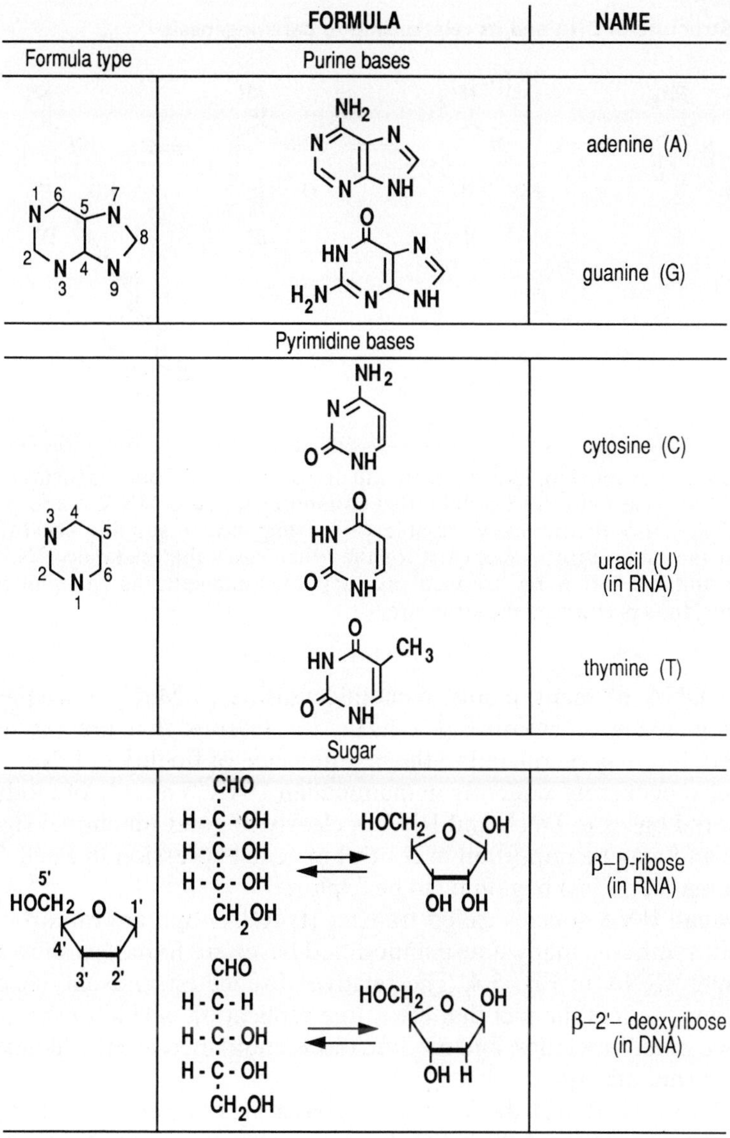

FORMULA		NAME
Formula type	Purine bases	

NH_2

adenine (A)

$$\begin{array}{c} 1\ 6\quad 5\quad 7 \\ N \qquad N \\ \ \ \ \ \ \ \ \ \ \ \ 8 \\ 2 \quad N^4 \quad N \\ \ \ \ \ 3 \quad 9 \end{array}$$

O

guanine (G)

HN

H_2N N NH

Pyrimidine bases

NH_2

cytosine (C)

$$\begin{array}{c} 3\quad 4 \\ N \qquad 5 \\ 2 \quad N \quad 6 \\ \ \ \ \ 1 \end{array}$$

O

HN

uracil (U)
(in RNA)

O CH_3

HN

thymine (T)

Sugar

CHO
H - C - OH
H - C - OH
H - C - OH
CH_2OH

$HOCH_2$ O OH

OH OH

β–D-ribose
(in RNA)

5'
$HOCH_2$ O 1'
4'
3' 2'

CHO
H - C - H
H - C - OH
H - C - OH
CH_2OH

$HOCH_2$ O OH

OH H

β–2'– deoxyribose
(in DNA)

Fig. 5.2 Structures, numbering systems, and abbreviations of the heterocyclic (nitrogenous) bases, and five carbon sugar moieties that are found in DNA and RNA. The numbering system currently in use for the six membered pyrimidine ring is to be regretted, since the original system wherein the corresponding rings in both purines and pyrimidines were numbered alike is a simpler system. Further, the latter was used in the classical hydrolytic studies of Chargaff where he showed that, regardless of the system used for isolation of DNA, or its overall base content, there was a conserved correspondence between the ratios of A:T and G:C. That is, the 'oxy' function in the 6 position of one purine was matched by a 6 amino function in the pyrimidine, and vice versa. Chargaff's work provided the basis for the well-known Watson Crick hypothesis of complementary structures in double-stranded DNA and led ultimately to the 'double helix' model of DNA. The pentose sugar numbers carry a 'prime' designation when this moiety is linked to the bases, to distinguish them from numbers given to the latter.

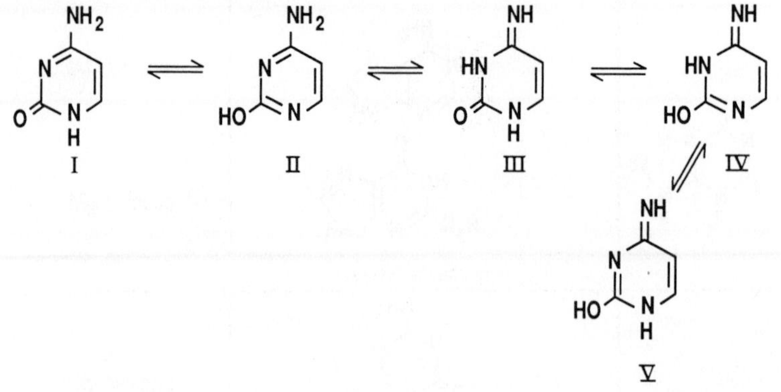

Fig. 5.3 An example of tautomerism, illustrated by the tautomeric forms of the pyrimidine base, cytosine. Form I is that commonly found in DNA, but forms III and V are also theoretically capable of existing under suitable conditions. Similar tautomeric forms can exist for the other bases that make up DNA (or RNA) and can often be 'trapped' as such by mutagenic and carcinogenic reagents, thus perturbing the structure.

in the DNA of many plants, 5-methylcytosine (5-MeC) is frequently found in place of cytosine (C), again for reasons that are yet to be defined, but may be related to the maintenance of fidelity of DNA. This base also occurs infrequently in mammalian DNA. The role of modified or altered bases in DNA and RNA is clearly of great functional significance and may be important in control of gene expression in a cell. This is an area that is just beginning to be explored.

In small RNA species called transfer (t) RNAs that are important in protein synthesis, many unusual modified bases are found, as shown for one such tRNA in Fig. 5.4. The relative absence of 'unusual' bases in DNA may reflect the fact that error free replication is vital to the maintenance of any specifics, and modifications could increase the likelihood of error (mutation).

In DNA (and RNA), the heterocyclic bases are covalently bound to the pentose sugar moiety, 2′ deoxyribose (or ribose), via an N-glycosidic bond. This link traps the carbohydrate in one of its tautomeric forms, that of a β-D-deoxyribose, see Fig. 5.2. (In RNA, the corresponding sugar moiety is β-D-ribose.) The combination of a heterocyclic base linked to the sugar is called a *nucleoside* or *deoxynucleoside*, as shown (Fig. 5.5, see also Table 5.1).

The third component of DNA, phosphoric acid, is covalently linked to the pentose sugars by a phosphate ester bond to produce, initially, a *nucleotide*. Nucleotides *per se*, especially in the cases where the sugar component is ribose, have many functions in cells. For example, they further combine with other molecules of phosphoric acid to produce

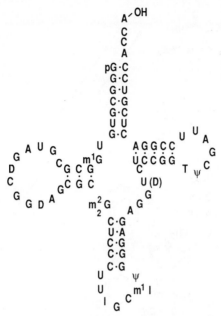

Fig. 5.4 The structure of one of the small transfer (t) RNAs, alanine tRNA, important in protein synthesis, showing some of the so-called 'minor bases' common to this kind of RNA, but not known to be ubiquitous in DNA or other RNA. Transfer RNAs carry amino acids to the sites of protein synthesis on ribosomes. It can be seen that more than 10 per cent of this particular molecule is composed of bases such as methylated guanines (m^1G, m_2^2G) or inosine (I) which are not normal components of nucleic acids. Similar types of modifications are found in other tRNAs and ribosome RNAs. (ψ is pseudouracil.)

compounds (such as adenosine triphosphate) that act as energy sources in many biochemical reactions. They also combine with other organic molecules to produce coenzymes, or they cyclize to produce signalling molecules important in regulating cellular functions.

A nucleotide, linked to another nucleoside via a phosphodiester bond, produces in an initial reaction a dinucleoside phosphate, which on phosphorylation gives a *dinucleotide,* part of the backbone of either DNA or RNA (see Fig. 5.5). The determination of the nature of the links involved in generating these species was one of the important contributions of the chemists of this field, since it provided the basis for our present understanding of the structure of DNA. It was found that the 5′ position of the deoxyribose in one nucleotide was covalently bound to the 3′ position of another by diester bonds with phosphoric acid (for numbering of the sugar, see Fig. 5.2). This linkage determines the polarity, sequence, and structure of the chain of molecules that compose nucleic acids, whether they be DNA or RNA (Fig. 5.5). It is interesting at this stage to note that

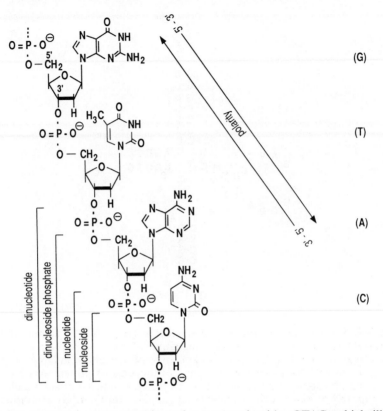

Fig. 5.5 Schematic representation of a tetranucleotide, GTAC, which illustrates the chemical linkages found in DNA, the terminology used to describe individual chemical entities, and the concept of polarity, important in any consideration of the double-stranded nature of DNA. The polarity must also be known before the structure of a polypeptide (or protein) can be predicted from DNA sequence.

Table 5.1 Standard nomenclature

Base	(deoxy)Nucleoside	Abbreviation	(deoxy)Nucleotide
cytosine	(deoxy)cytidine	C	(deoxy)cytidylic acid
thymine	thymidine	T	thymidylic acid
(uracil)	(uridine)	(U)	(uridylic acid)
guanine	(deoxy)guanosine	G	(deoxy)guanylic acid
adenine	(deoxy)adenosine	A	(deoxy)adenylic acid

for many years the macromolecular nature of DNA went unrecognized. Since only four major bases were evident from the hydrolysis of DNA, the compound was assumed to be a tetranucleotide and, as such, obviously lacked the capacity to be the genetic entity required even by the simplest of cells.

5.3 The genetic material

The DNA of most cells is double stranded. In the case of many viruses, it is also circular. A polymer composed of nucleotide components with a 5′—3′ polarity binds to another (complementary) polymer with a 3′—5′ polarity to create the double-stranded DNA (Figs 5.5 and 5.6). The

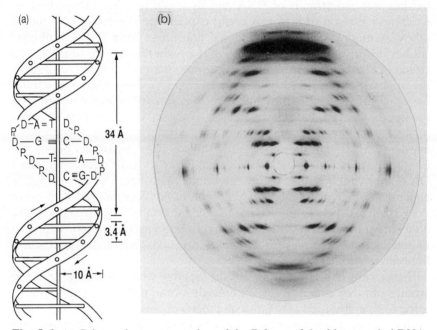

Fig. 5.6 (a) Schematic representation of the B form of double-stranded DNA, together with its dimension as derived from X-ray crystallographic analyses, and (b) its corresponding X-ray diffraction pattern (courtesy of Dr A. G. W. Leslie). Complementary bases (A and T, G and C) in opposing strands are held together by hydrogen bonds, as shown (Fig. 5.7). This structure produces grooves of two different sizes in DNA, designated 'major' and 'minor', which can act as sites of entry to DNA by chemicals, enzymes, etc. (D represents the deoxyribose moiety, P phosphate, and A, G, T, C the respective heterocyclic bases). The remarkable accuracy of the models of DNA has now been confirmed at the atomic level using a technique called scanning tunnelling microscopy (Driscoll *et al.* 1990).

recognition of the nature of the bonds that link one strand of a double-stranded DNA with its partner, and the fact that these bonds are very specific, provided the basis of the explanation of the mechanism by which DNA alone could encode genetic information. The groundwork for this important discovery came from analysis of the components released when DNA from a variety of sources was subjected to chemical hydrolysis. These experiments showed that although the base compositions, that is, the percentages of the various pyrimidines and purines, could vary enormously among species, a common relationship between bases was maintained such that the ratio of G:C or A:T always gave a figure that was about 1.0. These data, together with X-ray crystallographic evidence that showed the regularity of the structures of DNA, led not only to the very important suggestion of the nature of the base pairing between strands of DNA and its specificity, but recognition by Crick and Watson of how this could explain the key biological role of DNA. A model of the 'double helix' that arose from these combined studies is shown schematically in Fig. 5.6, wherein a C residue on one strand of DNA, wherever it occurs, is always 'paired' with a G residue on the opposite strand, likewise A with T (Fig. 5.7). The order in which the bases appear prescribes the genetic information. It is relevant to note that the bonds that link heterocyclic bases to sugars, and the latter to phosphates, are all covalent and as such very strong, requiring considerable energy to break. On the other hand, the so-called 'hydrogen bonds' (H bonds) that link one base residue to another to form double-stranded DNA are by their nature very weak bonds (less than 3 kcal of energy is generally sufficient to cleave a hydrogen bond as compared with more than 10 times this for the weakest covalent bond). The strength of the attachment between strands of DNA is in large part thus a consequence of the fact that many such H bonds are involved in the interaction between strands of DNA.

Biologically of great relevance is the fact that when, during mitosis, the strands of DNA separate and each single strand is then copied to reproduce double-stranded DNA, the specificity of base pairing ensures that a faithful, albeit complementary, replica of the coded DNA is made and the fidelity of the gene for future generations is maintained. If mistakes occur, however, as they do from time to time, normal cells have a variety of important functions that recognize individual errors and make the necessary repairs.

Following on the discovery of the mode by which fidelity of genetic information can be maintained was the elucidation of the mechanism by which the sequences of bases on any particular region of a strand of DNA could specify the sequence of amino acids in a corresponding protein, that is, how the genetic information is actually encoded within

(a)

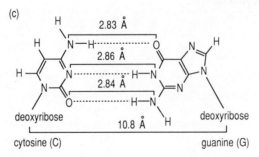

(b)

2.85 Å

H₃C

O ·············· H····N

2.90 Å

H— ... N—H ··············· N

deoxyribose   deoxyribose

11.1 Å

thymine (T)   adenine (A)

(c)

2.83 Å

N—H················ O

2.86 Å

N ···············H—N

2.84 Å

O················H—N

deoxyribose   deoxyribose

10.8 Å

cytosine (C)   guanine (G)

Fig. 5.7 Panel (a): a schematic representation of the links between the pyrimidine, thymine, and its purine complement, adenine. Panels (b) and (c): the hydrogen bonds formed between the base pairs T/A and C/G, respectively, and the distances found between the base links in double-stranded DNA. In general, the strength of a hydrogen bond is proportional to its distance. Here it is seen that not only are there three such bonds in the G/C partnership, but also that they are generally shorter. The energy required to separate these complementary bases is thus greater than that required for T/A base pairs. Hydrogen bonds determine the specificity that exists in DNA and can be considered the 'watchdogs' of fidelity during DNA replication, or transcription of RNA.

the DNA. The colinear relationship between DNA, RNA, and proteins is such that (except in the case of retroviruses) a gene containing an 'antisense' version of information maintained in DNA, is faithfully copied into a complementary 'sense' version of a species of RNA known as messengers (or mRNAs) using the specific base pairing discussed above (that is, for example, CAT in DNA would specify AUG in mRNA). RNA messenger, using blocks of trinucleotide sequences as its code, in turn specifies the amino acids and their order in a protein. But why a 'triplet' code? There are only 20 essential amino acids and four distinct nucleotides, so a doublet code would be inadequate, whereas a triplet could specify 64 amino acids, or more than enough. The precise nature of the code, as worked out with mixtures of synthetic oligonucleotides, is shown in Table 5.2. Certain amino acids, for example methionine (MET), are only encoded (specified) by one particular triplet (in this case AUG). In other cases, the coding is 'degenerate' and more than one triplet can specify a given amino acid. For example, proline (PRO) is encoded with two C residues and a third base which can be either C, U, A, or G. This degeneracy, together with the three triplets (UAG, UGA, and UAA) that specify the termination of translation of a nucleotide triplet into an amino acid, is such that all 64 potential triplet codons play some role in the specification of protein structures. Although the

Table 5.2 The genetic code

	U	C	A	G	
U	PHE	SER	TYR	CYS	U
	PHE	SER	TYR	CYS	C
	LEU	SER	STOP	STOP	A
	LEU	SER	STOP	TRP	G
C	LEU	PRO	HIS	ARG	U
	LEU	PRO	HIS	ARG	C
	LEU	PRO	GLN	ARG	A
	LEU	PRO	GLN	ARG	G
A	ILE	THR	ASN	SER	U
	ILE	THR	ASN	SER	C
	ILE	THR	LYS	ARG	A
	MET	THR	LYS	ARG	G
G	VAL	ALA	ASP	GLY	U
	VAL	ALA	ASP	GLY	C
	VAL	ALA	GLU	GLY	A
	VAL	ALA	GLU	GLY	G

The first letter of the triplet is in the left hand vertical column, the second in the horizontal axis, and the third in the right hand vertical column.

frequency of usage of individual codons appears to be species specific, all are used. The universality of this code has only been challenged fairly recently with the discovery that triplets which normally specify translational 'stops' are used as coding sequences in some species such as certain mitochondrial DNA. The exceptions would appear to be rare however.

Mitochondria are important organelles that provide the bulk of the ATP for eukaryotic cells and contain enzymes that catalyse many cellular reactions. It is generally speculated that this energy-converting organelle has been derived from a more primitive body, such as a virus, and has evolved together with its host, developing a symbiotic relationship with it. (The nearest equivalent to mitochondria in other species are the chloroplasts of plants.) Aside from its novel genetic code, mitochondrial DNA (16.5 kilobases in size) is also unique in being transmitted exclusively by maternal inheritance. The mitochondrial genome codes for at leaset 13 different proteins and it has long been suspected, but remains unproved, that defects in mitochondrial genes may be important in the genesis of at least some forms of cancer. On the other hand, the clinically and biochemically heterogeneous mitochondrial myopathies and encephalopathies—inborn errors of metabolism—are beginning to be better defined. In several recent reports, deletions or mutations (evidenced by restriction enzyme polymorphisms) in mitochondrial DNA have been specifically associated with human disease, in particular with the Kearns–Sayre syndrome and Leber's hereditary optic neuropathy. It remains to be seen whether there is a strong association with cancers, but as aberrant growth is a marker for these pathologies and mitochondrion provide cellular energy, with the newer molecular methods available for research into genetically inherited diseases, it would seem a fertile area for further exploration.

Space filling models of DNA usually represent it in its most stable (B) form. The dimensions of B DNA (Fig. 5.6) are derived from X-ray diffraction studies. (Similar studies suggest RNA exists in a less compact, or A form, type of helix.) As far as is known, B DNA structurally represents most of the DNA in a cell, and almost certainly that which is 'coding' (specifying proteins). However, for reasons yet to be resolved, much of the DNA in a mammalian cell would appear to be non-coding and has even been referred to as irrelevant (or 'junk') DNA. Biologically, this is a difficult concept to accept with regard to highly conserved, and conservative, organisms. It seems more probable that such DNA, although not directly related to coding, has a function yet to be recognized. In this regard, it is interesting that experimental data suggest that certain specific DNA sequences, such as regular repeats of purines and pyrimidines, may specify alternative structural forms of DNA, which in turn might play roles in regulation of gene expression or other cellular functions which might be modulated by DNA, as well as in intracellular DNA recombination.

The crystal structure of a hexadecanucleotide (C–G–C–G–C–G–T–T–T–T–C–G–C–G–C–G) shows this molecule to adopt a 'hairpin' configuration with the four T-residues forming a loop; hairpin structures have been postulated to be important in DNA replication. Recently also it has been shown that certain guanine-rich sequence in DNA can self-associate under physiological salt concentrations to form parallel four-stranded complexes (Sen and Gilbert 1988). It has been postulated that such sequences which occur in immunoglobulin switch regions, in gene promoters, and in chromosomal telomeres, may bring together the four homologous chromatids (see Glossary) during meiosis. Moreover, if the hypothesis that guanine-rich regions might be involved in meiosis is correct, such sites could be crucial in forming the (postulated) structures required for pairing homologous chromosomes during this stage of cell division.

Before turning to other aspects of DNA, two further topics should be briefly noted. One concerns the remarkable solubility of this highly polymeric species. Since water solubility is not a common property of most highly polymerized materials, the explanation for the great solubility of DNA must lie in its capacity to form specific interactions with water. The phosphodiester bond generated by the interaction of phosphoric acid with hydroxyl groups of sugar residues in nucleosides creates not only the backbone for DNA, but leaves a single acidic residue on the phosphate moiety that, at the normal pH of a cell, should be negatively charged (Fig. 5.5). *In vivo*, this charge is neutralized by cations, such as Mg^{++}, to generate a macromolecular version of a 'salt' which is capable of interacting both electrostatically or via hydrogen bonding with water. In support of this notion, DNA isolated from cells is neutral and contains many molecules of water of hydration. Moreover, once deprived of its water of hydration, DNA is remarkably difficult to redissolve.

The second point is that DNA in the nucleus of a cell is not 'naked'. Rather, it is found in association with histones and other proteins to produce a characteristic and fairly regular structure (Fig. 5.1). DNA in association with histones is known as 'chromatin'. Further organization of chromatin produces the highly ordered chromosome whose structures are specific and unique within each individual organism. A number of proteins have now been discovered which bind to regulatory regions of DNA adjacent to genes and allow transcription of these genes (Santoro *et al.* 1988). It is interesting to note that the DNA of some viruses also appears to be organized as 'mini-chromosomes', which nonetheless have regions that do appear to be 'naked' and act as origins of DNA replication. Whether the same is true for higher organisms with regard to areas relevant to the initiation sites of DNA replication remains to be seen.

5.4 'Informational' DNA

It is becoming increasingly apparent that not all DNA is used in the strictly genetic sense to 'specify genes'. Moreover, even within genes, there are signals that determine the function of a particular region (domain) within a protein. For example, some DNA sequences specify structural motifs in proteins. One, recently discovered, encodes a leucine at every seventh amino acid that allows the designated region in the protein to form an α-helix (Landschultz *et al.* 1988). It has been suggested that the leucine side chains from one protein interdigitate with those of another protein effectively forming a 'leucine zipper' that holds the molecules together. This dimer in turn can interact with DNA to regulate gene expression. Metal binding sites are also important components of many regulatory proteins and sequences specifying these have been located in a number of DNAs. The best characterized of these to date is that known as the 'zinc finger' motif (Klug and Rhodes 1987), which juxtaposes two cysteines, two histidines, two aromatic amino acid residues and one leucine with the appropriate spacing to bind zinc (See Fig. 5.8). These domains in turn attach the regulatory protein to recognition sequences in the appropriate gene. Yet other motifs in DNA determine the subcellular localization of proteins, the best studied being

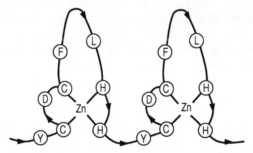

Fig. 5.8 A schematic folding arrangement for repeated zinc finger motifs in the *Xenopus* transcription factor IIIA (TF III A). Each domain is centred on a tetrahedral arrangement of Zn ligands. Ringed letters represent conserved amino acids; cysteine (C) and histidine (H) form bonds with the zinc, and other amino acids, leucine (L), phenylalanine (F), tyrosine (Y), and aspartate (D) may form a structural core. This protein mini-domain allows interaction with DNA (adapted from Klug and Rhodes 1987).

the signals that specify nuclear or nucleolar localization and the sequence G–C–C–A–A–T which has recently been implicated both in eukaryotic transcription and DNA replication.

Thus, little by little, the evidence is building up which allows patterns in DNA sequences to be recognized and the fine structure information in DNA to be understood. In the long run, such understanding should allow 'gene therapy' (see p. 145) to become a realistic proposition.

5.5 DNA damage

As earlier mentioned, many of the agents that produce mutations in DNA do so by altering the tautomeric form of a base (Fig. 5.3) such that inaccuracies occur during DNA replication which may be reflected in the transcription of DNA into RNA. In many cases, these mutations may be 'silent' in that they occur in the 'wobble' allowed at the third base (degenerate) position of some of the triplet codons (Table 5.2). In other positions within a codon, mutations could lead to an alteration of amino acid structure which, if protein function were not thus impaired, might be allowed. In fact, such alterations can even lead to mutants with selective advantages over the 'wild type' species. Other alterations could be positively harmful and, if not repaired, ultimately lethal either to the cell, or in the case of cancer causing lesions, if such there are, to the whole organism. In addition to mutations produced by exogenous agents, such as certain organic and inorganic chemicals, X-irradiation (see Chapters 6 and 7), etc., there is always a background level of mutation in a cell produced by the hydrolytic interaction of water itself with DNA. Such damages include hydrolytic cleavage of the glycosidic bond resulting in depurination (Fig. 5.9) or depyrimidination, and ultimately in strand breaks, or deamination of exocyclic amino groups. For example, cytosine converted to uracil (an abnormal base in DNA) by deamination would produce a 'mismatch' during DNA replication which could lead ultimately to the concomitant alteration of the structure of a protein. It has been calculated that the rate of release of purines from double-stranded DNA occurs at a rate of about $10^4/\text{day}/10^{10}$ bases (in rat liver cells), whereas depyrimidination occurs with slightly lower frequency (about $5 \times 10^2/\text{day}/10^{10}$ bases). Deamination proceeds marginally slower than depyrimidination. Were normal cells not endowed with a variety of mechanisms to combat such lesions, it can be seen that water alone could pose a serious threat to the accurate survival of genetic information. Hydrolysis could also lead directly to phosphodiester bond cleavage, disrupting the DNA backbone itself. Although theoretically possible, this mode of damage is not thought to be of great physiological relevance, although the alkylation of a phosphodiester (to a phospho-

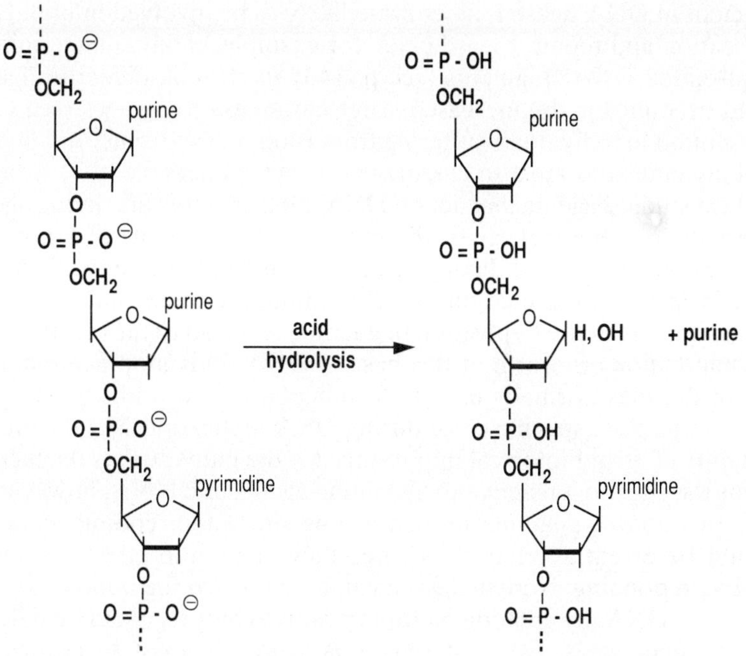

Fig. 5.9 A schematic representation of one of the well-characterized lesions of DNA, that of depurination. Loss of a purine (or pyrimidine) base, if not repaired, can lead subsequently to DNA chain breaks.

triester) could create a more labile substrate for hydrolysis. The role of phosphotriesters in DNA damage of mammalian DNA is only beginning to be investigated and its effect assessed.

Most cells have a limited capacity to correct lesions produced in DNA by exogenous or endogenous reagents. A variety of different repair processes have been identified, and some of them well characterized, at least in bacteria. There is no compelling evidence to suggest that higher organisms do not have repair processes comparable in type and in effect to those found in the better studied *E. coli*. In normal individuals, these processes must effectively compete with the background levels of mutations in DNA. It is obvious, however, that in individuals deficient (or defective) in one or more of the repair processes and probably in ageing populations in general, this delicate balance can be disturbed, with deleterious consequences. One of the most telling examples of this comes from recent studies on cells from patients with Bloom's syndrome (Willis and Lindahl 1987). Such individuals are known to have a greatly increased risk of developing cancer and have now been shown to be deficient in DNA ligase I, an enzyme likely to be involved in both DNA replication and repair. Ligase I can, for example, catalyse the joining of blunt-ended DNA fragments (see p. 143) *in vitro*. A defect in ligase I could account for the increased sister chromatid exchanges (see Glossary) found in individuals suffering from Bloom's syndrome.

Many mutagens are also carcinogens (see Chapters 6 and 7). Some of the best studied agents that act on DNA are the simple alkylating agents, whose biological effect can be directly related to their site of action (Fig. 5.10). Some of the most damaging agents, which can be shown in animal models *in vivo* to induce tumour formation, are those that modify the oxygen moiety at the 6–position of guanine and lead to the creation of an unusual tautomeric form of this base in DNA. This modification alters two of the sites normally used in forming base pairs with cytosine, and can lead to a site specific error during DNA replication. Significant, and certainly of great biological importance for the individual, is the fact that by far the site most reactive to alkylating agents in DNA is, however, the N–7 position of guanine; lesions at this site within coding sequences should be essentially harmless since they have little effect on normal hydrogen bonding. Nonetheless, it can be seen how the build up of such lesions in DNA, by creating multiple positive charges, could be deleterious in other ways. Efficient repair pathways do exist to remove N–7–methylguanine from DNA.

It has been shown by a variety of methods that during evolution trans-

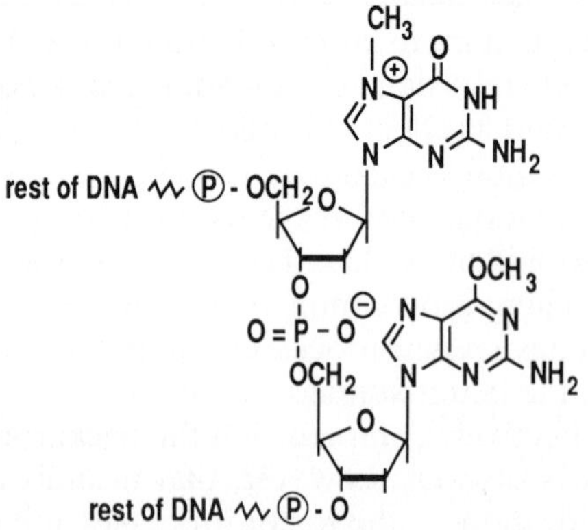

Fig. 5.10 Two of the well characterized lesions produced at guanine sites in DNA by simple alkylating agents. Modification of the N-7 position (on the 5 membered heterocyclic ring) generates a positively charged species, but is essentially 'silent' since it has no effect on base pairing in DNA. Conversely, methylation at the O-6 position of the 6 membered ring, if not repaired, is a dangerous lesion, being both mutagenic and carcinogenic.

forming (or tumorigenic) retroviruses arose by incorporation of cellular DNA into the genome of weakly or non-transforming viruses. A most interesting finding is that, in some instances, the difference between a viral oncogene and its normal cellular counterpart is the presence of mutations in the former, which lead to simple amino acid alterations in the protein specified by the respective genes (see Chapter 9). One of the most promising aspects of the high technology that now attends the analysis of DNA is that it provides the tools for investigating alterations in potentially normal functions that accompany malignancies. Note-worthy in this regard is the identification of the chromosomal deletions that give rise to (or result from) retinoblastoma and other tumours (see Chapters 4 and 10). It would be surprising if such changes were not fairly widespread.

5.6 DNA manipulation

The fact that mammalian DNA can be amplified by recombinant DNA technology and (potentially) expressed in both prokaryotic and eukaryotic systems makes many experiments possible that would hitherto not have been feasible. This should engender a sense of optimism, and even adventure, in those interested in exploring the cellu-lar and molecular biology of cancer. It has, for example, already allowed *inter alia*, a sequence homology to be observed between some of the normal growth factors of cells and those of viral and cellular 'oncogenes' implying a function for the latter in uncontrolled cell growth (see Chapters 9 and 11). Many vector systems for generating recombinant DNAs that can be amplified by replication in bacterial systems, and even expressed in bacteria or mammalian cells, have been generated.

A detailed description of nucleic acid molecular biology used for this purpose is beyond the scope of this chapter; however, a brief account may be given here to explain some of the terminology and methods employed. The applications are discussed elsewhere in this book. Perhaps one of the most important tools comes from the discovery of enzymes known as restriction endonucleases in many bacteria. The restriction enzymes recognize specific sets of DNA sequences in double-stranded DNA and cut both strands of DNA at or near these sequences. In some instances, the enzymes cut through both strands at a single point generating blunt-ended molecules, whereas with others, the cuts are at precisely spaced points along the two strands and generate overlapping ('sticky' or overhanging) ends. Various other enzymes are available for filling in or cutting back at these ends. These techniques allow unrelated DNA molecules to be joined together (ligated); in particular this is of use for propagating sequences from eukaryotic organisms in prokaryotic hosts such as bacteria. For example, the entire genetic information (genome) of human chromosomes can be ligated to bacteriophage (bacterial virus) DNA, giving rise to a 'library' of human DNA from which homogeneous clones can be isolated, amplified and manipulated further (e.g. for nucleotide sequence analysis, for mutagenesis experiments, for gene expression studies, etc.). One of the most common techniques is to isolate a specific clone of interest and label it with a radioisotope so that it can be used as a probe to examine the gene in various DNA or RNA preparations (see Chapters 4 and 12). In the example cited in Chapter 12, human DNA was cut with a restriction enzyme, the fragments of DNA separated by size by electrophoresis through a gel, and then transferred to a filter. The filter was exposed (hybridized) to a radioisotopically labelled probe for an immunoglobulin locus and the specific binding (hybridization) was measured by exposure of the filter to X-ray film. In the specific example shown, the study examined whether the immunoglobulin locus had been rearranged compared to the germ line sequences, during differentiation of B lymphocytes. Similar techniques have been applied to the study of cellular oncogenes (see Chapters 4, 9, and 11).

Such genetic manipulations are as yet in their infancy, and not wholly without pitfalls, as is beginning to be recognized. For example, genes expressed in heterologous systems may produce functions that are without activity in their normal hosts, possibly as a consequence of incorrect or incomplete protein modification that could result in aberrant folding or unusual instability. Further, gene functions expressed at abnormally high levels, even in homologous systems, can lead to unexpected effects, including cell death. 'Dose response' may prove a difficult problem to solve, at least in terms of defining underlying mechanisms of gene action within a cell. Another basic problem for biotechnology appears to be the expression of genes in the wrong cellular compartment following introduction of DNA into cells (transfection). The critical problems for gene manipulation now appear to be not how to 'clone' and express a particular part of DNA, or even a particular gene, but how to introduce it into a cell and regulate its expression so as to obtain data *in vitro* that are meaningful in terms of *in vivo* responses.

Essentially what is implicit in the above discussion is the fact that all the 'rules of the game' have not yet been determined, although great progress has been made. It is significant that the need for regulation of gene expression is now being widely recognized. With this in mind, vectors have been developed that allow control of DNA replication and many of them contain RNA polymerase promoters that can be additionally regulated by such external agents as temperature, hormones, or heavy metals.

One of the other basic problems in gene manipulation and expression arises from the fact that in mammalian cells mRNAs are often not colinear with respect to genomic DNA. Rather, they reflect the fact that enzymic processes have occurred in the cytoplasm, subsequent to transcription, which have 'spliced' together non-adjacent regions of RNA to produce the functional messengers for proteins. For expression of such genes *in vitro*, the messenger itself must first be isolated and reverse transcribed into DNA, before the latter can be introduced into suitable vector systems and studied. This process is both tedious and frequently unsuccessful, particularly in the case of mRNAs present in low copy numbers in cells, and/or unstable. In attempts to circumvent this, interesting new classes of vectors, have been developed. These are hybrid DNAs with elements derived both from plasmid and viral sources, such that for expression they can be 'shuttled' between bacteria and mammalian cells (Cepko *et al.* 1984), and even packaged as retrovirus particles, thus allowing recovery of input material. Since the latter contain all the signals for retrovirus transcription, these vectors provide the capacity for correctly splicing, in an *in vitro* system, the input DNA. Thus the latter should be re-isolatable as a reversible transcribed copy of its message. These vectors have, however, a size constraint that may limit their usefulness for many mammalian genes.

Gene manipulation is the basis of a new approach to medicine designated 'gene therapy'. For example, if 'cancer genes' can be identified and defined, they should be subject to manipulations (for example, controlled site specific mutagenesis) that could render them inactive and even possibly subjects for 'therapy', if corrected genes could be substituted for the aberrantly expressed ones. Such problems for the future deserve thought. An interesting alternative approach just beginning to be explored involves the use of complementary (anti-sense) sequence of mRNAs which, upon being introduced into cells, should at least theoretically be capable of binding to the messenger and rendering it inactive. This potentially fruitful avenue is at present being widely examined with particular regard to the prevention of expression of aberrant genes, but the approach has proved successful in only a few specific cases to date.

5.7 Conclusion

An essentially historical approach to DNA has been presented because it shows how many scientific disciplines have been, even indirectly, involved in taking us to a point where we can begin to approach the problem of human cancer in a non-empirical manner. At the moment it seems ironic, and paradoxical, that one of the few human cancers that has been firmly associated for over twenty-five years with a viral infection (e.g. Burkitt's lymphoma with Epstein Barr virus), is still preferentially treated with massive doses of cyclophosphamide, an alkylating agent and potent carcinogen. (Indeed, cyclophosphamide is a drug of choice in the therapy of many human tumours.) The aim of understanding at a molecular level the pathological process defined broadly as 'cancer' is obviously to be able to control it. Ideally such containment should come about by a less empirical manner than that presented as an example above, where although the disease is initially eradicated, the patient is left with many undesirable lesions and in a large proportion of cases, the tumour reappears and notably is no longer susceptible to treatment.

At least one success story can be cited as arising from application of the scientific method, that is the control of herpes simplex by the drug Acyclovir. Among others, patients immunosuppressed prior and subsequent to transplant therapy become immediately susceptible to the effects of reactivation of herpes viruses. Once the existence of a thymidine kinase gene was identified in herpes simplex virus, specific antagonists of the kinase enzyme were sought. The drug, Acyclovir, a nucleoside analogue, was developed; it blocks a vital step in the enzyme pathway and thus counters the reactivation of herpes simplex viruses, and provides protection for the patient. If cancer(s) can be related to specific sequences of DNA, it should be possible, in a similar fashion, to search for their control.

References and further reading

Alberts, B., Bray, D., Lewis, J., Raff, M., Roberts, K., and Watson, J. D. (1983). *Molecular biology of the cell.* Garland Publishing, New York.

Bramhill, D. and Kornberg, A. (1988). A model for initiation at origins of DNA replication. *Cell* **54**, 915–18.

Cepko, C., Roberts, B. E., and Mulligan, R. C. (1984). Construction and applications of a highly transmissible murine retrovirus shuttle vector. *Cell* **37**, 1053–62.

Chattopadhyaya, R., Ikuta, S., Grzeskowick, K., and Dickerson, R. E. (1988). X-ray structure of a DNA hairpin molecule. *Nature* **334**, 175–9.

Driscoll, R. J., Youngquist, M. G., and Baldeschwieler, J. D. (1990). Atomic-scale imaging of DNA using scanning tunnelling microscopy. *Nature* **346**, 294–6.

Glover, D. M. (ed.) (1985). *DNA cloning. A practical approach.* IRL Press, Oxford.

Gluzman, Y. (ed.) (1982). *Eukaryotic viral vectors.* Cold Spring Harbor Laboratory, Cold Spring Harbor, New York.

Klug, A. and Rhodes, D. (1987). 'Zinc fingers': a novel protein motif for nucleic acid recognition. *Trends in Biochemical Sciences* **12**, 464–9.

Landschultz, W. H., Johnson, P. F., and McKnight, S. L. (1988). The leucine zipper: a hypothetical structure common to a new class of DNA binding proteins. *Science* **240**, 1759–64.

Laskey, R. A. (1987). The cell nucleus. *British Medical Journal* **295**, 1121–3.

Lindahl, T. (1979). DNA glycosylases, endonucleases for apurinic/apyrimidinic sites, and base excision-repair. *Progres in Nucleic Acid Research and Molecular Biology* **22**, 135–92.

Messer, W. and Noyer-Weidner, M. (1988). Timing and targeting: The Biological Functions of Dam methylation in *E. coli. Cell* **54**, 735–7.

Radman, M. and Wagner, R. (1988). The high fidelity of DNA duplication. *Scientific American* **August**, 24–30.

Richards, B., Pardon, J., Lilley, D., Cotter, R. and Wooley, J. (1977). The substructure of nucleosomes. *Cell Biology International Reports.* **1**, 107–15.

Santoro, C., Mermod, N., Andrews, P. C., and Tjian, R. (1988). A family of human CCAAT-box-binding proteins active in transcription and DNA replication. *Nature* **334**, 218–34.

Sen, D. and Gilbert, W. (1988). Formation of parallel four-stranded complexes by guanine-rich motifs in DNA and its implications for meiosis. *Nature* **24**, 364–6

Singer, B. and Grunberger, D. (1983). *Molecular biology of mutagens and carcinogens.* Plenum Press, New York.

Siomi, H., Shida, H., Nam, S. H., Nosaka, T., Maki, M., and Hatanaka, M. (1988). Sequence requirements for nucleolar localization of human T cell leukaemia type I px protein which regulates viral RNA processing. *Cell* **55**, 197–209.

Willis, A. E. and Lindahl, T. (1987). DNA ligase I deficiency in Bloom's syndrome. *Nature* **325**, 355–7.

6

Chemical carcinogenesis and precancer
CAROLINE WIGLEY and ALLAN BALMAIN

6.1 The role of chemical carcinogens and mutation in human cancer

As outlined in Chapter 1 and considered in detail in this and subsequent chapters (7, 8, and 9), carcinogenesis is a multistage process. The first step(s) is known as initiation and is followed by one or more promoting events. These stages in the stepwise process of carcinogenesis have been detected in man and in animals and can now be analysed. This will be illustrated later (see p. 155). Each stage may be influenced by different factors. From epidemiological data and animal studies, and most recently from molecular analysis of tumours, there is little doubt that chemical agents are involved at some stage, although there are many other contributing factors. Many potentially carcinogenic agents are present in our diet and environment. There is convincing evidence that the site of action of most of these agents is the genetic material in cells; many known and suspect chemical carcinogens cause mutations. Even so, it is not entirely clear whether a change in DNA sequence is needed. We include in our definition of mutation gross DNA changes such as rearrangements and should bear in mind that heritable changes in gene expression, whilst not mutational events at the sequence level, might have similar effects on the cell phenotype.

6.1.1 *Epidemiological evidence*

This is discussed in Chapter 3 but a few examples illustrate the situation. The classic example was described in 1775 by Percival Pott who noted that chimney sweeps had a high incidence of cancer of the scrotal skin attributed, quite correctly, to chronic contact with soot—a mixture of chemicals including polycyclic hydrocarbons which were later shown to be carcinogenic in animals (see p. 152.) β-Naphthylamine and other aromatic amines have been linked to bladder cancer in workers in the dye industry, whereas industrial exposure to nickel and some chromates has been strongly implicated in the causation of cancers of the respiratory system. There are also naturally occurring carcinogens which may be present in the diet. A good example is a substance present in bracken fern which may cause tumours in the alimentary tract in animals and possibly in man in areas where fern hearts (fiddles) are eaten as a delicacy. Other intestinal carcinogens may be formed in the gut by the action of intestinal micro-organisms on substances in the diet or in the bile. Better known but less well defined is the chemical carcinogen(s) in tobacco smoke associated with lung cancer (see Chapter 3). A list of potentially carcinogenic chemicals is published by the International Agency for Research on Cancer (see Table 6.1).

We now know that almost all chemicals implicated by epidemiologists as human carcinogens can cause cellular mutations in the conventional sense, i.e. localized base changes in DNA, but there are a few exceptions. Asbestos causes cancer of the pleural cavity (see Chapter 3) but is not mutagenic in classical mutagenesis test systems, e.g. the Ames test (see p. 169). However, it has been shown to induce gross chromosomal changes such as non-disjunction in mammalian cells, probably because of binding to the spindle apparatus during mitosis. The resulting gene imbalance may be an important step in the generation of aneuploidy which is frequently observed during tumour progression. Diethylstilboestrol, a synthetic steroid hormone, was given in the past to prevent

Table 6.1 Chemicals with proven carcinogenic activity in humans

Chemical (or industrial process)[1]	Main type of exposure[2]	Main route of exposure[3]	Target organ(s)
Aflatoxins	Environmental, occupational	Ingestion, inhalation	Liver
4-Aminobiphenyl	Occupational	Inhalation, ingestion, skin contact	Bladder
Arsenic compounds	Occupational, medicinal, environmental	Inhalation, ingestion, skin contact	Skin, lung, liver[4]
Asbestos	Occupational	Inhalation, ingestion	Lung, pleural cavity, gastrointestinal tract
Auramine manufacture	Occupational	Inhalation, ingestion, skin contact	Bladder
Benzene	Occupational	Inhalation, skin contact	Haemopoietic system
Benzidine	Occupational	Inhalation, skin contact, ingestion	Bladder
Bis(chloromethyl)ether	Occupational	Inhalation	Lung
Cadmium-using industries (cadmium oxide?)	Occupational	Inhalation, ingestion	Prostate, lung[4]
Chloramphenicol	Medicinal	Ingestion, injection	Haemopoietic system
Chloromethyl methyl ether (associated with bis(chloromethyl)ether?)	Occupational	Inhalation	Lung
Chromium (chromate-processing industries)	Occupational	Inhalation	Lung, nasal cavities[4]

Cyclophosphamide	Medicinal	Ingestion, injection	Bladder
Diethylstilboestrol	Medicinal	Ingestion (acts transplacentally)	Uterus, vagina (in offspring)
Haematite mining (radon?)	Occupational	Inhalation	Lung
Isopropyl oils	Occupational	Inhalation	Nasal cavity, larynx
Melphalan	Medicinal	Ingestion, injection	Haemopoietic system
Mustard gas	Occupational	Inhalation	Lung, Larynx
2-Naphthylamine	Occupational	Inhalation, skin contact, ingestion	Bladder
Nickel (nickel-refining industries)	Occupational	Inhalation	Nasal cavity, lung
N,N-bis(2-chloroethyl)-2-naphthylamine	Medicinal	Ingestion	Bladder
Oxymetholone	Medicinal	Ingestion	Liver
Phenacitin	Medicinal	Ingestion	Kidney
Phenytoin	Medicinal	Ingestion, injection	Lymphoreticular system
Soots, tars, and oils	Occupational, environmental	Inhalation, skin contact	Lung, skin (scrotum)
Vinyl chloride	Occupational	Inhalation, skin contact	Liver, brain,[4] lung[4]

[1] The precise chemical(s) responsible may not be known.
[2] The main types of exposure mentioned are those by which the association has been demonstrated; other exposures may occur.
[3] The main routes of exposure given may not be the only ones by which such effects could occur.
[4] There is indicative evidence for these organs.
Adapted from Tomatis et al. (1978). Cancer Research 38, 877–85.

miscarriage in pregnant women; later it was found to cause vaginal tumours in female offspring at puberty, by a mechanism which does not seem to involve DNA mutation in the strict sense.

There are some instances where there may be an increased tissue susceptibility to the tumour-inducing activity of carcinogens. In some rare conditions an inherited trait predisposes affected members of a family to develop a particular cancer (see Chapter 4). This led Knudson and others, as has already been mentioned, to propose that the first event in carcinogenesis involved DNA mutation and that this mutation could, in rare instances, be transmitted in the germ cell line from parent to offspring. As a rule, this first or initiating mutation occurs after birth, in a target somatic cell. One heritable cancer-predisposing condition, xeroderma pigmentosum (XP), is one of the best understood of a group of conditions, where something is known about the defect. Individuals with XP suffer from a deficiency in their ability to repair DNA damaged by u.v. light in particular; this is demonstrated in cell cultures prepared from small biopsies of a patient's skin. The skin exposed to sunlight in XP patients is at risk of developing cancer, providing a very strong argument linking DNA damage (and, by implication, its faulty repair or lack of repair), mutation, and cancer.

6.1.2 *Evidence from animal studies*

Considerable support for the evidence from epidemiological studies came when Japanese workers showed for the first time early this century that a number of potent hydrocarbon carcinogens cause cancer when applied to the skin of rabbits. Subsequently, many other suspected chemicals were shown to be carcinogenic by similar techniques. Nowadays, most drugs, food additives, etc. are tested on laboratory animals, usually rodents, for general toxicity and long term carcinogenicity (see p. 167). Very many substances have now been shown to be tumour-producing in experimental animals. Some produce tumours at the site of application, e.g. on the skin. Others may produce tumours at the site of absorption, e.g. the bowel if given by mouth, or at the site of the breakdown (metabolism), e.g. in the liver, or in the excretory organ, e.g. kidneys or bladder (see Table 6.1). In some instances, a carcinogen may produce tumours in an entirely unexpected organ. For example, the carcinogens dimethylbenz(a)anthracene and nitrosomethylurea, when given by mouth, cause breast cancer in female rats, but have to be given at a particular time in the development of the breast during puberty. There seems to be a critical sensitive period which depends on the hormonal status of the cells (see Chapter 13). If nitrosomethylurea is given at an earlier stage to a pregnant animal it acts transplacentally to produce tumours of the nervous system in the offspring. Interestingly, tumours arising in these two tissues are specifically associated with activation of different proto-oncogenes, e.g. *ras* in mammary carcinomas and *neu* in Schwannomas of the nervous system. Similar complex factors may operate in other tissues also, so that exposure to a carcinogen is necessary, but not sufficient, for cancer induction in these cases. Intensive long-term administration by several routes is essential if possible toxic and carcinogenic effects are to be excluded by animal screening tests.

6.1.3 *Molecular evidence*

If cancer can be caused, partly or wholly, by chemical carcinogens inducing mutations in DNA, the powerful techniques of modern molecular biology can be used to detect differences between the DNAs of normal and tumour tissue from a single individual cancer. The cellular oncogenes (c-*oncs*) are one group of genes for which comparisons of DNA sequence have shown mutations in human tumours (Chapters 9 and 10). Members of the *ras* oncogene family, in particular, have been shown to be mutated, usually at around codon 12 or 61, leading to the production of proteins with specific amino acid substitutions at these positions. Examples of these mutations have been found in a proportion of virtually all human tumour types, including carcinomas of the bladder, colon, pancreas, and skin, and also sarcomas, lymphomas, and leukaemias. We know that these highly specific mutations must affect the function of the c-*onc* because the mutant gene, but not its normal homologue, can dramatically alter the phenotype of cells in culture transfected with the tumour DNA (see Chapter 9).

It has been found that different oncogenes tend to be activated preferentially in different tissues after carcinogen treatment. This may be related to the observation that carcinogen interaction with DNA is non-random throughout the genome and may depend on the degree of condensation of the chromatin or on the extent of DNA methylation at particular loci. This will, of course, vary with the differentiated phenotype of the target cell and may explain why a particular oncogene is activated more often in one type of tumour than another.

Since many of the oncogene protein products which have been identified are linked with the process of growth factor response, either as analogues of a factor itself, a receptor, or part of the intracellular signalling mechanism (see Chapter 11), it is likely that a mutated oncogene product would show altered interactions with other molecules in the finely tuned system which regulates the growth of the cell.

Agents other than chemicals can cause mutations. Irradiation is known to damage human DNA (see Chapter 7), and some viruses may also be mutagens when they integrate into host cell DNA (see Chapter 8). We must turn to experimentally induced cancer in animals to find direct evidence that chemical carcinogens can cause mutations such as those in c-*onc* genes that have been linked to cancer causation in humans (see Chapter 9).

One of the first examples of this came from the induction of breast tumours in female rats by nitrosomethylurea, a DNA alkylating carcinogen. All nine tumours in one experimental series contained an 'activated' *ras* gene (Ha-*ras*-1) and one of the genes which was isolated and analysed in detail showed that the same amino acid codon (at position 12 of the protein) was mutated as in the human cancers of unknown causation mentioned earlier. Evidence that the carcinogen causes the mutation directly, via interaction with DNA (see p. 160), is also provided by sequence analysis of the mutations. The change in nucleotide sequence observed is precisely that predicted from the known reaction between the particular carcinogen and a specific DNA base. For example, nitrosomethylurea methylates the O-6 position of guanosine residues and this adduct is known to mispair with thymidine during DNA replication (Chapter 5). Thus the observed G:C to A:T transition at position 12 of the protein is generated. A wide spectrum of different *ras* mutations is found in tumours of the breast, skin, and liver which are induced by carcinogens with different DNA-binding properties. In many, but not all, cases, the specific mutations observed can be correlated with the particular base involved in adduct formation.

Convincing as this might seem, many questions remain to be answered: only a minority of human tumours contain mutant oncogenes of the sort we have just described, whereas in other cases (e.g. colon or pancreatic carcinomas) the frequency is as high as 40–95 per cent. Moreover, one member of the *ras* family appears to be the preferred 'target gene' in each case. For example, Kirsten-*ras* (Ki-*ras*) is the most frequently mutated gene in colon or pancreatic tumours, whereas Harvey-*ras* (Ha-*ras*) or neuroblastoma-*ras* (N-*ras*) are more often mutated in bladder tumours and leukaemias respectively. The reasons for this apparent specificity, which is also observed in animal model systems, are unclear but may be related to the route of exposure or tissue-specific metabolism of certain carcinogens. The evidence that the mutant genes are causal in the development of the cancers which contain them is persuasive; the tumours are usually clonal in origin and the mutation is present in every cell, suggesting that the first cell in which the molecular change occurred had a selective advantage. In some cases, *ras* mutations may be an early event or even the initiating event in tumour formation, whereas in others there is evidence that mutations occur during tumour progression.

A great deal of attention has been devoted to *ras* mutations simply because this was the first human oncogene to be described in some detail. There is therefore a natural tendency to 'search where there is light', but the impression should not be given that these are the only genes involved. Many human and animal tumours have apparently normal *ras* genes, indicating that such mutations are not necessary for tumours to develop. Other genes must be involved in generating the complex cancer phenotype. Some of these may be alternative proto-oncogenes which can be activated in a positive sense by mutation, but there is strong evidence that recessive mutations can also play an important role. Wilms' tumour of the kidney (see Chapter 10) and retinoblastoma (see Chapters 4 and 10) are rare, childhood cancers associated with the loss or inactivation of genes on chromosomes 11 and 13 respectively. There is increasing evidence that loss of 'tumour suppressor genes' is also implicated in the generation of the more common human cancers which have no clear hereditary pattern (Chapter 4). Since in principle it is easier to inactivate a gene functionally by mutation or deletion than to activate a proto-oncogene in a specific way, the tumour suppressor genes may constitute important targets for carcinogenic chemicals at some stage of tumori-genesis.

6.2 Experimental approaches to the study of carcinogenesis by chemicals

6.2.1 *The biology of cancer induction in animals: precancer and multistage models*

Studies on the experimental induction by chemicals of cancer in laboratory animals introduced several new concepts, particularly the multistage theory (see Fig. 6.1). Carcinogens fall into two groups, complete and incomplete. The former can produce tumours on their own, whereas the latter cannot and require subsequent exposure of the treated (initiated) cells to promoting agents, which are not carcinogenic in themselves. Promoting agents can also lead to the development of tumours when applied to tissue previously treated with a subthreshold dose of a complete carcinogen which would not in itself produce tumours. Polycyclic aromatic hydrocarbons and the nitrosamines are complete carcinogens and can act as initiators and promoters. The complexity of the situation is illustrated by urethane, an incomplete carcinogen when applied to the mouse skin but a complete carcinogen in the fetal lung. Promoting agents are less well defined but seem to be specific for particular tissues. One group, the phorbol esters extracted from some plants, act as tumour promoters in skin. Other classes of promoting

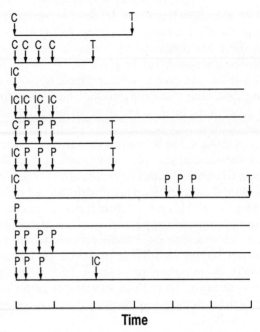

Fig. 6.1 Initiation and promotion in carcinogenesis. Schematic representation of various schedules of treatment of mouse skin with a complete or incomplete carcinogen and tumour promoter. Different combinations and sequences in time are shown horizontally. Tumours result after different latent periods only from schedules where 'T' is indicated. C, carcinogen; P, promoter; T, tumour; IC, incomplete carcinogen.

agents have been identified more recently, such as the teleocidin and aplysiatoxin classes. These were first identified from fungal and algal sources respectively. Aplysiatoxins were found to be responsible for outbreaks of a condition known as 'swimmer's itch' in waters off the Japanese coast. Sporadic algal blooms produced a highly irritating substance in the water which caused dermatitis. An important finding is that most promoting agents appear to act in a way similar to the phorbol ester class, binding to the same specific receptor molecule, protein kinase C, and activating the same intracellular signalling mechanism (see p. 166). There are a few exceptions to this, but the non-phorbol-like chemicals are less well characterized.

All promoting agents appear to act, at least in part, through their ability to cause irritation and inflammation. Many of them also interfere with intercellular communication through gap junctions between cells. This has led to theories which involve the response of initiated cells to the 'normalizing' influences of surrounding unaffected cells, as a critical component of promoter function. There is experimental evidence for this in studies on cellular transformation in tissue culture, but this has not yet been confirmed *in vivo* in studies on the effects of tumour promoters on gap-junctional communication between skin cells. In culture, at least, isolated cells show tumour-like characteristics after carcinogen treatment at a much higher frequency than cells in dense cultures. It is likely that the effects of promoters are multiple and that different cell types will respond differently to the same agents. This is currently an active area of research.

During experimental carcinogenesis in various tissues, macroscopic and microscopic changes in the affected tissues which precede the appearance of tumours have been defined (see Chapter 1) as altered discrete focal areas within the carcinogen treated region. These focal areas of abnormal tissue are intermediate in character between normal and malignant. In time, further changes occur in these foci culminating in the development of overt cancer. The cells in the precursor foci have an increased risk of cancer development compared with normal tissue, i.e. they are precancerous. The process will be illustrated by describing three experimental animal systems for inducing cancer of epithelial tissues in skin, liver, and large bowel (colon and rectum). In the colon system, in particular, there are similarities to tissue changes in humans that, from epidemiological evidence, are considered to be precancerous. This suggests that the experimental animal models are a valid way of studying the disease and may provide clues to possible means of medical intervention.

Carcinogenesis in mouse skin is the classic model system in which two stages in the process of cancer development, initiation and promotion, were first described. A single application of a chemical carcinogen is applied to the shaved back skin and this results in the initiation of an unknown number of cells which, if they are left without further treatment, will persist for a very long time without showing any apparent changes. If a second class of chemical agent, a tumour promoter, is subsequently applied to the same area at any time, even a year later, and the treatment repeated regularly, benign tumours (papillomas) appear. They are believed to arise from some of the initiated cells in the carcinogen treated skin. A small proportion of these papillomas may develop into fully malignant tumours with or without further applications of promoter. The two classes of chemical agent used appear to have different mechanisms of action (see Fig. 6.1).

Initiating agents and complete carcinogens are almost always DNA-damaging (genotoxic) and it seems very likely that the initiating event involves some form of carcinogen–DNA interaction and subsequent damage (see p. 160). Initiated cells persist in the tissue long after the initiating agent has disappeared and the lesion produced is both stable and heritable, i.e. it has the characteristics of a mutation. Conversely, most promoting agents are not mutagenic although in some cases they may modify gene expression. It is now thought that promotion itself consists of several steps, and that promoting agents too may be either complete and able to perform all functions, or incomplete, and active at only one or a few stages. A third term, progression, is usually reserved for the process by which cells of a benign or malignant tumour acquire more and more aberrant characteristics—the bad to worse principle of tumour evolution.

Similar sequences of events occur in other tissues. Cancer of the liver can be induced by chemical carcinogens fed to rats. For example, aflatoxin B_1, a mould product which may contaminate certain foods in the tropics, is one of the most potent liver carcinogens known. In all probability it contributes to the high incidence of human liver cancer in the tropics; it seems to act in combination with hepatitis B infection (see Chapter 8). One of the earliest effects of chronic aflatoxin B_1 treatment is the appearance of nodules of hyperplastic, probably precancerous liver cells. These nodules have a range of enzyme abnormalities distinguishing them from surrounding normal tissue. Iron is lost from the precancerous nodules but glycogen stores increase. In the liver, there seems to be an absolute requirement for cell proliferation before nodules can be induced. Aflatoxin is toxic and kills many liver cells. The tissue then regenerates to restore the lost mass. [The powers of regeneration in the liver are remarkable; in rats three-quarters of the organ can be removed surgically (partial hepatectomy) and regenerative proliferation will restore the original tissue mass within weeks.] Partial hepatectomy can in fact act as a promoting stimulus in rat liver carcinogenesis, and tumours arise in the regenerated liver after an initiating, prehepatectomy treatment with a carcinogen. Neither chemical promotion nor surgical treatment alone is sufficient for cancer induction in the adult animal. Interestingly, in young weanling rats, where the liver is growing rapidly during normal development, there is no need for a proliferative stimulus and an initiating dose of carcinogen alone will induce cancer.

Indeed, stimulation of cell division appears to be a necessary component of the promotion stages of carcinogenesis in most if not all tissues, but it is not usually sufficient in itself. Some types of hyperplastic stimuli are more effective than others. Promotion is obviously a complex process which is only recently becoming better understood, particularly with the development of cell culture model systems for studying this aspect of carcinogenesis (see p. 163).

Cancer induction in the colon induced by dimethylhydrazine injected into rats or mice shows well defined precancerous stages. A sequence of pathological changes (Fig. 6.2) can be observed before overt carcinomas develop, and the type and amount of altered colon epithelium depends on carcinogen dosage as well as length of treatment. Submucosal glands become abnormal (dysplastic), and benign polyps (adenomas) arise with increasing incidence with both dosage and time. Histologically, carcinomas can be shown to develop directly from polyps and, more rarely, from abnormal glands, indicating that these are precancerous stages in colon carcinogenesis. This is precisely the conclusion reached by pathologists from observations on human colon cancer. In families with familial polyposis coli (see Chapter 4), affected members develop multiple polyps of the colon and rectum at an early age and at least one polyp will almost certainly become carcinomatous within about 10 years. Usually, the entire colon is removed surgically before this time. This offers the pathologist a unique opportunity to study the precancerous lesions which in other circumstances would escape clinical detection. Most pathologists agree that, in polyposis patients, focal areas of carcinoma develop almost invariably from pre-existing polyps (or more rarely from microscopic glandular abnormalities) and that the polyps represent a precancerous stage. The adenomas themselves can be classified according to their potential for malignant change. Size (above 1 cm^2) and the presence of a particular histological pattern (villous) is accompanied by a statistically increased chance of cancer developing from a particular polyp. Thus there may be additional, more advanced precancerous stages in the multistage sequence, but these are less well defined (see Fig. 6.2). It is thought that, in the general population, precancerous polyps may also occur but are few in number and arise sporadically later in life, preceding the cancer by a similar 10 or 15 year interval. The animal carcinogenesis model of colon cancer is thus especially suitable for the study of factors (including dietary components) suspected from epidemiological studies of being involved in the adenoma (or polyp) to carcinoma progression sequence. For instance, bile acids and their derivatives may have promoter like activity (see earlier); this has already been shown in animal experiments. In addition,

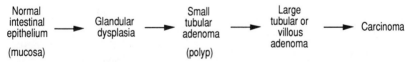

Fig. 6.2 Flow diagram showing the presumptive precancerous stages between normal and malignant tissue, identifiable histologically in human colorectal epithelium. Note: arrows indicate the direction of increasing potential for malignant change, not that cells necessarily pass through all of the precancerous stages between normal and malignant tissue.

a few laboratories are beginning to make use of cell culture techniques to investigate precancerous cells from polyps *in vitro* (see p. 165).

6.2.2 *Mechanisms of carcinogen activation and action*

Until the late 1960s, some of the most potent carcinogens, the polycyclic aromatic hydrocarbons, were unable to mutate cells in culture. We now know that this is because many carcinogens need to be metabolized by cellular enzymes to a reactive derivative before they can be effective, and the test cells used were deficient in one or more metabolic functions and were unable to activate the chemical. Similarly, these (and most other) chemicals must be metabolized to electrophilic derivatives before they are carcinogenic. In animals, metabolism usually happens in the target cells from which the cancer will develop; most reactive metabolites have short half lives in solution in body fluids. Occasionally, activation may occur in the liver. The capacity for metabolism is genetically determined and may be species specific. For instance, the guinea pig lacks a critical enzyme for the activation of acetylaminofluorene (AAF) to its active metabolite, N-hydroxy-AAF, and is thus resistant to its carcinogenic effects. There are also differences between tissues in the extent of metabolic activation of a particular chemical and in the relative extents of deactivation (or detoxification) and activation to the ultimate carcinogenic derivative.

Figures 6.3 and 6.4 show the main routes of chemical activation and sites of binding to DNA of two potent carcinogens. These polycyclic compounds, suspected of being human carcinogens, are both activated initially by a complex of enzymes associated with intracellular membranes, the cytochrome P450-associated mixed function oxidases (MFO). Aflatoxin B_1 is metabolized by MFO to several products including the 2,3-epoxide derivative shown in Fig. 6.3. Here an oxygen bridge

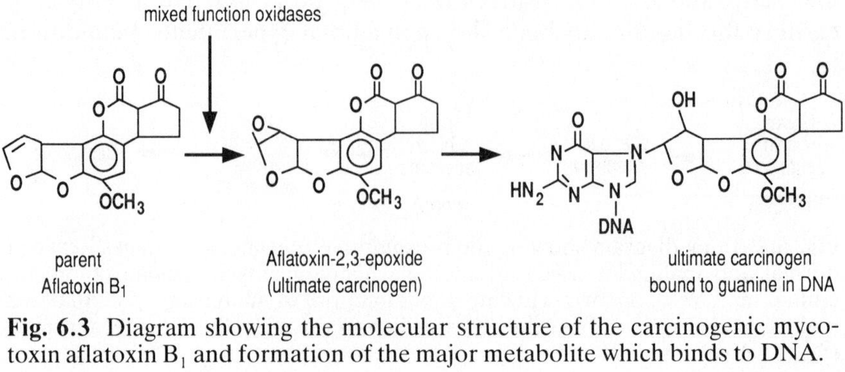

Fig. 6.3 Diagram showing the molecular structure of the carcinogenic mycotoxin aflatoxin B_1 and formation of the major metabolite which binds to DNA.

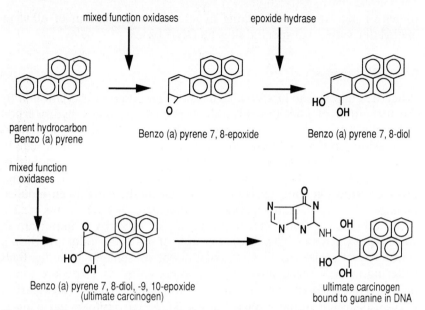

mixed function oxidases   epoxide hydrase

parent hydrocarbon
Benzo (a) pyrene

Benzo (a) pyrene 7, 8-epoxide

Benzo (a) pyrene 7, 8-diol

mixed function
oxidases

Benzo (a) pyrene 7, 8-diol, -9, 10-epoxide
(ultimate carcinogen)

ultimate carcinogen
bound to guanine in DNA

Fig. 6.4 Diagram showing the molecular structure of the aromatic hydrocarbon carcinogen benzo(a)pyrene and formation of the major metabolite which binds to DNA.

has been introduced enzymatically across a carbon–carbon double bond, producing an unstable intermediate proximate metabolite. This then reacts preferentially with the 7-position of guanine residues in DNA and forms one of the two major carcinogen–DNA adducts found in rat liver. It should be remembered though that alternative, or even very minor, carcinogen–DNA adducts may be the critical ones in carcinogenesis. The MFO are also responsible for metabolizing the polycyclic hydrocarbon carcinogens; benzo(a)pyrene is a good example. The 7,8-position double bond is opened enzymatically and the epoxide is formed, as with aflatoxin B_1. The compound is a substrate for a number of enzymatic and non-enzymatic reactions but the important one for carcinogenicity involves soluble enzymes (not bound to membranes) called epoxide hydrases. The 7,8-epoxide is converted to the 7,8-diol and this now forms a good substrate for, amongst other reactions, a second epoxidation at the 9,10-position as shown in Fig. 6.4. DNA binding occurs through this epoxide to the 2-amino group of guanine and this is the major adduct found in mouse skin under conditions where carcinomas would be expected after an appropriate latent period. Metabolism of other recognized or suspected carcinogens, the aliphatic N-nitrosamines, for instance, may involve less complicated metabolic routes of activation, culminating in alkylation (formation of ethyl or methyl derivatives) at specific sites on DNA bases.

6.2.3 *DNA repair*

What then happens to DNA which has been modified by alkylation or by the formation of nucleic acid adducts with large bulky hydrocarbon molecules? Are there mechanisms by which a cell detects and repairs such lesions so that its DNA sequence of bases is faithfully restored? In bacteria, this is certainly so and there is a considerable amount of knowledge about the precise pathways of repair and the enzymes involved in recognizing, excising, and repairing the carcinogen-induced damage. Much less is known about mammalian cells but it seems certain that inefficient or faulty (error prone) repair of DNA is important in some types of cancer. The classic example of this is the induction of skin cancer by exposure to the u.v. component of sunlight in individuals suffering from XP (see above and Chapters 4 and 5). Exposure to u.v. light also leads to skin cancer in normal individuals but only if excessive, i.e. in Caucasians in the tropics, but the XP patients' skin is particularly sensitive because of the underlying biochemical defect in their cells which is genetic in origin. This is known to involve part of the DNA repair mechanism so that their cells are deficient in repairing u.v. induced DNA damage. This can also be shown in cells cultured from patients' skin biopsies. There are other cancer prone conditions which are thought to involve defects in DNA repair, but these are less easily explained. For instance, ataxia telangiectasia (AT), Fanconi's anaemia, and Bloom's syndrome are thought of as chromosome breakage syndromes. AT cells are sensitive to agents such as X-rays which cause gross chromosome breakage due to deficient repair of this type of damage. As well as having many other clinical defects, AT patients are susceptible to cancers of lymphatic tissues and leukaemias, but the relationship between this susceptibility, DNA damage, and faulty DNA repair is still unclear. In Bloom's syndrome, the genetic defect involves an enzyme ligase which normally stitches DNA breaks together again (Willis *et al.* 1987). Such patients also develop various types of cancer.

In summary, most carcinogens need to be activated metabolically to be converted to the ultimate carcinogen that binds to DNA, modifying accessible DNA bases in a precise way throughout the genome, to an extent which depends on the dose and extent of metabolic activation. It is likely that small errors in repairing this damage or, on a larger scale, complete chromosome breakage, perhaps due to lesions on both DNA strands in the same vicinity, are important. Thus, mutations at the DNA sequence level or those involving gross changes such as large deletions, translocations, and mechanisms leading to a functionally homozygous state at particular loci (see Chapter 10) are strongly implicated in the mechanism of carcinogen action.

6.2.4 *Cellular transformation* in vitro

Transformation is a term used for changes seen in tissue culture, whereby more or less normal cells become altered to resemble cancer cells. This can happen spontaneously as a rare event whose frequency depends on a variety of factors and on the species. Cells from some rodents, e.g. mouse, transform spontaneously in culture whereas human and avian cells rarely (if ever) do. Physical agents (see Chapter 7), chemicals, and viruses (see Chapter 8) can transform cells *in vitro*. In many cases where the cell transformation system is well defined, near normal, diploid cells can be converted with reasonable efficiency into cells which can grow into invasive tumours if they are put back into a suitable animal host. The converted or transformed cells in culture are then said to be tumorigenic and the process by which they became so can be studied as a model for carcinogenesis *in vivo* (spontaneous or induced).

Many different culture systems for studying transformation have used mesenchymal or 'fibroblast' cells (see Chapter 1) which rarely give rise spontaneously to malignant tumours in man or laboratory animals. They probably do not provide a very good model system for studying the relationship between cell differentiation and neoplasia, but they are easy to grow and manipulate in culture and they have certainly provided us with ways of studying some fundamental aspects of carcinogenesis. For instance, the system devised by Heidelberger and his group used a clone of mouse embryo fibroblast cells called $C3H/10T\frac{1}{2}$ (Reznikoff *et al.* 1973). One parameter of transformation that correlates well with tumorigenicity in this cell system (as it does in many others, but not invariably) is the appearance of a property known as anchorage independent growth (AIG). Most normal cells need to be anchored to and spread on a solid substrate before they can divide and form a colony or clone from a single cell. Some cancer cells are able to grow and form colonies when suspended in a semi-solid medium such as soft (0.33 per cent) agar. If untransformed $C3H/10T\frac{1}{2}$ cells are treated with a chemical carcinogen, a small proportion of the cells will grow in soft agar and these cells are also usually tumorigenic. The frequency of transformation to this phenotype can be measured after correction for the proportion which survived the carcinogen induced toxicity. The efficiency with which certain carcinogens transform cells to AIG has been used by some scientists as a rapid screening test for chemicals which might cause cancer (see p. 168) but it is not as reproducible as other tests and is not widely used. There are many other changes which can be induced in cell culture by carcinogens and for which there is evidence of a link with the cancer cell phenotype. Changes in components of the filamentous cytoskeleton of cells is one such example. Since none of these markers of transformation in culture is invariably associated with cancer cells, their role in carcinogenesis remains an area of active investigation and dispute (see Chapter 1).

As mentioned earlier, some aspects of carcinogenesis cannot be studied in simple fibroblast cell systems, namely those concerned with differentiation and tissue homeostasis (the balance between cell renewal by division and cell death in a defined population)—a key abnormality in cancer. Some aspects of these properties can be investigated in specialized systems such as in differentiated epithelial cell cultures. Methods have been established for growing some of these more fastidious cell types from rodent and human tissues. It has been shown conclusively that chemical carcinogens, such as the hydrocarbons, benzo(a)pyrene, and dimethylbenzanthracene, can induce transformation and eventually tumorigenic potential in epithelial cells treated with the carcinogen in primary culture, i.e. cells grown directly from animal tissues. Various types of rodent cells have been used, including skin keratinocytes, salivary gland duct cells, epithelium from the respiratory system, and urinary bladder cells. All of these epithelial systems demonstrate one feature particularly clearly: transformation, like carcinogenesis, is a multistage process. There is a relatively long period of time between treating normal cells with a chemical, sometimes just for a single short exposure, and the eventual emergence of cells which will grow as tumours in an appropriate animal host. During this long latent period, more or less discrete precancerous stages occur wherein the cells appear altered in a characteristic way but are not yet capable of forming tumours in an animal. Some of these intermediate stages are probably equivalent to the precancerous stages in carcinogenesis observed *in vivo* (see above) but a comparison of the molecular changes *in vivo* and *in vitro* will be necessary to determine whether the sequence of events is the same. Amongst the properties which frequently alter during the precancerous stages in transformation are chromosome number (which usually increases, and often nearly doubles, the normal complement), loss of dependence on growth stimulating factors in serum (see Chapter 11), increased ability to grow clonally from single cells (clonogenicity) or at least at a reduced cell density, and the acquisition of a prolonged or indefinite lifespan in culture (immortality—an escape from senescence). We know very little about the factors which govern progression through these precancerous stages. Further studies will investigate the effects of tumour promoters, identified from animal experiments, and other known modifiers of gene expression which do not conform to the carcinogen/mutagen category of initiating agents.

6.2.5 *Precancerous cells* in vitro

Human cells of all types and, in particular, normal epithelial cells which give rise to the common human cancers, are extremely resistant to transformation induced by chemicals in culture. The reasons for this are poorly understood. It is now possible to culture epithelial cells directly from some tissues which are already precancerous. The early stages of transformation in these cells, especially the stage(s) leading to immortality and the capacity for indefinite propagation *in vitro*, may have already taken place. We can now study such cells and try to identify the factors which lead to the development of more malignant cell properties or conversely, those that induce reversion to a more normal state.

This approach is most useful where the precancerous tissue is readily available, usually through surgical procedures. In patients with familial polyposis coli, diseased tissue is removed surgically. Polyps from these surgical specimens have been cultured successfully in our laboratory and immortal cell lines have been derived from about one specimen in five which survived the initial preparation procedure and remained uncontaminated by intestinal micro-organisms.

The uterine cervix, oral tissues, oesophagus, and trachea also provide suitable precancerous tissue from biopsies, but work using these tissues is in its early stages.

6.3 The role of tumour promotion in human cancer

6.3.1 *Epidemiological evidence*

Whilst most of the known human chemical carcinogens (Table 6.1) and those which cause cancer in experimental animals are positive in screening tests designed to detect mutagenic activity (see p. 168), some do not fall into this category. It has been known for some time that there are many components with biological activity in cigarette smoke and that not all of these have the hallmarks of classical mutagenic/carcinogenic chemicals. We now believe that some constituents act to promote the development of cancers from cells of the respiratory epithelium which harbour covert mutations, perhaps in oncogenes. Such promoters (see Fig. 6.1 and related text) must be applied regularly over a long period of time to achieve the eventual completion of the carcinogenic process. This may explain partly the striking relationship between lung cancer incidence and duration of smoking (Chapter 3).

Other agents whose probable role in carcinogenesis is to promote rather than initiate tumour development include those contained in, or influenced by, the diet. High dietary fat has been linked epidemiologically with increased bowel cancer risk. This may be a reflection of consequent high levels of bile acids, some of which have been shown to promote colorectal cancer in experimental animal model carcinogenesis systems. Hormones can also act as growth promoting factors, which may in some circumstances create an environment in which initiated cells have a selective advantage. Obviously, the situation is complex and much experimental work needs to be done to correlate cancer risk statistics in human populations with identification of the causative agents and analysis of their modes of action.

6.3.2 *Mechanisms of tumour promotion*

Evidence for the role of some chemicals as promoters of tumour development comes from data on experimental animal model systems. Some of these systems are described in the preceding section. The generally applicable conclusion is that activation of an oncogene by mutation, which can be seen as an initiating event, does not in itself lead to cancer development. Promoting influences, such as the activity of hormones on a hormone-dependent tissue, for instance the mammary gland at puberty, are necessary for complete carcinogenesis. An activated oncogene may be linked experimentally to DNA sequences which, under hormonal control, promote expression of the gene, and then introduced into early embryos (to produce transgenic animals) (Stewart *et al.* 1984). Tumours develop, as predicted, predominantly in the tissues which respond to the hormonal stimulation. Presumably this is a result of activated oncogene expression under hormonal control. However, the situation is more complex than this, since expression of the activated oncogene can be detected in a wider range of hormone dependent tissues than those which give rise to tumours. This indicates that the hormone, in addition to stimulating expression of the oncogene, has other effects on the target tissue which promote tumour formation.

The primary cellular target for tumour promoters is thought to be an intracellular kinase known as protein kinase C (PKC). The endogenous mechanism of PKC activation is by binding diacylglycerol which is in turn generated by growth factor stimulated turnover of membrane bound phosphatidyl inositol (see Chapter 11). Some tumour promoters are chemically similar to diacylglycerol, and can consequently bind directly to PKC causing its 'translocation' (receptor translocation, see Glossary) to the cell membrane and the activation of pathways leading to DNA synthesis and cell division. This picture however represents a vast over-simplification of the true situation. Several members of the PKC family exist and it is not known which of these is critical for tumour promotion. In addition, some promoting agents that do not resemble phorbol esters do not bind PKC and conversely, other agents which bind PKC are either inactive as tumour promoters or, paradoxically, can inhibit promotion. It is therefore unlikely that the mechanism of promoter action involves only this pathway.

Although tumour promoters do not bind to DNA and are non-mutagenic in bacterial test systems (see p. 169), there is now evidence that promoters nevertheless have effects at the DNA level. The phorbol ester promoters have clastogenic effects on some mammalian cells, i.e. they cause single-strand breaks in DNA. This probably involves the generation of 'active oxygen' species within the cells which are known to cause such breaks. Indeed, some chemicals which induce active oxygen molecules can act as tumour promoters, for example, benzoyl peroxide. One possible explanation of this phenomenon is that the non-specific DNA damage caused by these agents induces a regenerative response which accounts for their tumour promoting ability but this is far from being understood at the moment.

Other documented effects of tumour promoters are the induction of gross chromosomal changes in epidermal cells, both *in vivo* and *in vitro*. Whilst these events may lead to changes in the function of critical genes, alterations in controlling elements or development of homozygosity at certain chromosomal loci, there is as yet no evidence that such changes are critical for the selection of initiated cells rather than a result of the toxic effects of the promoting agents used.

6.4 Screening for carcinogens

6.4.1 *Animal carcinogenicity tests*

Most known human carcinogens and many other chemicals will produce tumours in experimental animals under appropriate, although some-times very artificial, conditions. In fact, it is required by law that any new chemical introduced for use in or by humans (as medicines, cosmetics, food additives, weed and pest killers, agricultural fertilizers, household cleansing products, to give a few examples) must be tested in laboratory animals for their long-term effects. This is enormously costly in time and expenditure so it is vital that tests should be carefully planned and informative. This requires a knowledge of the way in which chemicals are metabolized (see above) and excreted in different species and whether these characteristics are appropriate to the human situation. For instance, the guinea pig would be inappropriate for testing AAF since it lacks the enzyme necessary to convert this chemical to its active, carcinogenic form. In practice, most tests are done on rats and mice, male and female, which are exposed for a long period, often one to two years, to the maximum tolerated dose. The route of administration usually depends on the likely mode of human exposure, by inhalation, in the diet or drinking water, or via skin contact. After the necessary length of time, animals still surviving are examined for tumours which can be confirmed as malignant by a pathologist.

As discussed earlier, human cancers are thought to take many years to evolve after the initiating event in a susceptible target cell. This can range from about 10 years up to almost the total lifetime of an individual. This fact alone poses a very real problem for the experimenter who wants to confirm the safety of a potentially useful chemical. Animal studies still form the main acceptable evidence for Food and Drug Safety authorities throughout the world, and yet they are very time consuming and expensive. As an example, to test one compound thoroughly for carcinogenicity in two species (rats and mice) costs approximately £500 000 at 1989 prices and takes up to three years. As it became apparent that most compounds with proven carcinogenic activity were genotoxic, a number of more rapid tests were developed as first-order screens for large numbers of chemicals at between 1 and 10 per cent of the cost of animal experiments, depending on the type and number of rapid tests used.

6.4.2 Rapid screening tests

A large number of different tests have been put forward over the last few years as potentially useful indicators of carcinogenic activity. After a series of comparative trials, the authorities in most countries reached agreement on the type of evidence which would be acceptable. This requires a compound to have been tested in assays measuring mutation (both in bacteria and in mammalian cells in culture) and for their ability to cause chromosome breakage both *in vivo* (in the animal) and *in vitro* (in cell culture). The results of a battery of about four rapid screening tests selected from within these categories should, if the results are unequivocal, provide information on the potential carcinogenicity of a chemical which would be accepted by safety authorities in lieu of long-term animal data. There are several problems in evaluating the usefulness of such rapid screening tests and in extrapolating from mutagenesis assays *in vitro* to carcinogenicity in animals and thence to the potential human cancer risk incurred by exposure to a suspect chemical. It is wise to assume that any chemical which is capable of causing mutation is a potential carcinogen. However, the expression of carcinogenic potential varies markedly between species, even those which are as close (in evolutionary terms) as rats and mice. Regulatory authorities are continually updating and revising test protocols to minimize the chance of undue exposure to carcinogens.

The most well known and widely publicized rapid test for chemical carcinogens takes the name of its originator, Bruce Ames (Ames *et al.* 1975). This test assesses the mutagenicity of chemicals, with or without metabolism by activating enzymes (from a crude subcellular fraction, S9, of rat liver, a rich source of membrane-bound enzyme activity including the MFO) in a range of specially selected strains of *Salmonella* bacteria. These bacterial strains have each been constructed in the laboratory to detect mutations of a specific kind. For instance, one strain is able to indicate a particular nucleic acid substitution which alters one base pair to another in bacterial DNA after exposure to a DNA damaging agent. Another strain detects chemicals able to cause frameshift mutations (whereby addition or deletion of one or more nucleic acids other than a multiple of three causes the whole reading frame of the triplet code to be thrown out of phase), when the encoded protein is drastically altered. The bacteria used as test strains are themselves mutants and are unable to make a particular amino acid essential for their growth, such as histidine. The test then detects whether a particular chemical can cause the specific base pair substitution or frameshift mutation needed to revert the mutant bacteria to their wild type capacity for synthesizing the essential amino acid. This is done because the investigator can most easily score the number of bacterial colonies which *do* grow on an incomplete nutrient agar substrate (one which does not supply the particular amino acid). Most carcinogenic chemicals induce a wide range of mutations which may kill the bacteria, so it is important to test a chemical at a dose giving an acceptable level of toxicity that can be measured separately under non-selective conditions. Obviously a positive result is one in which a chemical induces a significant increase in reverted (auxotrophic) bacteria capable of forming colonies on a selective (deficient) nutrient agar substrate where the unaffected bacteria cannot grow (Fig. 6.5).

The Ames test is probably one of the least expensive and most rapid tests available (taking only a couple of days) to detect a property, namely mutation, common to most human carcinogens. However, there are some drawbacks to the test which mean that the results obtained with it should be considered suggestive rather than conclusive evidence for or against the carcinogenicity of a suspect chemical compound. First, the complete spectrum of enzymes involved in metabolizing a variety of carcinogens is not present in the particular liver microsome fraction (S9) component of the assay mixture. This means that although mutagenic derivatives of a chemical may be generated by the microsomal enzymes, they may not be the ones produced by an intact cell or in the body, and so

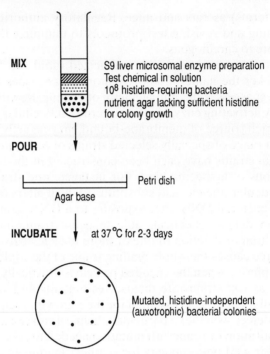

MIX

S9 liver microsomal enzyme preparation
Test chemical in solution
10^8 histidine-requiring bacteria
nutrient agar lacking sufficient histidine
for colony growth

POUR

Petri dish

Agar base

INCUBATE   at 37°C for 2-3 days

Mutated, histidine-independent
(auxotrophic) bacterial colonies

Fig. 6.5 Diagram showing the basic Ames test procedure for detecting mutagenicity of chemicals in bacteria. A panel of similar tests is generally used, each
designed to detect different chemical activities and types of mutation in order to
predict carcinogenic potential.

present a different picture of the chemical's potency. It is even possible
that a false positive or negative result might be obtained for similar
reasons. In practice, the Ames test, using a battery of four *Salmonella*
strains each designed to detect a particular type of mutation, has
achieved greater than 90 per cent accuracy in predicting both carcinogens and non-carcinogens in 'blind' trials. However, the relative
potency of the chemicals was predicted less accurately in quantitative
comparisons between different classes of chemicals.

Because mammalian cell DNA is more complex than bacterial DNA,
mutation tests in mammalian cells *in vitro* must also be performed, even
though they are rather more complicated to do and take longer to
produce results. The mutations most commonly used are deficiencies
due to DNA sequence modification in either hypoxanthine phosphoribosyl transferase (HPRT) or thymidine kinase (TK), enzymes involved
in nucleic acid synthesis (see Fig. 6.6) or to a mutation which reduces the
capacity of the drug ouabain to bind at the cell surface and block

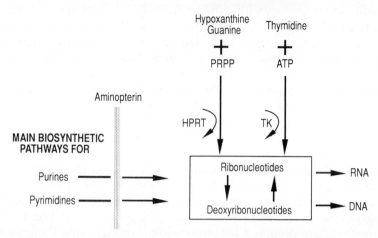

Fig. 6.6 Diagram showing the biochemical pathways utilized in the synthesis of nucleic acids. The biosynthetic (*de novo*) pathway can be blocked by aminopterin and the alternative salvage pathway can be abolished by mutation in hypoxanthine phosphoribosyl transferase (HPRT) or thymidine kinase (TK) genes. This is detected by cellular resistance to the cytotoxic base analogue drugs 6-thioguanine (or 8-azaguanine) and bromodeoxyuridine respectively. Mutation in a forward direction (HPRT⁻ or TK⁻ phenotypes) or reversion of mutants to the wild type (resistance to aminopterin) are increased in frequency by many chemical carcinogens.

membrane transport. The most commonly used mutation is HPRT deficiency (HPRT⁻) because, although recessive, it can be induced at relatively high frequency due to its location on the X chromosome, i.e. inactivation of the single gene copy in male cells is sufficient for expression of the mutant phenotype. The assay procedure involves treating suitable mammalian cells (often Chinese hamster fibroblasts) with a suspect chemical, with or without activation by the S9 rat liver cell enzyme fraction. The treated cell population is assessed (i) for its ability to grow from single cells to form clones as a measure of the degree of toxicity of the chemical (reduction in clone formation compared with untreated cells), and (ii) for the number of mutations in a given number of cells cultured in the presence of the drugs 6-thioguanine or 8-azaguanine. Unaffected cells take up the selective drug, incorporate it into nucleic acid, and are thus killed. Mutated cells fail to do so because they lack HPRT of the salvage pathway and cannot utilize base analogue drugs. They depend solely on the alternative biosynthetic pathway and are spared. They grow to form clones of mutant cells which can be stained and counted after one to two weeks, and their numbers expressed as a mutation frequency within the surviving fraction (after correction for toxicity of the suspect chemical). Relative to the control 'background' level of mutation in untreated cells, an elevated mutation frequency would indicate that the suspect chemical has potential carcinogenic activity.

Chromosome mutations, i.e. aberrations such as chromatid breaks (see Chapter 10) caused by chemicals, are also good indicators of carcinogenic activity and this property forms the basis of the second pair of rapid screening tests. The *in vivo* assay relies on the living animal to activate the test chemical, if this is necessary, after it has been fed or injected, and metabolism may occur in the liver or in the target cells. In this test, target cells are usually from the bone marrow where blood cell precursors are dividing rapidly and are highly sensitive to DNA damaging agents. At several times after injecting the chemical, samples of bone marrow cells are prepared for chromosome analysis (see Chapter 10). Chromosomes in the metaphase part of the cell cycle, just before cell division, are condensed and relatively easy to see. Many metaphase chromosome spreads are analysed and the average numbers of breaks, or discontinuities, in the chromatid arms of each chromosome are counted. Significant increases above the spontaneous background level indicates that the chemical may be carcinogenic. The same kind of analysis may be performed on cells in tissue culture, which can either be human blood lymphocytes in short term culture or permanent cell lines of human or rodent origin. These are treated with the chemical *in vitro*, with or without the metabolising enzyme S9 preparation, and chromosomes prepared at two or three times thereafter to find the peak of chromosome breaking activity; this may vary, but is generally at 24–48 hours.

6.5 Prospects

6.5.1 *Early diagnosis of precancerous conditions*

With most precancerous lesions, there are no problems for the patient, who is probably unaware of their presence. In rare circumstances, a large adenoma of the bowel for instance may cause obstruction or bleed chronically and require surgery, but generally the lesions are asymptomatic. These situations are distinct from some other clinical conditions, such as ataxia telangiectasia and Down's syndrome, where obvious multiple abnormalities exist, including an increased risk of particular cancers to which the clinician will already be alerted.

The main clinical problems in cancer usually arise when metastasis to distant parts of the body occurs (see Chapter 2) so that local surgical excision or radiation therapy is no longer feasible. By definition, pre-cancerous tissues of epithelial origin have not invaded the underlying stroma (see Chapter 1) although the individual cells may be highly abnormal in other respects. There can only be the possibility of metastasis once invasion has taken place. In some tissues, such as breast, this may take place when the cancer is very small but cancer cells must at least have penetrated vessels in the stroma. Is there any way of detecting abnormal precancerous tissues before invasive properties are acquired? In some cases there may be. The haemoccult test used to detect cancer of the colon and rectum relies on the fact that many cancers ulcerate and bleed chronically and blood can be detected biochemically in the faeces. Thus a simple screening test can often help in diagnosing cancer in individuals with bowel problems, when malignancy is suspected. In fact, many precancerous adenomas of the large bowel, particularly the more advanced ones, will also be detected with this test, and can be removed surgically. In the near future, it should be possible to use a similar approach to detect in faecal samples more specific products of pre-malignant colon cells either actively secreted into the bowel lumen or shed from dead cells. The problem is whether this type of test will be practical for the population as a whole, perhaps over a certain age, or just for high-risk individuals with a family history (see Chapter 4). Similar principles could theoretically apply to products of abnormal tissues released into other body fluids. Screening for cervical precancerous lesions using a smear from the cervix to look for abnormal cells is a well known example of a different technique used to identify potential cancers before the risk of invasion and metastases arises. Apart from these examples, screening for overt cancer of various tissues and organs presents a great problem, with little chance at present of finding pre-cancerous lesions by current, insensitive and, for the most part, non-specific methods. It is in this area of clinical cancer research that much effort is needed, particularly where clinicians and scientists in the laboratory can combine efforts and devise sensitive diagnostic pro-cedures. Monoclonal antibody technology will undoubtedly have a major influence in this area (see Chapters 18 and 19).

6.5.2 *Understanding and preventing tumour progression*

Very little is known about the factors that govern the fate of pre-cancerous lesions and determine whether or not a cancer develops and progresses. Some clues have come from epidemiological studies (see Chapter 3) but, information is sparse. To take the example of cigarette smoking, where components of the smoke are thought to promote development of lung cancer in the later stages of carcinogenesis, this

tumour progression phase can be retarded by stopping smoking—
however long ago the habit was established (see Chapter 3). In the case
of liver cancer, hepatitis B virus infection and chronic hepatitis probably
act as promoting influences, particularly in the Third World (see
Chapter 8). It is likely that this occurs through the creation of tissue
damage, which in turn stimulates regeneration, a situation which is
known from animal experiments to allow expression of chemical
carcinogen-induced genetic changes, culminating in cancer. Improved
availability of preventative measures, viral vaccines particularly, would
almost certainly reduce the overall incidence of this cancer. It may also
be effective to modify the diet, or hormonal status, in individuals with
precancerous lesions of particular tissues or in those in a high risk group
for cancer development.

Much more needs to be done to identify potential promoting agents for
the common cancers and thus the means for intervention in the process
of tumour progression. It may be more practicable to arrest the development of the disease or slow its progress sufficiently for it to cease to be
life-threatening, than to prevent its initiation.

References and further reading

Ames, B. N., McCann, J., and Yamasaki, E. (1975). Methods for detecting
carcinogens and mutagens with the Salmonella/mammalian-microsome
mutagenicity test. *Mutation Research* **31**, 347–64.

Balmain, A. and Brown, K. (1988). Oncogene activation in chemical carcinogenesis. *Advances in Cancer Research* **51**, 147–82.

DeCosse, J. J. (1983). Precancer. *Cancer Surveys* **2**, 347–518.

Farber, E. (1986). Some general principles emerging in the pathogenesis of
hepatocellular carcinoma. *Cancer Surveys* **5**, 695–718.

Franks, L. M. and Wigley, C. B. (ed.) (1979). *Neoplastic transformation in
differentiated epithelial cell systems in vitro.* Academic Press, Orlando.

Freeman, A. E. (1980). Induction of mammalian cell transformation by
chemical carcinogens: basic considerations. In *Mammalian cell transformation by chemical carcinogens* (ed. N. Mishra, V. Dunkel, and M. Mehlman),
pp. 37–45. Senate Press, Princeton, New Jersey.

IARC Monographs. Suppl. 2 (1980). *Long term and short term screening assays
for carcinogens: a critical appraisal.* IARC, Lyon.

Moolgavkar, S. H. and Knudson, A. G. Jr. (1981). Mutation and cancer: a model
for human carcinogenesis. *Journal of the National Cancer Institute* **66**,
1037–52.

Reznikoff, C. A., Bertram, J. S., Brankow, D. W., and Heidelberger, C. (1973).
Quantitative and qualitative studies of chemical transformation of cloned
C3H mouse embryo cells sensitive to post-confluence inhibition of cell
division. *Cancer Research* **33**, 3239–49.

Stewart, T. A., Pattengale, P. K., and Leder, P. (1984). Spontaneous mammary adenocarcinomas in transgenic mice that carry and express MTV/myc fusion genes. *Cell* **38**, 627–37.

Willis, A. E, Weksberg, R., Tomlinson, S., and Lindahl, T. (1987). Structural alterations of DNA ligase 1 in Bloom's Syndrome. *Proceedings of the National Academy of Sciences, USA* **84**, 8016–20.

Yuspa, S. and Poiret, M. C. (1988). Chemical carcinogenesis: from animal models to molecular models in one decade. *Advances in Cancer Research* **50**, 25–70.

7

Radiation carcinogenesis
G. E. Adams and R. Cox

7.1 Introduction

7.1.1 *The problem*

Although mankind has always been exposed to natural background ionizing radiation, there is considerable doubt whether in the past such exposures have had any significant role to play in the aetiology of human cancer. Radiation carcinogenesis is a twentieth century problem, as indeed are the problems of carcinogenic risk from other hazards to which society is exposed. Cancers induced by ionizing radiation are indistinguishable from most cancers arising from other causes and their occurrence can only be identified by a statistical analysis of excess incidence over the 'natural' incidence. Much of our information on human radiation carcinogenesis is derived therefore from epidemiological sources. Studies of occupational exposure of diagnostic radiologists, uranium miners, and workers in the nuclear industries, for example, have provided some information. Much more, however, has come from analyses of cancer incidence in patients exposed to radiation for medical purposes, either for diagnosis or for treatment of non-malignant conditions.

Another major source of information has been the long-term follow up of survivors of the atomic bombs dropped in 1945 on Nagasaki and Hiroshima. This Life Span Study has been in progress since 1950 and has achieved a remarkable level of precision particularly in regard to the dosimetry. In many cases, the precise location of the individuals and the shielding effect of buildings and other structures is now known accurately.

Studies on radiation carcinogenesis in experimental animals have addressed problems such as the pathogenesis of the various cancers that have been identified, inter-species variation, the relationships between cancer induction and cell mutation and other cellular phenomena, dose relationships, and, most important of all, the validity or otherwise of animal experiments for assessing radiation risk in human populations. While such studies have provided much information on the biology of radiation carcinogenesis, estimates of radiation risk in humans still rest heavily on the data from epidemiological studies. As in other fields of carcinogenesis, knowledge of events at the cellular and molecular level is essential to an understanding of radiation carcinogenesis, a complex multistage process that extends from the very early physical, chemical, and cellular changes initiated by the absorption of radiation to the delayed effects that only appear many years later.

The energies of photon or particulate radiations emanating from radionuclides, X-ray sets, and particle accelerators are vastly in excess of those of the chemical bonds in biological molecules. Ionization, i.e. electron ejection from atoms with which the radiation interacts, is therefore the primary initial event. The time-scale over which energy is imparted to the atom is governed by the speed of the particle (usually at or near the velocity of light), the dimension of the atom, and the extent of energy loss. A quantum of γ radiation, or an energetic α particle, will pass through a small molecule and deliver energy to it, in a time between 10^{-17} and 10^{-18} s. The subsequent physical, chemical, and biological processes that follow this event are only expressed as an induced cancer perhaps 30 years or more later. Thus, it is not surprising that the interpretation of radiation carcinogenesis in terms of the primary physical and molecular events is a complex undertaking.

7.1.2 *The temporal stages of radiation action*

It is convenient, though not rigorously precise, to classify the many processes of radiation action into four stages, namely, physical, chemical, cellular, and tissue effects (Table 7.1).

The physical stage. Radiation deposits its energy in discrete packages. Their magnitude and spatial distribution depend on factors such as the

Table 7.1 The temporal stages of radiation action

1. *The physical stage*	
10^{-18}–10^{-17} (s)	Fast particle traverses small atom or molecule
10^{-16}	Ionization: $H_2O \rightarrow H_2O^+ + e^-$
10^{-15}	Electronic excitation: $H_2O \rightarrow H_2O^*$
10^{-13}	Molecular vibrations: dissocation
10^{-12}	Rotational relaxation: $e^- \rightarrow e^-aq$
2. *Physico-chemical and chemical stage*	
10^{-10}–10^{-7} (s)	Reactions of e^-aq and other free radicals with solutes in radiation tracks and spurs
10^{-7}	Homogeneous distribution of free radicals
10^{-3}	Free radical reactions largely complete
Seconds, minutes, hours	Biochemical changes (enzyme reactions)
3. *Cellular and tissue stage*	
Hours	Cell division inhibited in micro-organisms and mammalian cells; reproductive death
Days	Damage to gastrointestinal tract (and central nervous system at high doses)
Months	Haemopoietic death; acute damage to skin and other organs; late normal tissue morbidity
Years	Carcinogenesis and expression of genetic damage in offspring

energy of the radiation, the nature of the absorbing medium, and particularly the type of radiation. The 'densely ionizing' radiations (i.e. α particles, protons, and neutrons) lose energy over a much shorter distance than do the 'weakly ionizing' X and γ rays. The biological effectiveness of the particulate radiations in cell killing, mutagenicity, cell transformation, and carcinogenic potential are substantially greater than the weakly ionizing or low 'LET' radiations (LET or 'Linear Energy Transfer' is a measure of the rate at which energy is imparted to the absorbing medium per unit distance of track length).

Interaction of radiation with an atom ejects an electron, or electrons, in less than $\approx 10^{-16}$ s. The ejected electrons have energy greatly in excess of atomic ionization potentials and cause many more of the secondary ionizations responsible for the subsequent chemical changes which lead to biological damage. The secondary electrons lose energy by collision and eventually undergo dipole interaction. This involves electrostatic interaction between the negative charges of the secondary electrons and the slight positive charge associated with the hydrogen atoms in water. This polarization in water is due to the higher electronegativity of the oxygen atom compared with that of hydrogen. The interaction in aqueous media is complete in about 10^{-12} s ('the dielectric relaxation time'). The trapped electron or, as it is called 'the hydrated electron' (e⁻aq), has, in many respects, the properties of many free radicals. It can diffuse considerable differences and can undergo rapid reaction with many diverse types of chemical structures including those present in most biological molecules. Its formation marks the transition to the chemical stage of radiation action.

The chemical stage. The chemical stage of radiation action is concerned mainly with the formation and reaction of molecular fragments such as free radicals and excited molecules. Roughly speaking, radiation energy deposited in the cell is partitioned according to the relative proportions of the constituent atoms (at least for the low atomic number elements normally represented in biological tissue). This is known as 'the principle of equipartition of energy' and implies that about 80 per cent of the overall energy deposited in the cell by ionizing radiations initially occurs in the aqueous component. This is why so much attention in the past has been devoted to the study of the radiation chemistry of water and aqueous solutions.

The hydroxyl radical (OH), an oxidizing species of high reactivity, is formed very quickly from the ionized water molecule, H_2O^+, by inter-action with neighbouring water molecules and by rapid dissociation of excited H_2O molecules. The reactive hydrogen atoms and hydrated electrons are the corresponding reducing equivalents so that, overall, water radiolysis is described by the simple equation

$$H_2O \rightarrow H(+e^-aq) + OH$$

Some of the radicals interact together to form molecular hydrogen and hydrogen peroxide. The remaining radicals diffuse away from the radiation track and react with other molecules in the environment. There is much evidence that damage to biological molecules caused by free radicals contributes to loss or change of cellular function following irradiation. The problem that remains is to identify those reactions that are relevant to the observed cellular response to radiation. The time scales for these reactions are very short and most will be complete in times much less than a millisecond. Others, however, will take longer.

The cellular stage. Ultrastructural changes in cells can sometimes be observed a short time after irradiation. Local protrusions of the plasma membrane, for example, can be observed within minutes of exposure of cells to a relatively high dose of radiation. Within a few hours these changes are followed by membrane distention and later by invagination of the nuclear membrane. These effects are accompanied by changes in the permeability of the membrane and loss of essential enzymes. However, the more important effects (i.e. the loss of, or changes in, cellular function that occur at much lower radiation doses) can only be observed after longer periods. Loss of reproductive capacity is only evident when the cell fails to divide and chromosomal changes or cellular mutations are observable only after sufficient cell divisions have taken place to allow the analyses of aberrant cells in the total population. Similarly, measurement of changes in the repair capacity of irradiated cells, a process which is normally complete within a few hours of irradiation, usually requires clonal analysis of irradiated populations. Nevertheless, the stages in the cell cycle when radiation damage is most critical are now known. Mammalian cells are usually most radiosensitive during mitosis and very early in the G_1 phase and usually at their most resistant in early S phase, although this is very dependent on radiation quality. Cellular sensitivity to low LET radiation is usually much more variable than it is to high LET radiation.

The tissue stage. The response times of mammalian tissues to radiation exposure vary widely as do their sensitivities. The observation that mammalian cells show maximum sensitivity to radiation during mitosis predicts firstly that the fertilized mammalian egg cell (zygote) would be highly sensitive to radiation, and secondly that, in the animal, the most radiation-sensitive tissues would be those that turn over rapidly. This is indeed the case. The rapidly dividing proliferating stem cells of the haemopoietic system and the intestinal epithelium are particularly sensitive and respond more quickly than do the moderately sensitive tissues such as lung and the basal layer of the skin. Cells that do not normally divide except after an appropriate stimulus, i.e. parenchymal cells of the liver and connective tissue, are less sensitive still, and cells that divide only during embryonic development are the least radiosensitive. The time of onset for normal tissue damage and mortality depends on the radiation dose.

7.2 Radiation and human cancer

7.2.1 *Radiation dose and radiation risk*

Radiation dose to tissue is expressed in terms of absorbed energy per unit mass as the gray (Gy), which is 1 joule/kg. The older unit, the rad, still in common use, is equivalent to 100 ergs absorbed per gram of tissue and is equal to 0.01 Gy. Carcinogenic potential depends upon absorbed dose and is greater per unit dose for high LET radiations than for low LET radiations. The latter type becomes less effective per Gy as the dose falls which is not the case for high LET radiation. This means that, for example, neutrons that are five times more effective than gamma rays at a given dose may be relatively much more effective than gamma rays over a lower dose range. This dose dependence of relative biological efficiency

(RBE) is a problem in assessing risk following exposure to a mixture of radiation qualities. This has been encountered in the analysis of the atomic bomb data from Japan.

Estimates of cancer risk may be made in various ways. *Additive* risk expresses the number of *excess* cases per unit of time per unit of dose in a given number of exposed individuals. The *multiplicative* or relative risk model expresses the ratio of the risk in the irradiated population to that in a non-irradiated control group. Additive risk has the advantage of specifying the number of individuals involved and is the approach favoured by UNSCEAR (United Nations Scientific Committee on the Effects of Atomic Radiations) in the Absolute Risk Model. For example, a risk of 10^{-4} implies one excess cancer over a given period, in a population of 10 000 individuals each of which has received an average dose of 1 Gy.

7.2.2 *Radiation epidemiology*

Numerous long-term studies are in progress on human populations that have been exposed to radiation. These studies ask questions about overall cancer incidence, excess of individual cancers, latency periods, and, where possible, dose response relationships in groups of individuals exposed to radiation arising from occupation, the environment, and diagnostic or therapeutic medical procedures (UNSCEAR 1988, BEIR 1980, 1988).

Occupational exposure. Excess lung cancers have been observed in underground workers including uranium miners, fluorspar miners in Newfoundland, and some Swedish zinc and iron ore miners. The cancers are due to α particle radiation from inhaled radon gas emanating from radium present in the ores. Risk estimates are complicated by the long average latency period of 20 years, difficulties of assessing dose, and the evidence that heavy cigarette smoking substantially increases the risk.

Painters of luminous watch dials have been exposed to radiation through ingestion (by brush licking) of substantial quantities of radium[226] and radium[228]. These isotopes concentrate in the bone matrix and emit short range α particles. Results from a large US study have revealed 62 bone sarcomas and 32 carcinomas of the mastoid and paranasal sinuses in a total of about 2000 female dial painters. No more than one case would have been expected.

Environmental exposure. Exposures to natural sources of radiation are very low. It has been estimated, for example, that the average US citizen has accumulated by the age of 65, the equivalent of only 0.12 Gy from all natural sources and 0.04 Gy from man-made sources. However, concern is increasing in several countries, including the UK, that there is a small, but significant risk to health arising from exposure inside dwellings to the radioactive gas radon. This element has three naturally occurring isotopes, radon[219], radon[220], and radon[222] of which only the last two are of radiological consequence. Of these, radon[222] is the more important because of its longer half-life (3.824 days). Radon[222] is formed in the uranium[238] decay series and is environmentally important because it decays (by α particle emission) to form products or 'daughters' that are themselves radioactive. Radiation doses to the individual arise predominantly from irradiation of lung epithelium by α particles from the decay products. Radon generated in rock, soil, and building materials, used for example as in-fill, diffuses readily and can attain relatively high levels, particularly in poorly ventilated rooms. Radon decay products form small ion clusters with water molecules or react chemically with vapours present in air. These decay products can become attached to aerosol particles and ultimately enter the body by inhalation.

A recent survey has estimated that in the UK the total average annual effective dose-equivalent from radon is about half of the radiation dose received from natural sources. This is misleading, however, since there is much regional and individual variation in domestic exposure to radon. In some cases, annual doses may be 1–2 orders of magnitude higher than the average and it is estimated that about 20 000 dwellings in the UK, mostly in the south-west of the country, require action to reduce radon levels indoors. The calculation of excess lung cancer risk to the population arising from exposure to radon is fraught with uncertainty because of the assumptions involved. However, a recent calculation using current risk estimates (National Radiological Board 1988) indicates that about 1500 persons may die annually from lung cancer associated with exposure to radon and its decay products. To put this in context, however, this figure is somewhat less than 4 per cent of the total annual death rate from lung cancer in the UK and about a quarter of the death rate from domestic accidents.

Although the *average* individual dose from man-made sources of environmental exposure is very low, substantial exposure has occurred in some select groups of individuals. For example, exposure to short- and long-lived radioisotopes of iodine in the fall-out from atmospheric weapons testing is responsible for excess thyroid cancer found in a small group of 240 female Marshall Islanders. Estimates of thyroid doses vary between 0.15 and 15 Gy.

The main source of data in the field of exposure to man-made environmental radiation is the Life Span Study on survivors from the atomic bombs dropped on Japan. This study has followed about 80 000 survivors who were briefly exposed to the radiation. Cancer incidence in this population is compared with that in 26 500 age matched, non-irradiated controls. A total of 180 excess cancer deaths occurred over the period 1950–74, corresponding to an increase over control of about 5 per cent, although this figure is misleading since a large proportion of the irradiated population received low doses relative to the remainder. First, excess cancer has occurred mainly in the groups receiving a whole body dose of 1 Gy or greater and, secondly, the excess will undoubtedly increase after total lifetime follow up. A further 4-year follow-up has shown that the *total* cancer mortality from all causes has increased by 24 per cent. Figure 7.1 shows the *relative* risk of cancer mortality for various sites over the 1950–78 period. The excess leukaemia prominent over the first 20-year period is obviously still present although analysis has shown that it is no longer significant over the last 4-year period. Conversely, there is now evidence of increased rates for cancer of the stomach, lung, breast, and urinary tract and, for the first time, excess risk of colonic cancer and multiple myeloma. These data clearly illustrate the remarkably long latent period in some types of radiation-induced cancers. Final risk estimates will have to await the entire post-irradiation life span. Calculation of risks is also complicated by factors such as age at irradiation, sex, variation in dose response relationships for different cancers, and differences between incidence and mortality which may be influenced over a protracted period by improvements in therapy.

There is considerable debate in the UK over reports of increased

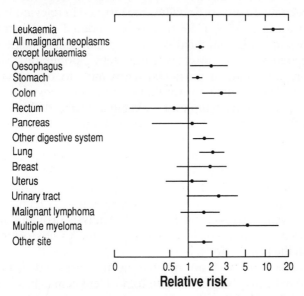

Fig. 7.1 Life Span Study: relative risk of mortality (90 per cent confidence intervals) for specific types of cancer (1950–78) (Kato and Schull 1982).

incidence of childhood leukaemia in the vicinity of some nuclear installations. In particular, small but statistically significant excesses of
leukaemia have been identified in the populations living near the two
nuclear-reprocessing plants at Sellafield in Cumbria and Dounreay in the
north of Scotland. The question as to whether or not these leukaemia
clusters arise from man-made sources of radiation, or even radiation at
all, is too complex to discuss here. (It is recognized for example, that
leukaemia clusters also exist in areas quite remote from nuclear installations.)

Background levels of radiation in the environment rose substantially in
the early sixties due to fall-out from nuclear weapons testing. At their
peak, these levels were greatly in excess of the measured background
levels in the vicinity of the reprocessing plants. It is argued (Darby and
Doll 1987) that if the leukaemia clusters *were* attributable to these
discharges then the relatively much larger doses from nuclear fall-out
received by the general population during the weapons testing period
should have caused a material increase some years later in the risk of
childhood leukaemia elsewhere in the country. Analysis of temporal
fluctuations of childhood leukaemia incidence in the UK as a whole and
in Scandinavia showed no convincing evidence of an increase in
incidence attributable to fall-out let alone any increase comparable with
that found around Dounreay. There are, however, some reservations and
uncertainties concerning the estimate of radiation dose arising from
ingestion of radionuclides from reprocessing waste, particularly if such
nuclides become disproportionally concentrated in relevant tissues.

Cook-Mozaffari *et al.* (1989) have considered the distribution of
cancer mortality in England and Wales. Estimates were made of relative
risk associated with social class, rural status, population size, health
authority region, and, in particular, proximity to one of 15 nuclear
installations. The results confirm that there is an excess mortality from
leukaemia, particularly lymphoid leukaemia, in the 0–24 age group in
districts with some of the population resident within 10 miles of a nuclear
installation, during 1969–78. The excess risk of persons under 25 years
is small but statistically significant for all leukaemias (relative risk,
$RR = 1.5$, $P = 0.01$) lymphoid leukaemia ($RR = 1.21$, $P = 0.01$), and
Hodgkin's disease ($RR = 1.24$, $P = 0.05$). There is also, however, a
significant deficiency of mortality from lymphoid leukaemia for persons
aged 25–64 years ($RR = 1.24$, $P = 0.05$). Analysis provided no positive
evidence that the increase in leukaemia is due to environmental pollution from the installations.

A hypothesis, in which an infective agent, unrelated to radiation, is
considered to be involved in the aetiology of leukaemia, has been investigated by Kinlen (1988). The nuclear-reprocessing plants at Dounreay

and Sellafield were built in unusually isolated place where, it is postulated, herd immunity to widespread viral infections tended to be lower
than average. The hypothesis is based on the premise that influxes of
populations into previously isolated areas lead to epidemics of certain
infections and that leukaemia is a rare response to some viral infections.
Sellafield and Dounreay were identified as extreme examples of such
communities. The two of Glenrothes in the rural district of Fife in
Scotland was identified as a unique test of the hypothesis. Glenrothes,
which is not near any nuclear installation, underwent a rapid expansion
during the fifties and sixties. The community remained relatively
isolated until the opening of a major road bridge in 1964. For the period
1951–67, there is a significant excess of leukaemia deaths below age 25
(10 observed and 3.6 expected). This excess is mainly accounted for by
deaths at ages below 5 in the relevant period 1954–9. No excess was
found for the period 1968–85. It is concluded that at least some of the
excess leukaemia cases near the nuclear installation at Dounreay and
Sellafield could be attributed to an infective agent mechanism since
these locations represent even more extreme degrees of isolation combined with population influx.

Medical exposure. The evidence for cancer induction by radiation used
for various medical treatments for non-malignant conditions is too substantial to review here, but a few general points can be made. Excess
thyroid cancers have been observed in children and young adults treated
with X-rays for enlarged tonsils, enlarged thymus, and nasopharyngeal
disorders, and ring worm of the scalp. There is equivocal evidence of
excess cancer in patients treated for hyperthyroidism with iodine[131]
(which is concentrated by the thyroid) probably because the epithelial
cells are killed by the high local doses of radiation.

Excess cases of leukaemia and cancers of the uterus, kidney, and
bladder have occurred in women treated with pelvic irradiation for non-
malignant gynaecological disorders. Leukaemia incidence, for example,
in one group of patients had a two- to threefold increase over that
expected. Excess breast cancer has also been reported in women who
received fairly high doses of X-rays for treatment of mastitis during the
period 1940–55, or during fluoroscopic examinations.

Two large current epidemiological studies are particularly significant.
During the period 1935–54, some 14 000 patients with ankylosing
spondylitis were given a course of X-rays to ameliorate pain associated
with the disease. A recent follow-up of earlier studies by Darby and
colleagues (1987) has shown that mortality from neoplasms other than
leukaemia or colon cancer increased by 28 per cent compared with that
in the general population. The proportional increase reached a maximum

(71 per cent) between 10 and 12.4 years post-irradiation and declined to only 7 per cent after 25 years. For leukaemia there was a threefold increase in relative risk overall which was at a maximum between 2.5 and 4.9 years after treatment. The risk was still nearly double that of the general population more than 25 years after treatment. There was evidence of increased risk for acute lymphatic and both acute and chronic myeloid leukaemia. The relative risk appeared to be greatest for acute myeloid leukaemia. There was also a 30 per cent increase in relative risk for colon cancer although this disease is associated with spondylitis through a common association with ulcerative colitis.

There is much concern over the risks arising from the ingestion of radionuclides such as radium and plutonium. These isotopes, which concentrate in the bones, decay by emission of energetic, short-range α particles which can irradiate various regions of the bone matrix. Relevant to this are the results of studies on patients all of whom were given multiple intravenous injections of radium[224] over a period of months to treat tuberculosis and ankylosing spondylitis. This isotope is short-lived (half-life of 3.6 days) and therefore decays while still on bone surfaces. In one study involving 681 adults and 218 juveniles (0–20 years) 18 and 35 cases of bone sarcoma, respectively, were reported, where only an 0.2 incidence rate would have been expected (Gössner et al. 1985). Five cases of leukaemia have been reported in the higher age group but none occurred in the younger group. Two cases would have been expected in a normal population of comparable size, age, distribution, and follow-up time. It is pointed out, however, that the excess cases may not have been caused by the radiation in view of the possible leukaemogenic effect of the pain-killing drugs taken by spondylitic patients during the relevant period.

The induction of acute myeloid leukaemia (AML) has also been observed in experimental animals treated with such radionuclides. Deposition of the isotope on bone surfaces can lead to heavy irradiation of the marrow. Figure 7.2 is a neutron autoradiograph of plutonium[239] in trabecular bone and marrow of the mouse lumbar vertebrae. A thin section of bone is mounted on plastic sheet and placed in a nuclear reactor. Neutron induced fission fragments of the plutonium and other fragments damage the plastic. After etching of the plastic, the fission fragment tracks show as black lines. Irradiation of the bone marrow is clearly evident.

7.2.3 Dose response relationships

The epidemiological data on radiation-induced leukaemia generally, its relatively short latency period, and the substantial evidence available

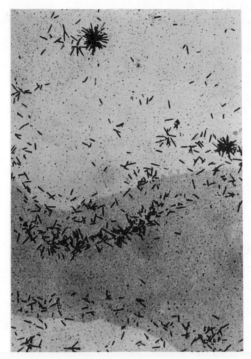

Fig. 7.2 Neutron induced autoradiograph of α particle tracks from plutonium[239] in mouse bone (see text).

from experimental laboratory studies, combine to make leukaemia the most suitable model for studying dose response relationships. Such information is essential for an understanding of the molecular and cellular aspects of radiation carcinogenesis as well as the calculation of risk for radiological protection purposes.

Figure 7.3 compares the dose response for mortality from all forms of leukaemia as observed in the ankylosing spondylitis study with that for induction of acute myeloid leukaemia in male CBA/H mice. The more precise experimental data clearly show an initial rise with increasing dose followed by a decrease at even higher doses. The human data are consistent with this type of response, as indeed are data for several other radiation-induced human and animal tumours. The overall response curve is influenced by two factors. The probability of a malignant transformation at the *cellular* level rises with increasing dose. However, when the dose is sufficient to sterilize (kill or prevent cell division) some cells, the number that survive, and are therefore capable of transformation, *falls* with increasing dose. The counterbalancing of these two effects results in the overall dose response for cancer induction.

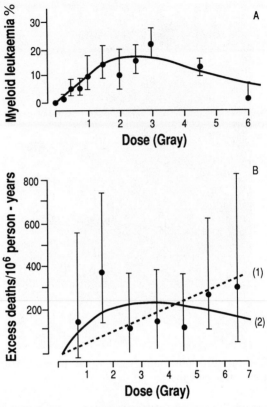

Fig. 7.3 Comparison of dose response data for induction of myeloid leukaemia in male CBA/H mice after brief exposures to 250 kVp X-rays (A) (Mole *et al.* 1983) with excess mortality rate from leukaemia (as a function of mean bone marrow dose) in irradiated spondolytic patients (B) (Smith and Doll 1982). Line 1, linear dose response relationship; line 2, linear dose response relationship with cell killing component included.

Various empirical expressions have been used in attempts to describe the overall dose response relationships for human cancer induction although the lack of accurate data on radiation dosage is often a problem. For the experimental AML data in CBA/H mice, the curve fits reasonably well the expression $P = \alpha D^2 e^{-\lambda D}$ where P is the probability of induction, D is the radiation dose, and α and λ are constants. Knowledge of the response relationship for inactivation of the cells at risk would permit derivation of the response relationship for neoplastic transformation.

7.3 Cellular and molecular processes

7.3.1 *Cell inactivation*

A radiation dose of about 1 Gy leads to about 2×10^5 ionizations within the mammalian cell, of which approximately 1 per cent occur in the genomic material. A primary consequence is breakage of DNA strands. Of the 1000 or so strand breaks that occur, almost all disappear within a few hours, either by spontaneous rejoining, or by enzyme mediated repair (see Chapter 5). Some breaks remain, possibly as aligned double-strand breaks, and these are the major cause of loss of viability of some cells. Nevertheless, at this dose, some 40 per cent of the irradiated cells retain the capacity for growth despite the large amount of chemical damage sustained by the cells. Much of the initial chemical damage caused by the radiation is therefore of no consequence to the fate of the cell. Only a very small part of the damage, caused perhaps by the fairly rare local deposition of a large amount of energy near critical molecular sites, is important.

Useful information has been obtained from studies of the relative cytotoxic, mutagenic, and transforming abilities of radiations of different LET. Figure 7.4 shows typical dose response 'survival' curves for cells in tissue culture irradiated with either low LET X-rays or high LET α

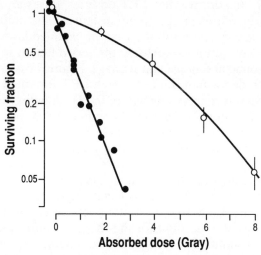

Fig. 7.4 Survival curves for V79 cells irradiated with low LET γ rays ($\bigcirc$) and high LET γ particles ($\bullet$) (from Thacker *et al.* 1982).

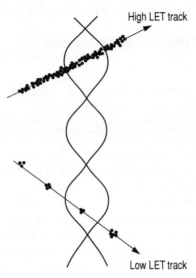

Fig. 7.5 Simple schematic illustration of the different patterns of energy depositions arising from low LET and high LET radiation tracks traversing a section of the DNA helix.

particles. High LET radiations invariably are more effective and frequently show an exponential dose relationship (linear on the semi-log plot in Fig. 7.4). In contrast, low LET dose responses are curved when plotted similarly although, in some instances, the initial curvature observed at lower doses is followed by an exponential response. The LET differences in response are due to the very different spatial patterns of the initial atomic and molecular damage. Figure 7.5 shows schematically the energy deposition tracks for high LET α particles and typical low LET X-rays traversing a segment of the DNA helix. Some typical α particles deposit energy at the rate of about 100 keV per micron track length, which means that many ionizations will occur when the particle track passes through the DNA. The energy deposition rate is very much less for X-rays. There is a greater probability of double strand break formation from a single α particle track than there is from either a single X-ray track or from the alignment of two single strand breaks arising from two X-ray tracks. Many physical models based on such processes have been proposed to account for the shapes of dose response curves based on the 'accumulation' of damage or 'interaction' of sub-lesions. There are, however, alternative explanations.

Cells irradiated at low dose rates or with intermittent radiation show reduced inactivation. The effect of such treatments on the shape of dose response curves has been explained on the basis of enzyme mediated 'restoration' or 'repair' processes. There is direct evidence that single-strand breaks are much more readily repaired than double-strand breaks and therefore repair processes should be more efficient for low LET irradiation. The presence of a shoulder on low LET and survival curves is consistent with the existence of repair mechanisms which become inactivated, or more probably saturated, at higher doses.

7.3.2 *Chromosome damage and cell mutation*

Chromosomal aberrations. Radiation causes a variety of structural aberrations in mammalian chromosomes only some of which are lethal. Most of these aberrations appear to result from interactions between two or more lesions and can be conveniently grouped as exchanges (interchanges, intra-arm intrachanges, and inter-arm intrachanges) and breaks or discontinuities. Not all such aberrations lead to cell death nor do those that are transmissible necessarily have detectable genetic consequences. The position of the cell in its cycle (or its status in regard to DNA duplication) leads to two types of aberration observed at metaphase (see Chapter 10). These are *chromosome type*, where both sister chromatids are involved in exchange for a given locus, and *chromatid type* where only one is affected. Radiation produces both types of aberration in contrast to most chemical chromosome damaging agents which cause only the latter type of aberration, as a rule.

Since cell sterilization and chromosome aberrations involve damage at the DNA level, one might expect some common types of behaviour in radiation dose response relationships. The efficiency of aberration induction for low LET radiations falls at low dose rates and for intermittent, or fractionated, radiation. Figure 7.6 compares the efficiencies for the formation of asymmetrical interchanges (dicentric aberrations) in human lymphocytes irradiated with single doses of either γ rays or fission neutrons. As for cell inactivation, the high LET neutrons are substantially more effective than the low LET γ rays on an equi-dose basis.

Single gene mutation and DNA repair. The scale of the initial radiation chemical lesions and that of the chromosomal aberrations observed at the light microscope level are separated by several orders of magnitude. The considerable length of nuclear DNA associated with the nucleosomes is duplicated and packaged in a very precise configuration in 25 nm fibres which are themselves folded and coiled into the structure of the metaphase chromosome (see Chapter 5). The small scale molecular changes which reveal themselves as major changes in chromosomal structure must involve considerable amplification. Further, it is likely

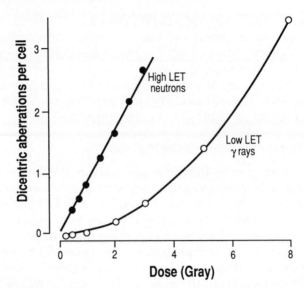

Fig. 7.6 The different efficiencies of high LET fission neutrons and low LET γ rays in causing dicentric chromosome aberrations in human lymphocytes (combined data of Lloyd *et al.* 1975, 1976).

that modification of some of the initial molecular changes will occur long before these chromosomal changes are observable. Studies at the single gene level provide a much higher degree of resolution.

Radiation-induced mutation in specific genes in target somatic cells is one plausible explanation for the initiation of cancer. It is useful therefore to consider briefly some of the molecular characteristics of mutations induced by radiation in somatic cells cultured *in vitro*. An end-point frequently used in radiation mutagenesis is the induction of thioguanine resistance in various mammalian cell lines. Resistance is due to the induced reduction in activity of the enzyme hypoxanthine phosphoribosyl transferase (HPRT) whose normal function is to enable cells to incorporate purines from the growth medium (see Chapter 6). The coding gene is located on the X chromosome and therefore only one gene alteration may be required to show mutant behaviour. This allows elevation of mutation frequencies to be observed at fairly low radiation doses. Indirect evidence from various biochemical, cytogenetic, and immunological studies (Thacker 1986) have indicated that the majority of radiation-induced mutations at the HPRT locus are associated with deletions and/or rearrangement of DNA sequences within the HPRT region. In general terms, the evidence indicates that radiation-induced *point* mutation at this locus is either infrequent or masked by large rearrangements and deletions. This contrasts with mutation at HPRT by both ultraviolet (u.v.) light and by chemicals where point mutation appears to be the dominant mechanism. It is premature, in any case to consider HPRT mutations to be at all representative mechanistically of radiation induced mutations elsewhere in the genome. Evidence, for example, that radiation induces *point* mutation at the autosomal *APRT* locus suggests that the particular characteristics of a genomic domain may influence mechanisms of radiation mutagenesis and hence onco-genesis also. This underlines the caution necessary in extrapolation of mechanisms of mutagenesis uncovered in model genetic systems *in vitro* to mechanisms of oncogenesis *in vivo* particularly involving mutations in classes of cellular genes that are still poorly understood.

It is interesting that the efficiencies of radiation in causing cellular inactivation and mutation often appear to be quantitatively related. An example is shown in Fig. 7.7, again for mutation at the HPRT locus. The plot of mutation frequency corrected for cell lethality, against the logarithm of the fraction of the irradiated population that survives the radiation, is linear. This relationship is not affected by variations in the shapes of the individual radiation dose response curves for both mu-tation and cell kill between the different cell types. It is difficult to reconcile this behaviour in terms of two independent lesions leading either to cell death or to mutation. Thacker has argued that the repair process used by the cell in attempting to deal with the initial radiation damage is not entirely error free. If the probability of the repair system for *changing* the genetic material relative to that for *elimination* of the lesion in order to give non-mutant survivors is fixed, this would explain the constancy of the slope for the data in Fig. 7.7. To what extent such behaviour holds for high LET radiations is not fully explored but it is

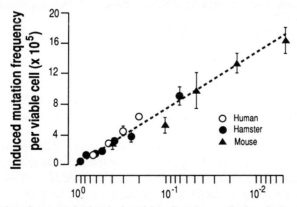

Fig. 7.7 Mutation survival relationships for the radiation induction of thio-guanine resistant mutants of cells from different species (from Thacker 1979).

significant that, as is the case with radiation lethality, the efficiency of radiation mutagenesis increases at high LET.

Ionizing radiation acts fairly non-specifically to produce a very broad spectrum of types of molecular damage in DNA. The question as to which of these many different molecular lesions are biologically important remains largely unanswered. However, applications of gene transfer and recombinant DNA techniques to problems of mammalian cell radiosensitivity and DNA-repair has yielded important information, particuarly in regard to the role of DNA strand scissions in cellular radiosensitivity.

Studies on the autosomal recessive human genetic disorder, ataxia telangiectasia (AT) are particularly relevant to the rôle of DNA repair in oncogenesis. AT is a disease which is characterized mainly by neuro-motor dysfunction, immunodeficiency, proneness to T-cell neoplasia, and high radiosensitivity both *in vitro* and *in vivo*. Transfer of plasmid-encoded genes containing inactivating DNA strand scissions into both normal human and AT cell lines have shown that AT cells exhibit greatly elevated *mis-repair* of these scissions. This readily explains both the radiation sensitivity and apparent repair deficiency associated with the disorder (Cox *et al.* 1986). It is suggested that such mis-repair might thereafter affect sequence specific DNA recombination processes associated with maturation of the B-cell immunoglobulin and T-cell receptor systems and be implicated therefore in the immunodeficiency and the specific chromosomal rearrangements that characterize T-cell neoplasia in AT. There is evidence in support of a close link between repair deficiency, enzyme (recombinase) function, and T-cell neoplasia, firstly in that a chromosomal translocation (t7; 14) observed in a T-cell leukaemia in an AT patient involved aberrant recombination of the T-cell receptor β-gene and secondly, reduced DNA topoisomerase II activity in AT cells.

7.3.3 *Radiation transformation and oncogenesis*

There are clear dose response interrelationships for radiation-induced cell inactivation, chromosomal rearrangements, and mutagenesis. One may ask: Are such interrelationships of value in understanding mechanisms of radiation carcinogenesis?

There is now much research directed towards developing cellular systems *in vitro* that can be used to describe events associated with oncogenic transformation induced by radiation (see Hall 1988). In some established cell lines, neoplastic transformation can be induced by various agents, including some chemicals and radiation. Such trans-formed cells can induce tumours when re-implanted into the animal from which they were originally derived, or when transplanted into immuno-logically compatible hosts. Transformed cells *in vitro* usually display characteristics such as loss of anchorage dependence *in vitro*, loss of contact inhibition, and changes at the DNA level.

Radiation transformation *in vitro* was first observed by Borek and Sachs in 1966 using short-term cultures of Syrian hamster embryo cells. Transformation is indicated by the appearance of 'piled up' colonies on the culture plates. The $C3H/10T_{1/2}$ cell line (see Chapter 6) is also used frequently for radiation transformation studies. These fibroblast-like cells show normal contact inhibition when grown in confluent monolayer but radiation treatment causes morphological changes. Colonies appear that overgrow the confluent layer and cells from these colonies can give rise to fibrosarcomas after inoculation into the appropriate animal host. Transformation efficiency can be expressed as a frequency per irradiated cell or per surviving cell. A disadvantage of this system is that the cells are less stable than the cells of the short-term hamster embryo cell system and spontaneous transformation may occur. Studies with this and other transforming systems show decreased transformation efficiency with decreasing radiation LET. This is in line with the trends for cell lethality, chromosomal aberration, and gene mutation described earlier. Dose response curves for γ-irradiated $10T_{1/2}$ cells show the 'bell-shaped' curves sometimes observed in experimental and human carcinogenesis (Section 7.2.2 and Fig. 7.3). This is illustrated by the transformation data in Fig. 7.8 for $10T_{1/2}$ cells exposed to γ radiation. Correction of the bottom curve, which is expressed as transformants per exposed cell, to take account of those cells that do not survive the treatment, gives the dose response curve for the remaining cells that are at risk.

7.3.4 *Molecular studies in radiation oncogenesis*

Although studies on DNA repair and mutagenesis provide valuable information on processes that may underlie the initiation of oncogenesis, the principal approach to the problem now centres largely on direct molecular analysis of somatic cells carrying lesions that may be associated with the neoplastic phenotype.

In the $10T_{1/2}$ cell line, utilization of DNA-mediated gene transfer techniques have shown that the transformed phenotype in $10T_{1/2}$ cells can be transferred to murine NIH-3T3 cells (Borek *et al.* 1987) (see Chapter 9). The identity of the specifically oncogenic DNA sequences in $10T_{1/2}$ cells remains to be established.

A major problem in the study of radiation oncogenesis at the cellular and molecular level remains the limited available cell lines. Mostly these are immortalized rodent fibroblast lines and the relevance of single cell

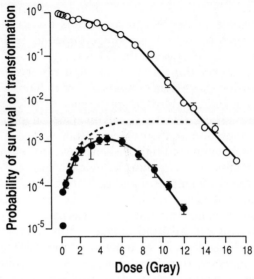

Fig. 7.8 Radiation survival (○) and transformation (●) data for $10T_{1/2}$ cells exposed to cobalt[60] γ rays (1 Gy/min) (from Elkind *et al.* 1983). Transformation frequency corrected for cell loss due to inactivation is represented by -----.

responses to the more complex 'whole organ' responses that influence oncogenesis *in vivo* is a problem that is not just confined to radiation research. Nevertheless, these problems are recognized and much effort is currently devoted to the development of better model systems *in vitro*.

Radiation can activate oncogenic viruses under some circumstances. The so-called radiation leukaemia virus (RadLV) is a member of a class of retroviruses (see Chapter 8) first isolated as a leukaemogenic activity in cell-free extracts from radiation-induced thymic lymphomas in mice. Fractionated whole body X-irradiation can produce up to 100 per cent lymphoma incidence in C57BL/Ka mice. These tumours express the virus in the primary tumour and in serial syngeneic transplants. Injection of cell-free extracts intrathymically gives rise to identical tumours in the same strain of mouse. The induction of myeloid leukaemia in RFM/Vn mice following irradiation is also associated with virus infection.

Ras oncogene activation occurs in some murine thymic lymphomas induced by radiation. Whilst some degree of codon specificity has been reported to be associated with Ki-*ras* activation, it is still not possible to relate these events directly to the initiation of neoplasia. It is feasible that they are associated with some late event in neoplastic development (virus and oncogene activation are discussed in detail in Chapters 8 and 9).

The problem of the interpretation of specific molecular changes in overt neoplasms is compounded in radiation oncogenesis because radiation is so non-specific in its capabilities for damaging DNA. Radiation causes many types of chemical damage in DNA of which only a few may be relevant to oncogenesis. An alternative approach to the problem of initiation of radiation oncogenesis lies in the search for chromosomal markers in preneoplastic cells. In essence, if such neoplasia-specific chromosomal effects exist, it may be possible to track these back in animal systems to the immediate post-irradiation phase in the target organ.

Trisomy of chromosome (ch) 15 is known to occur in a high proportion of spontaneous and induced murine thymic lymphomas and there is evidence for duplication of a specific region of ch15 containing the *myc* and *put-1* oncogenes. Trisomy 15 may be detected also in thymocytes from irradiated mice 12–18 weeks after exposure. In some animals, however, a translocation marker (tl; 5) is observed at earlier times. It is suggested that this translocation is an early event in lymphomagenesis whereas trisomy 15 may only contribute to neoplastic progression (McMorrow *et al.* 1988).

Rearrangement of a specific interstitial region of ch2 is a common feature of radiation-induced actute myeloid leukaemia (AML) in various mouse strains. It has been shown (Silver *et al.* 1987) using in-vitro X-irradiation and bone marrow transplantation, that rapidly proliferating multi-potential haemopoietic cell clones carrying such ch2 rearrangements can be observed in the marrow of recipient mice within five days of transplantation. It was also found that the haemopoietic growth factor gene interleukin (IL1) is located close to the critical ch2 breakpoint and that this gene is over-expressed in cases of AML carrying the characteristic ch2 rearrangement. In this respect it is interesting that a high proportion of human AML patients similarly overexpress IL1 suggesting that changes in this gene may play an important part in the early phase of the genesis of AML.

7.3.5 *Some free radical aspects of radiation carcinogenesis*

Radiation carcinogenesis is a multistage process conveniently divided into *initiation* and *promotion* phases. Free radical processes are involved in both. The initial intracellular chemical damage caused by radiation must be mainly free radical in nature since the local energy deposition is greatly in excess of the normal bond energies of all the affected molecules. Many radiation chemical studies on DNA systems *in vitro* have led to a fairly complete knowledge of the various types, structures, and reactions of DNA free radicals produced by radiation action. The fundamental problem that remains to be solved, a problem common to the molecular action of other types of carcinogens, is the identification of the *specific types* of free radical chemical damage critical to the onset of the multistage process of carcinogenesis.

Radioprotectors. It has long been established that various types of molecules containing the reactive sulphydryl group, –SH, influence radiation response. The effects include protection against: (i) cell lethality and mutagenesis *in vitro*, (ii) acute radiation morbidity *in vivo*, and (iii) late effects including radiation-induced life shortening. There is much evidence that these protective effects are directly due to the relative weakness of the sulphur–hydrogen bond in the sulphydryl compounds. Free radicals produced either by the radiation or in subsequent chemical reactions can be 'restored' or 'repaired' by transfer of a hydrogen atom from the sulphydryl group of the protector. Such processes have been observed directly in various radiation chemical model systems using fast response pulse radiolysis techniques i.e.

$$X^{\cdot} \quad + \quad RSH \quad \rightarrow \quad XH + RS^{\cdot}$$
$$\text{(free radical)} \qquad \text{(protector)}$$

Sulphydryl compounds can also inhibit radiation-induced carcinogenesis. The effect is complicated, however, by the protective effect of these agents against cell lethality and life shortening. The reduction in the number of potential tumour cells and the shortening of the time in which late tumours can be expressed influences their radiation-induced incidence. Protection against both these effects can lead to an apparent *increase* in incidence and this has been observed for some tumours. However, there is more consistency in the data for radiation-induced thymic lymphomas. Figure 7.9 shows the protective effect of a mixture of radioprotective compounds (most containing sulphydryl groups) on the incidence of thymic lymphomas in irradiated C57 mice.

Superoxide dismutase. Evidence that free radicals are involved in the promotional phases of radiation carcinogenesis comes from studies with the enzyme superoxide dismutase (SOD). This enzyme is a powerful catalyst for the removal of the superoxide radical-anion (O_2^-) ultimately formed in many cellular electron transfer processes. It has been proposed that this radical (or one of its reaction products, the hydroxyl radical) is highly damaging to the cell and that SOD has evolved as a natural cellular defence mechanism against free radical injury. There is evidence that this enzyme inhibits radiation transformation in the $10T_{1/2}$ and in the hamster embryo systems but, interestingly, does not need to be present at the time of irradiation. It does require, however, *prolonged* post-irradiation exposure. The hypothesis that free radicals are involved in the promotional phase of radiation carcinogenesis is supported by

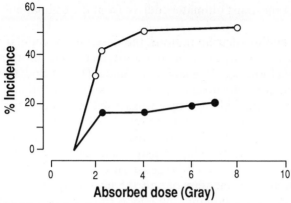

Fig. 7.9 Incidence of thymic lymphoma in C57BL male mice irradiated with 650 R X-rays (○). Protective effect of a mixture of radioprotective agents (●) (from Fry 1983).

transformation experiments using chemical promoters and SOD. The phorbol ester tumour promoter, TPA for example, substantially increases radiation-induced cell transformation but the effect is completely eliminated by treatment of the cells with SOD. The mechanisms of inhibition of free radical processes involved in tumour promotion are obscure but a plausible explanation may lie in the ability of SOD to inhibit free radicals involved in lipid peroxidation in cellular membranes.

7.3.6 *Interaction of radiation and chemical carcinogens*

Of particular concern in the field of radiological protection and the assessment of risk is the question of possible interaction between radiation and environmental carcinogens. Many experimental in-vivo studies have been carried out to assess interactions between radiation and chemical carcinogens. Overall, the evidence is equivocal. A complicating factor is the life shortening effect of radiation which may mask any interactive influence of a chemical carcinogen. Nevertheless, there are experimental data indicating positive interactions particularly after fetal exposure. The carcinogen ethylnitrosourea (ENU) can act transplacentally in inducing various tumours in the offspring of mice treated with the chemical before birth. Leukaemia incidence is greater after both ENU and X-irradiation than with each agent alone. Interestingly, the incidence of radiation induced ovarian tumours is also increased by exposure to ENU even though no such tumours were observed when ENU was administered without radiation. While such studies may highlight possible additional risks associated with combined exposure to radiation and chemical carcinogens, progress in understanding the mechanisms of such interactions must ultimately derive from the cellular and molecular approach.

Notwithstanding their limitations, the *quantitative* cellular transformation systems currently available have provided some information. For a general discussion see Nygaard and Simic (1983) and Cerutti *et al.* (1987). Radiation transformation *in vitro* can be potentiated by various chemical agents including some drugs used in cancer chemotherapy, tumour promoters such as TPA, thyroid hormone extracts and particularly, pyrolysates of protein foods such as Try-P-2 (3-amino-1-methyl-5H-pyrido-4,3-b-indol). The transforming efficiency of 1.5 Gy of X-irradiation in hamster embryo cells, for example, is increased 20-fold by the addition to the medium of only 0.5 µg/ml Trp-P-2.

Information on the potentiating effects (and indeed inhibiting effects) of various other chemical agents on radiation-induced cellular transformation is steadily accumulating. The major problems in applying such information to the problems of human radiation carcinogenesis are the limitations of the *quantitative* transformation assay systems currently available. Established rodent cell lines of fibroblastic origin may not be appropriate models for investigating human radiation carcinogenesis since human tumours arise mainly from epithelial cells. Transformed human epithelial cells can behave quite differently from transformed cells of rodent origin. Nevertheless the rapid accumulation of knowledge on the various stages involved in the transition of a normal primary cell to a frankly malignant phenotype will eventually lead to the establishment of an assay system that is sufficiently quantitative for studies in radiation carcinogenesis. When that point is reached, there will be a much firmer basis for assessing radiation risk in the human population.

References and further reading

BEIR (1980). National Academy of Sciences. National Research Council. Committee on the Biological Effects of Ionizing Radiations. *The effects on populations of exposure to low levels of ionizing radiation: 1980.* National Academy Press, Washington, DC.

BEIR (1988). National Academy of Sciences. National Research Council. Committee on the Biological Effects of Ionizing Radiations. *Health risks of radon and other internally deposited alpha-emitters.* BEIR IV. National Academy Press, Washington, DC.

Black, D. (1984). Report of independent advisory Group: *Investigation of the possible increased incidence of Cancer in West Cumbria.* HMSO, London.

Borek, C., Ong, A., and Mason, H. (1987). Distinctive transforming genes in X-ray transformed mammalian cells. *Proceedings of the National Academy of Sciences, USA.* **84**, 794–8.

Borek, C. and Sachs, L. (1966). In vitro cell transformation by X-irradiation. *Nature* **210**, 276.

Cerutti, P. A., Nygaard, O. F., and Simic, M. G. (ed.) (1987) *Anti-carcinogenesis and radiation protection.* Plenum.

Cook-Mozaffari, P. J., Darby, S. C., Doll, R., Forman, D., Hermon, C., Pike, M. C., and Vincent, T. (1989). Geographical variation in mortality from leukaemia and other cancers in England and Wales in relation to proximity to nuclear installation. 1969–78. *British Journal of Cancer* **59**, 476–85.

Cox, R., Debenham, P. G., Masson, W. K., and Webb, M. B. T. (1986). Ataxia-telangiectasia: a human mutation giving high frequency misrepair of DNA double strand scissions. *Molecular Biology and Medicine* **3**, 229–44.

Darby, S. and Doll, R. (1987). Fallout, radiation doses near Dounreay, and childhood leukaemia. *British Medical Journal* **294**, 603–7.

Elkind, M. M., Han, A., Hill, C. K., and Buonaguro, F. (1983). Repair Mechanisms in radiation-induced cell transformation. In: *Proceedings of the 7th International Congress of Radiation Research* (ed. J. J. Broerse, G. W. Barendsen, H. B. Kal, and A. J. van der Kogel), pp. 33–42. Martinus Nijhoff, Amsterdam.

Fry, R. J. M. (1983). Radiation carcinogenesis: radioprotectors and photo-sensitizers. In: *Radioprotectors and Anticarcinogens* (ed. O. F. Nygaard, and M. G. Simic), pp. 417–36. Academic Press, New York.

Gray, L. H. (1965). Radiation biology and cancer. In: *Proceedings of the XVIII Annual Symposium on Fundamental Cancer Research*, pp. 7–25. Williams and Wilkins, Baltimore.

Gössner, W., Gerber, G. B., Hagen, U., and Luz, A. (ed.) (1985). *The radiobiology of radium and thorotrast.* Urban and Schwarzenberg.

Hall, E. J. (1988). *Radiobiology for the radiologist* (3rd edn). Lippincott.

Kato, H. and Schull, W. J. (1982). Studies of the mortality of A-bomb survivors. 7. Mortality, 1950–78, Part 1, Cancer mortality. *Radiation Research* **90**, 395–432.

Kinlen, L. (1988). Evidence for an infective cause of childhood leukaemia. Comparison of a Scottish new town with nuclear reprocessing sites in Britain. *Lancet* **i**, 1323–6.

Lloyd, D. C., Purrott, R. J., Dolphin, G. W., Bolton, D., Edwards, A. A., and Corp, M. J. (1975). The relationship between chromosome aberrations and low LET radiation dose to human lymphocytes. *International Journal of Radiation Biology,* **28(1)**, 75–90.

Lloyd, D. C., Purrott, R. J., Dolphin, G. W., and Edwards, A. A. (1976). Chromosome aberrations induced in human lymphocytes by neutron irradiation. *International Journal of Radiation Biology* **29(2)**, 169–82.

McMorrow, L. E., Newcomb, E. W., and Pellicer, A. (1988). Identification of a specific marker chromosome early in tumour development in γ-irradiated C57BL/6J mice. *Leukaemia* **2**, 115–19.

Mole, R. H., Papworth, D. G., and Corp, M. J. (1983). The dose-response for X-ray induction of myeloid leukemia in male CBA/H mice. *British Journal of Cancer* **47**, 285–91.

National Radiological Board (1988). Bulletin No. 89.

Nygaard, O. F., and Simic, M. G. (ed.) (1983). *Radioprotectors and anticarcinogens* Academic Press, New York. (Collected papers.)

O'Riordan, M. C. (1988). *Notes on radon risks in homes,* Radiological Protection Bulletin No. 89, February 1988, National Radiological Protection Board, Chilton, UK.

Silver, A. R. J., Breckon, G., Masson, W. K., Malowany, D., and Cox, R. (1987). Studies on radiation myeloid leukaemogenesis in the mouse. In *Radiation research,* Vol. 2 (ed. E. M. Fielden *et al.*) pp. 494–500. Taylor and Francis, London.

Smith, P. G., and Doll, R. (1982). Mortality among patients with ankylosing spondylitis after a single treatment course with X-rays. *British Medical Journal* **284**, 449–60.

Thacker, J. (1979). The involvement of repair processes in radiation-induced mutation of cultured mammalian cells. In: *Radiation Research Proceedings of the 6th International Congress of Radiation Research* (ed. S. Okada), pp. 612–20. Japanese Association for Radiation Research, Tokyo.

Thacker, J. (1986). The use of recombinant DNA techniques to study radiation-induced damage, repair and genetic change in mammalian cells. *International Journal of Radiation Biology* **50**, 1–30.

Thacker, J., Stretch, A., and Goodhead, D. T. (1982). The mutagenicity of α particles from plutonium-238. *Radiation Research* **92**, 343–52.

UNSCEAR (United Nations Scientific Committee on Effects of Atomic Radiation) (1988). Sources, effects and risks of ionizing radiations. 1988 Report to the General Assembly, UN, New York.

8

Viruses and cancer

J. A. Wyke

8.1 Introduction

Different forms of human cancer show marked geographic variations in incidence that mainly reflect social rather than genetic differences in the populations at risk. This suggests that variable environmental factors are responsible for a great deal of cancer (Chapter 3). These environmental risk factors comprise three categories: (i) physical agents (such as X-rays or u.v. light, Chapter 7), (ii) chemical agents (either directly carcinogenic or converted to carcinogens in the body, Chapter 6), and (iii) infectious agents. Of the infectious agents, bacteria, fungi and parasitic animals have all been considered as potential carcinogens, but it is viruses that have received the most attention as risk factors in neoplasia of verte- brates. This has been justified for two major reasons. First, viruses are important causes of cancer in certain animals and they are being increas- ingly implicated in human neoplasia. Secondly, the laboratory study of tumour viruses has led to important insights into the mechanisms of carcinogenesis. These, however, are accolades conferred only with the benefit of hindsight and before we consider our present knowledge of virus associated cancer in more detail it is worth surveying the develop- ment of the subject. It is an instructive tale, demonstrating the stimula- tory effects on research of advances in apparently unrelated areas as well as the stultifying influence of prejudice on the one hand and uncritical oversimplification on the other.

8.2 The history of tumour virology

8.2.1 *Pioneer days*

The first viruses were discovered towards the end of the nineteenth century as very small infectious agents, pathogenic for both plants and animals, that passed through filters capable of retaining the smallest bacteria. It was not long before comparable filtrable agents were found to cause tumours in animals. The first of these, discovered by Ellermann and Bang in 1908, induced erythroblastosis (erythroid leukaemia) in chickens. Interest in this finding was muted by contention over whether the disease was a true neoplasm or a hyperplasia, but it was soon shown by Rous in 1911 and Fujinami and Inamoto in 1914 that viruses could induce true tumours, sarcomas, in fowls. This too created relatively little interest in the scientific community at large, perhaps because the diseases of chickens were thought to have little relevance for those of man, and for 20 years the study of chicken tumour viruses was an esoteric pursuit.

Nonetheless, this period saw advances in other fields that would later become very significant. Selective breeding of laboratory mice produced some with high incidences of various cancers. This enabled Bittner in 1936 to show that the high incidence of mammary carcinomas in some strains was due to transmission of a filtrable virus. In the same decade, Rous, Shope, and others produced interesting studies on the virus induced papillomas and carcinomas of rabbits, and work with chicken viruses gathered momentum. The next major advance occurred in 1951 when Gross discovered the first mouse leukaemia virus. This early post-war finding was not the beginning of a new era in tumour virology but the end of an old one. By the beginning of the 1960s our views on tumour viruses were changing rapidly. The momentum behind this change had several sources, the main ones being our growing understanding of both viruses and neoplastic growth and the development of techniques to advance this understanding by studies in tissue culture.

8.2.2 *The nature of viruses*

Viruses vary enormously in structure and complexity but they all share certain features that distinguish them from other forms of life. They do not have a cellular organization that is propagated by division of the whole entity, but can multiply only by replication of their genetic material (genome) (Brown and Wilson 1984). The genome comprises either RNA or DNA and the proteins it encodes usually serve in genome replication or as structural components that protect the genome and facilitate its spread from host to host. Depending on the complexity of the virus, host functions may or may not be required to aid genome replication and transcription, but all viruses require host ribosomal functions for translation of their messenger RNA and thus they are all obligatory intracellular parasites. This enforced intimacy with their host can take many forms, ranging from cytolytic viruses that overwhelm host functions, replicate rapidly, and kill the cell, to latent forms that seldom express their own functions and whose genome is replicated, in concert with that of the cell, by the host's own machinery. As we shall discuss later, this close symbiosis is also the reason why some viruses can cause cancer.

Some important human diseases are caused by viruses and there have been strong incentives to understand, and act against, these pathogens. Such stimuli led to the development of laboratory systems for detecting, growing, and quantifying viruses, notably the use of tissue cultures to examine virus cytopathic effects (CPE). An important development was a virus assay, based on CPE, in which virus spread was limited and localized plaques of dead cells were thus produced on a layer of tissue culture cells. When similar assays were tried during the 1950s with certain tumour viruses, it was found that, instead of plaques of CPE, the viruses induced focal areas of piled up and morphologically altered cells (Fig. 8.1). Such behaviour was not entirely unexpected, for in the late 1930s it was shown that chicken sarcoma viruses could produce comparable changes in organ explants and on the chorioallantoic membrane

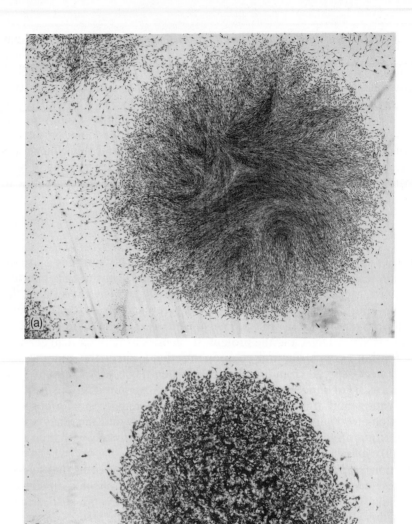

Fig. 8.1 Cell transformation by tumour viruses. (a) Shows a colony of un-infected cells of the Syrian hamster cell line BHK21/C13. The colony is flat and the cells are aligned in parallel array. In (b) a colony of the same cell line is shown after transformation with the small DNA virus, polyomavirus. The cells are piled up and disorientated.

of eggs. This tissue culture cell 'transformation' by tumour viruses provided a ready means to examine and quantify the effects of tumour viruses on cells, and it proved a great impetus to further research.

8.2.3 *Cell transformation: 'tumours' in* vitro

The full significance of the discovery that some (by no means all) tumour viruses can transform cells *in vitro* can be appreciated by considering concepts then current on the nature of cancer. The availability of inbred laboratory animals, which facilitated the identification of some tumour viruses, also permitted experiments on the transplantation of tumours, and it was possible to show that with some tumours a single transplanted cell could cause a tumour in a previously normal animal. This focused attention on cancer as a disease of the single cell, a concept that accorded with a clonal origin of many cancers. Transplantation studies also demonstrated the stable heritability of the tumour phenotype and this, together with the knowledge that many carcinogens were also mutagens, emphasized the possibility that the initiating defect in cancer may often be a mutation in the genome of a somatic cell. The problem with this concept was demonstrating its validity. Every cell in a complex vertebrate contains tens of thousands of genes, so identifying the one or a few that are altered in cancer seemed a hopeless task.

Cell transformation by tumour viruses seemed to offer a way round this impasse. The speed and efficiency of this event suggested to workers in the 1960s that it resulted directly from the functioning of a virus gene. Thus, to find a gene whose activity led to cancer one had only to sift through the genome of a virus rather than that of a cell. Since some tumour viruses contained only enough nucleic acid to encode three or four genes, this simplified the quest over 10 000-fold, bringing its potential achievement within the ambit of the techniques then available. This reasoning encouraged many researchers to investigate the genetics and biochemistry of virus induced cell transformation and not only has this led to the discovery of a number of tumour inducing genes in viruses but studies on the nature of these genes have had enormous conceptual implications for the whole of basic cancer research (see p. 226, and Chapter 9). It should be remembered, however, that these important advances had an element of serendipity in them for, as we shall discuss below, there are ways in which a virus might induce cancer that do not result directly from the action of one of its genes inside the tumour cell.

Work with tumour viruses is only one aspect of the use of tissue culture to study the behaviour of normal and neoplastic cells. Indeed, studies with cells explanted into the culture dish have pervaded all areas of modern biology and the very facility with which these techniques are applied have, from the outset, raised questions about the validity of the results they provide. Reservations about their application in investigating neoplasia centre around the problem of reproducing in culture all the cellular and humoral responses and topological restraints that influence tumour evolution in the whole animal. Conversely, some cells adapt readily to culture and the artificial environment of the culture dish may select characteristics of the cell that have little importance in the host. These considerations have stimulated many studies to identify the features of in-vitro transformation that correlate most precisely with the ability to form a tumour in the animal. Unfortunately, no single characteristic of cells *in vitro* permits such a correlation. As we might have expected, the phenotypes of tumour cells and in-vitro transformed cells differ in a complex, but not fully congruent, fashion from their normal counterparts. Studies *in vitro* will help to dissect this complexity but, for the time being, their full significance can only be assessed by continual reference to the behaviour of the cells in the whole animal.

8.2.4 *Modern concepts and questions*

Tumour virology still concerns itself with two broad questions: the role of viruses in clinical neoplasia (Klein 1985), and the use of viruses in probing the mechanisms of carcinogenesis (Tooze 1980; Weiss *et al.* 1984). We now understand enough about viral genomes and host cell functions to formulate ideas about how the former topic influences the latter (see p. 222) and to relate these concepts to specific diseases. We also appreciate better the role of viruses as one component in what may be a multifactorial disease process, and this concept is taken into account when examining viruses as risk factors in human and animal cancer (see Section 8.3). Both these considerations are underpinned by the premise that cancer results from alterations in the structure or activity of a certain number of genes in the cell (see Chapter 9). Tumour virology, as a whole, can thus be regarded as a branch of molecular genetics.

8.3 Implicating viruses in experimental and natural cancers

How do we decide that an infectious agent is responsible for causing a given disease? The classic criteria were embodied in Koch's postulates of 1876 which stated that (i) the agent should be found in all cases of the disease, its location corresponding with the observed lesions; (ii) the agent should be capable of isolation from the lesions and growth in pure culture outside the body; and (iii) the culture, when inoculated back into an animal, should produce the identical disease.

It is apparent that many pathogenic bacteria, let alone viruses, do not fulfil Koch's postulates but, when satisfied, these criteria provide convincing proof of causality. Many viruses have been shown to cause tumours in laboraory and domestic animals in this way, with the proviso that 'culture outside the body' has, perforce, been done on living cells, not inert media. Any doubts this raises about purity of the organism could, in theory, now be answered by molecular cloning of viral genomes in bacteria, but this has not yet been considered a necessary precaution.

In contrast to the ease with which viruses have been implicated in animal tumours, their role in human disease has been extremely difficult to investigate. This explains why major efforts, during the 1970s, to identify viral causes for human cancer produced few clear leads, and convincing causal associations are only now becoming apparent. A major problem, of course, is the impossibility of ethically fulfilling Koch's third postulate; other animals can be used to test the oncogenicity of human viruses but the results they yield can be misleading. Attempts have been made to modify Koch's postulates by including evidence for the presence of viral genomes or proteins in the tumours (when infectious virus cannot be found) and by asking whether the person with a tumour shows a specific immune response against the candidate virus. However, even if positive, these tests only show that a virus is associated with a tumour, not that it is a cause. Good evidence for causality would, of course, be provided if elimination of virus infection (for instance by immunization) also eliminated the incidence of the tumour, but such evidence has not yet been provided for any major virus associated human neoplasm.

A further problem in implicating viruses in human neoplasia is the complexity of the disease in man. Many virus induced tumours in animals occur in a high proportion of the infected population, often at a relatively early stage of the animal's short lifespan and, in the case of laboratory animals, selective breeding has enhanced these features. The virus is clearly a major risk factor in the disease. Human populations, in contrast, are usually outbred with no marked inherited predispositions to cancer (but see Chapter 4) and the disease usually occurs towards the end of a long lifespan with a pattern of incidence that suggests a multifactorial causation (Chapter 3). Indeed, the role of a virus in human neoplasia is often first hinted at by epidemiological data, the same data also showing that only a minority of the infected population develop tumours. Virus infection is thus usually one of a complex of interrelated risk factors that operate at different stages of development of a tumour, as shown in Figure 8.2 (Wyke and Weiss 1984). This figure depicts three main groups of risk factors: those relating to virus infection, those concerning the cellular alterations involved in neoplasia, and those affecting the host response to both viruses and tumour cells. Let us consider how they may operate at three arbitrary stages in the evolution of a virus associated neoplasm.

8.4 Risk factors predisposing to virus associated neoplasia

The risk factors are shown diagrammatically in Figure 8.2. In Stage 1 a virus must infect the host and, depending on the mechanism by which it instigates neoplasia (see p. 222), infection must lead to persistence of virus genetic material in the host or to some stable virus induced alteration in certain cells. Infection-related risk factors important at this stage are exposure to the virus (dose of virus and route of infection) and whether or not the host cells are susceptible to virus penetration and

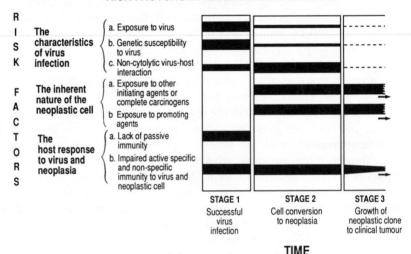

RISK FACTORS IN VIRAL CARCINOGENESIS

Fig. 8.2 Diagram of the relationship between the arbitrary stages of virus associated neoplasia, the component features of the neoplastic process, and the risk factors that influence each of these features. Only the stages of neoplasia have a clear temporal relationship to one another, as shown on the abscissa. The breaks in the time axis indicate that these stages vary in length and they are not shown to scale (in general Stage 1 is likely to be relatively short and Stage 2 is usually the longest period). The risk factors affecting either of the three features characterizing virus associated neoplasia are shown on the ordinate and, as shown by the bars, they may operate at different, and often more than one, stages of the disease process. The thickness of the bars indicate the relative importance of each risk factor at each stage of neoplasia, broken lines indicating periods at which the risk factors are of uncertain significance.

growth. The acquired resistance of the host, either by passive or active immunity (see Chapter 15), is also crucial, as in any other virus infection.

The requirement for Stages 2 and 3 distinguishes the pathogenesis of virus associated neoplasia from that of acute cytolytic virus diseases. In Stage 2, the conversion of an infected cell to neoplastic growth, the important characteristic of virus infection is the need to establish a non-cytolytic virus cell association, since, obviously, only live cells can form tumours (an exception to this requirement may be those instances where the mechanism of virus oncogenesis is indirect, see on). Some tumour associated viruses, such as the retroviruses and hepatitis B virus, can replicate without causing massive cell death, whilst herpesviruses can persist in latent, inactive forms in living cells. Other tumour viruses, however, can only convert a cell to neoplasia if full replication of the virus is blocked, either because the host cell is unusual or the virus has some defect. In either case, the infection that leads to neoplasia is unusual and may be inapparent in other ways, yet another factor making it difficult to link virus infection to neoplasia.

Further complicating factors important at Stage 2 are the requirements for (i) additional carcinogens or promoting agents (see Chapter 1); or (ii) the action of agents that affect host immune responses. In many virus associated tumours of man, such additional factors seem as or more important than virus infection itself.

Stage 3 in this sequence is the multiplication and progression of a neo-plastic cell to form a clinical tumour. Infection related risk factors are probably of little importance to this late stage in tumour evolution, although if expression of the virus persists it may serve as an antigenic target for host immune responses. Co-carcinogens, on the other hand, may remain important as tumour progression occurs (Chapter 6). Impairment of the host immune response is also likely to remain important. There is, in fact, a pervasive connection between host immune deficiency and the development of virus-associated tumours, particularly in man, which may be summarized as follows.

1. Immune mechanisms, such as T cell mediated cytotoxicity, natural killer cell activity, and interferon, can reduce the growth of virus induced tumours in laboratory animals (see Chapter 15).

2. Immune impairment is seen in many virus induced tumours in animals. In some instances, as with agents that infect T cells like Marek's disease virus and feline leukaemia virus, virus infection itself is immuno-suppressive. In other cases a co-carcinogen has this effect, an example being the role of bracken fern in papillomavirus associated alimentary carcinomas of cattle.

3. Human patients receiving immunosuppressive therapy show an

increased incidence of a limited range of tumours, many of which are tumours in which virus infection has also been implicated. Indeed, so striking is this overlap that it has been suggested that viruses may play a role in all tumours whose incidence is increased in immune deficient individuals. However, in only one type of therapy induced tumour has virus been directly implicated so far: the immunoblastic 'lymphomas' associated with Epstein Barr virus (EBV) (see Chapter 12).

4. Finally, immune impairment seems a risk factor in the natural history of a number of virus associated human neoplasms. Notable among these are tumours linked to infection with the human T cell leukaemia virus (a retrovirus), certain papillomaviruses, hepatitis B virus, and the herpesvirus, Epstein Barr virus.

8.4.1 *Searching for new viruses*

This discussion points to a recurring triplet motif in naturally occurring viral cancers in animals and man: (i) chronic viral infection, (ii) immune impairment, and (iii) co-carcinogens. These factors must be considered when asking where and how to implicate viruses in neoplasia.

The most promising tumours to examine are those whose incidence shows a clustering that can be either familial, geographic, or social/ occupational, implying genetic or environmental factors, both of which can point to viruses, as can an increased prevalence in immuno-suppressed individuals. Once candidate tumours have been identified, whole individuals and tumours should be examined for evidence of virus infection. Virus isolation may require *in vitro* culture of tumour cells, employing various stratagems to unmask latent viruses, including a search for appropriate cell types in which to propagate virus.

Isolation of a candidate virus will provide tools for a more detailed study of its association with the tumour but, failing isolation, a great deal may be learned by comparison with known tumour viruses, screening tumour cells for molecular evidence of virus infection, and examining hosts for seroepidemiological clues. If evidence is found for infection with a virus common in the population, then features of the infection peculiar to the tumour-bearing hosts should be sought. Is the infection unusually persistent? Is the serological response abnormal? Is the virus in the tumour defective? Are unusual viral antigens expressed? Even when a virus cannot be identified, a specific tumour antigen may be an important clue to a viral aetiology.

In man epidemiological evidence is crucial, since it is impossible to test a causal role for the virus by animal inoculation. Case control studies will further associate the virus with the tumour and laboratory studies will indicate features of high and low risk groups. Retrospective and, ideally, prospective studies of high risk groups should establish the temporal relationship between virus infection, other risk factors, and tumour development, providing a basis to tackle the management of the tumour.

8.5 Important tumour viruses of animals and man

The oncogenic viruses are a very diverse group. They include members of all the major families of DNA viruses that infect vertebrates, with the exception of the very small parvoviruses and the very large poxviruses. On the other hand only one family of RNA viruses, the retroviruses, can cause tumours. The tumour viruses vary greatly in the complexity of their genomes, in the types of neoplasms they induce, and in the requirement for cofactors in tumorigenesis. What do they have in common?

One universal feature is the importance of a DNA stage in the replication of the viral genome. The retroviruses are unique among viruses whose free infectious forms contain RNA in that this genome is copied soon after infection into double-stranded DNA by an RNA dependent DNA polymerase enzyme carried in the virus particle. Moreover, this DNA 'provirus' is then inserted ('integrated') by a covalent linkage into the host cell DNA (Weiss *et al.* 1984). Such integrations seem another frequent, but not invariant, hallmark of tumour viruses: whole or partial viral genomes are very often detected in tumour cell chromosomes. It is not clear why integration occurs so commonly since it is obligatory in only one of the mechanisms by which tumour viruses cause neoplasia (see p. 227). However, it may serve mainly to ensure a stable association between viral and host genomes during the lengthy development of a neoplastic cell lineage. In this context it is interesting that a number of poxviruses can induce cellular proliferation and, since some encode proteins related to growth factors (see Chapters 1 and 11), this growth may be truly neoplastic, yet none of them induces stable progressing tumours. Could this be because they complete their life cycles in the cytoplasm of the cell and do not enter the nucleus?

Various groups of tumour viruses are shown in Tables 8.1, 8.2, and 8.3. These lists are representative and not comprehensive and for a fuller account the texts listed at the end of this chapter should be consulted. Entries in the tables have been chosen because of (i) their historical or research interest; (ii) an intrinsically interesting pathogenesis; (iii) their importance as pathogens of man and domestic animals; or (iv) their inclusion elsewhere in this Chapter.

8.5.1 *Retroviruses (Table 8.1)*

This is a family of small viruses with RNA genomes of 5–10 000 nucleotides. Most tumorigenic members are in the oncovirus subfamily and, of

Table 8.1 Some oncogenic retroviruses of animals and man

Virus classification[1]	Viruses	Associated tumours	Other risk factors[2]
Oncovirus subfamily, Type B genus	Mouse mammary tumour virus	Mammary adenocarcinoma	Pregnancy (altered hormone levels), genetic susceptibility affecting endogenous virus production
Oncovirus subfamily, Type C genus	Avian sarcoma-leukosis virus complex	Various sarcomas, some carcinomas, lymphomas, and leukaemias	Genetic susceptibility affecting virus penetration, replication, and spread
	Avian reticuloendotheliosis virus complex	Lymphomas and leukaemias	
	Mouse leukaemia and sarcoma viruses	Various sarcomas, lymphomas, and leukaemias, some carcinomas	Genetic susceptibility affecting virus penetration, replication, and spread
	Feline leukaemia virus	Leukaemia, lymphosarcoma (mainly T-cell)	Possible genetic susceptibility, concurrent infections
	Bovine leukosis virus	Lymphosarcoma, leukaemia (B-cell)	Genetic susceptibility to development of lymphocytosis
	Primate leukaemia and sarcoma viruses	Fibrosarcoma, myeloid leukaemia	
	Human T-cell leukaemia lymphoma viruses	Adult T-cell leukaemia/lymphoma	
Oncovirus subfamily, Type D genus	Simian immunodeficiency viruses	Fibrosarcomas	
Lentivirus subfamily	Human immunodeficiency viruses	Kaposi's sarcoma, lymphoma	Concurrent infections with HTLV-1, Human herpesvirus 6

[1] Classification is based upon virus particle structure and limited biological criteria. Evolutionary relationships being revealed by genetic comparisons may not accord with this taxonomy.

[2] Where known or suspected. The absence of listed risk factors does not imply that such factors are unimportant.

the genera in this subfamily, by far the most important is the Type C virus genus.

The retroviruses of chickens were the first tumour viruses to be discovered and, together with comparable viruses of mice, they have played a crucial role in the history of tumour virology. They are still important in our attempts to understand the molecular basis of neoplasia (see p. 226 and Chapter 9), but only one group of avian retroviruses, the causal agents of fowl leukosis, is of commercial importance.

Two other important retrovirus pathogens of domestic animals are feline leukaemia virus and bovine leukosis virus (Onions and Jarrett 1987). The latter is of commercial importance in many parts of the world where it causes the most common malignancy in cattle. Leukaemia and lymphosarcoma are also the most frequent tumours of domestic cats but they account for only a fraction of the deaths attributable to feline leukaemia virus, which also causes anaemia, immunosuppression, and related diseases, making it now the most frequent non-traumatic cause of death among cats in developed countries.

The diseases caused by these cattle and cat retroviruses provide interesting parallels for a disease complex in man associated with infection by a retrovirus group called human T cell leukaemia virus (HTLV). With all three agents the prevalence of infection in the population is greater than the incidence of neoplasia. Indeed, in the case of HTLV the associated tumour, adult T cell leukaemia/lymphoma, is rare and, even in areas where HTLV infection is widespread, only about one in 80 of the infected population develop the malignancy (presumably, unidentified cofactors are important in oncogenesis). Nevertheless, as with feline leukaemia virus, HTLV may induce diseases other than neoplasia. Two strains, HTLV-1, and -2, are associated with T cell malignancies whilst a third virus, identical to or closely related to HTLV-1, is strongly implicated in tropical spastic paraparesis, a chronic degenerative disease of nervous tissue.

An even more important group of human retroviruses, human immunodeficiency viruses (HIV), together with an almost identical simian agent (SIV), are grouped in the lentiviruses, a subfamily they share with feline T lymphotropic lentivirus, equine infectious anaemia virus, visna virus of sheep, and others. Like many lentiviruses, HIV causes degenerative diseases of the immune and nervous systems, the most frequent manifestation being a T cell deficiency leading to acquired immunodeficiency syndrome (AIDS). However, unlike other lentiviruses, the HIV isolates earn a place in this chapter because AIDS patients frequently develop otherwise rare mesenchymal tumours, such as Kaposi's sarcoma, which may affect connective tissues in many parts of the body (Fauci 1988).

8.5.2 *Small DNA viruses (Table 8.2)*

The most important human oncogenic viruses appear to be in this group (Klein 1985). It is estimated that 200 million people, mainly in Third World countries, are chronically infected with hepatitis B virus and are at risk of developing cirrhosis (fibrosis of the liver) and primary liver cancer, a tumour that causes about 500 000 fatalities per annum. However, even at this level it is clear that only a minority of those infected with the virus develop the tumour, and other risk factors, which may vary from area to area, must be important. Postulated factors include smoking, superinfection with another virus (the Delta agent), and consumption of alcohol or food contaminated with aflatoxin B_1 derived from the fungus *Aspergillus flavus*. Very similar viruses cause liver tumours in rodents and these should provide a good model for studying this important human disease.

One genus of the Papovavirus family contains two important tools of the molecular biologist, polyoma and SV40, and two agents closely related to these that commonly infect man, BK and JC virus. These latter have not been associated with any human tumour but they are frequently detected in immunosuppressed individuals, they transform cells in tissue culture, and they can induce tumours in rodents, so they are clearly still viewed with suspicion.

The papillomaviruses, in contrast, have long been known to cause tumours in many animals—benign warts (papillomas). However, it is now clear that these lesions can become malignant, but they have a natural history that may be very complicated. For instance, molecular biology techniques have revealed over 50 different papillomaviruses that infect man alone, and those types commonly associated with benign growths may not be the same as the types found in malignant lesions (the most important of which is carcinoma of the uterine cervix). Moreover, the progression from benign to malignant growth can depend on several other predisposing factors (see Table 8.2) making it difficult to implicate a virus by epidemiological data alone. One example gives a flavour of this complexity. The very rare human disease epidermodysplasia verruciformis shows some familial clustering (suggesting a genetic component) and usually arises in young patients with congenitally defective cell mediated immunity. It is characterized by disseminated skin warts of two main types from which can be isolated many different papillomaviruses. Carcinomas may later develop from one type of wart associated with a subset of these papillomavirus types. The carcinomas arise mainly in areas exposed to sunlight. Thus specific papillomaviruses, immune deficiency, and a co-carcinogen (in this case u.v. light) all combine in the disease process. Factors of these three classes also

Table 8.2 Some oncogenic small DNA viruses of animals and man

Virus family	Virus	Genome size[1]	Host of origin	Associated tumours	Other risk factors[2]
Hepadnavirus	Hepatitis B group	3 kb	Man, apes, rodents, ducks	Liver cancer	In man: alcohol, smoking, fungal toxins, other viruses
Papovavirus	Polyoma	5 kb	Mouse	Various carcinomas and sarcomas	
	SV40	5 kb	Monkey	Sarcomas (in rodents)	
	BK and JC	5 kb	Man	None in man; neural tumours in rodents and monkeys	
	Papilloma	7–8 kb	Man	Genital, laryngeal, and skin warts, may progress to:	
				Cervical carcinoma	Smoking, herpes simplex viruses, immune suppression
				Laryngeal carcinoma	X-irradiation, smoking
				Skin carcinoma	Sunlight, genetic disorders possibly affecting immunity
			Cattle	Genital, alimentary, and skin warts, may progress to:	
				Alimentary carcinoma	Possibly carcinogens and immune suppressants in bracken fern
				Skin carcinoma	Sunlight, genetic predisposition (lack of pigmentation)
			Other mammals	Papillomas, may progress to carcinomas	Experimentally, carcinogens such as methylcholanthrene

[1] In kilobases (kb); one kilobase is 1000 base pairs of nucleic acid.
[2] Where known or suspected. The absence of listed risk factors does not imply that such factors are unimportant.

operate in the genesis of alimentary carcinomas in cattle, emphasizing the triad of virus, co-carcinogen and immune impairment as a common motif in many tumours.

8.5.3 *Large DNA viruses (Table 8.3)*

Of the two families considered here, the adenoviruses cause mild non-neoplastic diseases in man but the same strains can cause tumours in hamsters and, as a consequence, they have been studied intensively. There is no evidence that they play a role in human cancer, but these studies have been enormously fruitful for those interested in the basic molecular mechanisms of gene expression and regulation.

The herpesviruses are more significant pathogens. Marek's disease virus causes a commercially important disease of chickens characterized by a T cell proliferation that infiltrates nervous tissue (hence the description, neurolymphomatosis). Three other herpesvirus types have been linked with human neoplasia. For two of these, the Herpes simplex viruses and cytomegalovirus, the evidence that they play a causal role in certain cancers (Table 8.3) is intriguing but not conclusive (Macnab 1987). For the lymphotropic herpesviruses the causal link is more persuasive and, although some of these agents have only recently been described, EBV has received considerable attention. Not only does EBV cause the non-neoplastic infectious mononucleosis (glandular fever), a disease particularly common in young adults, but it is also a major risk factor in nasopharyngeal carcinoma, a common malignancy in some heavily populated parts of the world such as southern China. Once more, however, we see the familiar pattern: infection is far more widespread than the incidence of the tumour, and other factors are clearly important. This is even more evident in the case of Burkitt's lymphoma, a B cell tumour of children in West Africa and New Guinea linked jointly to EBV infection and endemic malaria. A large proportion of children in these areas have experienced both known risk factors, yet only a small minority develop the tumour. What else is required for tumorigenesis? Specific chromosomal rearrangements seem an important prerequisite, but what favours such events and what are their biological consequences? These are discussed further in Chapters 9 and 10 but the answer can be given here—we do not know.

8.6 Prophylaxis and therapy of virus associated neoplasia

Although we are clearly ignorant of many aspects of the pathogenesis of virus associated tumours, the knowledge that a virus is implicated can be used in attempts to manage the disease at any of the three arbitrary stages

Table 8.3 Some oncogenic large DNA viruses of animals and man

Virus family	Genome size[1]	Virus	Host of origin	Associated tumours	Other risk factors[2]
Adenovirus	30–50 kb	Types 2, 5, 12	Man	None in man; sarcomas in hamsters	
Herpesvirus	130–250 kb	Frog herpesvirus Marek's disease	Frog Fowl	Adenocarcinomas Neurolymphomatosis (T cell)	Ambient temperature Genetic predisposition of unknown basis
		H. ateles and *H. saimiri* Epstein Barr virus	Monkey Man	Lymphoma, leukaemia (T cell) Burkitt's lymphoma Immunoblastic lymphoma Nasopharyngeal carcinoma	 Malaria Immune deficiency Salted fish in infancy, histocompatibility antigen genotype
		H. simplex types 1 and 2 Cytomegalovirus	Man Man	Cervical neoplasia (?) Kaposi's sarcoma (?) Cervical neoplasia (?)	Papillomaviruses, smoking Immune deficiency, histocompatibility antigen genotype

[1] In kilobases (kb); one kilobase is 1000 base pairs of nucleic acid.
[2] Where known or suspected. The absence of listed risk factors does not imply that such factors are unimportant.

of virus associated neoplasia described above and in Fig. 8.2. In general these measures are more effective in veterinary medicine (where the health of the herd can override the survival of the individual) than in human practice.

8.6.1 *Preventing virus infection*

The level of oncogenic viruses in the hosts' environment can be reduced by hygiene and husbandry techniques, and by detecting and, if necessary, eliminating carriers. The latter approach has been used in managing diseases caused by the avian, feline, and bovine retroviruses and it has some limited applications in man, for instance in detecting agents like hepatitis B virus and HIV in donor blood intended for transfusion.

Increased genetic resistance to infection can be bred into domestic animals, an approach successful in developing chicken strains relatively resistant to both avian leukosis and Marek's disease.

However, in most cases acquired immunity to infection is the only option. Vaccines made from ground-up wart tissue have long been used as a prophylactic and, indeed, therapeutic measure against papillomas in animals, but the first successful commercial vaccines against a neoplasm were produced against Marek's disease and are a great boon to the poultry industry. A useful, but far from ideal, vaccine now exists against feline leukaemia but, in contrast to veterinary problems, proposals to produce vaccines against human oncogenic viruses have long been controversial, for several reasons. (i) In many instances the role of the virus in the tumour is uncertain; (ii) tumour production may be a rare outcome of infection by a widespread and not very pathogenic virus; and (iii) since tumours may result from an aberrant virus cell interaction, a classic vaccine, based on inactivated or attenuated virus, may itself pose a health risk. This last objection may be obviated by using purified immunogens produced by genetically manipulated portions of viral genomes or synthesized in the laboratory.

However, the first two considerations suggest that the returns (in terms of improved health of the population) may not justify the outlay unless the virus causes significant disease in addition to its oncogenic potential (as is the case with HIV and hepatitis B virus), unless there are clearly defined, small, high risk populations or unless the virus is the only clearly defined risk factor in the genesis of a common tumour (as with EBV associated nasopharyngeal carcinoma).

These reservations, together with the problems of developing a safe and effective vaccine and delivering it to a susceptible population before they contract the virus, provide scope for other approaches. An alternative, popular with the pharmaceutical industry, is to limit infection, usually after it is clinically evident, with antiviral chemotherapy. This, too, poses problems inherent in the close symbiosis between viruses and their host cells, since drugs must be devised that are preferentially toxic for the virus. One approach focuses on processes essential to the virus and dispensable by the cell, such as reverse transcription by the retroviral RNA dependent DNA polymerase. Within these constraints, some widely used and moderately effective drugs have been developed against, for instance, herpesviruses and retroviruses.

8.6.2 *Preventing cell conversion to neoplasia*

The problems of tackling virus infection suggest that other potentially avoidable risk factors, which tend to operate at this stage, might provide easier targets for preventive measures. Such hopes, however, seem largely misplaced. In man these risk factors include habits (notably smoking), dietary factors (which may be even harder to eliminate than smoking unless, like aflatoxin contamination, they are obviously undesirable), and other diseases. The most striking example of the latter is malaria, a risk factor in Burkitt's lymphoma in certain tropical areas but also a crushing disease problem in its own right in many parts of the world. Indeed, reducing or eliminating some of these risk factors would have benefits far beyond the reduction in cancer incidence, and this seems so evident that one doubts that the incentive of reducing cancer will work where other imperatives have so signally failed.

8.6.3 *Tackling clinical neoplasia*

Can a knowledge of tumour virology contribute to tumour diagnosis and therapy? So far such applications have been on a small scale but in principle a detailed understanding of the role of any given virus in cancer should aid diagnosis and prognosis by screening for characteristic viral genes or proteins, or a host response to them. It is harder to see how such knowledge could be applied to therapy but, there is the tantalizing example of interferon, a general antiviral agent produced by host cells after infection with many different viruses, that also has an antitumour effect in a few cases that is, as yet, poorly understood. There have also been instances of regression of virus associated tumours in response either to viral antisera or to a 'vaccine' extracted from homogenized tumour. The latter effect, well known for papillomavirus infections, helps to justify attempts to produce vaccines against HPV16 and HPV18, the two papillomaviruses most implicated in human cervical carcinoma. By the criteria mentioned in Section 8.6.1 above, venereally transmitted human papillomaviruses are not strong candidates for management by vaccination. Although there are high risk groups such as prostitutes, a very large proportion of the population in a sexually permissive society is at some risk of infection, which suggests the need for an extensive vaccination programme. Set against this is the fact that neoplasia is a relatively rare consequence of sexual promiscuity that might better be tackled by a screening programme allied to an education campaign (that would also help to reduce other venereal diseases). However, if vaccines against HPV can promote regression of cervical carcinoma, then they may play a greater role in its therapy than in its prevention.

Clearly, we will not advance much further with these considerations until we unravel the details of how a tumour virus subverts cell growth. Fortunately, as the next section describes, such understanding is beginning to take shape from the amorphous complexity of tumour virus behaviour.

8.7 The mechanisms of virus induced neoplasia

We have already seen that workers in the 1960s set out to identify viral genes that directly converted normal cells to neoplastic growth. However, the many variations of intracellular parasitism exhibited by viruses could, in theory, permit neoplasia by a number of other mechanisms. Indeed, as we identify more tumour viruses so we are compelled to invoke an ever widening spectrum of pathogenic mechanisms. In this section we will survey, with examples, these different modes of virus induced neoplasia (Table 8.4).

The two major categories of the disease process are: (i) that in which the tumour cell ancestry must at some stage have been infected by the tumour inducing virus; and (ii) that in which the tumour cell ancestry need not be infected by the virus. Neoplasia in the first category is a direct consequence of infection whilst in the second category it results indirectly, from infection of other cells (the terms intrinsic and extrinsic have also been used to describe these two mechanisms).

8.7.1 *Indirect mechanisms: suppression and stimulation of cell proliferation*

Most virus infections kill cells and if the cells of the host's immune system are targets for viral infection then immune deficiency, particularly an impairment of cell mediated immunity, can result. The role of immune surveillance in cancer has been controversial (see Chapter 15) but, as we saw above, many virus associated neoplasms occur in hosts with immune deficiency and it is possible that tumour cells arising in these hosts by unknown mechanisms are simply not eliminated. Viruses such as avian

Table 8.4 Mechanisms of virus induced neoplasia

Pathogenic mechanism	Examples
Indirect	
1. Suppression of host immune system, impaired elimination of tumour cells	Avian reticuloendotheliosis viruses, feline leukaemia virus, HTLV-1, HIV-1, Marek's disease virus, cytomegalovirus
2. Stimulation of cell proliferation, increased 'targets' for other neoplastic changes	
(a) tissue regeneration after virus cytolysis	None yet known
(b) mitogenesis of immune competent or other cells	Possibly some mouse leukaemia viruses, HTLV-1, HIV-1
Direct	
1. 'Hit and run': no crucial virus gene or structure whose persistence is essential Viral DNA or viral functions act transiently	Possibly some herpesviruses
2. Crucial parts of viral genomes persist in tumour cells	
(a) virus carries a gene whose product directly or indirectly initiates and/or maintains neoplasia (oncogene)	
(i) oncogene descended from normal cell counterpart (*proto-oncogene*) in past evolution	Rous sarcoma virus, acute leukaemia viruses
(ii) oncogene directly transduced from proto-oncogene during virus infection	Feline leukaemia virus, avian leukosis virus
(iii) oncogene with no related normal cell counterpart	SV40, polyomavirus, papillomaviruses, adenoviruses, EBV, bovine leukosis virus, HTLV-1
(b) Insertional mutagenesis: virus DNA inserted in the host chromosome augments or destroys normal gene expression	Avian leukosis viruses, mouse mammary tumour virus, possibly hepatitis B virus and some papillomaviruses

reticuloendotheliosis viruses, Marek's disease virus, feline leukaemia virus, HTLV, and cytomegaloviruses are known to have immune suppressive effects sometimes, but not always, associated with cytotoxicity. However, these agents also seem to have another, possibly direct, tumorigenic effect. One likely exception is HIV-1 which does not appear to act directly in tumour causation. Its effect seems due partly to immune impairment and a consequent inability to kill tumour cells arising by other mechanisms, and partly to other indirect mechanisms (see p. 224).

Another possible consequence of virus infection is a reactive cell proliferation, the expansion of a cell population, increasing the chance of occurrence of a neoplastic change. In principle this proliferation can have several causes, one of which, regeneration of tissue damaged by virus cytolysis, has been invoked for liver carcinogenesis associated with hepatitis B virus infection. It is also possible that the frequently observed production of growth factors by virally transformed cells might stimulate hyperplasia of uninfected cells but the only good example of this plausible phenomen is the induction of Kaposi's sarcoma in HIV-1 infected AIDS patients. There is better evidence for chronic retroviral infection stimulating immune competent cells into proliferation. For example, in HTLV-1 associated B-cell chronic lymphocytic leukaemia, the tumour cells are uninfected but recognize viral antigens (Mann *et al.* 1987). A similar mechanism also operates in HIV-1 associated B cell lymphomas and variants have also been implicated in murine leukaemia virus-induced T cell lymphomas, although in this latter case the lymphoma cells are infected by the viruses, which additionally have a direct insertional mutagenesis component (see p. 227). Furthermore, virion constituents may mimic ligands for receptors on the surface of other cell types (see Chapters 9 and 11). If these receptors modulate normal cell growth and behaviour, chronic unscheduled binding of ligand analogues may lead to hyperplasia and then neoplasia.

8.7.2 *Direct mechanisms: the 'hit and run' hypothesis*

It is conceivable that transient cell infection induces a heritable neoplastic change in the cell lineage and the infecting virus is then lost. In practice this mechanism is indistinguishable from the indirect modes described above. It can also resemble the direct mechanisms detailed below if specific viral functions are required only for the earliest steps of a multistage tumour evolution, as may occur with bovine papillomavirus. However, in other instances, notably herpes simplex virus and cytomegalovirus induced tumours, portions of the viral genome frequently persist. If, neither the fragment of persisting virus, nor its location in the cell genome, show any discernible pattern, it is suspected that they represent the 'footprints' of an earlier 'hit and run' event whose significance cannot now be assessed (Macnab 1987). Since they are difficult to investigate, 'hit and run' and indirect mechanisms are usually only considered after failing to demonstrate tumour formation by one of the direct mechanisms we now describe.

8.7.3 *Direct mechanisms: transforming genes with cellular ancestors*

The viruses that most readily transform cells in culture (Fig. 8.1) and

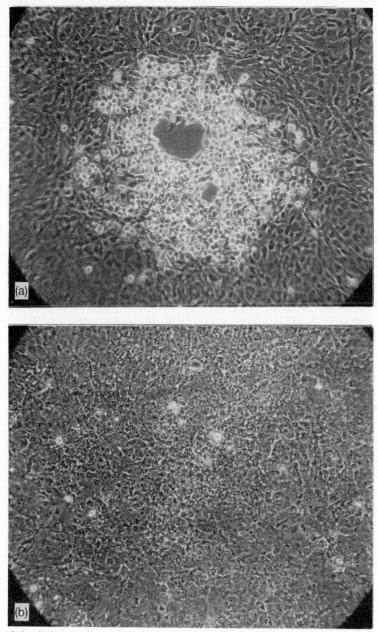

Fig. 8.3 Cell transformation by viruses with mutations in transforming genes. Some mutants of Rous sarcoma virus have 'temperature sensitive' defects in transformation. Infected cells grown at the 'permissive' temperature are transformed because the viral transforming gene functions normally. At the 'restrictive' temperature the gene is inactive and the infected cells are normal. (a) A focus of transformed cells on a chicken embryo cell monolayer induced by infection with a Rous sarcoma virus temperature sensitive mutant and incubation at permissive temperature. The cells are rounded and detaching from the substrate, so that holes are appearing in the cell sheet. After photography the culture was incubated at restrictive temperature and four days later was photographed (b). The cells are now flatter and more indistinct and the gaps in the cell sheet have been filled in. (Magnification ×100 in both)

most rapidly induce tumours in animals are the chicken and mouse retroviruses that cause sarcomas and 'acute' (rapid) leukaemias and the
polyomavirus group of the papovaviruses. Using cell culture, mutants of
these viruses with defects in transformation were obtained (Fig. 8.3) and
analysis of these mutants defined transforming genes in the viruses.
These genes, as predicted by the pioneer workers in this field, encode
proteins necessary to initiate and/or maintain transformed cell growth,
and they comprise two classes.

The retroviruses mostly contain transforming genes that play no part
in virus replication and have, indeed, usually replaced portions of viral
replicative genes in the virus genome (Weiss *et al.* 1984). These genes are
related to sequences in normal host cells from which they are believed to
have evolved after 'capture' of the cell gene by the virus in some ancestral
infection (indeed, this evolution is apparently sometimes recapitulated
when tumours are induced by certain chicken and cat leukaemia viruses).
These transforming genes were named viral oncogenes and the cellular
counterparts became known as cellular oncogenes, or since they presumably serve some crucial non-neoplastic function in the host, proto-
oncogenes. These oncogenes, first brought to our attention by the
retroviruses, are now central to studies on the molecular biology of
cancer (see Chapters 9 and 11).

8.7.4 *Direct mechanisms: transforming genes without cellular ancestors*

The second class of transforming gene is exemplified by those in
polyoma, SV40, papillomaviruses, adenoviruses, and herpesviruses such
as EBV. These genes play a part in the virus life cycle (albeit sometimes a
rather peripheral role) and they are not obviously derived from cellular
ancestors (Tooze 1980). Protooncogenes are thought to function in
normal cell growth and behaviour, with the viral oncogenes representing
perverted forms of these activities that tip the cell into neoplasia. Since
the actions of papovavirus transforming genes have similar consequences for the cell as the functioning of the retrovirus oncogenes, we
can conclude that the former mimic functions in comparable regulatory
pathways. They are 'analogues' to the retroviral oncogene's 'homologues'
of cellular functions.

A few retroviruses also seem to possess genes whose provenance and
activity are more akin to the DNA virus transforming genes than to retroviral oncogenes. Despite their morphological differences, bovine
leukosis virus and HTLV on the one hand, and HIV on the other, have
similarities in their pathogenesis (see above) and it may be significant
that they have evolved markedly analogous means of regulating their
intracellular life cycles. Both virus groups encode proteins that deter-

mine the activity of their own genes but may also regulate host cell genes. It seems probable that these phenomena contribute to the cell proliferation or cell death characteristic of infection by these viruses. There may, furthermore, be parallels in the pathogenesis of HTLV-1 and EBV, for in each case the transforming genes may induce a cell-specific growth factor that acts in an autocrine fashion to promote hyperplasia (Yoshida and Seiki 1987).

8.7.5 *Direct mechanisms: insertional mutagenesis*

It appears, however, that many retroviruses without oncogenes induce neoplasia by a different means. The majority of retroviruses contain only the genes needed for their own replication, they cannot transform cells in tissue culture, they induce tumours after a long latency, and tumour induction results from a direct effect of the viral genome rather than from the action of a virus-coded protein.

Avian leukosis viruses (ALV) are retroviruses of this type and studies on the lymphomas they induce provided the following clues to this mode of pathogenesis. All or most of the cells in a lymphoma contain ALV proviruses integrated at the same site in the host DNA. There are two explanations for this: either the provirus is obliged to integrate at one location or the tumour is a clone, all its cells deriving from a single ancestor. The former possibility is eliminated by showing that ALV can insert itself at many sites in the normal cells of the host. It follows that, although many cells in the bird are infected, in only a very small minority of these cells does an event occur that promotes the clonal neoplastic growth of the cell. When lymphomas from different birds are compared, ALV proviruses are found integrated in the same region of the host DNA in over 90 per cent of them, although many of these proviruses are incomplete. The conclusion, that the site of provirus integration is vital to tumorigenesis, led to the postulate that the provirus acts as a mutagen whose insertion disrupts host gene expression in that region (Weiss *et al.* 1984). You should note that this is the only one of these postulated pathogenic mechanisms in which virus DNA integration is obligatory.

This concept of 'insertional mutagenesis' has been supported by studies very similar to those described above on other virus induced chicken tumours, on comparable tumours induced by some mouse leukaemia viruses, and on adenocarcinomas caused by mouse mammary tumour virus. It has also been invoked to explain oncogenesis by hepatitis B virus and some papillomaviruses. How does the virus exert its mutagenic effect? Does it destroy host gene functions or stimulate them? In most cases host genes seem to be stimulated. Increased transcription of host DNA in the vicinity of an integrated provirus has often been detected, and this is generally ascribed to the action of elements in the provirus that increase transcription (such elements are required by the provirus to transcribe its own genes—see Chapter 9—and they are themselves subject to modulation by tissue specific host cell factors, such that they are more efficient in some tissues than others). Moreover, in a number of instances it has been shown that the proviruses have integrated in the vicinity of, and increased transcription of, host proto-oncogenes. This further implicates such genes in neoplasia, and where a provirus has not integrated near a known protooncogene it is suspected that its integration site pinpoints other genes that are important in tumorigenesis. We thus hope that new oncogenes will be identified in this 'guilt by association' process and this reasoning is pursued further in Chapters 9 and 10.

8.8 Conclusions

We have seen that viruses are important environmental carcinogens. In domestic animals they can be the predominant risk factors in some common and commercially important cancers. They are also implicated in some major human malignancies, although in man the interplay between viruses and other risk factors may be very complex. Prophylactic procedures against infectious diseases have been used efficiently for over a century and it was hoped that similar approaches might prove successful in eliminating virus associated cancer. However, the complexity of the disease process has frequently complicated such attempts and to improve management of these diseases we clearly need to understand more about the pathogenesis of virus induced neoplasia.

Such basic studies have, in fact, already spurred advances in other directions. Genes in small tumour viruses directly mediate the neoplastic growth of the cells they infect and such genes are analogous or homologous to genes in host cells. Tumour virology has thus provided the first glimpses of genes that become altered in cancer cells, no matter what the precipitating cause of the disease may be. Much modern cancer research now aims to widen this window, to identify the full panoply of these oncogenes, to detect the functions of their products, and to determine how these functions affect normal and neoplastic cell growth and behaviour. The development of this work is reflected in the next three chapters and we will leave it with one final thought. At various stages in its history, tumour virology has been dominated by the precepts and techniques of different scientific disciplines. Beginning with observations on whole animals, the field has become sequentially the preserve of the pathologist, the cell biologist, and the molecular geneticist. Now with increasing emphasis on the products of oncogenes, the biochemist and protein chemist are coming to the fore. It is important that each of these disciplines realizes that it is looking at aspects of one biological question, and that it appreciates the accretion of understanding provided by its predecessors in the field. Only by such a breadth of view are we likely to close the circle and return to answer the questions that stimulated these studies—how do we prevent, diagnose, or treat malignant disease in the whole organism?

References and further reading

Brown, F., and Wilson, G. S. (ed.) (1984). *Topley and Wilson's principles of bacteriology, virology and immunity.* Vol. IV. *Virology* (7th edn). Edward Arnold, London.

Editorial (1988). HTLV-1 comes of age. *Lancet* i, 217–19.
An epidemiological survey.

Fauci, A. S. (1988). The human immunodeficiency virus: infectivity and mechanisms of pathogenesis. *Science* **239**, 617–22.
Short and up-to-date.

Klein, G. (ed.) (1985). Viruses as the causative agents of naturally occurring tumours. *Advances in viral oncology,* Vol. 5, Raven Press, New York.

Macnab, J. C. M. (1987). Herpes simplex virus and human cytomegalovirus: their role in morphological transformation and genital cancer. *Journal of General Virology* **68**, 2525–50.
Tackles the difficult question of oncogenesis by herpesviruses that lack a specific oncogene.

Mann, D. L. *et al.* (1987). HTLV-1-associated B-cell CLL: indirect role for retrovirus in leukemogenesis. *Science* **236**, 1103–6.
A primary source with useful references on indirect modes of neoplasia.

Onions, D. E. and Jarrett, O. (ed.) (1987). Naturally occurring tumours in animals as a model for human disease. *Cancer Surveys* **6**, 1–181.

Tooze, J. (ed.) (1980). *The molecular biology of tumour viruses: DNA tumour viruses* (2nd edn). Cold Spring Harbor Laboratory, Cold Spring Harbor, New York.

Weiss, R. A., Teich, N., Varmus, H. and Coffin, J. (ed.) (1984). *The molecular biology of tumor viruses: RNA tumor viruses* (2nd edn and Appendices). Cold Spring Harbor Laboratory, Cold Spring Harbor, New York.

Wyke, J. and Weiss, R. (ed.) (1984). Viruses in human and animal cancers. *Cancer Surveys* **3**, 1–218.

Yoshida, M. and Seiki, M. (1987). Recent advances in the molecular biology of HTLV-1: *trans*-activation of viral and cellular genes. *Annual Review of Immunology* **5**, 541–59.

9

Oncogenes and cancer

NATALIE M. TEICH

9.1 What is an oncogene?

The term oncogene has been used in several of the preceding chapters and will arise again in subsequent chapters. In this section, we shall try to amalgamate current knowledge and hypotheses of this fascinating subject from a more functional view. The generic name 'oncogene' was coined to delineate a gene capable of causing cancer. Obviously, this may be an oversimplified concept as the previous chapters have presented evidence that the genesis of a tumour is a complex issue involving multifactorial and/or multistep processes.

9.1.1 *First encounter: retroviruses*

As discussed in Chapter 8, the family of RNA tumour viruses, Retroviridae, comprises a vast number of members from many animal species, with a wide range of pathogenic properties including neoplastic and non-neoplastic diseases. The most common oncogenic viruses (such as long latency leukaemia viruses and the murine mammary tumour virus) contain genes for their replication only, and cause tumours by mechanisms grouped as insertional mutations which may often take many months for clinical manifestation (see Chapter 8, Fig. 9.1, and below).

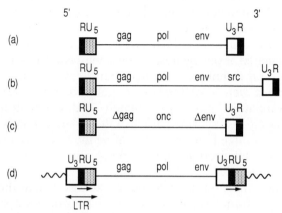

Fig. 9.1 Retroviral genomes. (a) Typical genome structure of a long latency leukaemia virus. The three replicative genes are shown: *gag*, the gene encoding the internal core structures; *pol* encoding the RNA dependent DNA polymerase enzyme known as reverse transcriptase; and, *env* encoding the envelope proteins inserted into the plasma membrane which becomes incorporated into mature virions. 5′ and 3′ denote the polarity of the molecule, as the RNA can serve directly as a mRNA molecule and be translated into protein (a positive strand genome). R denotes a short segment of RNA repeated at each end of the molecule. U_5 and U_3 denote non-coding unique sequences at each end that contain the regulatory elements (the promoter and enhancer sequences) for viral transcription. (b) The structure of the Rous sarcoma virus of chickens, the only retrovirus that contains all replicative genes and additionally an oncogene (*src*). (c) Structure of the acutely leukaemogenic or transforming retroviruses. These genomes generally lack all or part of one or more of the replicative genes, thus rendering them defective for replication. The deleted sequences have been replaced by cellular sequences that confer the oncogenic potential (generically called *onc* genes). The extent of replicative gene deletion and location of the *onc* gene are distinct for each viral isolate. (d) Structure of the integrated virus (DNA provirus). The mode of replication of retroviruses generates duplicated ends known as long terminal repeat (LTR) structures in which the U_5 region is duplicated at the extreme 3′ end and U_3 at the 5′ end. The LTRs are important for the integration of the provirus into chromosomal DNA and always generate a complete proviral DNA colinear with viral RNA. Another important feature is that the regulatory elements for viral transcription are now found also at the 3′ end of the molecule and thus can serve as promoters or enhancers for adjacent cell genes (denoted by the wavy lines).

However, about 50 virus isolates have the interesting property of being able to induce tumours in infected animals after very short latency periods (generally days or weeks) and furthermore often are capable of causing morphological alterations in cells grown *in vitro* (transformation, see Chapter 8). Dissection of the genomes of these viruses showed that most of them had lost genetic information coding for their replicative genes (and hence were known as replication defective, albeit transformation competent, viruses). New genetic information was inserted in place of the deleted material (Fig. 9.1). By a variety of genetic techniques, this new set of sequences was shown to be responsible for the short latency and transformation inducing capacities. Hence, the new sequences were designated as oncogenes, or *onc* genes. Nucleic acid sequencing and hybridization techniques revealed that the *onc* gene sequences of these 50 viruses were sometimes essentially identical and sometimes completely unrelated. Thus, approximately 20 different viral oncogenes (v-*onc* genes) were distinguished. Each of the separate v-*onc*s was given a different name, a three letter word to define the virus from which it was isolated which is italicized in type like other genes; this nomenclature is illustrated in Table 9.1.

The first important question to be asked was: where did the v-*onc* sequences come from? Again nucleic acid hybridization studies were used to demonstrate unambiguously that the v-*onc* sequences were almost identical to sequences in the cellular DNA of the animal species from which the virus was isolated so that cellular DNA contains genes which when 'transplanted' (transducted) into retroviruses are cancer causing genes. Moreover, it was shown that the *onc* sequences could be found in the DNA of every cell from virtually all higher vertebrate orders ('evolutionary conservation'). Thus, these sequences related to viral oncogenes are inherited from one generation to the next in the same way as one inherits genes for eye colour. It is clear, however, that these genes cannot be operating as cancer genes in cells (and indeed our own cells). How can we envisage this subtle but extremely important dichotomy? First, one could imagine that the genes were never turned on (expressed) in animals. This is clearly not so, as messenger RNA (mRNA) molecules and also protein products of the genes can be found in different cell types, sometimes expressed in particular stages of the cell division cycle. Second, one could speculate that *onc* gene products were overexpressed when under viral regulatory signals, i.e. in their viral form, compared to levels observed normally in cells. This possibility too has been ruled out for many of the cellular oncogene counterparts, but not for all. A third hypothesis is that the gene has cancer inducing activity if expressed in the wrong place or the wrong time. This theory is possible, there being at this time no evidence to rule it out definitively. A fourth alternative is that there are actual changes between the viral and cellular homologues. There is some evidence that such changes do occur, but again this does not apply to all oncogenes. Thus, both qualitative and quantitative changes may be responsible for oncogenic activity. For clarity, the term protooncogene is used for the cellular species; sometimes, the cellular counterpart may also be called a c-*onc* gene, but this term has a connotation that the gene is an oncogene, and it is best to distinguish it instead as one with only a *potential* oncogenic activity.

There is currently very active and widespread research to sort out the properties and functions of v-*onc* genes and protooncogenes, both similarities and differences, to understand what may cause the potential carcinogenic activity to become an actual one. Many of the problems have been tackled initially with v-*onc* genes which are more amenable to manipulation and dissection in the laboratory, and this will form the basis of the following sections.

The v-*onc* genes are useful for analysis of expression of the related protooncogenes in normal and tumour tissues. Messenger RNA preparations from different tissues, including embryonic tissues at different stages of development, were examined (in general, the technique known as Northern hybridization was used wherein mRNA molecules are fractionated by size during electrophoresis through agarose gels and then annealed to specific *onc* gene 'probes') to look for both the level of expression and the size of the specific mRNAs (transcripts). The most consistent finding was that most of these genes are expressed at some stage in some tissue; that is, they are regulated genes and participate in the cell's normal differentiation programme. It also became evident that the cellular *onc* genes may be abnormally expressed in certain tumours due to alterations in regulation, mutation, gene amplification, or chromosomal translocation (see also Chapter 10).

9.1.2 *Second encounter: cellular oncogenes*

While the RNA tumour viruses have given us a handle on nearly 20 oncogenes, there are probably at least 20 others that have been determined from studies on tumours. These have been found by three major approaches (i) gene transfer; (ii) insertional mutagenesis mapping of virally induced tumours; and (iii) analysis of known chromosomal translocations or amplifications.

Gene transfer. The primary method of gene transfer is often known by the name DNA transfection. The DNA is isolated from tumour cells and introduced into recipient cells. It may be added as a calcium phosphate precipitate to assist uptake, taken up following an electric shock (electroporation), or microinjected; the first is the most commonly used

Table 9.1 Retroviral oncogenes

onc	Retrovirus isolates [1]	v-onc origin [2]	v-onc protein [3]	Virus disease	v-onc product activity [4]	v-onc product location	Human chromosome
src	RSV	Chicken	pp60^{src}	Sarcoma	PK(tyr)	Inner side of plasma membrane	20q12-q13
	rASV	Quail	pp60^{src}	Sarcoma	PK(tyr)	Inner side of plasma membrane	
fps	FuSV-ASV	Chicken	P130$^{gag-fps}$	Sarcoma	PK(tyr)	Plasma membrane	15q24-q25
	PRCII-ASV	Chicken	P105$^{gag-fps}$	Sarcoma	PK(tyr)	Plasma membrane	
	PRCIV-ASV	Chicken	P170$^{gag-fps}$	Sarcoma	PK(tyr)	Plasma membrane	
	UR1-ASV	Chicken	P150$^{gag-fps}$	Sarcoma	PK(tyr)	Plasma membrane	
	16L-ASV	Chicken	P142$^{gag-fps}$	Sarcoma	PK(tyr)	Plasma membrane	
fes	ST-FeSV	Cat	P85$^{gag-fes}$	Sarcoma	PK(tyr)	Plasma membrane	
	GA-FeSV	Cat	P110$^{gag-fes}$	Sarcoma	PK(tyr)	Plasma membrane	
	HZ1-FeSV	Cat	P100$^{gag-fes}$	Sarcoma	PK(tyr)	Plasma membrane	
yes	Y73-ASV	Chicken	P90$^{gag-yes}$	Sarcoma	PK(tyr)	Plasma membrane	18
	Esh-ASV	Chicken	P80$^{gag-yes}$	Sarcoma	PK(tyr)	Plasma membrane	
fgr	GR-FeSV	Cat	P70$^{gag-actin-fgr}$	Sarcoma	PK(tyr)	?	1p36
ros	UR2-ASV	Chicken	P68$^{gag-ros}$	Sarcoma	PK(tyr)	Plasma membrane	6q22
abl	Ab-MLV	Mouse	P90-P160$^{gag-abl}$	Pre B cell leukaemia	PK(tyr)	Plasma membrane	9q34-qter
	HZ2-FeSV	Cat	P98$^{gag-abl}$	Sarcoma	PK(tyr)	Plasma membrane	
ski	Skn-rASV	Chicken	P110$^{gag-ski-pol}$ P125$^{gag-ski-pol}$	Squamous carcinoma	?	Nucleus	1q12-qter

Oncogene	Virus	Species	Product	Disease	Function	Location	Chromosome
erbA	AEV-ES4 AEV-R	Chicken Chicken	P75$^{gag\text{-}erbA}$ P75$^{gag\text{-}erbA}$	?	Thyroid hormone receptor	Cytoplasm Cytoplasm	(1): 17p11-q21 (2): 17
erbB	AEV-ES4 AEV-R AEV-H	Chicken Chicken Chicken	gp65^{erbB} gp65^{erbB} gp65^{erbB}	Erythroblastosis and Sarcoma	Truncated EGF receptor PK (tyr)	Plasma membrane	7pter-q22
fms	SM-FeSV	Cat	gP180$^{gag\text{-}fms}$ gp140^{fms} gp120^{fms}	Sarcoma	?PK(tyr) M-CSF receptor	Intermediate filaments, membranes	5q34
kit	HZ4-FeSV	Cat	P80$^{gag\text{-}kit}$	Sarcoma	*W* allele PK(tyr)	Plasma membrane	4q11-q21
sea	AEV-S13	Chicken	gP155$^{env\text{-}sea}$	Sarcoma, granulocytic leukaemia, and erythroblastosis	PK(tyr)	Plasma membrane	?
fos	FBJ-MSV FBR-MSV NK24-ASV	Mouse Mouse Chicken	pp55^{fos} P75$^{gag\text{-}fos\text{-}fox}$ P100$^{gag\text{-}fos}$	Osteosarcoma Osteosarcoma Nephroblastoma	DNA binding; complexes with *jun*	Nucleus	14q21-q31
mos	Mo-MSV Gz-MSV MPV-MSV	Mouse Mouse Mouse	P37$^{env\text{-}mos}$? p34^{mos}	Sarcoma Sarcoma Sarcoma, erythroleukaemia, and myelo-proliferation	PK(ser, thr); oocyte maturation factor	Cytoplasm	8q22

Table 9.1—*continued*

onc	Retrovirus isolates[1]	v-onc origin[2]	v-onc protein[3]	Virus disease	v-onc product activity[4]	v-onc product location	Human chromosome
sis	SSV	Monkey	$P28^{env\text{-}sis}$	Sarcoma	Truncated PDGF (B chain)	Membranes ? secreted	22q11-qter
	PI-FeSV	Cat	$P76^{gag\text{-}sis}$	Sarcoma			
	FT-FeSV	Cat	?	Sarcoma			
	TP2-FeSV	Cat	?	Sarcoma			
myc	MC29	Chicken	$P110^{gag\text{-}myc}$	Sarcoma, carcinoma, and myelocytoma	DNA binding	Nucleus (P100 cytoplasmic)	8q24
	MH2	Chicken	$P100^{gag\text{-}mil\text{-}myc}$ $P58^{myc}$				
	CMII	Chicken	$P90^{gag\text{-}myc}$	Myelocytoma			
	OK10	Chicken	$P200^{gag\text{-}pol\text{-}myc}$	Carcinoma			
	FeLV-*myc*	Cat	?	Granulocytic leukaemia			
crk	CT10-ASV	Chicken	$p47^{gag\text{-}crk}$	Sarcoma	Phospholipase C-related; regulator of PK(tyr) activity	Cytoplasmic	?
maf	AS42-ASV	Chicken	?	Sarcoma	?	Nucleus	?
cbl	CasNS-MSV	Mouse	$P100^{gag\text{-}cbl}$	Pre B leukaemia	?	?	?
jun	S17-ASV	Chicken	$P65^{gag\text{-}jun}$	Sarcoma	Complexes with *fos*	Nucleus, transcription factor?	1p31-p32

	Virus	Species	Protein	Tumour	Function	Localization	Chromosome
myb	AMV-BAI/A AMV-E26	Chicken Chicken	p45^{myb} P135$^{gag\text{-}ets\text{-}myb}$	Myeloblastosis Myeloblastosis and erythroblastosis	?	Nucleus	6q22-q24
rel	REV-T	Turkey	p64^{rel}	Reticulo-endotheliosis	?	Nucleus	2
raf	3611-MSV	Mouse	gP90$^{gag\text{-}raf}$ P75$^{gag\text{-}raf}$	Sarcoma	PK(ser, thr)	?	(1): 3p25 (2): 4[5]
mil	MH2	Chicken	P100$^{gag\text{-}mil\text{-}myc}$	See above for *myc*			
Ha-*ras*	Ha-MSV	Rat	pp21^{ras}	Sarcoma and erythroleukaemia	GTP binding GTPase	Plasma membrane	(1): 11p13 (2): X[5]
	RaSV	Rat	P29$^{gag\text{-}ras}$	? Sarcoma			
	BALB-MSV	Mouse	pp21^{ras}	Haemangiosarcoma			
Ki-*ras*	Ki-MSV	Rat	pp21^{ras}	Sarcoma and erythyroleukaemia	GTP binding GTPase	Plasma membrane	(1): 6p32-q13[5] (2): 12p12-pter
	NY-FeSV	Cat	?	Sarcoma			
ets	AMV-E26	Chicken	P135$^{gag\text{-}ets\text{-}myb}$	See above for *myb*		Nucleus	(1): 11q23-q24 (2): 21q22

[1] RSV, Rous sarcoma virus; ASV, avian sarcoma virus; FeSV, feline sarcoma virus; FeLV, feline leukaemia virus; MLV, murine leukaemia virus; AEV, avian erythroblastosis virus; SSV, simian sarcoma virus; AMV, avian myeloblastosis virus; REV, reticuloendotheliosis virus; MSV, murine sarcoma virus. The other notations indicate specific viral strains.

[2] The origin denotes the species of animal from which the *onc* gene was transduced by a retrovirus. Note that in some cases the same *onc* gene has been transduced by retroviruses from different animals.

[3] The nomenclature has been standardized as follows: p, protein; pp, phosphoprotein; P, fusion protein between a retroviral replicative gene (*gag, pol,* or *env*; see Fig. 9.1) and the *onc* sequence (note that this may influence the intracellular localization); gp or gP, glycoprotein. The numerals denote the molecular weight in kilodaltons, generally deduced from polyacrylamide gel electrophoresis.

[4] PK, protein kinase; tyr, tyrosine; ser, serine; thr, threonine; EGF, epidermal growth factor; PDGF, platelet derived growth factor; M-CSF, macrophage colony stimulating factor. See text for further details.

[5] These genes lack introns and are probably not transcribed (pseudogenes).

technique. The recipient cells are most usually the 'normal' immortalized mouse fibroblast line, NIH/3T3. The NIH cells have several advantages in this system: (i) they are flat, contact inhibited cells in monolayer culture; and (ii) their DNA can be distinguished in hybridization experiments from DNAs of other species (thus allowing identification of donor DNA). The first property is the most important as the general goal is to examine the transfected cultures for the appearance of morphologically altered (transformed) foci due to the expression of an introduced oncogene. This procedure is represented diagrammatically in Fig. 9.2. Although a lot of DNA is taken up by an individual cell, the majority of this material is degraded with time, and only a minority becomes stably incorporated into the cellular genome. To identify the gene responsible for the cell's transformed phenotype, transfection of DNA from the transformed cells is carried out a second and third time; a procedure that eventually whittles down the amount of non-murine DNA carried over.

One of the surprises from such analyses was that cellular genes related

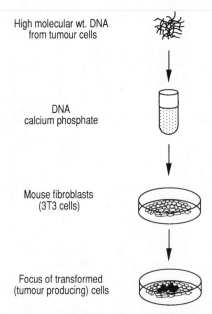

High molecular wt. DNA
from tumour cells

DNA
calcium phosphate

Mouse fibroblasts
(3T3 cells)

Focus of transformed
(tumour producing) cells

Fig. 9.2 DNA transfection. DNA isolated from tumour cells is introduced into recipient NIH/3T3 mouse cells. After several weeks, the monolayer is observed for foci of morphologically transformed cells. Repeated cycles of DNA transfection from transformed foci are generally performed. The transformed cells may be tested for tumorigenicity in nude mice. DNA from transformed cells is also examined for retention and identification of donor cell sequences.

to the already known *ras* genes of retroviruses were found to be responsible for the transformed phenotype in about 20 per cent of human tumours tested. This could be due, at least in part, to the ability of NIH cells to respond morphologically to a given oncogene product, to the need for the gene to be genetically dominant (e.g. lack of expression of an inhibitor gene in the NIH cell), and to the need for the gene to be sufficient for the expression and maintenance of the transformed state. On second thought, perhaps it is not unusual to find that few genes other than *ras* related genes have been identified as being 'activated' (see p. 242, 251) in tumours. This may be due to the NIH cells being of the wrong lineage or species to respond to the effects of other genes. Or one may consider them to have a preneoplastic phenotype (see Chapter 6) as they already have the ability to grow continuously in culture, thus they may be more susceptible to genes whose effects are manifest during the later stages of tumour progression. Nonetheless, new oncogenes have been detected by this technique (Table 9.2).

Insertional mutagenesis. Retroviruses lacking v-*onc* gene sequences carry only the genes required for replication; after infection, a DNA copy of their genome is synthesized and becomes integrated within chromosomal DNA of the cell (the provirus, see Fig. 9.1). This event by definition is an insertional mutation and can have several consequences. The long terminal repeat (LTR) structures do not code for proteins but do contain the regulatory elements ('promoters' for initiation of transcription of mRNA molecules from the DNA, and 'enhancers' that may influence the levels of transcription of adjacent or distant genes) which permit transcription of the viral genes into mRNA molecules so that viral proteins are produced and incorporated into progeny virus particles. However, as shown in Fig. 9.1, the replication and insertion of a provirus leads to the generation of two LTRs; one may be used for the production of progeny virus whereas the other may be used for promotion of an adjacent cellular gene ('downstream promotion'). The first example of this phenomenon came from the examination of B cell tumours induced in chickens by avian leukaemia viruses (ALVs) lacking *onc* genes. Hybridization of LTR probes to tumour RNA revealed not only transcripts expected from the proviral genome but also some that contained an LTR but no viral sequences; the remaining sequences were derived from cellular genetic material. Further analysis showed that the cellular gene was in fact the counterpart of the *myc* gene already identified in several avian viruses (Table 9.1 and Fig. 9.3). The integration event resulted in the use of the LTR as a promoter for transcription of the cellular *myc* gene, leading to inexorable production of *myc* RNA and protein. Upon analysis of many of these B cell tumours, it was found

Table 9.2 Non-retroviral oncogenes

Name	Tumour of origin	Method of detection	Human chromosome
hst	Stomach tumour		11q13
N-*myc*[1]	Neuroblastoma	Amplification	2p23-p24
L-*myc*[1]	Lung carcinoma		1p32
neu[1,4]	Mammary carcinoma		17
N-*ras*[1]	Neuroblastoma[2,3]	Transfection	1cen-p21
dbl	B cell lymphoma		
K-*fgf* (= *hst*)	Kaposi's sarcoma		11q13
*mcf*2	Mammary carcinoma		
*mcf*3	Mammary carcinoma		
met	Chemically transformed osteosarcaoma cells		7p11-p14
neu	Neuroblastoma[4]		17
onc-D	Colon carcinoma		
ras related	Melanoma		
Tlym-1, -2	T cell lymphoma		
trk	Colon carcinoma		
tx-1	Mammary carcinoma		
tx-2	Pre B cell leukaemia		
tx-3	Plasmacytoma		
tx-4	T cell lymphoma		
bcl-1	B cell leukaemia	Translocation[5]	11q13
bcl-2	B cell lymphoma		18q21
bcr	Chronic myelocytic leukaemia		22q11
tcl-1	T cell lymphoma		14q32
$T_k NS$-1	Plasmacytoma[6]		
int-1	Mammary carcinoma	Insertional mutation[6]	12q14-pter
int-2	Mammary carcinoma	(all rodent)	11q13
Mlvi-1, -2, -3	T cell lymphoma		
Pim-1	T cell lymphoma		
Pvt-2	Plasmacytoma		
RMO-*int*-1	T cell leukaemia		
fim-1, -3	Myeloid leukaemia		
evi-1, -2	Myeloid leukaemia		
Spi-1	Erythroid leukaemia		
fis-1	Myeloid and lymphoid leukaemia		

[1] Related to *onc* genes already known from retroviruses (see Table 9.1). The *neu* gene is related to the EGF receptor gene (c-*erbB*). Also called c-*erb*B2 or HER2.
[2] Also in rodent tumours.
[3] See also Table 9.5 for mutated N-*ras* genes in other human tumours.
[4] Rodent tumour.
[5] These genes represent breakpoint regions; no oncogenic potential has yet been demonstrated for any genes in this region.
[6] Targets of proviral insertion in rodent tumours. No oncogenic potential has yet been demonstrated for the activated gene *per se.*

(a)

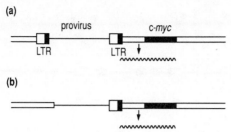

(b)

Fig. 9.3 Promoter insertion. In the example shown here, DNA from a B cell tumour (bursal lymphoma) of chickens induced by an avian leukaemia virus (ALV) has been examined for the site of provirus insertion. The provirus (shown as a single line bounded by LTRs) is located 'upstream' of the cellular gene known as *myc* and the viral LTR is being used as a promoter for the transcription of high levels of *myc* gene messenger RNA (mRNA), shown as a wavy line. For simplicity, the *myc* gene is shown as a contiguous element whereas it is actually composed of a 5′ untranslated exon and 3 coding exons separated by regions of non-coding sequences known as introns (the introns are also removed, 'spliced out', during the processing to mature mRNA molecules). (a) Provirus is intact. (b) Provirus shows deleted structure, lacking the 5′ LTR and some coding sequences.

that there was no specific integration of the ALV provirus at a unique site upstream of the *myc* gene; the integration sites in individual tumours were often several kilobases different from one to another. ALV proviruses can integrate at a multitude of sites, even perhaps totally at random; however, the integrations near to *myc* presumably led to some selective advantage for the particular cell in proliferation, which might prime such cells for other neoplastic events to occur, or which might be oncogenic *per se*. Interestingly, the involvement of *myc* promoter insertions has also been observed in murine T cell leukaemias induced by murine retroviruses and in feline T cell leukaemias induced by feline retroviruses. Further, the previously known *erbB* oncogene (known as the counterpart of an oncogene isolated from avian erythroleukaemia viruses, AEV) has also been activated by promoter insertion by other strains of ALV in situations where they induced erythroleukaemia instead of B cell leukaemia. In many of these instances, the integrated provirus is no longer intact, i.e. one of the LTRs, as well as coding sequences, may be lost, suggesting that the provirus itself is no longer required for maintenance of the neoplastic state (Fig. 9.3).

A different phenomenon occurred when the integration sites for another retrovirus group, mouse mammary tumour viruses (MMTV, see Chapter 8) were mapped in a large number of tumours. As illustrated in Fig. 9.4, there were apparently clusters of regions where proviruses inserted. However, the LTRs in most cases could not be used to promote transcription of downstream genes. In the majority of tumours, it turned

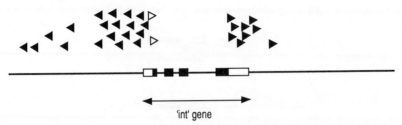

'int' gene

Fig. 9.4 Proviruses in mouse mammary tumour virus (MMTV) induced breast tumours. The integration sites of MMTV proviruses in a large number of tumours have been mapped chromosomally. Each site is denoted by an arrow-head and the direction of the arrow shows the direction of transcription of the provirus (5' to 3'). Note that the activated gene is in a central region (designated *int* for integration) around which the proviruses cluster and that the polarity of the provirus is in an orientation such that downstream promotion cannot occur. There are at least two well characterized *int* loci, *int-1* and *int-2*; others may also exist.

out that there were regions bounded by the provirus insertions and that this central region contained a gene that was activated in the tumour and silent in normal mammary tissue. The proviruses were oriented in a polarity that proscribed transcription of the central gene from the viral LTR, but presumably the enhancer element within the LTR was able to activate this cellular gene. In fact, (at least) two cellular genes may become expressed in this manner in different murine mammary carcinomas and have become known as the *int-1* and *int-2* loci. Thus, a second mechanism of insertional mutagenesis has been identified. Integration sites in other tumours have delineated further activated genes (Table 9.2). However, it must be stressed that these genes have not yet been shown to be oncogenes in a definitive test.

Chromosomal abnormalities. Various chromosomal abnormalities have been detected in tumours by karyotype analysis of metaphase chromosomes; some abnormalities seem specific for a distinctive tumour type (see Chapter 10). Because the *onc* genes mentioned above, whether originally identified in retroviruses or by gene transfer, all have cellular counterparts, each one can be mapped to a distinct region of a particular chromosome (Tables 9.1 and 9.2, and Chapter 10). These two observations were linked to ask two questions. First, did any of the known *onc* genes map in or near regions involved in abnormal tumour karyotypes? Second, could the abnormalities be used to discover new oncogenes?

The foreknowledge of promoter insertion of *myc* in ALV induced B cell tumours quickly played a role. Burkitt's lymphoma is a B cell tumour of humans in which both Epstein Barr virus and malaria are known cofactors (see Chapter 8). Additionally, it was previously known that tumour cells contained one of three chromosomal translocations (trans-locations refer to movement of genetic sequences from the normal chromosome site to another chromosome, see Chapter 10), involving part of the long arm of chromosome 8 becoming translocated to chromo-some 14, 2, or 22 (the first occurring in approximately 90 per cent of tumours examined). The *myc* locus mapped to the long arm of chromo-some 8 and it became obvious to examine whether it was translocated in these tumours. The simple answer was yes. Even more exciting was the finding that the *myc* locus translocated to regions containing different immunoglobulin loci (the heavy chain, κ and λ light chains, respectively, on chromosomes 14, 2, and 22). Mechanistically, this meant that the *myc* locus could now be found adjacent to a highly transcribed gene (an important function of mature B cells being that of immunoglobulin production, see Chapter 15). The generation of the translocation also tended to have another effect: the non-coding regulatory sequences of the *myc* gene were frequently 'decapitated' from the gene, thus leaving it to the controls of its new chromosomal milieu. In most instances the altered *myc* allele was shown to be transcriptionally active while the non-altered one was silent. Moreover, the *myc* allele was sometimes located near the immunoglobulin enhancer element which could affect the level of *myc* transcription. Similar sorts of rearrangements also occur in immunoglobulin producing tumours (plasmacytomas) of mice and rats.

Another particularly well known translocation was the formation of the 'marker' Philadelphia chromosome in chronic myelocytic leukaemias (CML, see Chapters 10 and 12) which represents a translocation between chromosomes 9 and 22 (although other variants are known which may involve segments from three different chromosomes). The net result is that the *abl* locus (known from retroviruses) becomes trans-ferred from chromosome 9 to sequences on chromosome 22. The *abl* gene is located rather a long distance away from the site of the breakpoint (and hence is presumably not dissociated from its upstream regulatory sequences), but transcription of the gene is abnormal in that the *abl* mRNA molecules are much greater in size than normal.

Many other tumours contain recognizable translocations (see Chapter 10) and the translocation breakpoints of some of these have already been molecularly cloned. Although often listed as cellular oncogenes (e.g. some are included in Table 9.2), this remains a tenuous assignment until the sequences are shown experimentally to have neoplastic potential.

Another type of chromosomal abnormality seen in tumours is related to gene amplification and generally involves huge (100–1000 kilobases or more) DNA segments spanning many genes. The amplification may be observed as a homogeneously staining region (HSR) in which it occurs as a contiguous element in a chromosome, or as the formation of double minute (DM) chromosomes which are additional tiny mini-chromosomes (although they lack centromeres and therefore may be lost, or segregate, during cell division). The availability of v-*onc* gene probes proved useful in demonstrating that HSRs and DMs observed in cell lines derived from various tumours contained *onc* gene amplifications (e.g. Ki-*ras* in a mouse adrenocortical tumour cell line, *myc* in a human colon carcinoma line). Needless to say, these particular types of abnormalities require large degrees of gene amplification (large DNA segments repeated perhaps 50 or 100 times) to be detected by micro-scopic analysis of chromosomes. More sensitive techniques such as Southern blotting and hybridization (in which DNA is cut by endo-nucleases known as restriction enzymes, separated by size in gels, and transferred to nitrocellulose filters for hybridization to DNA probes) allow the detection of low levels of gene amplification or indeed the disruption or rearrangement of a single gene (Table 9.3).

One particularly interesting finding is the N-*myc* amplification in neuroblastomas. This property is observed only in those tumours histo-logically defined as grades III and IV and thus may be a reflection of tumour progression toward a more advanced malignant state. Similarly,

Table 9.3 Oncogene amplification

Gene	Human tumour	Degree of amplification
c-*myc*	Promyelocytic leukaemia[1]	20
	Colon carcinoma	40
	Small cell lung carcinoma	5–30
N-*myc*	Neuroblastoma[1]	5–1000
	Retinoblastoma[1]	10–200
	Small cell lung carcinoma[1]	50
c-*abl*	Chronic granulocytic leukaemia	5–10
c-*myb*	Colon carcinoma	10
	Acute myeloid leukaemia[1]	5–10
c-*erbB*	Vulval carcinoma	30
c-Ki-*ras*-2	Lung carcinoma[1]	
	Colorectal carcinoma[1]	4–20
	Bladder carcinoma[1]	
N-*ras*	Mammary carcinoma	5–10

[1] Primary tumour material.

myc amplification in cell lines derived from small cell lung carcinomas is correlated with an enhanced growth of the tumours when transplanted into nude mice.

Nonetheless, it should be remembered that large segments of DNA are reiterated in these amplifications and the *onc* gene may be merely a passenger in this event. Thus, other genes in the amplified regions could have significant consequences.

9.2 Identification of viral and cellular oncogene products

It is important to ascertain the physiological functions of oncogene products and to determine (or at least speculate) on how these products may lead to neoplastic changes with a cell. Such analyses include the sub-cellular localization of the gene product and deciphering biochemical functions associated with the expression of the product (see Table 9.1). To a large extent, these experiments rely on recombinant DNA technology, specific antisera, and luck.

9.2.1 *The tyrosine kinase family*

src. The Rous sarcoma virus (RSV) of chickens was one of the earliest retroviruses isolated and has turned out to be one of the most interesting from a genetic and biochemical standpoint. It is the only naturally occurring retrovirus with all replicative genes as well as an oncogene (see Fig. 9.1). The oncogene, v-*src*, was transduced from cellular DNA in such a way that the replicative genes of its presumptive parent virus (an avian leukaemia virus) remained intact with the new sequences appearing appended to the viral *env* gene. Early genetic experiments showed that inactivation of the v-*src* gene by mutations or deletions left a long latency leukaemia virus that could no longer cause morphological transformation of cells in culture.

With recombinant DNA technology, it was possible to obtain molecular clones of proviral DNA and the v-*src* (and subsequently cellular *src*) gene was characterized down to its last nucleotide. Comparisons showed that the nucleotide sequence of the cellular and viral genes were essentially identical. Two major differences could be noted however: (i) the v-*src* gene lacks introns (some time during the process of transduction this occurs with all oncogenes picked up by retroviruses wherein the introns are 'spliced' out); and (ii) the last (carboxy terminal) 12 coding amino acids of v-*src* were different from those of cellular *src* (which is also 7 amino acids longer). Thus, transduction leads to splicing, truncation, and perhaps other sorts of recombinational events—a motif seen frequently in the derivation of the *onc* containing retroviruses.

The next important breakthrough came with the development of an antiserum that reacted specifically with the *src* gene product, obtained from an animal bearing an RSV induced tumour. The serum was used to show that the *src* gene product was located largely on the cytoplasmic side of the plasma membrane and was approximately 60 000 daltons in molecular weight. By metabolically labelling transformed cells with ^{32}P, it was shown that the protein was phosphorylated, and hence became known as pp60$^{v\text{-}src}$. The protein of the cellular gene also shared these properties and was called pp60$^{c\text{-}src}$.

One of the most interesting discoveries from this study was that the protein was apparently responsible for its own phosphorylation (auto-phosphorylation) and also that it had the ability to phosphorylate the immunoglobulin (Ig) molecule in the antiserum that was used to identify it. Further studies showed that the two proteins were phosphorylated on tyrosine residues and thus the oncogene product was identified function-ally as a tyrosine protein kinase (cellular enzymes that phosphorylated serine or threonine residues were previously known, but this was the first time that a tyrosine kinase was characterized).

Several of the proteins phosphorylated by *src* are of interest. One is the protein vinculin, a component of the cytoskeleton network of mesen-chymal cells which is situated at areas of contact between cells or between a cell and a substrate (such as tissue culture dishes) known as focal adhesion plaques. The change in the phosphorylation pattern of vinculin could be responsible for some of the observable changes in transformed cells (rounding up, decreased adhesion to solid substrates, ruffling of the plasma membrane). Another interesting target is a protein known as p90 (so designated because it is 90 kilodaltons in molecular weight) which with another protein (p50) serves to transport pp60^{src} from its site of synthesis to the plasma membrane. While complexed with these two proteins, the *src* protein is neither phosphorylated nor does it exhibit kinase activity. The *src* product is also capable of phosphorylat-ing lipids (i.e. it is a phospholipid kinase) such as phosphatidylinositol and diacylglycerol. These molecules are believed to function as 'second messengers' in response to growth factor stimulation, an important aspect of cell metabolism (see Chapter 11). Several enzymes of the glycolytic pathway are also targets for *src* kinase activity. Despite these identifiable changes, we are still a long way from demonstrating a direct link between these effects and the ultimate heritable change necessary for a cell to transform and develop into a neoplasm.

Guilt by association. As further viral and cellular oncogenes were isolated, they were tested for a similar kinase function, and a number were found to be positive (Table 9.1). Each represented the transduction of a distinct cellular gene, as the cellular homologues mapped to different chromosomes. Nucleotide sequence data confirmed this and also pointed out that there was a small region of sequence homology presumably related to the shared biochemical activity. The oncogene *erbB* also shares this sequence, but definitive evidence for tyrosine kinase activity is more difficult to establish (see Chapter 11).

Bearing in mind that a rather large class of genes were turning out to share this biochemical activity, oncogenes themselves were used as probes against genomic or cDNA libraries to pick up genes bearing sequence homology. As a number of these were isolated, it also became clear that the tyrosine kinases could be divided into two classes: (i) those that were membrane associated without clear transmembrane or extracellular domains (like *src*); and (ii) those that had apparent extracellular domains (like *erbB*) (Table 9.4). The latter class contained many members of the growth factor receptor category, with particular interest for studying proliferative stimuli.

There were two obvious questions at this stage. Do the cellular *onc* counterparts also display kinase activity? Could specific proteins in the cell be identified as target substrates? To the first question, the answer was generally yes, in that the *onc* proteins were capable of autophosphorylation, but often did not show any ability to phosphorylate Ig molecules. Answering the second question was more difficult. Levels of phosphotyrosine containing peptides in normal non-transformed cells

Table 9.4 Tyrosine protein kinases

Membrane-associated	Transmembrane
abl[1]	*eph*
arg	*erbB*[1] (epidermal growth factor receptor)
fes[1]	*erbB2*
fgr[1]	*fms*[1] (monocyte colony stimulating factor receptor)
fyn	*kit*[1] (*W* locus)
hck	*met*
lyk	*neu*
lyn	*ret*
src[1]	*ros*[1]
tkr	*sea*[1]
yes[1]	*trk*
	Insulin receptor
	Insulin-like growth factor type I receptor
	Platelet-derived growth factor receptor

[1] First isolated as a retroviral oncogene.

are extremely low; thus, despite demonstrated kinase activity for c-*onc* gene products *in vitro*, their activity is virtually undetectable in cells. However, there is often a considerable elevation of phosphotyrosine in cells transformed with the kinase v-*onc* genes, suggesting that the v-*onc* products have greater activity. However, many different species of proteins seem to be affected rather than one or two specific target molecules.

Lastly, one is interested in asking if any of these growth factor receptors can be turned into oncogenes by mutations similar to those of the viral counterparts [generally 5′ (ligand binding) or 3′ (catalytic) subunit truncation]. Of these, only the insulin receptor (when truncated and inserted into a provirus) and the EGF receptor (when truncated in nature as the viral oncogene *erbB*) have been proved to be true oncogenes; the others remain to be tested.

kit. The oncogene *kit*, first discovered in a cat sarcoma virus, is a member of the transmembrane protein kinase like *fms* and *erbB* (see Table 9.4 and Chapter 11). What makes it stand out from the rest at this time is its apparent identity with the gene called *W* in mice. The initial relationship was hinted at by the finding that *kit* maps in the human genome at chromosome 4q11-q21, a region containing the genes for phosphoglucomutase-2, α-fetoprotein, and albumin which are conserved together on a region of mouse chromosome 5 (synteny) where *W* maps. The *W* gene (or locus as it is called more familiarly) has been a gene of much interest since the 1950s because of its pleiotropic effects on the stem cells of at least three distinct lineages: haematopoieses (the animals suffer from severe macrocytic anaemia and mast cell defects), melanogenesis (the *W* stands for dominant *w*hite spotting which is an early discernible phenotype), and gametogenesis (the animals are infertile). In some of the mutants of the *W* family, one or another of these lineages may appear normal (e.g. the animals have normal red blood cells or are fertile), and the white spotting may occur to greater or lesser extents; these differences may represent the degree of penetrance of the gene or extent of the mutant protein's kinase activity.

Although the initial experiments only suggested that *W* and *kit* might be the same gene, recent work has added more definitive evidence of their identity: (i) the W^{19H} mutant which visibly lacks part of mouse chromosome 5 also lacks the *kit* gene; (ii) other *W* mutants have been shown to have point mutations in their *kit* gene sequence; and (iii) all *W* mutants lack functional (tyrosine kinase) expression of *kit* although *kit* RNA may still be present in tissue wherein *kit* is normally active (e.g. erythroid cell progenitors, mast cells and testes). Interestingly, *kit* is also expressed at a high level in the brain, and is lacking in brain tissue of *W* mutant mice, although no noticeable defect in neurological function has yet been described! Taking these data further, there is interest in examining patients with two gene defects that map to this region of human chromosome 14q: those with the piebald trait (showing streaks of different hair colour and sometimes with mental retardation as well), and the haematological disorder known as Diamond–Blackfan syndrome. Further, it would certainly be of interest to search for a *Drosophila* homologue of *kit* (see p. 265) to discover if any mutations have been found giving rise to concordant phenotypes (e.g. infertility).

The *W/kit* relationship is the first example of a germ-line mutation in a mammalian protooncogene.

crk: *an oncogene related to phospholipase C.* Another oncogene for which a biochemical function has been postulated is *crk* (for *C*T-10 virus *r*egulator of *k*inase) now known to have been transduced from chickens into two avian sarcoma viruses. Initial studies on the viral gene product showed that it did not have intrinsic tyrosine protein kinase activity like many other viral oncogenes, although cells transformed by the virus had elevated levels of intracellular phosphotyrosine proteins. When the putative amino acid sequence of *crk* was entered into a protein database, it was observed to have a sequence related to the enzyme phospholipase C which is postulated to play a key role in signal transduction mechanisms, particularly in the activation of protein kinase C (see Chapter 11). These relationships suggest that the *crk* oncogene might function as a regulator of the endogenous non-receptor (membrane-associated) class of tyrosine protein kinases. Thus, once again an oncogene has been implicated as a potential disrupter of the cell receptor-growth factor recognition pathway.

9.2.2 *The other kinases:* raf *and* mos

Two other oncogenes, *raf* (also called *mil*) and *mos*, show kinase activity but, in these instances, serine and threonine are the substrate sites rather than tyrosine. Expression of *raf* (of which there are also several related genes) is fairly ubiquitous in mammalian tissues, which makes it difficult to speculate on a function. On the other hand, the oncogene *mos* was originally thought to be a pseudogene because the gene lacked introns and no expression could be detected in adult tissues. Then in the last two years, its expression was detected at very low level in testes and oocytes, and the pursuit was on for determining some physiological function in gametogenesis or early development. By far the most compelling evidence at this stage is its pattern of expression in the oocytes of the frog *Xenopus*. Although the cellular *mos* RNA is expressed from early oogenesis and persists through gastrulation, the protooncogene product, pp39^{mos}, rapidly appears following progesterone treatment (which stimulates maturation) and disappears extremely rapidly following fertilization. Investigations with *mos* RNA and monoclonal antibodies suggest that the *mos* protein is the key to holding the egg cell in metaphase arrest during meiosis and subsequently it becomes a relatively 'silent' gene. It is a little difficult to extrapolate from this scenario to the situation when the gene becomes oncogenic (as in the murine leukaemia virus transductions—Table 9.1—and in insertional mutagenesis in some rodent plasmacytomas) in which its role is as a proliferative signal rather than an anti-proliferative one. Of course, it may be simply a question of its expression in the wrong cell type at the wrong time, or the altered gene product may phosphorylate different target substrates.

9.2.3 ras *and GTPase activity*

The *ras* oncogene was originally discovered in two murine retroviruses, known as Harvey and Kirsten murine sarcoma viruses (Ha-MSV and Ki-MSV). By nucleic acid hybridization, the *onc* genes of the two viruses were non-homologous. This turned out to be due to the degeneracy of the triplet code of DNA because the *onc* gene products from the two viruses were virtually identical, p21^{ras}. Moreover, it was found that both Ha-*ras* and Ki-*ras* were each found at two separate chromosomal loci, although in each case, one locus appeared to be a pseudogene (a gene sequence lacking introns and thought to be untranscribed and thus not translated into protein, Table 9.1). Later, a gene showing homology to Ha-*ras* and Ki-*ras* was detected in neuroblastoma DNA by transfection (see Table 9.2); this became known as N-*ras* and was mapped to yet another chromosome. These five different *ras* genes became known as the *ras* multigene family. By using conserved parts of the *ras* gene, a number of other *ras*-related genes have been detected in mammalian genomic DNA as well as in lower species, such as yeast.

The viral p21^{ras} proteins were shown to have binding activity for guanosine triphosphate (GTP) and in fact enzymatically converted GTP to its di- and mono-phosphate forms. Thus, *ras* too is an enzyme, a GTPase. Comparisons of enzyme activity showed that p21$^{v\text{-}ras}$ has less activity than p21$^{c\text{-}ras}$.

How did the change in activity come about? Nucleotide sequence analysis showed that in viral transductions (three separate isolates, the third being another capture of Ha-*ras*, see Table 9.1), the *ras* gene was intact, i.e. it was not truncated. But in each instance, there was one nucleotide change which in turn led to the substitution of a different amino acid compared to the cellular *ras* protein. The same event occurred in the neuroblastoma N-*ras* gene as well. These results, as well as those from other tumour transfections or site specific mutagenesis *in vitro*, showed that changes in the *ras* protein at amino acid residues 12 or 13, or residues 59–61 almost invariably led to a change in the oncogenic potential (Table 9.5, see also Chapters 10, 11, and 12). The easiest explanation is that these changes lead to major alterations in the conformational (three-dimensional) state of the protein thereby diminishing its activity.

Mutant *ras* sequences are the most frequently detectable alterations in some types of human tumours (Bos 1988). In all cases tested to date, the single point mutations listed above are responsible for the changes in biological activity, keeping the protein in the active GTP form.

Table 9.5 Mutated human *ras* genes

		Codon	
		12	61
c-Ha-*ras*-1[1]	Normal	gly	gln
	Bladder carcinoma	val	
	Lung carcinoma		leu
	Mammary carcinoma	asp	
c-Ki-*ras*-2[2]	Normal	gly	gln
	Lung carcinoma	arg	
	Lung carcinoma	cys	
	Lung carcinoma	lys	
	Lung carcinoma		his
	Colon carcinoma	val	
	Bladder carcinoma	arg	
	Neuroblastoma	cys	
N-*ras*[3]	Normal	gly	gln
	Neuroblastoma		lys
	Teratocarcinoma	asp	
	Fibrosarcoma		lys
	Melanoma		lys
	Lung carcinoma		arg
	Leukaemia	asp	
	Rhabdomyosarcoma		his

[1] Mutated c-Ha-*ras*-1 also detected in a human melanoma.
[2] Mutated c-Ki-*ras*-2 also detected in human pancreatic carcinoma, gall bladder carcinoma, rhabdomyosarcoma, ovarian carcinoma, and acute lymphoblastic leukaemia.
[3] Mutated N-*ras* also detected in Burkitt's lymphoma, acute promyelocytic leukaemia, T cell leukaemia, acute myelocytic leukaemia, and chronic granulocytic leukaemia.

Another clue for *ras* function comes from studies on the yeast *Saccharomyces* which has two *ras* genes that stimulate adenylate cyclase. This finding led to the deduction that *ras* could be a regulator of adenylate cyclase similar to the known G proteins that transmit signals from cell surface receptors ('second messages', see Chapter 11). Binding of GTP is necessary for activation of G proteins and thus GTPase activity is a regulatory factor. The diminished GTPase activity of mutated *ras* proteins could lead to sustained effects on adenylate cyclase and the ensuing events in cell metabolism. The location of the *ras* protein on the cytoplasmic side of plasma membranes adds credibility to this hypothesis. Once again, how this leads to oncogenesis is unknown.

Another piece of evidence suggesting the involvement of *ras* proteins in signal transduction comes from studies using microinjection of anti-*ras* neutralizing antibodies into cells *in vitro*. The net result is a block to cell proliferation normally induced by a variety of mitogenic stimuli such as growth factors and activators of protein kinase C (see Chapter 11). Conversely, microinjection of *ras* protein into quiescent cells stimulates DNA synthesis. However, if such cells are treated so as to down-regulate protein kinase C, *ras* is unable to induce DNA synthesis. Interestingly, mutant *ras* protein-induced DNA synthesis also has an absolute requirement for insulin-like growth factor type I (IGF-I, see Chapter 11) whereas morphological transformation is independent of this factor (Lloyd *et al.* 1989).

While the precise nature of the *ras* effector system remains unknown, recent findings implicate a protein known as GAP (*G*TPase *a*ctivating *p*rotein) in the pathway by regulating the level of bound GTP.

9.2.4 *Oncogenes with products related to growth factors*

sis and platelet derived growth factor activity. The *sis* oncogene was originally found in a simian sarcoma virus and has a protein product, p28^{sis}, apparently found in the cytoplasm. The use of a computer database was the key to deciphering its physiological function. When the amino acid sequence of several tryptic peptide digests of the growth factor known as platelet derived growth factor (PDGF) was entered into the database, it was immediately obvious that the amino acid sequence was highly homologous to that of the putative protein product of v-*sis*. Further experiments verified this observation, showing that v-*sis* represented a truncated form of the B chain of PDGF (this is discussed in greater detail in Chapter 11).

This observation was very exciting in regard to a central theme of tumours: continuous cell proliferation. Normally, PDGF is packaged in the subcellular blood components known as platelets which are released by the disintegration of blood megakaryocytes. The release is generally triggered as a response to a wound and the local delivery of PDGF stimulates, for example, endothelial or epithelial cells around the wound to proliferate and fill in the gap. Once healing has begun, platelets are no longer released, thus decreasing local PDGF concentrations; the net result is that the cells become quiescent again. However, if PDGF were to be continuously produced, the cells would be constantly stimulated to divide. So the function of v-*sis* in virally induced tumours might be envisaged in one of two ways: (i) as an intracellular product that can still stimulate the specific PDGF receptor such that its 'signal' for cell division is transmitted to the cell nucleus (an autocrine mechanism); or (ii) as a secreted product interacting with specific PDGF receptors on its own, neighbouring (paracrine), or distant (exocrine) cells, thus stimulating cell growth.

Although PDGF (cellular *sis*) is overexpressed in at least one human osteosarcoma cell line, this does not seem to be a general feature of bone or other tumours so far examined.

int-2, hst, *and K*-fgf: *genes related to basic fibroblast growth factor.* As mentioned above, the *int-2* gene was first identified as a target of MMTV proviral insertion in mouse mammary tumours. The effect of integration was the activation of this previously silent gene. In fact, when a variety of adult mouse tissues were examined, no *int-2* gene expression could be detected. Further, when whole mouse embryos were surveyed, the only traces of expression were in early (6.5 day) embryos with a limited temporal and spatial distribution.

The amino acid sequence of the *int-2* gene product bears striking homology to the growth and angiogenic factor known as basic fibroblast growth factor (bFGF) which acts by stimulating the phosphorylation of the ribosomal 40S subunit (see also Chapter 11) and it is worth recalling at this stage that vascularization (angiogenesis) is required for development of a solid tumour (see Chapter 2).

Another gene, known as *hst*, was isolated from the amplified DNA of a human stomach tumour. Later, transfection techniques identified a gene from Kaposi's sarcoma (see also Chapter 8), called K-*fgf* due to its relatedness to bFGF. Moreover, K-*fgf* and *hst* were found to be the same gene. An interesting finding arose once the *hst* gene was mapped chromosomally: it maps on chromosome 11q13 in humans within 35 kb of the *int-2* gene (similarly, the two genes are close together on mosue chromosome 7). Following the finding that *int-2* is amplified in about 10–20 per cent of human mammary carcinomas, it was also detected that *hst* was co-amplified in many of the same tumours (including some of the murine tumours wherein *int-2* is expressed due to the integration of an

MMTV provirus nearby). In contrast to the pattern found in mouse mammary tumours, however, in no case yet has detection of *int-2* or *hst* mRNA been found in any human tumour regardless of whether the genes are amplified. This result casts some doubt as to the significance of either of these genes in the mammary carcinogenic process. For this reason, there has been some interest in the *bcl-1* gene (see Table 9.2), detected as a breakpoint of translocation in human leukaemias, which maps about 1000 kb away from *int-2/hst* chromosomally. The fact that *bcl-1* is generally not found to be within the amplified DNA segment (the 'amplicon') suggests that there may be another gene of major importance somewhere between the two genes.

9.2.5 *Oncogenes with products related to growth factor receptors*

erbB *and the epidermal growth factor receptor.* The computer database was again fundamental in showing high levels of homology between the putative protein product of the v-*erbB* oncogene (from avian erythroleukaemia viruses) and amino acid sequence derived from the receptor for the epidermal growth factor (EGF). As with v-*sis* and PDGF, the transduction of the cellular gene was a truncation event involving both the carboxyl and amino terminal coding portions of the gene. The v-*erbB* gene product is a faulty receptor; it lacks most of the external domain, including the binding site for ligand (EGF) and is also missing part of the intracytoplasmic domain (see Chapter 11). Interestingly, the EGF receptor is a protein that becomes modified by phosphorylation and possesses tyrosine kinase activity; both of these functional regions are retained in the v-*erbB* sequence (although as mentioned above, it has been difficult to demonstrate kinase activity in the viral product). Going back to the central *sine qua non* of cell proliferation, we can suggest that these modifications to the EGF receptor protein may be sufficiently important to its conformational state so that a continuous proliferation signal may be generated in the absence of the appropriate ligand (which it cannot bind in any case).

Malfunction of the EGF receptor molecule also occurs in nonviral human tumours. For example, a truncated version is produced in a cell line derived from an epidermoid carcinoma of the vulva due to gene amplification (and gene mutation presumably) as well as in several neurological tumours (for further discussion, see Chapter 11).

The oncogene *neu*, a transfectable gene from a carcinogen induced rat neuroblastoma, is related though non-identical to *erbB* (see Chapter 11). Its ligand is as yet unknown.

fms *and the macrophage colony stimulating factor.* The pattern of mimicry of a receptor is also observed with the oncogene v-*fms* (from a feline sarcoma virus). Although the normal cell gene has not yet been isolated, studies on ligand and antibody binding suggest that v-*fms* has been derived from the receptor gene for macrophage colony stimulating factor (M-CSF or CSF-1).

And others. As mentioned above in detail, the protooncogene *kit* is a cell surface receptor for an as yet unidentified, though quite interesting, ligand. Likewise, the protooncogene *mas* is apparently the angiotensin receptor. Also as discussed above, many of the tyrosine kinase oncogenes are cell surface proteins and are structurally related to known growth factor receptors. Hence it has been speculated that *abl* and *ros* may similarly represent growth factor receptors.

Lastly we must not forget that cytosolic or nuclear receptors, rather than cell surface proteins, exist for some of the hormones. One interesting example comes from the study of *erbA*, an oncogene originally detected in the avian erythroleukaemia virus in which *erbB* was also first discovered (Table 9.1). It is now clear that *erbA* is the cellular thyroid hormone (triiodothyronine, T_3) receptor.

9.2.6 *Oncogenes with nuclear products*

myc, myb, ets *and* ski. Oncogene products located in the nucleus might be expected to have functions quite distinct from the *onc* proteins mentioned above. One of the most facile hypotheses would be that these oncogenes might be involved in direct regulation of gene expression, perhaps serving as DNA binding proteins to activate or inactivate transcription of a particular gene or set of genes.

The *myc* oncogene, isolated from four separate avian sarcoma viruses which induce sarcomas, carcinomas, and myeloid leukaemias, is perhaps the best characterized. One reason for heightened interest is its involvement in tumours caused by retroviruses lacking oncogenes as well as its translocation and increased expression in tumours lacking retroviruses (e.g. Burkitt's lymphoma, see above). The *myc* gene is also a multigene family with the cellular *myc* homologue of v-*myc* and two distinct but related genes detected in neuroblastomas and retinoblastomas (N-*myc*), and small cell lung carcinomas (L-*myc*) (see Table 9.2).

The *myc* gene is expressed in many cell types suggesting that its normal function is in a pathway shared by most cells. There is often more *myc* product in cells actively proliferating, although levels are also high in the most mature macrophages whose differentiation programme is nearing the terminal phase. It has been speculated that increased levels of *myc* transcription in tumour cells (in situations where the regulation is governed by new promoter or enhancer elements) may represent deregulation of a negative feedback inhibition between the *myc* product and the *myc* gene, in situations where non-coding sequences have been removed or when the gene is amplified.

The *myb* gene, first detected in two avian myeloid leukaemia viruses, appears to be the best example of a cell cycle dependent gene with levels highest during the G_1 phase of growth. Whether this portends a role as an initiator or effector of cell division is as yet unknown, but intensive research in this area is under way.

Little is known about the *ski* oncogene detected in an avian carcinoma virus other than its size, chromosomal site, and nuclear location. However, the latter would suggest that it may also be important in gene regulation.

The v-*ets* gene product is also located in the nucleus; however, because it is found in the virus as a protein fused to *myb* (see Table 9.1), it is as yet unclear whether the normal *ets* product is nuclear or not.

The fos–jun *(and AP-1) connection.* The oncogene *fos* was originally discovered in 1982 to be the transforming region of the FBJ mouse osteosarcoma virus. Later it was also shown to be the oncogene of the FBR mouse osteosarcoma virus as well, and still later it was found in an avian virus (NK24) which gave rise to nephromas and fibrosarcomas with osteoid deposition *in vivo* (Table 9.1); these findings certainly imply a role for this gene in osteogenesis (formation of bone) although we now know several facts about *fos* that implicate its major role in tissues of nearly all types. The *fos* gene product is a nuclear phosphoprotein of 55–62 kilodaltons (kDa). It is expressed constitutively in a few cell types, notably mature monocytes and cells of the placenta. However, in nearly all other cells, the *fos* gene becomes activated shortly after treatment with agents that induce differentiation (e.g. specific growth factors) or those that induce cell division (e.g. serum *in vitro*, PGDF, and the phorbol ester promoter, TPA). Increased levels of *fos* mRNA can be detected as early as 5 min following induction, making this one of the earliest nuclear events observed in response to mitogenic stimuli—a so-called 'immediate early' gene. However, the induction is only transient and diminishing levels of specific RNA are found by about 45 min after stimulation. Because of the broad range of signals to which *fos* regulation responds, it is worthwhile to pursue those elements that control *fos* expression. On the one hand, it has been shown that the oncogenic conversion in the retroviruses containing *fos* relies on the truncation of the 3′ end of the gene (coding and non-coding sequences) as well as the substitution of the gene's own promoter by the viral (or another strong) promoter. In the former situation, a motif (ATTTATTT . . .) in the 3′ non-coding region is removed; this motif is associated with relative mRNA instability, and hence the altered (viral) *fos* mRNA would have a longer half-life. In the latter situation, a region designated SRE (for *s*erm *r*esponse *e*lement) is no longer juxtaposed to the *fos* gene and hence the specificity of response to mitogenic elements of serum, for example, is lost. It is this short sequence, located about 300 bp upstream of the c-*fos* mRNA cap site, which mediates transcriptional activation of the gene. A 62 kDa binding protein (SRF, serum response factor) has been detected which binds to the SRE specifically. The c-*fos* protein itself has been shown to be DNA binding, a fact which suggests that it might act as a regulator of gene expression *per se*. Some evidence for this has been obtained during the induction of differentiation of adipocytes. Interestingly, even the earliest papers on *fos* mentioned its association with a cellular protein of 39 kDa (termed p39); this association has been repeatedly detected by immunological and biochemical means. However, it has only been two years since p39 has been recognized to be the cellular homologue of the viral oncogene *jun* first discovered in avian sarcoma virus S17 (see Table 9.1). The *jun* oncogene is highly related to a known mammalian transcriptional factor termed AP1 (it is also related to the yeast transcription factor known as GCN4). Hence the formation of the *fos–jun* complex can be seen to be involved in the regulation of transcription. Thus, a new motif for transformation might be the induction of transcriptional factors specific to different programs of differentiation or cell division in various cell types. Since changes in gene expression following growth factor stimulation still occur in the presence of inhibitors of protein synthesis, the activity of transcription factors must be regulated at the post-translational level.

One other interesting feature of the *fos–jun* complex is the so-called 'leucine zipper'; regions of the proteins contain five leucine moieties at regular intervals of seven residues forming a helix that would allow intercalation with similar regions on another protein bearing the same motif. Thus, the three-dimensional configuration of the two proteins allows for very specific binding between them, and accounts to some extent for their affinity to one another. This motif has since been found to occur in other nuclear protooncogenes (e.g. *myc*) as well as other possible transcription factors.

fos, like *myc*, may be involved in self-regulation by virtue of its carboxyl amino acid sequences interacting with the 3′ non-coding regulatory sequences. How this effect modulates expression of other genes is yet to be discovered.

9.2.7 Anti-oncogenes or tumour suppressor genes

Definition and conceptual views. The terms anti-oncogene and tumour suppressor gene are often used synonymously, although some investigators would use them in the negative or positive sense, respectively. Hence an anti-oncogene would be a gene whose repression, inactivation, loss, or dysfunction would lead to cell transformation or neoplastic conversion, whereas a tumour suppressor gene would be one whose activation, expression, or introduction would result in the suppression or inhibition of a tumorigenic phenotype. In addition, the term recessive oncogene has also been coined and used synonymously with tumour suppressor gene (given that oncogenes are thought to act in a dominant genetic fashion—or have a dominant rather than a recessive phenotype). Nonetheless, we shall use the term tumour suppressor gene throughout as its introduction into the literature has priority.

Tumour suppressor genes: the story of Rb-1. One of the genetic highlights of the last few years is the identification of a gene involved in the formation of retinoblastomas. As expounded in Chapters 4 and 10 the inheritance of this tumour, in families displaying bilateral tumours, is often associated with the visible loss of part of chromosome 13 (specifically region 13q14). Sporadic cases also occur and sometimes also exhibit obvious abnormalities of this chromosome. The pattern of inheritance suggests that the disease is caused by a two-hit phenomenon, as first postulated by Knudson: one of these changes occurs in the germ-line and hence is genetic whereas the second presumably occurs early in life or perhaps during embryogenesis. In some instances where no visible deletion was apparent, when the DNA from tumours was compared to the DNA from the normal tissue of the same donor, one often detected the loss of heterozygosity (or a reduction to homozygosity) of loci near 13q14 (these changes are often detected by the use of restriction fragment length polymorphisms, RFLP). Taken together, these findings suggested that it was more likely a case of the *absence* or loss of a normal gene rather than the presence of an abnormal one. The gene identified to this critical chromosomal region by Weinberg's laboratory (Friend *et al.* 1986) is called *Rb-1* and is found to be expressed in all cells and tissues tested with the exception of retinoblastomas, whereas normal retinal cells do express it. Thus, in the case of this tumour it would appear that absence of a gene product is a prerequisite to the development of the tumour; thus one is tempted to call the *Rb-1* gene a tumour suppressor gene or an anti-oncogene in relation to the discussion of this chapter.

What is so far known about the *Rb-1* gene product? It seems to be universally expressed as a nuclear protein of about 105–108 kDa and, interestingly, is often found in a complex with the nuclear proteins of several oncogenic viruses (e.g. the T antigens of SV40 and polyoma viruses, the E1a gene product of adenoviruses, and the E6 gene product of papillomaviruses) in cells/tissues infected and tranformed by these viruses. Thus, the level of expression of the *Rb-1* gene product is also being measured in other tumour types, based on the speculation that it might function coordinately with other as yet unidentified tumour-associated genes.

Although the biochemical function of the *Rb-1* gene product is unknown, if one considers that it has a tumour suppressing activity, then it is logical to speculate that the amount (dosage) of this product might be important to its role whereby too little of the normal product (below the threshold value for function) would lead to expression of the oncogenic phenotype. It is also conceivable that the *Rb-1* gene product expressed in most tissues might have another physiological function as yet un-deciphered or one might hypothesize that the normal levels act as a con-stitutive precautionary safeguard against tumorigenic insults to the cell. Introduction of several copies of a normal *Rb-1* gene into tumour cells (a human osteosarcoma cell line) which were found to lack expression due to gene mutation had the ability to confer a normal-looking, non-neoplastic phenotype to the cells. Further knowledge is necessary to elucidate the precise role of the *Rb-1* gene.

Another aspect of retinoblastomas that deserves mention is the fact that the tumours arise early in the lifespan of the individual (most fre-quently within the first two years of life) suggesting that the second muta-tion involved in the tumour phenotype most likely occurs during embryogenesis. Although one might argue that all tissues of the body are undergoing rapid and complex development during this period, it leads one to speculate that other tumours of the newborn might also arise from similar mechanisms (see p. 266 and discussions in Chapters 4 and 10).

p53. A last gene to mention at this stage is p53, a gene whose product of 53 kDa was originally immunoprecipitated from cells transformed by SV40 virus as part of a complex with the viral T antigen (see Chapter 8). The p53 product can be detected in normal cells as well, although it appears to have a shorter half-life. This result suggests that the viral pro-teins stabilize p53 by formation of the complex. Further studies showed that p53 was present at above normal amounts in many different trans-formed and tumour cell lines including those not involving viruses. This nuclear protein has been observed in many different human tumours, and specific antibodies are detectable in a small but significant propor-tion of breast cancer patients. It also is capable of participating in co-operative transformation described below.

Two exciting results have been reported. First, microinjection of cultured cells in G_1 with anti-p53 antibodies inhibits DNA synthesis; however, during other phases of the cell cycle the cells proliferate normally. This suggests that p53 could be a requisite rate-limiting protein for progression through the cell cycle. Second, a series of mouse leukaemias induced by the Abelson murine leukaemia virus (v-*abl*) have been established as cell lines; all were tumorigenic, but only one failed to metastasize. This exception was due to the integration of a 'helper' virus within the coding sequences of the p53 gene. Upon introduction of an intact p53 gene into these cells by transfection, the ability to metastasize was re-established.

As with the *Rb-1* gene product, p53 can also be found to complex with the tumour antigens of several oncogenic viruses. Morever, it is now observed that alteration in the gene which lead to complete absence of p53 product or to the appearance of an altered product is the observed phenotype in a number of tumours (e.g. small cell lung carcinoma and mammary carcinomas in humans, leukaemias of mice and man, etc.). It is also interesting that in an animal model where an altered p53 gene has been introduced into the germ line (the transgene of a so-called 'transgenic' mouse), tumours of lung, bone, and haematopoietic cells (which exhibit high levels of expression of the altered gene) have been observed in the progeny animals. Such data argue for the role of p53 as another tumour suppressor gene.

Others. Interestingly, another situation for which the existence of a tumour suppressor gene is invoked is that of Wilms' tumour, a tumour of the kidney also found in children under two years of age. As with retinoblastoma, visible deletions have been recorded—this time in chromosome 11p—and a search for a gene in that region is currently underway in several laboratories. See Chapter 10 for new developments.

A further corollary to mention at this point is the finding that in somatic cell hybrids created *in vitro* by fusing a tumorigenic cell type with a non-tumorigenic one (usually human×mouse hybrids from which the human chromosomes are selectively lost in a random fashion with the continued passage of the cells in culture), the retention of human chromosome 11 is frequently associated with a normal morphological phenotype whereas loss of this chromosome correlates with the neoplastic morphology. More recently (Weissman *et al.* 1987) the introduction of a normal chromosome 11 into a tumourigenic cell led to a reversion of phenotype to normal morphological appearance; this situation is analogous to the *Rb-1* experiment discussed above (p. 259).

As discussed earlier in Chapter 4, the aetiology of colorectal tumours has also recently been linked to the absence of a normal gene product. The story begins with the discovery of a mutant karyotype involving chromosome 5 in one patient who exhibited familial adenomatous polyps (FAP, formerly the disease syndrome was known as polyposis coli)—involving hundreds of tiny polyps in the large intestine and colon, some of which may convert to malignant papillomas with time—and in addition was mentally retarded (a circumstance which suggested that this person might have a larger genetic abnormality than people with FAP alone). By using probes from different regions of chromosome 5, the defective regions was discovered to be at 5q21. Taking these data a few steps farther, there were apparent rearrangements in this region in cells derived from polyps of FAP patients and in patients with colorectal cancer of a presumably non-familial (sporadic) origin. As colorectal tumours represent the second most common tumour in western countries, it would seem that such tumour suppressor genes may exert a very important and frequent modality of carcinogenesis. More recent data on genetic abnormalities in colorectal carcinomas have brought to light the importance of chromosomes other than 5q, notably chromosomes 17 and 18. The chromosome 17 abnormality involves the p53 gene, a gene already implicated in the genesis of osteosarcomas and haematological malignancies (see p. 260). Once again, the type of abnormality observed appears to be the absence of a functional p53 gene product (e.g. due to point mutations, deletions, or other rearrangements leading to no expression or expression of a defective protein). Interestingly, it has been speculated that p53 dysfunction occurs in all adult tumours but this is yet to be proven.

9.2.8 *Other oncogenes*

We have already seen that the cellular oncogenes may encode proteins that are growth factors, growth factor receptors, enzymes, and perhaps *trans*-activating DNA binding proteins. Each new oncogene is examined for these properties but oncogenes with other functions may yet be discovered. The major theme is that the normal gene counterparts are active, and presumably vital, to the functioning of cells, and abnormal properties may be attributed to mutation, truncation, amplification, or deregulation.

9.3 Interactions between oncogenes

9.3.1 *Tripartite retroviruses*

Perhaps one of the most unexpected results from studying v-*onc* genes has been the discovery that a single virus genome often contained genetic information derived from two distinct cellular genes (often from two different chromosomes). The v-*src* gene of Rous sarcoma virus, as mentioned above, derives its carboxyl terminal sequences from outside the cellular *src* gene; these sequences are apparently from about 1 kb downstream of the gene. The FBR strain of MSV in addition to *fos*, contains sequences called *fox* from another location which confer additional properties. Similarly, the AEV strain, ES4, differs from strain H; the former contains *erbA* and *erbB*, the latter *erbB* alone. Both strains cause erythroleukaemias but erythroid cells transformed in culture show differences in growth requirements and differentiation capacity depending on the specific virus strain. The GR strain of feline sarcoma virus shows fusion of part of an actin gene with the transforming *fgr* gene. Finally, two viruses have shown acquisition of sequences from two genes, both of which have oncogenic potential on their own: the avian myelocytomatosis virus MH2 contains the oncogenes *myc* and *raf* (also called *mil*), and the E26 avian myeloblastosis virus contains *myb* and *ets*. The neoplastic potential of these viruses is expanded with regard to the types of tumours induced compared to viruses containing only one of the genes (see Table 9.1). In the case of v-*src* mentioned above, the unusual structure might have arisen by a large splicing event, but the others required more bizarre recombination. (However, it is still speculation whether transduction occurs at the DNA level between integrated provirus and adjacent cellular DNA, or at the RNA level between viral and cellular species, or by both mechanisms.)

9.3.2 *Co-operative transformation*

An old observation that primary rodent embryo fibroblasts, but not later passaged or established cell lines, were generally refractory to morphological transformation by transforming viruses was the basis of a new assay for oncogene activity. For example, the dual expression of the genes encoding the nuclear (large T) antigen and the membrane associated (middle T) antigen of polyomavirus (see Chapter 9) cause transformation whereas neither alone has this capacity, although large T confers longevity ('immortalization' to continued cell growth). A similar relationship was found between unrelated oncogenes, e.g. the mutated alleles of Ha-*ras*-1 (whether viral or cellular in origin) and *myc*. Eventually a pattern emerged of co-operation between a nuclear oncogene product and a cytoplasmic oncogene product (Table 9.6); the former also generally can immortalize non-established cells.

Co-operative transformation is not as stringent as described above; in a few cases, a single oncogene will suffice. Despite this, the discovery of co-operating oncogenes was particularly satisfactory given the multistage character of carcinogenesis, interweaving pathways of biochemical and physiological functions. But we have determined only the earliest step without generating clues to distinct targets or later cascades. Elucidation of the intermediate and final stages is necessary for both understanding and devising possible modes of intervention.

Table 9.6 Co-operating oncogenes[1]

Non-nuclear	Nuclear
Ha-*ras*-1[2]	v-*myc*
N-*ras*[2]	c-*myc*[3]
Polyoma middle T	Polyoma large T
Adenovirus E1b	Adenovirus E1a
	v-*myb*
	v-*fos*
	p53

[1] Manifestation of the transformed phenotype requires one oncogene from the nuclear group and one from the non-nuclear group.
[2] Mutated allele.
[3] Normal and rearranged forms when supplemented with a strong promoter.

9.4 Protooncogene expression in normal development and differentiation

Although molecular clones to probe for oncogene expression are available for many of the genes described above, it is a rather daunting task to examine many different tissues. In some cases, it is also difficult to obtain sufficient amounts of homogeneous cell populations and thus one is not always able to make direct comparisons between a neoplasm and its normal counterpart. This has left large gaps in our knowledge of oncogene expression during embryonic development, differentiation stages, or cell cycling (Table 9.7), although eventually this knowledge will be available.

One well exploited system is the examination of mouse embryos at different days of gestation. Genes such as *raf*, Ha-*ras*, Ki-*ras*, and *myc* are expressed throughout. The *abl* gene is highly expressed at day 10, corresponding to a period of fetal liver haemopoiesis, whereas *fos* peaks at later stages due to high levels in the extraembryonic membranes, particularly the amnion, and *fms* expression is observed in the chorion. Other genes are not present at detectable levels. It is not facetious to speculate that protooncogenes are vital to embryogenesis and that we may eventually know their specific roles.

To study changes with regard to cell cycling and proliferation, several methods have been used. Most rely on selecting cells in the same phase of the cycle using a cell sorter which measures levels of DNA or by synchronizing the cycle, usually by blocking the cells in a specific phase of the cycle by deprivation of specific nutrients or by chemicals. On removing the block, the cycle continues with many of the cells in the

Table 9.7 Protooncogene expression

Gene	High expression[1]	Abnormal expression[2]
abl	Mid-gestation embryo, testes, spleen, thymus	Chronic myelocytic leukaemia
erbA	Mid to late gestation embryo	
erbB	Mid-gestation embryo	Squamous cell carcinomas, glioblastomas
fes	Bone marrow	Myeloid and lymphoid leukaemia
fms	Macrophage, placenta	Mammary and renal carcinoma
fos	Amnion, chorion, mature macrophages, growth factor stimulated cells	Choriocarcinoma
mos	Testis, oocytes	Plasmacytoma[3]
myb	Yolk sac, bone marrow, thymus	Myeloid and lymphoid leukaemia
myc	Ubiquitous	B cell lymphomas, promyelocytic leukaemia
raf	Ubiquitous	
ras	Ubiquitous	Numerous (see Table 9.3)
rel	Spleen	
ros	Kidney	
sis	Platelets	Osteosarcoma
ski	Cartilage, muscle, skin	
src	Spleen, macrophages, brain	Brain tumours
yes	Kidney	

[1] Most studies from experimental animals, principally mice.
[2] Human tumours or tumour cell lines.
[3] Rodent tumour only, gene activated by insertional mutagenesis (see Table 9.2).

same phase. In-vivo systems such as partial hepatectomy (where one lobe of the liver is removed) can be used to stimulate rapid regeneration of cell number. Cell sorting has shown *myb* expression to be cycle dependent. Stimulation from quiescence shows *fos* to be activated within minutes followed by a rapid decline, whereas *myc* expression appears later but is less transient. The liver regeneration experiments show that the major nuclear oncogenes, members of the kinase family, and *ras* are expressed during the highly proliferative phase. Thus, these results recapitulate our awareness that oncogenes are involved in normal proliferative responses, but as yet do not explain how or why.

9.5 Lessons learned from oncogenes

It should be clear from the discussion above the oncogenes are expressed in a plethora of cell types. Even the truly tumorigenic v-*onc* genes do not answer our queries. For example, the murine *abl* containing virus causes immature B cell tumours in mice whereas the feline *abl* virus causes fibrosarcomas (Table 9.1). Is this due to differences in the nature of the fusion proteins (between *abl* and viral genes) or to some inherent properties of the two animal species? Site specific mutagenesis has shown that the *gag* portion of the mouse virus fusion protein is required for efficient transformation of cultured B lymphocytes though it is dispensable for fibroblast transformation which may explain in part this discrepancy. Furthermore, many strains of mice are genetically resistant to leukaemogenesis by the virus by a mechanism as yet unknown. We have to know more about the normal *abl* product. The highest level of *abl* mRNA in normal tissues occurs in the testes. Haemopoietic tissues also express high levels; is this significant in regard to aberrant transcription due to translocation in CML? Other examples of these phenomena exist (see Tables 9.1, 9.2, and 9.7).

Understanding the role of the *myc* gene is certain to prove of central interest. The various v-*myc* genes are responsible for the induction of carcinomas, fibrosarcomas, and myeloid leukaemias, whereas hyperexpression due to translocation or promoter insertion is observed in T and B cell leukaemias. Additionally, *myc* expression is apparent at low levels in most tissues, becomes elevated in cells responding to mitogens, and is very high in the most mature macrophages. There is also intriguing information from experimental manipulation of the gene. First, a new 'virus' was made by putting the cellular *myc* gene between viral LTRs from the MMTV genome and introduced into fertilized mouse ova. The use of LTRs ensured that the gene could be incorporated into chromosomal (and germ line) DNA and the mice (known as transgenic mice) were examined for abnormalities of development. However, it was only in adult female mice that a change was seen: the animals developed mammary carcinomas as a heritable trait. In a similar way, a construct containing *myc* and the enhancer element from an immunoglobulin gene (from a rearranged gene in a mouse plasmocytoma, see above) induced a high incidence of B cell leukaemias in the transgenic mice, whereas other enhancers from a leukaemia virus LTR or SV40 virus were less efficacious in inducing tumours and showed different target specificities (a T cell lymphoma with the former, and lymphosarcoma, fibrosarcoma, and kidney carcinoma with the latter). These results emphasize the importance of specific promoter and enhancer elements of gene transcription in different cells and in neoplasia.

As discussed previously, oncogenes are relatively highly conserved evolutionarily in vertebrates. Indeed, structural and biochemical (functional) counterparts have been detected in yeast, slime moulds, fruit flies (*Drosophila*), and frogs (*Xenopus*). At the moment, the genes related to oncogenes in *Drosophila* provide a lot of interest because there exist many mutant flies which could throw light on the physiological function of oncogenes. Further, it is also possible to induce mutations or to introduce mutant genes into the animals during embryonic development. As can be seen in Table 9.8, there already are identifiable mutations in *Drosophila* genes which are apparently related to mammalian oncogenes. While it is difficult to say that one such as *wingless* (the fly equivalent of *int-1*) has a direct functional homologue in primates, for example, it is interesting that a number of the oncogene counterparts are involved in overall body pattern formation. As another example, the mutation *sevenless* has been postulated to be a receptor defect and this would fit in very well with the idea that *ros* is a receptor based on its structural counterparts in the tyrosine kinase family (see p. 254). Moreover, another gene called *bride of sevenless* encodes the ligand of *sevenless*; once purified and identified, this product should elucidate the nature of the ligand for mammalian *ros* and hence its functional role.

In the past few years, a number of new techniques for introducing DNA into cells, other than calcium precipitation, have been devised and discussed above; some have proved to be of greater value in cells of small size or those very sensitive to the chemical treatment. One such improvement is the use of electroporation in which uptake of exogenous materials is facilitated during transient changes to cell membrane electrical potential. This method has been particularly useful in cells known as embryonic stem (ES) cells derived from the inner cell mass of early embryonic mice. ES cells are the closest type to pluripotential stem cells that can be cultured *in vitro* and *in vivo*; indeed they have the capacity to contribute to all three germ layers (endoderm, mesoderm, and ectoderm) of a developing embryo.

Homologous recombination is a term applied to the exchange of allelic, though not necessarily identical, genetic sequences from an exogenous source with the endogenous sequences of an organism. In this way, one might hope to substitute—with specificity—the allelic gene: (i) from a different species but having the same physiological function (a

Table 9.8 Oncogene homologues in *Drosophila*

Oncogene	Homologue	Effect of mutation
rel	dorsal	Pattern development
ros	sevenless	Absence of eye segment
abl	abl	Lethal mutation (90% larvae and pupae)
ras	ras	Reduced viability and fertility; poor wing development; disordered eye structure
int-1	wingless	Absence of wings
erbB (EGFR)	egfr	Disordered eye structure

specific type of transgenic manipulation); (ii) from an intact gene copy to replace a defective endogenous gene (i.e. gene therapy); or (iii) from a defective gene to replace the endogenous normal gene. In the last case, if the defective gene was incapable of being expressed, the result would be the loss ('knock out') of the endogenous gene and hence absence of its biochemical function. All of the above alternatives are 'transgenic' experiments *per se*, and the chances are more likely that recombination at a non-homologous region should occur at higher frequency. It has thus been surprising that the actual frequency of homologous recombination is often quite high (up to 1 per 30 transgenic events), certainly much greater than anticipated; this result makes the experimental system much more attractive and efficacious for studying the importance of proto-oncogenes, oncogenes, and their regulatory sequences on viability, development, and oncogenesis.

We are looking through a small window at what is going on in a cell. The viral oncogenes have called our attention to the cellular counter-parts and transfection assays and chromosomal aberrations have helped to identify further potentially important genes for neoplastic growth.

Genetic studies suggest that recessive changes or deletions may be contributory, but our present technology limits the means of detecting them.

Our sojourn into oncogenes has led to the conclusion that normal cellular genes can be diverted into neoplastic pathways by somatic mutations as diverse as single nucleotide substitutions to gross altera-tions involving translocations or amplifications. Underscoring these perturbations, there may be more subtle effects on expression in inappropriate cells, alterations in transcription rates or mRNA turnover, or stabilization of the gene products and, in turn, their target substrates within the cell.

Most of the changes observed have led to overexpression of the gene. But taking a cellular *onc* gene and putting it under the control of a strong promoter does not necessarily lead to oncogenic potential (indeed this works for *mos* but not for *src*). Overabundance of an *onc* gene product might also uncover cryptic activities normally absent when the protein occurs at some subthreshold dose.

Overall, cell metabolism and proliferation is an intricate and delicately balanced response to extracellular signals (e.g. mitogens), involving regulated transmission by effector molecules to the nucleus. Some of the known oncogenes already recapitulate a few stages of these mechanisms in their mimicry of growth factors and growth factor receptors. Others display functions suggestive of second messages or transmitters. And the nuclear oncogenes might be the ultimate effector molecules. Despite the formidable task ahead of identifying these interweaving pathways step by step, we have gained considerable insight within the last five years. Indeed, the studies on the first identifiable retrovirus oncogenes quickly aroused the interest of scientists in many areas of diverse research and showed how interrelated their endeavours had become.

In the final analysis, c-*onc* genes are just protooncogenes that march to the tune of a different drummer.

References and further reading

Bishop, J. M. (1985). Viral oncogenes. *Cell,* **42**, 23–38.

Bishop, J. M. and Varmus, H. (1982). Functions and origins of retroviral trans-forming genes. In: *RNA tumor viruses*, Vol. 1. (eds. R. Weiss, N. Teich, H. Varmus, and J. Coffin) pp. 999–1108. Cold Spring Harbor Laboratory, Cold Spring Harbor, New York.

Bishop, J. M. and Varmus, H. (1985). Functions and origins of retroviral trans-forming genes. In: *RNA tumor viruses*, Vol. 2 (eds. R. Weiss, N. Teich, H. Varmus, and J. Coffin), pp. 249–356. Cold Spring Harbor Laboratory, Cold Spring Harbor, New York.

Bodmer, W. F. *et al.* (1987). Localisation of the gene for familial adenomatous polyposis on chromosome 5. *Nature* **328**, 614–16.

Bos, J. L. (1988). The *ras* family and human carcinogenesis. *Mutation Research* **196**, 255–71.

Friend, S. H. *et al.* (1986). A human DNA segment with properties of the gene that predisposes to retinoblastoma and osteosarcoma. *Nature* **323**, 643–6.

Glover, D. M. and Hames, B. D. (ed.). (1989). *Oncogenes.* IRL Press, Oxford.

Huang, H. S. *et al.* (1988). Suppression of the neoplastic phenotype by the replacement of the RB gene in human cancer cells. *Science* **242**, 1563–6.

Lloyd, A. C., Paterson, H. F., Morris, J. D. H., Hall, A., and Marshall, C. J. (1989). p21$^{H\text{-}ras}$ induced morphological transformation and increases in c-*myc* expression are independent of functional protein kinase C. *EMBO Journal* **8**, 1099–104.

McCormick, F. (1989). Ras GTPase activating protein: signal transmitter and signal terminator. *Cell* **56**, 5–8.

Marshall, C. (1985). Human oncogenes. In: *RNA tumor viruses* Vol. 2 (eds. R. Weiss, N. Teich, H. Varmus, and J. Coffin), pp. 487–558. Cold Spring Harbor Laboratory, Cold Spring Harbor, New York.

Solomon, E. *et al.* (1987). Chromosome 5 allele loss in human colorectal carcinomas. *Nature* **328**, 616–18.

Varmus, H. E. (1984). The molecular genetics of cellular oncogenes. *Annual Review of Genetics,* **18**, 553–612.

Weissman, B. E. *et al.* (1987). Introduction of a normal human chromosome 11 into a Wilms tumour cell line controls its tumorigenic expression. *Science* **236**, 175–8.

10

Chromosomes and cancer
Denise Sheer

10.1 Introduction

The genetic basis of cancer was discussed in Chapter 4 in relation to both hereditary predisposition to cancer and DNA changes in tumours themselves. In this Chapter I shall describe chromosome changes in tumours.

Most tumours have structural and/or numerical chromosome aberrations, many of which are consistently associated with particular types of tumours. It is now apparent that sites of consistent chromosome rearrangements pinpoint genes which may be critically involved in malignant transformation, and that the rearrangements themselves can subvert the normal functioning of these genes.

Human somatic cells normally have 46 chromosomes; they are said to be diploid. Chromosome analysis is usually carried out on cells in mitosis (cell division), when the chromosomes become visible as distinct entities. After identifying each chromosome in a cell by its characteristic size (the largest chromosome was named 1 and others numbered in descending order of size), shape, and staining properties, a karyotype displaying the full chromosome complement of the cell can be prepared.

The first specific chromosome abnormality observed in a human tumour was seen in Philadelphia in 1960 by Nowell and Hungerford who found an unusually small chromosome in the leukaemic cells of patients with chronic myeloid leukaemia (CML). This small chromosome was named the Philadelphia chromosome. The discovery of the Philadelphia chromosome aroused considerable interest in cancer cytogenetics as it gave the first direct evidence for a consistent DNA-associated change in a tumour. The search for similar abnormalities in other tumours was hampered by the great variation in chromosome abnormalities from one patient to the next and by the finding of aneuploidy (abnormal chromosome numbers) and multiple rearrangements (abnormal breakage and rejoining of chromosomes) in many tumours. Furthermore, chromosome rearrangements could not be defined using staining methodology available then.

The introduction of chromosome banding techniques in 1970 revolutionized cancer cytogenetics. Consistent chromosome aberrations which are uncommon or extremely rare in normal tissues were found in the different cells of a tumour, and further karyotypic changes were shown to occur during tumour progression. Short term culture methods to improve yields of dividing cells, and high resolution banding of elongated chromosomes now allow a more precise definition of re-arrangements as well as the identification of previously undetected rearrangements. Using these techniques, most tumour cells can be shown to have some chromosomal defect.

Evidence that these aberrations are non-random has come from cytogenetic analysis of large numbers of human tumours (Heim and Mitelman 1987). Haemopoietic tumours (tumours of blood forming cells) usually have rearrangements involving only a few chromosomes in an otherwise normal diploid karyotype. Many solid tumours have numerous chromosome rearrangements with gross aneuploidy and the technical quality of the chromosome preparations is often poor. For this reason, consistent or specific aberrations have been more easily detected in the haematological neoplasms, which are also generally easier to sample and have more dividing cells than solid neoplasms.

Both dominantly and recessively acting genes are known to be directly affected by chromosome aberrations in tumours. For example, the oncogenes c-*abl* and c-*myc* are activated by specific translocations in CML and in B and T cell tumours, respectively, and are unaffected by the presence of normal alleles. These oncogenes are said to be dominantly acting. On the other hand, both alleles of recessively acting genes, such as the retinoblastoma (RB) gene, need to be deleted or mutated in order for neoplastic transformation to occur. These genes are believed to suppress tumour formation even when only one functional allele is present.

10.2 Methodology

Cytogenetic analysis of leukaemias is carried out on a bone marrow aspirate, or on a peripheral blood sample if the white cell count is high, by immediate processing (direct method) or after in-vitro culture for 24–72 hours. Solid tumours are processed directly or after culturing for variable times. The cells in solid tumours have to be separated from each other before chromosomes can be prepared. This problem may be circumvented by analysing pleural or ascitic fluid containing metastatic tumour cells, or cell lines derived from tumours. These types of samples have the disadvantage, however, that extensive chromosome rearrangements may be present which make it difficult to distinguish primary from secondary changes (which could have occurred during metastatic progression or during in-vitro culture).

As a rule, chromosomes are studied during or just prior to the metaphase stage of mitosis when they become condensed and can easily be seen under the microscope (Rooney and Czepulkowski 1986). DNA replication occurs before mitosis so that each chromosome consists of two identical sister chromatids held together at the centromere. When making chromosome preparations, colchicine or a related agent is added to the tumour material to arrest cells in metaphase by disrupting the formation of the mitotic spindle fibres which normally separate the chromatids. The cells are then swollen in a hypotonic solution to make it easier to disrupt them and release the chromosomes. The cells are fixed in methanol-acetic acid, and metaphase 'spreads' are prepared by dropping fixed cells onto microscope slides. Chromosomes are identified by staining techniques which produce a characteristic series of bands along the chromosomes. The two techniques most often used, G- and Q-banding, result in essentially the same banding patterns on the chromosomes. One method of producing G-banding is to treat the chromosomes with a weak solution of trypsin prior to staining with Giemsa stain; the preparation is then examined using conventional light microscopy. Q-banding is produced by treating the chromosomes with quinacrine-dihydrochloride or quinacrine mustard; the preparation is then examined using fluorescence microscopy.

Until recently, cytogenetic techniques produced karyotypes of normal human cells with approximately 300 bands. It is now possible to obtain less condensed chromosomes earlier in mitosis. One method used is to synchronize the cultures using the folic acid analogue methotrexate which inhibits the production of thymidine and therefore prevents DNA synthesis. Cells collect at the G1/S interphase of mitosis and can be released by thymidine. After a brief exposure to colchicine, high resolution karyotypes can be obtained with up to 1200 bands, thus allowing the delineation of previously undetected chromosome aberrations. In the standard method these bands would normally coalesce as condensation proceeds during mitosis into fewer and thicker bands. Another method for high resolution banding is to synchronize the cells by incorporating

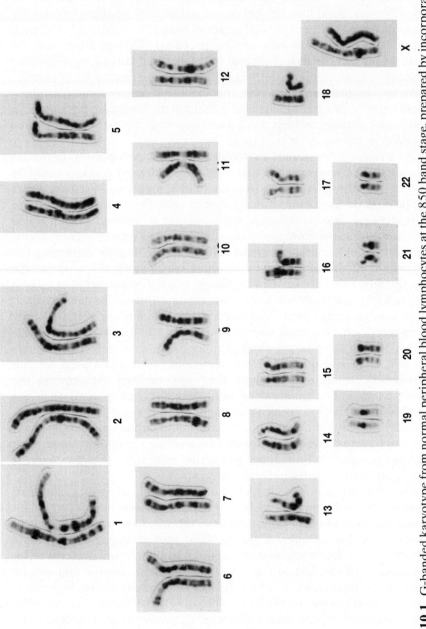

Fig. 10.1 G-banded karyotype from normal peripheral blood lymphocytes at the 850 band stage, prepared by incorporating BrdU into the chromosomes.

5-bromodeoxyuridine (BrdU) into DNA in place of thymidine. This dramatically alters the staining characteristics of the chromatin. High quality chromosome banding can be achieved using this method, provided the BrdU pulse is carefully timed in relation to the cell cycle, since the DNA in light G-bands replicates before that in dark G-bands (Fig. 10.1).

Our understanding of the molecular events occurring in chromosome rearrangements in cancer has been greatly facilitated by advances in gene mapping (localization of genes on chromosomes). This is classically done by analysing human genetic markers present in human–rodent somatic cell hybrids (produced by whole cell fusions) which after passage in-vitro, retain only a few human chromosomes. More precise mapping to single chromosome bands is done by in-situ hybridization of radiolabelled molecular probes directly to complementary sequences in fixed metaphase chromosomes. Hybridization is visualized on the chromosomes by conventional autoradiography, and chromosomes identified by either pre- or post-hybridization banding. Techniques have recently been developed for non-isotopic labelling of probes, for example, with biotin. Hybridization can then be visualized in various ways, such as reaction of the biotin-labelled probe with avidin or streptavidin coupled to a fluorochrome. These techniques dramatically improve the resolution, efficiency and safety of the in-situ hybridization procedure.

10.3 Terminology and types of chromosome aberrations in tumours

Karyotypes are described according to an International System for Human Cytogenetic Nomenclature (ISCN 1985). The total chromosome number is listed first, then the sex chromosomes, then gains and losses of whole chromosomes, and finally structural rearrangements. The short and long arms of chromosomes are represented by 'p' and 'q', respectively. Gains or losses of whole chromosomes are identified by a '+' or a '−' before the chromosome number. 46,XY and 46,XX represent normal diploid male and female karyotypes. 47,XX, + 8 represents a female karyotype with an extra copy of chromosome 8, which is the most common aberration in myeloproliferative disorders. The most common structural rearrangement found in tumours is a translocation, in which segments of two or more different chromosomes are interchanged. Translocations are signified by a 't'; the chromosomes involved are enclosed within a first set of brackets and the translocation breakpoints are enclosed within a second set of brackets. For example, t(9;22)(q34;q11) represents a reciprocal translocation between chromosomes 9 and 22, with the breaks in the long arms of chromosomes 9 and 22 in bands q34 and q11, respectively. This translocation is present in most cases of chronic myeloid leukaemia, and will be described in detail below. Other structural chromosome rearrangements seen in tumours are inversions, deletions, and insertions. For example, inv(14)(q11q32) represents an inversion within chromosome 14, with the breakpoints occurring in bands q11 and q32. This inversion is present in some T lymphocyte malignancies and will also be described in detail below. 46,XY,20q− represents a male karyotype with a deletion of the long arm of chromosome 20, which is present in the affected cells of 10–20 per cent of patients with polycythaemia vera. Finally, an isochromosome is derived from the loss of either the long or short arm of the chromosome with the duplication of the other arm. Thus, an i(17q) consists of the duplicated long arm of chromosome 17. This aberration is frequently present in the acute phase of CML.

Only clonal chromosome aberrations will be reviewed below. These are defined as occurring within at least two cells in a tumour for structural rearrangements or the same extra chromosome, or within at least three cells for the same missing chromosome.

10.4 Leukaemias and lymphomas

10.4.1 *General survey*

Leukaemias and lymphomas usually have few chromosome rearrangements and chromosome numbers in the diploid range (other than Hodgkin's lymphomas which usually have chromosome numbers in the triploid–tetraploid range). These rearrangements have therefore been relatively easy to define (Rowley 1990), especially with high resolution chromosome banding techniques. These techniques have also enabled the definition of subtle rearrangements in different subtypes of acute myeloid leukaemia (AML). Nomenclature of acute myeloid leukaemias, published by the French, American, and British (FAB) Cooperative group is of particular relevance to cytogenetic studies, and is summarized in Table 10.1. A summary of consistent chromosome defects in leukaemias and lymphomas, and in the myeloproliferative disorder polycythaemia vera in which there is an increased production of red blood cells and their precursors, is presented in Table 10.2.

Trisomy (three copies of a chromosome) is in some cases the only aberration observed in leukaemias and lymphomas, particularly trisomy 8 in all subtypes of AML and trisomy 12 in B cell chronic lymphocytic leukaemia (CLL). However, translocations are the most common struc-

Table 10.1 FAB classification of AML

Description	Code
Acute myeloblastic leukaemia without maturation	M1
Acute myeloblastic leukaemia with maturation	M2
Acute promyelocytic leukaemia	M3
Acute myelomonocytic leukaemia	M4
Acute myelomonocytic leukaemia with increased basophils	$M4_{EO}$
Acute monoblastic leukaemia (poorly differentiated)	M5a
Acute monocytic leukaemia (well differentiated)	M5b
Erythroleukaemia	M6
Megakaryocytic leukaemia	M7

tural aberration. Translocations involving the same pairs of chromosomes recur within particular malignancies to varying extents. For example, t(9;22) which generates the Philadelphia chromosome is present in the leukaemic cells of virtually all patients with CML, whereas t(8;21) is present in the leukaemic cells of 40 per cent of patients with AML-M2. The degree of specificity of different translocations is also variable. For example, a t(9;22) which appears cytogenetically to be identical to that found in CML also occurs in AML-M1 and is the most frequent rearrangement in non-B, non-T adult acute lymphocytic leukaemia (ALL) (although we now know that the exact breakpoints in ALL are different to those in CML). The t(15;17) associated with acute promyelocytic leukaemia (APL-M3) has virtually never been seen in any other malignancy other than in a few very rare cases of promyelocytic blast crisis of CML.

10.4.2 *Myeloid malignancies*

Myeloid malignancies—chronic myeloid leukaemia (CML) and the acute myeloid or non-lymphocytic leukaemia (AML or ANLL)—are located primarily in the bone marrow. AML may also be present in the peripheral blood. The molecular analysis of the Philadelphia chromosome in CML provides one of the paradigms for oncogene activation by chromosome translocation, and will be discussed in detail below (see p. 000).

Chronic myeloid leukaemia. CML is characterized by an initial chronic phase where mature granulocytes proliferate, followed by a blast crisis which has features similar to acute leukaemia in which either myeloid or lymphoid cells can proliferate. Bone marrow cells from the majority of patients with CML contain a reciprocal translocation, t(9;22)(q34;q11), which gives rise to a small 22q− chromosome called the Philadelphia

Table 10.2 Consistent chromosome aberrations in leukaemias and lymphomas

Malignancy	Chromosome aberration
Leukaemias	
Chronic myeloid leukaemia	t(9;22)(q34;q11)
Acute myeloid leukaemia	
M1	t(9;22)(q34;q11)
M2	t(8;21)(q22;q22)
M3	t(15;17)(q22;q12)
M4 with abnormal eosinophils	inv(16)(p13q22)
M5a	t(9;11)(p22;q23)
M1, M2, M4 with increased basophils	t(6;9)(p23;q34)
M1, M2, M4, M5, M6	Monosomy 5/del(5q)
	Monosomy 7/del(7q)
	Trisomy 8
Chronic lymphocytic leukaemia	t(11;14)(q13;q32)
	Trisomy 12
Acute lymphocytic leukaemia	t(9;22)(q34;q11)
	t(4;11)(q21;q23)
Acute B cell leukaemia	t(8;14)(q24;q32)
	t(2;8)(p12;q24)
	t(8;22)(q24;q11)
Acute T cell leukaemia	inv(14)(q11q32)
	t(14;14)(q11;q32)
	t(8;14)(q24;q11)
	t(10;14)(q24;q11)
	t(11;14)(p13;q11)
Lymphomas	
Burkitt's lymphoma	t(8;14)(q24;q32)
	t(2;8)(p12;q24)
	t(8;22)(q24;q11)
Small non-cleaved cell lymphoma, large cell immunoblastic lymphoma	t(8;14)(q24;q32)
Follicular small cleaved cell lymphoma	t(14;18)(q32;q21)
Small cell lymphocytic lymphoma	Trisomy 12
Small cell lymphocytic transformed to diffuse large cell lymphoma	t(11;14)(q13;q32)
Polycythaemia vera	del(20q)

(Ph′) chromosome (Fig. 10.2). As a result of this translocation, the oncogene c-*abl* becomes transferred from chromosome 9 at band 9q34 to the *bcr* gene at band 22q11. The translocation breakpoints in different patients are restricted to a 5.8 kb region of DNA (M*bcr*, Major breakpoint cluster region) of the *bcr* gene on chromosome 22, but can occur over a relatively large distance of DNA (200 kb) in the 5′ region of the c-*abl* gene on chromosome 9 (Bernards *et al.* 1987).

CML cells contain an unusually long 8.5 kb c-*abl* mRNA synthesized from the fusion gene in which *bcr* sequences replace part of the 5′ end of the c-*abl* gene (Fig. 10.3). All CMLs examined so far have the 8.5 kb transcript. The protein translated from the fusion mRNA has a relative mass (M_r) of 210 kDa. It has tyrosine kinase activity (see Chapters 9 and 11) that is greatly increased when compared with the normal c-*abl* protein which has a M_r of 145 kDa (Konopka *et al.* 1985).

It is not known how the *bcr–abl* gene product promotes myeloid tumorigenesis. The normal *bcr* gene product is a phosphoprotein of M_r 160 kDa. It is expressed in various cell types as is the c-*abl* gene product. The structure of the fusion protein suggests that it may function as a receptor for a growth factor (see Chapter 11), therefore, substitution of the amino terminus of the molecule might result in constitutive phosphorylating activity thus promoting cell proliferation.

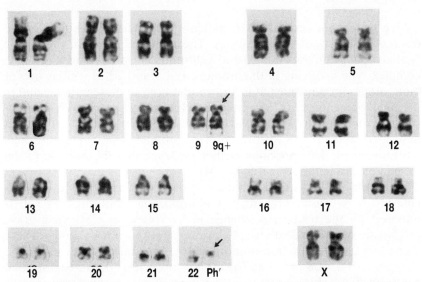

Fig. 10.2 G-banded karyotype showing the Philadelphia chromosome.

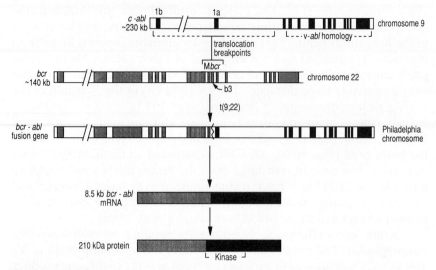

Fig. 10.3 Schematic representations of the *bcr-abl* fusion arising from the t(9;22) in CML. Exons (coding regions) are represented by filled boxes. Exons 1b and 1a are alternative first exons of c-*abl*. If present in the *bcr-abl* fusion gene, they are spliced out of the fusion mRNA. If exon b3 of M*bcr* is present in the fusion gene, it is alternatively spliced out of or included in the fusion mRNA (Shtivelman *et al.* 1986).

Five to ten per cent of CML patients with classical features of the disease have variant translocations. These can be either of the simple type involving chromosome 22 and a chromosome other than 9, or of the complex type involving both chromosomes 9 and 22 as well as one or more other chromosomes. It is known that both in these patients and in the small percentage of CML patients without any apparent Ph chromosome, a *bcr–abl* fusion protein is produced that is identical to that described above, presumably from a submicroscopic chromosomal rearrangement. This provides further evidence that the *bcr–abl* fusion protein plays a pathogenic role in CML.

Leukaemic cells from approximately 80 per cent of CML patients entering blast crisis show further chromosome abnormalities superimposed on the t(9;22). A change in karyotype in CML is a grave prognostic sign, with death usually occurring within a few months. These secondary changes are non-random with at least one of three particular changes, + Ph, + 8, or + i(17q) usually occurring. This suggests that genes on these chromosomes confer a proliferative advantage to cells carrying them.

Acute myeloid leukaemias. The most frequent numerical aberrations observed in AML are trisomy 8 (13 per cent of all cases), monosomy 7 (9 per cent), and monosomy 5 (6 per cent). Recurring structural chromosome rearrangements have now been identified in subtypes of AML representing the various lineages of myeloid cell differentiation (Tables 10.1 and 10.2).

An (8;21)(q22;q22) translocation is present in the leukaemic cells of 10–15 per cent of patients with AML-M2, and in a small number of patients with AML-M4. A (15;17)(q22;q12) translocation is present in virtually all cases of acute promyelocytic leukaemia, AML-M3.

Subtle structural rearrangements of chromosome band 16q22 have been demonstrated in AML-M4 with abnormal eosinophils, using high resolution chromosome banding. These rearrangements are either an inversion or translocation with band 16p13, various other translocations involving band 16q22, or a deletion, del(16)(q22). The metallothionein gene cluster which is normally located in band 16q22 has been shown to be split by inversions and translocations involving this band. It has been suggested that metallothioneins function in the intracellular binding and storage of zinc in eosinophils and granulocytes, and therefore tumori-genesis might occur by combining the control sequences of these genes with other genes.

A translocation t(6;9)(p23;q34) has been observed in AML-M1, -M2, and -M4 in which there is an increased basophil count. The basophils appear morphologically normal. Using high resolution chromosome banding, the breakpoint in chromosome 9 has been shown to be slightly more proximal than that in the t(9;22) in CML.

Structural rearrangements, translocations, or deletions involving the long arm of chromosome 11, particularly bands 11q23-24 but also 11q13-14, have been observed in acute monoblastic leukaemia. In one study, six of eight children and five of 16 adults with poorly differentiated or monoblastic leukaemia, M5a, had abnormalities of 11q, whereas only one of three children and none of seven adults with well differentiated or monocytic leukaemia (M5b) had these abnormalities. The breakpoints on 11q can vary, and where the abnormality is a translocation, the other chromosome may be variable, although the most common translocation is t(9;11)(p22;q23). Deletions of chromosome 11, usually occurring as del(11)(q23) are also found in AML-M5.

Secondary acute myeloid leukaemia. Radiation or cytotoxic chemotherapy for a first malignancy can give rise to a myelodysplastic syndrome (MDS) (an abnormal overgrowth of some bone marrow cells) terminating in a secondary acute myeloid leukaemia (S-AML) that is generally unresponsive to therapy. Characteristic chromosome aberrations are present in most patients with these diseases.

In one study, 61 of 63 patients with secondary AML had visible chromosome rearrangements and, of these, 55 patients had deletions of the long arm of either or both chromosomes 5 and 7, as shown in Table 10.3 (Le Beau *et al.* 1986). These figures should be compared with those in the same study for apparently *de novo* AML where only 22 out of 140 patients had deletions or losses of chromosomes 5 and/or 7. Regions 5q23-32 and 7q32-36 are commonly deleted, although the amount of deleted material varies. In an earlier study, patients with AML who had been previously exposed to insecticides or petroleum products were also found to have deletions of 5q and 7q in their leukaemic cells.

Table 10.3 Deletions of 5q and 7q in secondary AML

Number of patients	Chromosome aberrations
6	Monosomy 5
8	Deletion of 5q
22	Monosomy 7
2	Deletion of 7q
17	Abnormalities of both 5 and 7

Several different models can be envisaged for the role of these deletions in tumorigenesis, i.e. either that a tumour suppressing gene needs to be deleted, followed by mutation or loss of the remaining allele (see Chapter 4), or that a specific combination of genes or sequences is brought about by loss of chromosomal material. An alternative possibility is that control sequences of particular genes on chromosomes 5 and 7 need to be deleted or disrupted to deregulate transcription. The fact that the breakpoints tend to be variable (although there appear to be a few preferential breakpoints) favours the first model.

In a different study of 115 patients with S-AML whose first malignancy was Hodgkin's or non-Hodgkin's lymphoma, the most common numerical abnormality was loss of chromosome 7. The most common structural abnormalities involved various different aberrations of chromosome 3 with breakpoints around 3p14 and 3q21, and deletions of 5q.

10.4.3 *Lymphoid malignancies*

Acute lymphoblastic leukaemia (ALL) occurs more commonly in children. The disease is characterized by the accumulation of immature malignant lymphoid cells in the bone marrow and blood. Chronic lymphoproliferative diseases, such as chronic lymphocytic leukamia

(CLL) and prolymphocytic leukaemia, are characterized by the accumulation of relatively mature lymphoid cells in the bone marrow and blood.
Lymphocytes are the main cellular elements of the immune system. B
lymphocytes are responsible for immunoglobulin-mediated immunity,
while T lymphocytes are responsible for cellular immunity via T cell
receptors (Chapters 15 and 18). Specific translocations result in activation of the oncogene c-*myc* by immunoglobulin and T cell receptor
genes in B and T cell tumours, respectively.

At the Third International Workshop on Chromosomes in Leukaemia
in 1980, two-thirds of 330 ALL patients were shown to have clonal
chromosome aberrations. Of the 213 aneuploid cases, 35 per cent were
pseudodiploid, 25 per cent were hyperdiploid, and 7 per cent were
hypodiploid. Numerical aberrations were non-random, with chromosomes 6, 8, 18, and 21 being preferentially gained, and chromosomes 7
and 20 being preferentially lost. The structural abnormalities which
occurred most frequently were t(9;22) in 17.9 per cent of patients with
structural abnormalities, t(4;11) in 8.2 per cent, t(8;14) in 7.3 per cent, a
14q + chromosome in 6.9 per cent, and a 6q − chromosome in 6.0 per
cent of patients. Half the patients with the t(4;11)(q21;q23) were children, most of whom were less than one year old.

The t(9;22) is the aberration found most frequently in adult non-B,
non-T ALL. It is cytogenetically identical to that which gives rise to the
Ph′ chromosome in CML. However, only some Ph′-positive ALLs have
breakpoints within M*bcr* resulting in a *bcr-abl* fusion protein of 210
kDa. Others have breakpoints within the first intron of *bcr* resulting in a
fusion protein of 145 kDa, while a few do not appear to have breakpoints within the *bcr* gene at all suggesting the involvement of different
sequences on chromosome 22 (Heisterkamp *et al.* 1989).

Karyotype is an important correlate of survival in ALL (Williams *et al.*
1986). Patients with leukaemias showing modal chromosome numbers
greater than 50 respond best to treatment, with children in this group
frequently being cured. Patients without apparent chromosome abnormalities in their leukaemias also do well, while those with the t(4;11) or
t(8;14) respond particularly poorly and have the shortest survival times.

B cell malignancies. The specific translocations found in B cell
leukaemias and lymphomas have been subjected to rigorous molecular
analysis (Croce *et al.* 1987). These translocations show several features
that differ radically from the t(9;22) described above.

Burkitt's lymphoma is a tumour characterized by massive enlargement
of lymph nodes due to infiltration of malignant B lymphocytes, the
immunoglobulin-producing cells. Two categories of the disease are
defined, the African or endemic form in which children often present

with jaw masses, and the non-African or sporadic form. An (8;14)(q24;q32) translocation is present in the malignant cells of approximately 90 per cent of patients with both categories of Burkitt's lymphoma (Fig. 10.4). One of two variant translocations, t(2;8)(p12;q24) and t(8;22)(q24;q11), is present in the remaining patients with the disease. All three translocations have also been observed in B cell ALL. Interestingly, chromosomes 14, 2, and 22 carry the genes for the immunoglobulin heavy (IgH), and the kappa (κ) and lambda (λ) light chains (discussed below), respectively. In the typical t(8;14), the oncogene c-*myc* which is normally present on band q24 of chromosome 8, translocates to the IgH locus on chromosome 14. In the variant translocations, c-*myc* remains on chromosome 8 while sequences from the κ and λ light chain loci translocate to the vicinity of the c-*myc* gene. Analogous translocations between c-*myc* and immunoglobulin loci are

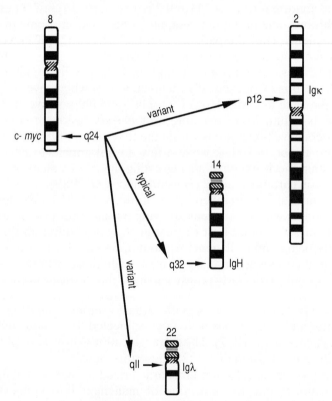

Fig. 10.4 Schematic representation of the typical and variant translocations in Burkitt's lymphoma and B-cell ALL.

present in rat and mouse plasmacytomas which are also derived from B cells. All three translocations in human tumours, and those in rat and mouse plasmacytomas, result in constitutive transcription of the trans-located c-*myc* gene. It is believed that c-*myc* activation occurs by physical separation of the gene from its normal regulatory elements. However, it is also important that c-*myc* is in these circumstances physically adjacent to genes that are highly transcribed in mature B lymphocytes, i.e. the immunoglobulins.

A brief introduction to the immunoglobulin (Ig) and c-*myc* genes helps in understanding the mechanisms involved in c-*myc* activation. B lymphocytes are responsible for antibody-mediated immunity. Antibody molecules are composed of two identical Ig light chains and two identical Ig heavy chains (see Chapters 15 and 18). The light chains each have a variable region and one constant region domain, whereas the heavy chains each have a variable region and three or four constant region domains. The Ig loci have a large number of variable (V) region genes located upstream from the constant (C) region genes (Fig. 10.5). These

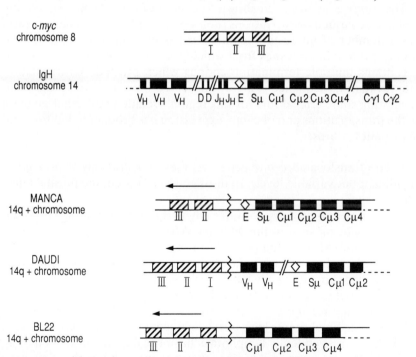

Fig. 10.5 Schematic representation of c-*myc* and IgH genes, and the arrange-ment of these genes at the translocation breakpoints in three cell lines derived from Burkitt's lymphoma with the typical t(8;14) (see text). Boxed regions: exons (coding regions). Arrows indicate direction of transcription of c-*myc*.

loci have a very unusual property: during B lymphocyte differentiation they undergo recombinations that are crucial for the generation of antibody diversity. The first step is a rearrangement that joins one of the V-region segments with a C-region segment, separated by the small D (diversity) and J (joining) segments in the case of the heavy chains or by a J segment only in the case of the light chains. This is brought about by recombinase enzymes which recognize specific heptamer-nonamer signal sequences in these regions. The promoter of the V-segment that has been moved is thus brought under the influence of an enhancer sequence immediately upstream of the C-segment, which leads to the production of a complete Ig transcript. The enhancer is, by definition, tissue specific and can enhance transcription of genes on either side of it. In the case of the heavy chain loci, further rearrangements occur later that involve a switch in the class of the C gene expressed with a given V segment. In any one B cell, these recombinations occur at only one allele of each Ig locus, resulting in the production of only one type of light chain and one type of heavy chain.

The c-*myc* gene has been shown to consist of three exons (i.e. coding regions) separated by two introns (non-coding) (Fig. 10.5). The first exon has a number of interesting features: it has two active promoters with transcription initiation sites from which two transcripts are synthesized; it contains multiple termination codons which preclude its translation into protein; its sequence is highly conserved between mouse and man, suggesting some crucial role for this region. Detailed molecular analysis of the translocations and of c-*myc* expression has brought to light several important features:

1. The translocated c-*myc* gene becomes constitutively expressed at significant but variable levels. In most cell lines tested, the normal allele remains silent.

2. The breakpoints vary considerably in different tumours, with slight differences being seen in the endemic African type lymphomas and in sporadic non-African tumours. Most cases of endemic lymphomas appear to have breakpoints upstream of the first exon of c-*myc*, whereas sporadic lymphomas usually have breakpoints within the first exon or in the intron between exons 1 and 2. However, in all known cases the regions coding for the c-*myc* protein (second and third exons) are unaffected. The breakpoints have also been located in different positions in the Ig loci. Most endemic lymphomas have breakpoints in the V_H or J_H regions, whereas in sporadic cases they appear to be in the μ, γ, or α switch regions (Fig. 10.5). These findings reflect different mechanisms whereby in endemic lymphomas the translocations occur as a result

of mistakes in V-D-J joining, while in sporadic lymphomas they occur as a result of mistakes during heavy chain class switching at a later stage in B cell development.

3. The orientation of c-*myc* and Ig genes in the t(8;14) is head-to-head, or 5′ to 5′, whereas in the t(2;8) and t(8;22) variants it is head-to-tail, or 5′ to 3′. This variation results from the fact that IgH sequences are oriented with the constant region facing the centromere with the variable region placed more distally, while this is reversed in both Igκ and Igλ. No effects have been noted from this variation in orientation.

4. The translocation always involves the non-functional Ig allele in which no recombination has occurred.

The variability in the locations of the translocation breakpoints makes it difficult to define one mechanism for c-*myc* activation in Burkitt's lymphomas. In one case, transcriptional activation has been shown to be due to the Ig enhancer element (see Fig. 10.5, MANCA). In other cases, however, the enhancer is either too far away to influence c-*myc* expression (DAUDI, Fig. 10.5) or else is not on the chromosome containing c-*myc* (BL22, Fig. 10.5). It has been proposed that sequences within the first exon of c-*myc* are involved in regulating expression and that removal of this exon in some tumours may be responsible for activation of the oncogene. Extensive mutation of the first exon has also been found in one case. The fact that the first exon is absent from the viral *myc* oncogene lends further support for its regulatory role in normal cells (see Chapter 9).

The direct participation of the deregulated c-*myc* gene in tumorigenesis has been elegantly demonstrated by the formation of transgenic mice carrying various constructions of the c-*myc* gene. One experiment in which the normal upstream control regions of the c-*myc* gene were replaced by an Ig enhancer region, 13 of 15 transgenic mice developed aggressive B cell lymphomas. Each of these lymphomas was shown to be of clonal origin, i.e. derived from a single progenitor cell, implying that at least one other event besides deregulation of the c-*myc* gene is necesary for neoplastic transformation to occur. Explanations for the mechanisms of c-*myc* activation in B cell tumours are for the most part speculative at this stage, but the general principle remains remarkably clear: c-*myc* becomes deregulated as a result of its relocation into the immediate vicinity of the Ig loci. Although the exact function of the *myc* protein is not yet known, there is some evidence linking it to DNA synthesis. The *myc* protein is nuclear and can bind to double-stranded DNA. It is expressed in dividing cells and switched off when the cells enter a resting state, suggesting a mechanism whereby deregulation of the gene maintains cells in a cycling state. Activation of the c-*myc* gene can occur by different mechanisms such as gene amplification leading to enhanced expression (see p. 000).

It is interesting that chromosome band 14q32, the site of IgH, is also involved in recurring chromosome translocations in other B cell malignancies, and that these have also been shown to directly involve the IgH genes. Small cell lymphocytic lymphomas and B cell CLLs have a t(11;14)(q13;q32), while follicular lymphomas (small cleaved cell lymphomas) have a t(14;18)(q32;q21). Sequences designated *bcl*-1 and *bcl*-2 (B cell lymphoma/leukaemia) have been shown to become translocated from their locations at bands 11q13 and 18q21, respectively, into the immediate vicinity of the IgH locus at band 14q32. No transcription unit has yet been found for the *bcl*-1 gene. However, the *bcl*-2 gene is expressed in B cell lines with high levels being found in cells showing the translocation. There is some evidence that in cells with the t(14;18), expression of *bcl*-2 is deregulated as a result of its being brought under the influence of Ig sequences. It appears that in some cases the translocation between chromosomes 14 and 18 occurs as a result of a mistake during immunoglobulin V-D-J joining at the pre-B stage of differentiation, so that there is an interchromosomal rather than an intrachromosomal joining. Heptamer-nonamer signal sequences have been found near the breakpoint on chromosome 18, and it has been suggested that these are mistakenly recognized by the recombinase enzyme(s) that normally facilitate Ig recombination during B-cell differentiation (Finger *et al.* 1986).

T cell malignancies. T cell malignancies exhibit chromosome translocations that result in rearrangement of cell type specific genes in a fashion analogous to that found in B cell malignancies (Croce *et al.* 1987). T cell receptors function as recognition molecules in cellular immune responses. There are marked structural similarities between the T cell receptor (TCR) and Ig loci, and both undergo somatic recombinations during lymphocyte differentiation. Two loci are involved in the production of TCR molecules, the α-chain locus at chromosome band 14q11 and the β-chain locus at band 7q35. The function of a third locus, the γ-chain locus at chromosome band 7p13, is largely unknown.

Chromosome band 14q11 is frequently involved in translocations and inversions in T cell malignancies (Table 10.2). T cell ALLs have translocations between chromosome 14 at band 14q11 and chromosome bands 8q24, 10q24, and 11p13, respectively, while chronic T cell tumours have a translocation between two chromosomes 14 with the breakpoints at 14q11 and 14q32, and also an inversion of chromosome 14 between bands 14q11 and 14q32.

In the t(8;14)(q24;q11), the translocation splits the TCRα locus in the Jα region between the V and C regions, relocating the Jα sequences downstream of the c-*myc* gene on chromosome 8. In all cases examined, the break on chromosome 8 is 3' (downstream) of the intact c-*myc* gene. Signal sequences for normal VJ joining are present at the breakpoints in chromosome bands 8q24 and 14q11, suggesting that the translocation occurred during TCR gene rearrangement (Finger *et al.* 1986). The translocation results in constitutive expression of the c-*myc* gene as a result of its juxtaposition with TCRα sequences. In the t(11;14)(p13;q11), the TCRα locus is again split between the V and C regions (although in one case the break is within the Vα region), resulting in the juxtaposition of TCRα sequences with sequences designated *tcl-2* on chromosome 11. In the t(14;14)(q11;q32) and the inv(14)(q11q32) TCRα sequences are believed to activate a gene designated *tcl-1* on chromosome 14 at band 14q32, located proximal to the IgH locus. The gene which is believed to combined with TCRα sequences in the t(10;14)(q24;q11) has been designated *tcl-3* (Fig. 10.6).

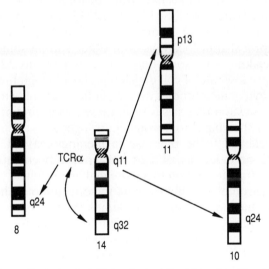

Fig. 10.6 Schematic representation of consistent translocations found in T-cell ALL involving the TCRα locus.

The TCRβ locus on chromosome 7 at band 7q35 can also be disrupted by chromosomal rearrangements in T cell malignancies. For example, in one case of T cell ALL, a $C_{\beta 1}$ gene from a rearranged TCRβ gene was found to be inserted into a region of chromosome 6.

10.5 Solid tumours

The identification of consistent chromosome aberrations in solid tumours has lagged far behind that in leukaemias and lymphomas, largely for technical reasons. It is difficult to obtain sufficient numbers of dividing cells from many solid tumours, and when chromosomes are obtained they are often of poor quality. Further, the complexity of chromosome aberrations observed in many solid tumours makes the distinction between primary and secondary changes difficult. Recently, the use of enzymes for disaggregating tumour tissue, the addition of feeder layers, cell adhesion substances and growth factors to short term cultures, and improved cytogenetic techniques have begun to allow the accumulation of data showing that consistent chromosome aberrations are present in several solid tumour types (Table 10.4).

One of the interesting facts to emerge from these data is that consistent chromosome breakpoints can be present in benign tumours. For example, a large proportion of mixed salivary gland adenomas which show chromosome aberrations have translocations and deletions involving chromosome bands 3p21, 8q12, and 12q13–15. Approximately half of karyotyped lipomas have chromosome bands 12q13–14 involved in balanced reciprocal translocations with other chromosomes. Meningiomas have been known for some time to have monosomy 22 or deletions of 22q. The very low rate of cell division in most benign tumours has so far precluded cytogenetic studies on greater numbers of benign tumours. However, the question that arises from these recent findings is whether benign tumours showing chromosome rearrangements are premalignant. This is not believed to be the case in meningiomas, at least, where monosomy 22 or deletions of 22q have never been reported to be associated with transformation to a more malignant state.

Consistent chromosome aberrations in solid tumours have recently been shown to have diagnostic applications (Sandberg and Turc-Carel 1987). For example, it may be difficult to distinguish histologically between various types of childhood small round cell tumours. The presence of a t(11;22) in Ewing's sarcomas and some peripheral neuroepitheliomas but never in neuroblastomas or embryonal rhabdomyosarcomas can therefore be an invaluable aid to classification and to deciding about the type of treatment. Similarly, the presence of a t(12;16) in myxoid liposarcomas can help to distinguish these from other myxoid tumours.

Prognostic applications have also emerged from these studies, as is clearly seen in bladder tumours (Sandberg 1986). In non-invasive bladder tumours, recurrence has been reported to occur in 90 per cent of

Table 10.4 Consistent chromosome aberrations in solid tumours

Malignancy	Chromosome aberration
Alveolar rhabdomyosarcoma	t(2;13)(q37;q14)
Bladder carcinoma	Structural changes of 1
	i(5p)
	Structural changes of 11
Breast carcinoma	Structural changes of 1
	t/del(16q)
Ewing's sarcoma, Askin's tumour/neuroepithelioma	t(11;22)(q24;q12)
Kidney carcinoma	t/del(3)(p11–21)
	t(5;14)(q13;q22)
Large bowel cancer	Structural changes of 1
	Trisomy 7
	Structural changes of 17
Lipoma	t(12)(q13–14)
Malignant melanoma	t/del(1)(p12–22)
	t(1;19)(q12;p13)
	t/del(6q)/i(6p)
	Trisomy 7
Meningioma	Monosomy 22
Mixed salivary gland adenoma	t(3)(p21)
	t/del(8)(q12)
	t/del(12)(q13–15)
Myxoid liposarcoma	t(12;16)(q13–14;p11)
Neuroblastoma	del(1)(p31–32)
Ovarian carcinoma	t(6;14)(q21;q24)
	Structural changes of 1
Prostatic carcinoma	del(7)(q22)
	del(10)(q24)
Retinoblastoma	Structural changes of 1
	i(6p)
	del(13)(q14)/-13
Small cell lung carcinoma	del(3)(p14p23)
Synovial sarcoma	t(X;18)(p11;q11)
Testicular teratoma/seminoma	i(12p)
Uterine carcinoma	Structural and numerical changes of 1
Wilms' tumour	Structural changes of 1
	t/del(11)(p13)

Adapted from Heim and Mitelman (1987).

cases with chromosome rearrangements but in only 5 per cent of cases with no visible rearrangements. Furthermore, in invasive bladder tumours, the extent of invasiveness correlates with the extent of chromosome rearrangement. Prognosis also correlates with amplification of cellular oncogenes in certain solid tumours, which may be visualized in

the form of homogeneously staining regions or double minute chromosomes (see below; p. 290).

The molecular analysis of chromosome translocations in leukaemias
and lymphomas described above suggests that activated oncogenes such
as c-*myc* and c-*abl* behave in a dominant fashion, but investigations of
solid tumours reveal that other genes behave recessively in tumorigenesis. These genes are believed to suppress tumorigenesis when
normally present in the cell. Only when both alleles are inactivated
(through deletion or mutation, for example) does a tumour arise.
Although these studies have yielded different types of information in
leukaemias and lymphomas on the one hand, and solid tumours on the
other, both dominant and recessive genes are believed to participate in
tumorigenesis in all these tumour types. Recessively acting genes in
tumours have been described in detail in Chapter 4 and will be
mentioned only briefly here.

The prototype tumour in these studies is retinoblastoma which occurs
in both dominantly inherited and spontaneous (sporadic) forms. Several
lines of evidence pointed towards the retinoblastoma gene being located
in chromosome band 13q14, the most important being that a small
proportion of the inherited cases carry a constitutional chromosome
deletion involving band 13q14. Both inherited and sporadic retinoblastomas show loss of an entire chromosome 13, deletions involving
chromosome band 13q14, or loss of genetic material from this band
when compared to normal tissue from the same patient. The retinoblastoma (RB) gene has now been cloned and shown to be deleted or to
have inactivating alterations in retinoblastomas. A candidate gene for
Wilms' tumour, a childhood kidney tumour, has recently been cloned
from chromsome band 11p13. This followed observations of constitutional deletions in band 11p13 in patients with inherited Wilms' tumour
and of loss of genetic material from this band in sporadic tumours.
However, other data suggest the presence of at least two further Wilms'
tumour genes elsewhere in the genome. Recessively acting genes have
now been implicated in other tumour types, such as small cell lung
carcinomas, meningiomas and colon carcinomas, through the identification of consistent chromosome deletions. Intensive efforts are being
made to clone these genes.

10.6 Gene amplification

Gene amplification has been recognized for some time as one way in
which cells gain resistance to selective drugs. Cytogenetic studies of drug
resistant cell lines demonstrated that amplified genes were found in
homogeneously staining regions (HSRs) on single chromosomes or

alternatively in double minutes (DMs). The term HSR comes from the unusually uniform staining of these chromosome regions, while DMs are small paired chromosome structures which lack centromeres and therefore do not segregate symmetrically at mitosis. HSRs and DMs are common features of human solid tumours, but they are only rarely seen in leukamias or lymphomas. Because they are never seen together within the same cell (although tumour cell lines may have both HSR- and DM-containing cells), HSRs and DMs are believed to represent alternate states of gene amplification. There are many examples of amplified cellular oncogenes in tumours, often in association with the presence of HSRs or DMs (Fig. 10.7).

Many neuroblastomas show extensive amplification of N-*myc*, a gene

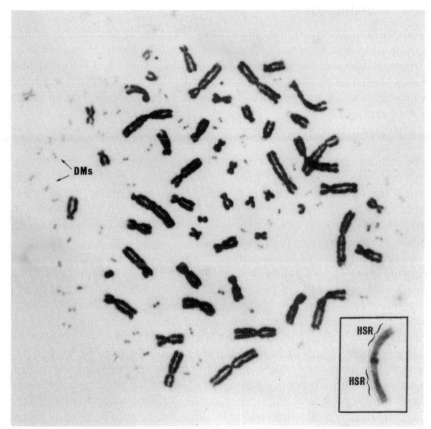

Fig. 10.7 Metaphase spread from neuroendocrine tumour cell line with multiple double minutes. Inset shows chromosome with homogeneously staining region.

related to c-*myc*, usually within HSRs and DMs. The extent of amplification of N-*myc* in neuroblastomas has been shown to correlate strongly with clinical stage of the disease (Brodeur *et al.* 1984). More importantly, the level of amplification indicated the probability of relapse-free survival. In patients with no amplification of N-*myc* the probability of relapse-free survival at 18 months was 70 per cent, in patients with 3 to 10 copies the probability was 30 per cent, and in patients with more than 10 copies the probability was 5 per cent. The prognosis of breast tumours has been similarly shown to correlate with amplification of the *neu* oncogene. A third tumour type with similar characteristics is small cell lung carcinoma where more highly malignant tumours have greater amplification of c-*myc*, L-*myc* or N-*myc* genes.

10.7 Conclusions

Non-random chromosome aberrations are closely associated with a variety of human (and animal) malignancies. The aberrations consist of structural abnormalities, most often translocations, and of gains and losses of part or all of certain chromosomes. They are believed to confer a proliferative advantage to cells carrying them and to be crucial in the pathogenesis of tumours.

Molecular analyses of these chromosome aberrations are now accumulating as the genes directly affected are cloned and identified. The cellular oncogenes c-*myc* and c-*abl* have been shown to be activated by consistent chromosome translocations so that they assume dominantly acting functions similar to their viral counterparts. Other genes which are recessively acting, such as the RB gene, need to be deleted or inactivated in order for tumorigenesis to occur and are believed to have a function in suppressing tumour formation.

Careful analysis of the exact nature of consistent aberrations can provide clues as to whether dominantly or recessively acting genes are involved. For example, if two particular chromosome bands are always brought together, one would surmise that dominant activation of one of the genes at the breakpoints is occurring. On the other hand, if a particular chromosome band is always lost from the genome, perhaps by different mechanisms including large and small deletions or isochromosome formation, one might speculate that recessively acting genes, perhaps 'tumour-suppressing' genes, are located within the chromosome band. One way in which the exact locations of recessively acting genes might be determined within normally large deletions, would be to isolate new polymorphic probes and use RFLP analysis (see Chapter 4) of tumour and corresponding normal tissue in situations where no visible deletions are found even though they would be expected for that particular tumour type. Obvious candidates for such studies are the deletions of chromosomes 5 and 7 in secondary AML.

The immunoglobulin and T cell receptor genes, which are expressed specifically by B and T lymphocytes, play a major role in translocations in tumours of these lineages. The specificity with which certain rearrangements are seen in various other tumour types, such as the various myeloid malignancies, suggests that tissue specific genes might be located at these sites. Alternatively, these rearrangements may be lethal or may not confer a proliferative advantage to other tumour types. An example which falls into this category is the t(15;17) in acute promyelocytic leukaemia, which has never been seen in any other malignancy other than the rare promyelocytic blast crisis of CML.

Malignant transformation is a multistage process involving genetic alterations at several sites in the genome. While consistent chromosome aberrations provide clear indications of some of these alterations, others such as the activation of cellular oncogenes by point mutation are not visible microscopically. Sequential karyotypic evolution has often been observed in experimental tumours and in human tumours where it has been possible to do serial sampling. These secondary changes, which usually appear more random than the primary changes, may affect the behaviour of cells carrying them and thus be selected for by leading the cells towards increased independence from host control. In tumours which have multiple aberrations, as is often the case in the solid tumours, it is difficult to distinguish the primary from the seconary changes. The finding of consistent chromosome rearrangements in benign solid tumours lends further weight to the need to develop techniques for karyotyping benign and obviously 'premalignant' lesions. Since carcinomas represent about 80 per cent of all human tumours, it is clear that studies should be focused also in this area to complement the extensive progress made on leukaemias and lymphomas.

The actual mechanisms for generating consistent chromosome rearrangements are unknown. Certain chromosome regions may be particularly vulnerable to breakage and rearrangement, as has been shown for the immunoglobulin and T cell receptor genes. No other genes are known to have the property of somatic recombination. Tumorigenic chromosome translocations involving these genes may arise from erroneous recombinase activity. Recombinase accessibility is believed to be facilitated by prior transcription of the genes or by the presence of negatively supercoiled regions or Z-DNA (Boehm *et al.* 1989). The role of fragile sites in chromosome rearrangement has been investigated, but without clear evidence for or against their direct involvement. Random chromosome breakage and rejoining may occur continuously at a low frequency, with only those conferring a proliferative advantage being ultimately observed. There is some evidence that spontaneous chromosome rearrangements observed in peripheral blood lymphocytes increase in frequency as a person ages, and these could contribute to the age-related incidence of cancer.

Direct applications of the work described above are already emerging. For example, prenatal diagnosis can now be offered for inherited retinoblastoma by using the cloned RB gene. A different application comes from the cloning of the translocation breakpoints of the t(14;18) in follicular lymphoma. The extent of residual disease in patients who have undergone treatment for follicular lymphomas can now be measured by using the polymerase chain reaction (this technique enables sequences, in this instance around the translocation breakpoint, to be selectively amplified and measured). Such new approaches towards understanding malignant transformation should facilitate advances towards the prevention and treatment of cancer.

References and further reading

Bernards, A., Rubin, C. M., Westbrook, C. A., Paskind, M. and Baltimore, D. (1987). The first intron in the human c-abl gene is at least 200 kilobases long and is a target for translocations in chronic myelogenous leukemia. *Molecular and Cellular Biology* 7, 3231–6.

Boehm, T., Mengle-Gaw, L., Kees, U. R., Spurr, N., Lavenir, I., Forster, A. and Rabbitts, T. H. (1989). Alternating purine-pyrimidine tracts may promote chromosomal translocations seen in a variety of human lymphoid tumours. *The EMBO Journal* 8, 2621–31.

Brodeur, G. M., Seeger, R. C., Schwab, M., Varmus, H. E., and Bishop, J. M. (1984). Amplification of N-*myc* in untreated human neuroblastomas correlates with advanced disease stage. *Science* 224, 1121–4.

Croce, C. M., Erikson, J., Tsujimoto, Y., and Nowell, P. C. (1987). Molecular basis of human B- and T-cell neoplasia. In *Advances in Viral Oncology,* Vol. 7 (ed. G. Klein), pp. 35–51. Raven Press, New York.

Finger, L. R., Harvey, R. C., Moore, R. C. A., Showe, L. C., and Croce, C. M. (1986). A common mechanism of chromosomal translocation in T- and B-cell neoplasia. *Science* 234, 982–5.

Heim, S. and Mitelman, F. (1987). *Cancer Cytogenetics.* Alan R. Liss, New York.

Heim, S. and Mitelman, F. (1989). Primary chromosome abnormalities in human neoplasia. *Advances in Cancer Research* 52, 2–45.

Heisterkamp, N., Jenkins, R., Thibodeau, S., Testa, J. R., Weinberg, K., and Groffen, J. (1989). The *bcr* gene in Philadelphia chromosome positive acute lymphoblastic leukaemia. *Blood* 73, 1307–11.

ISCN (1985). *An International System for Human Cytogenetic Nomenclature.* (ed. Harnden, D. G. and Klinger, H. P.). Published in collaboration with *Cytogenetics and Cell Genetics* (1985). Karger, Basel.

Klein, G. (1987). The approaching era of the tumour suppressor genes. *Science* **238**, 1539–45.

Konopka, J. B., Wanatabe, S. M., Singer, J. W., Collins, S. J., and Witte, O. N. (1985). Cell lines and clinical isolates derived from Ph positive chronic myelogenous leukemia patients express c-*abl* proteins with a common structural alteration. *Proceedings of the National Academy of Sciences, USA.* **82**, 1810–14.

Le Beau, M. M. *et al.* (1986). Clinical and cytogenetic correlations in 63 patients with therapy-related myelodysplastic syndromes and acute non-lymphocytic leukemia: Further evidence for characteristic abnormalities of chromosomes no. 5 and 7. *Journal of Clinical Oncology* **4**, 325–45.

Nowell, P. and Hungerford, D. A. (1960). A minute chromosome in human chronic granulocytic leukemia. *Science* **132**, 1197.

Rooney, D. E. and Czepulkowski, B. H. (ed.) (1986). *Human cytogenetics, a practical approach.* IRL Press, Oxford.

Rowley, J. D. (1990). Recurring chromosome abnormalities in leukemia and lymphoma. *Seminars in Hematology* **27**, 122–36.

Sandberg, A. A. (1986). Chromosome changes in bladder cancer: clinical and other correlations. *Cancer Genetics and Cytogenetics* **19**, 163–75.

Sandberg, A. A. and Turc-Carel, C. (1987). The cytogenetics of solid tumours. Relation to diagnosis, classification and pathology. *Cancer* **59**, 387–95.

Shtivelman, E., Lifshitz, B., Gale, R. P., Roe, B. A. and Canaani, E. (1986). Alternative splicing of RNAs transcribed from the human *abl* gene and from the *bcr-abl* fused gene. *Cell* **47**, 277–84.

Third International Workshop on Chromosomes in Leukemia, 1980 (1981). *Cancer Genetics and Cytogenetics* **4**, 95–142.

Williams, D. L. *et al.* (1986). Chromosomal translocations play a unique role in influencing prognosis in childhood acute lymphoblastic leukemia. *Blood* **68**, 205–12.

11

The role of growth factors in cancer
M. D. WATERFIELD

An understanding of the mechanisms responsible for the control of normal proliferation and differentiation of the various cell types which make up the human body will undoubtedly allow a greater insight into the abnormal growth of malignant cells. Particular attention is now focused on the role of polypeptide growth factors as molecules which may play a central role as both positive and negative regulators of normal and abnormal growth control and development.

The characterization of polypeptide growth factors began with the analysis of components of nutrient media necessary to support optimum growth of cells in tissue culture and with the search for the factors which could induce various biological responses in animals. The introduction of recombinant DNA techniques has now made it possible to characterize easily and produce large amounts of rare growth regulatory molecules, revolutionizing the ability to study their involvement in neoplastic diseases and to provide potential new routes to therapy. In this chapter the structure and function of growth regulators and their receptors will be described.

11.1 Structural and functional diversity of growth factors

The polypeptide growth factors are a class of molecules which can act as mitogens for target cells *in vivo* or *in vitro*, either alone or synergistically with other factors, as a result of a primary interaction with specific cell receptors. The expression of these receptors on target cells governs, at least in part, the cellular specificity of different growth factors. It is clear that particular cells may express several distinct growth factor receptors; responses are governed by the levels of factor(s) to which a cell is exposed and by interactions between the signal pathways stimulated by the receptor(s) following ligand binding. The study of such interactions provides clues to the synergistic effects mixtures of growth factors and other molecules have on the proliferation of target cells.

The types of cells that synthesize specific growth factors are often ill-defined and, in some cases, it seems that many different cell types may synthesize particular factors. This observation and others suggest that factor synthesis is not confined within particular endocrine organs, as is the case with the signal molecules grouped together as hormones (see Chapters 13 and 14), and has led to the hypothesis that three different levels or systems of signalling may exist (see Fig. 11.1).

The first system has been termed *endocrine*; synthesis takes place in a specific tissue or organ; the hormone is stored and secreted under the control of specific releasing factors whose action may in turn be controlled by the nervous system. Delivery of hormones is by the circulation. The second system is called *paracrine*, and here a cell may produce a signal molecule that can interact with nearby cells expressing appropriate receptors. In this case the cell synthesizing the factor would not necessarily express receptors capable of interpreting a signal from the factor. In the third type of system, *autocrine*, the cell responds to signal molecules which it generates itself if the appropriate post-receptor mechanisms are activated. Clearly the use of these systems by a diverse series of hormones and factors could generate a formidable array of regulatory controls.

Early work on the isolation of growth factors relied on biological

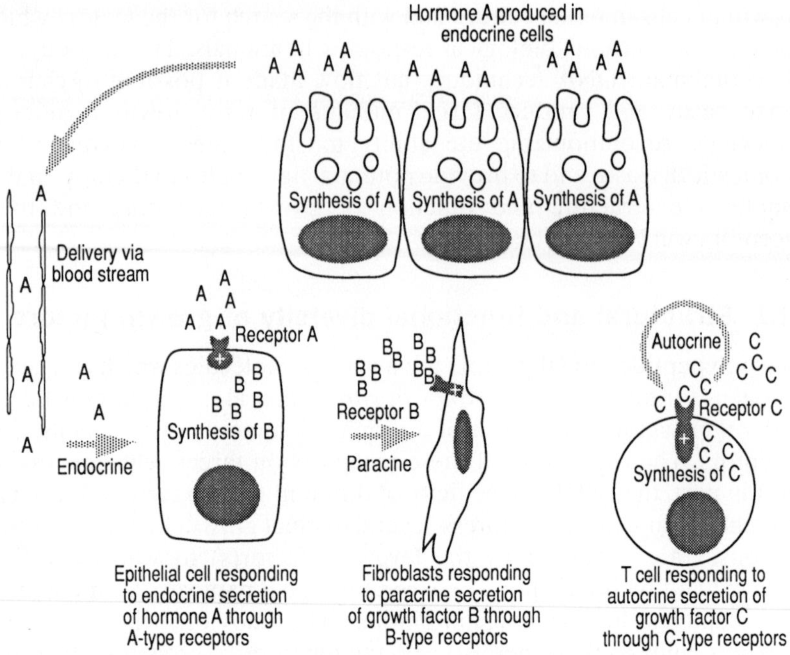

Fig. 11.1 Sites of release and action for hormones and growth regulators.

sources where expression was abnormally high and where there was a
suitable assay technique available for the factor. The pioneering work of
Cohen on epidermal growth factor (EGF) is a particularly good example
(reviewed in Waterfield 1989a). A factor which influenced the growth of
nerves had been isolated from the salivary gland of the snake and Cohen
sought a similar activity in the mouse salivary gland. Extracts injected
into newborn mice caused their eyes to open earlier and their incisor
teeth to appear faster—apparent effects on tissues of epidermal origin. The
factor was purified and was called epidermal growth factor. Receptor
binding radioimmunoassays and DNA synthesis assays on cells *in vitro*
now provide more convenient methods of measuring EGF levels. The
salivary glands of male mice for some unknown reason produce and
store large amounts of EGF and the discovery of this site of synthesis and
storage was critical in the isolation of EGF.

Other factors, such as platelet derived growth factor (PDGF), a potent
mitogen for cells of glial and connective tissue origin, were discovered
because serum, but not plasma, supported optimum growth of these
cells. (Plasma is the fluid component of blood; blood clotting, involving
the platelets, removes some proteins from the plasma which is converted

into serum.) In this case it was clear that the serum contained products released by platelets and thus the platelet was the obvious source for the purification of a serum growth factor.

Improved analytical methods have made it possible to analyse the amino acid sequence of minute amounts of proteins generating either complete sequences of the protein or providing information to make oligonucleotide probes predicted from partial protein sequence. Such probes can be used to isolate cDNA clones of the genes encoding the factors. The availability of cloning techniques which allow expression of biological activity (e.g. selection of biologically active mRNA species by microinjection into frog oocytes or the transient expression of cDNAs in tissue culture cells) has made it possible to characterize factors and receptors which could never be isolated from natural sources. The clones can then be used to synthesize the growth factor in bacteria, yeast, insect, or mammalian cells, providing large amounts of the factor for further study. These techniques have also been used to analyse the genes which encode the factors, to discover related but previously unknown factors, and to characterize possible defects in disease.

The structures of the receptors for many factors have now been elucidated at the molecular level. Characteristic primary structures for several families of cell surface receptors provide models for under-standing the processes of transmembrane signal transduction. Similarly, several growth factors can be grouped into families based on shared structural or functional features. Many factors have yet to be classified. The spectrum of tissue specificity and size of these polypeptides is presented in Table 11.1.

To link the structural data with studies in whole animals or in tissue culture systems, biological and molecular genetic studies of the abnormal proliferative properties of tumours and tumour derived cell lines provide information on the effects of normal growth factors and their receptors and the abnormal ones induced by certain oncogenes and tumour promoters resulting in subversion of their signal pathways (see also Chapters 6 and 9).

11.1.1 Haemopoietic cell growth factors

The continuous generation of mature circulating blood cells which can respond to environmental stress throughout life involves the action of a number of growth factors on a small pool of stem cells and committed progenitor cells. The stem cells themselves, which are in the bone marrow, can proliferate to generate new stem cells (self-renewal) and/or differentiate to generate specific cell lineages under the influence of factors which have restricted target specificities. The differentiation

Table 11.1 Growth regulators

Regulator	Tissue specificity	Source	Structure
Haemopoietic and lymphoid			
Erythropoietin	Modulates numbers of circulating erythrocytes	Kidney and liver of adult and fetal liver	166aa (34–39 000 K; human)
IL1 (Interleukin 1) α and β	Stimulation of IL2 synthesis by T cells, induces bone resorption and cartilage breakdown, central mediator of inflammation	Macrophages, T & B cells and fibroblasts	153aa (15 000 K; human)
IL2	Long term maintenance and growth of T cells (also B cells and macrophages)	Activated T cells	133aa (15 000 K; human)
IL3 (see multi-CSF)			
IL4 (also BSF1, BCGFγ, and BCGF$_I$)	Stimulation of anti-IgM B cells, of IgE and IgG1 production and T cell and mast cell activity	Activated T cells	124aa (15 000 K; human)
IL5 (also TRF [T cell replacement factor], BCGF$_{II}$, BCGFμ, and EDF)	Stimulation of IgA production and acts as an eosinophil CSF	Activated T cells	134aa (22 000 K; human)
IL6 (also TRF, BCDF, BSF2, and IFNβ2)	Stimulation of B cell differentiation and proliferation (also other non-lymphoid cells)	T cells, macrophages, endothelial cells, and fibroblasts	184aa (26 000 K; human)
IL7	Stimulation of B cell progenitors	Bone marrow stromal cells	(25 000 K; human)
M-CSF (macrophage colony stimulating factor, also CSF-1)	Stimulation of production of macrophages	Multiple cell types (e.g. endothelial, fibroblast, macrophages, and lymphocytes)	224aa (dimer of 14 000 K chains; human)
G-CSF (granulocyte colony stimulating factor)	Stimulation of production of granulocytes	Multiple cell types (as above)	178aa (24–25 000 K; murine)
GM-CSF (granulocyte and macrophage colony stimulating factor)	Stimulation of production of granulocytes, macrophages, and eosinophils and initiates (only) proliferation of erythroid and megakaryocyte precursors	Multiple cell types (as above)	124aa (23 000 K; murine)

Multi-CSF (also IL3)	Stimulation of production of most haemopoietic lineages and self renewal of multipotential stem cells	Antigen primed T cells	140aa (23–30 000 K; murine)
Nervous system			
NGF (nerve growth factor)	Prevents loss of neurons *in vivo*	Salivary gland (mouse), developing skin and Schwann cells	116aa (13 000 K; murine)
GGF (glial growth factor)	Stimulation of proliferation of Schwann cells	Pituitary gland	(31 000 K; bovine)
BDNF and CNTF (brain and ciliary neurotrophic factors)	Promotes survival of subsets of neurons	Unclear at present	(25 000 K; various species)
Specific for other cell types			
EGF (epidermal growth factor)	Stimulation of proliferation of cells from all 3 germ layers (e.g. fibroblasts, glial, epithelial, and endothelial cells)	Salivary gland (mouse), kidney, other sources unclear	53aa (6000 k; human)
TGFα (transforming growth factor α)	Specificity as EGF	Macrophages, keratinocytes, transformed cells, and in fetal development	50aa (6000 K; human)
VVGF (Vaccinia virus growth factor)	Specificity as EGF and TGFα	Virus infected cells	77aa
IGF-I (insulin-like growth factor I; also somatomedin C and insulin acting via IGF-I receptor)	Stimulation of cells of mesenchymal origin, controls skeletal growth with growth hormone	Mesenchymal cells	70aa (7500 K; human)
IGF-II (insulin-like growth factor II; also MSA)	Cells of mesenchymal origin stimulates fetal development and acts on brain cells	Mesenchymal cells	67aa (7500 K; human)
PDGF AA, BB, AB (platelet derived growth factors)	Cells of mesenchymal origin e.g. fibroblasts, smooth muscle cells, and Glial cells. Response to dimers varies	Platelets, endothelial cells, macrophages	104aa A chain and 109aa B chain, (Dimers 31 000; human
aFGF (acidic fibroblast growth factor; also ECGF)	Vascular endothelial cells and cells of mesodermal and neuroectodermal origin	Brain, retina, and kidney	134–140aa (16–17 000 K; human)

Table 11.1—*continued*

Regulator	Tissue specificity	Source	Structure
bFGF (basic fibroblast growth factor)	Specificity as aFGF	Brain, liver, and other tissues, capillary endothelial cells	131–146 and 156aa (18 000 K; human)
TGF beta family			
TGFβ1, β2, β3, Mullerian inhibiting substance (MIF), (transforming growth factor β related factors also isolated as BSC-1 and CIF-A), bone morphogenic proteins (BMP-2A, 2B, and -3)	Stimulation of proliferation of mesenchymal cells modulated by other factors. Inhibition of epithelial, endothelial, lymphocytic, haemopoietic, and neuroectodermal cells. Factors have a role in differentiation. MIF causes regression of the Mullerian duct. BMPs initiate bone and cartilage formation	Ubiquitous, platelets, and placenta

BMPs isoated from bone | 112aa (homodimer of 28 000 K; human)

(homodimers of 30 000 K; human) |
Neuromedins and neuroendocrine peptides			
Bombesin and neuromedins	Stimulation of proliferation of fibroblasts and bronchial epithelial cells	Gastrin releasing peptide (GRP) present in fetal lung	14aa bombesin (frog); 27aa GRP, and 9–10aa for neuromedins (human)
Substance P and substance K	Stimulation of proliferation of fibroblasts and smooth muscle cells	Central nervous system	
Vasopressin	Stimulation of proliferation of fibroblasts	Neurohypophyseal tissue	

process leads to commitment of cells to form progenitors for particular lineages each of which ultimately forms the different mature cell types seen in the circulation.

Growth factors controlling these processes have been elucidated as a consequence of the pioneering work of Metcalf (1987) and Sachs (1987) who developed in-vitro assays using cells suspended in semi-solid media which were able to respond to added colony stimulating factors (CSFs) by forming colonies of cells. In some cases, CSFs were also isolated as interleukins [multi-CSF, and interleukin 3 (IL3) are terms used to describe the same factor]. At least four distinct factors which control haemopoietic cells are now well characterized and, as a result of cloning, available in large amounts. This has allowed the evaluation of their role in disease and as potential therapeutic agents. Multi-CSF produced by activated T cells shows the widest cell specificity in its action, being able to support the proliferation of myeloid and mast cells and promote the survival, proliferation and differentiation of stem cells and of progenitor cells of the megakaryocyte, eosinophil, erythrocyte and granulocyte/ macrophage lineages (see Chapter 12). The specificity of action of granulocyte-macrophage (GM)-CSF which is produced by multiple cell types rather than just activated T cells, is more limited, acting mainly on the granulocyte/macrophage and eosinophil lineages. A further narrowing of target specificity is shown by granulocyte (G)-CSF which acts in a lineage specific fashion on granulocytes, and monocyte (M)-CSF (also called CSF-1) which acts on macrophages. The action of these four CSFs is also influenced by other growth factors which act synergistically. Of importance in this context are interleukins 1, 4, 5, and 6 which are also involved in lymphoid cell development and erythropoietin which acts to maintain circulating red blood cells.

The CSFs all appear to be encoded by distinct and unrelated single copy genes and have specific high affinity receptors which are expressed at low levels (50–500/cell) on target cells. For CSF-1 and GM-CSF the detailed structure of the receptors is known. The CSF-1 receptor was discovered because it has been identified as an oncogene (v-*fms*) carried by a feline sarcoma virus (see p. 322).

The ability to fractionate specific subpopulations of cells using the fluorescence activated cell sorter makes it possible to evaluate the role of the CSFs on leukaemic cells.

The proliferative response of leukaemic cells to CSFs has been found to be highly variable. The cells in primary myeloid leukaemia remain dependent *in vitro* on exogenous CSF and hence it is unlikely that they are proliferating through autocrine factor production. It is important to realize that some normal cells can produce as much CSF as leukaemic cells and that in the case of myeloid leukaemia, the cells are exposed in the circulation to abundant CSF.

Experiments wherein CSF production has been induced in pre-leukaemic or immortalized cells suggest that as yet unknown changes which alter the balance between self-renewal and differentiation are prerequisites to the generation of the leukaemic properties of a cell. Activated growth factor genes alone do not confer leukaemogenic properties on normal cells but may be able to produce the disease if certain cells are already immortalized.

11.1.2 *Immune system growth regulators*

The origin and operation of the complex series of cells which make up the immune system and provide a number of different defence mechanisms for the body are coordinated by a variety of interactions, many of which are mediated by peptide growth regulators. At the heart of the system is the T lymphocyte (see Chapters 12, 15, and 18) which can produce a wide range of regulators known as cytokines or lymphokines after activation by antigen presenting cells. At present, it is known that interleukins 2, 3, 4, 5, and 6, interferons α and β, tumour necrosis factors (TNF) GM-CSF and transforming growth factor β can be produced by T cells, and it is to be expected that further factors will be defined in the future. As with other cell types, the importance of synergistic effects induced by combinations of different factors is of fundamental importance.

The major factor which influences T cell growth is IL2. IL2 has been produced in quantity by molecular cloning techniques, and it has been crystallized alone and also in combination with its receptor. Since the primary structures of both receptor and factor are known and a full three-dimensional picture of the ligand–receptor interaction will soon be produced we may have a clear rational route to the design of potentiators or antagonists. The mechanism of signal transduction by the IL2 receptor is unclear.

IL2 has been used in cancer therapy by Rosenberg (1988) and others to help produce particular types of T cells which should recognize and destroy tumours; at present, results are disappointing for all but a limited number of cancers.

The T cell population includes cells which are cytotoxic to other cells. The elimination process involves cell–cell recognition and is less dependent on factors than other steps in the immune response but could still involve the TNFs and interferons. One of the major processes involving the immune regulators is the provision of T cell help to B cells which results in the production of antibodies. This process is extremely complex and involves IL2, 3, 4, 5, and 6 and the interferon and TNF molecules. It is clear that IL4 and IL5 can mediate the isotype regulation of antibodies thus determining the types of antibodies produced to mount a defence against a particular antigen (see Chapter 15). The functions of the other factors are complex and presently being un- ravelled.

Rapid progress is being made in defining the roles of these mediators because the factors and their receptors can be cloned and also because many distinct sub-types of cells can be recognized and defined using the fluorescence-activated cell sorter. Reference to a specialist review is essential to follow the latest ideas of immune function. The role of these regulators in T and B cell leukaemias is discussed in Chapter 12.

In cancer research and therapy, it is of paramount importance to understand the detailed role of these regulators. It seems clear that regulation normally involves a complex series of synergistic effects on many different processes, and factors cannot be assigned unique roles. Receptors for most immune regulators can be found on a variety of cell types, other than those of the immune system, throughout the body. Understanding how to confine intervention to a particular cell type or tumour remains a major goal.

11.1.3 *Nervous system growth factors*

The isolation of a factor with target cell specificity for cells of the nervous system was one of the pioneering events in the growth factor field. Levi-Montalcini, Hamburger, and Cohen in 1960 were responsible for purifi- cation of nerve growth factor (NGF) from snake venom and the mouse submaxillary gland. The male mouse provided a convenient source of glands for purification as it also did for EGF. NGF promotes the survival of embryonic sensory and sympathetic neurons in culture and prevents loss of such neurons in the whole animal as long as it is administered in the critical period when target innervation begins. Many of the neurons made in the developing nervous system die and this is part of the developmental 'wiring up' process. Neurons interact with specific target fields which contain post-synaptic neurons or effector cells. It seems that the target cells make NGF and the neurons with receptors for NGF which interact with the target are stimulated and survive and thus make permanent connections. This process can be interrupted with antibodies to NGF or by interfering with the target cells, and for example, by preventing axonal transport of the receptor/ligand complexes. The primary structure of the receptor for NGF has been deduced but there are no clues as yet to its mechanism of signal transduction. Two other factors, brain derived neurotrophic factor (BDNF), which may mediate similar functions on the survival of sensory neurons, and ciliary neuro- trophic factor (CNTF), which has a wider spectrum of action, have been characterized (see Table 11.1).

A number of factors isolated as mitogens from non-nervous system cells have been found to promote proliferation or differentiation of cells in the nervous system. Thus PDGF, fibroblast growth factors FGFs, and IL1 can promote proliferation of glial cells *in vitro*. PDGF has been implicated as one factor that mediates oligodendrocyte differentiation induced by astrocytes. A factor with specificity for Schwann cells, glial growth factor (GGF), has been purified from the pituitary gland.

It is clearly very early days in defining the roles of these factors. How these factors might be involved in tumours of the central and peripheral nervous system remains to be elucidated. The possible involvement of altered factor production and receptor function in brain tumours is described below (see p. 320 and 323).

11.1.4 *EGF and transforming growth factor α*

The pioneering studies of Cohen carried out on EGF have provided a general model for the development of methods suitable for investigating the structure-function relationships and the mechanism of action of most growth factors.

EGF, a small peptide of 53 amino acids (aa), derived from a large membrane glycoprotein precursor (1200 aa) by proteolysis was originally isolated from mouse salivary glands. It can act as a mitogen for target cells which express its specific cell surface receptor and these can be derived from all three embryonic germ layers; hence 'epidermal' growth factor is a term kept for tradition rather than accuracy. The receptors can be detected and their numbers and binding affinities measured using radioligand immunoassays with monoclonal antibodies.

DeLarco and Todaro (1978) studied tumour cell conditioned medium for the ability to confer properties of anchorage independent growth and proliferation, similar to those of transformed cells, on normal cells. The early studies revealed a novel growth factor related to EGF that was called transforming growth factor α (TGFα). It was eventually found that a second factor called TGFβ was also necessary to confer the transformed phenotype. The name TGF is also a misnomer because it is now clear that normal cells can produce TGFα and, although certain tumour cell lines (mostly carcinomas) do make the mitogen it is not a tumour specific factor. High level TGFα or EGF expression in immortalized cells can induce partial transformation (see p. 320).

TGFα (50 aa) is also produced as a transmembrane glycoprotein precursor, albeit as a much smaller protein than the EGF precursor. It is also known that viruses of the vaccinia virus family encode an EGF-like molecule which has been named VVGF (vaccinia virus growth factor). The structural features of EGF, TGFα, and various VVGF are beginning to be defined by biophysical study involving nuclear magnetic resonance

(NMR) spectroscopy and their partial three-dimensional structure can thus be deduced.

The normal roles of EGF and TGFα are unclear. The best evidence suggests EGF and TGFα are important in fetal development but the TGFα gene is also expressed in adult brain and pituitary, skin keratinocytes, and macrophages. Interestingly, EGF can act as a potent anti-ulcer drug by repairing damaged gut tissue. Both EGF and TGFα are potent angiogenic agents, i.e. they stimulate the growth of blood vessels, and may also play a role in regulating bone resorption and blood calcium levels.

TGFα and EGF both appear to act through the same receptor (see p. 313).

11.1.5 *Transforming growth factor β*

Despite its original discovery as an agent promoting proliferation, further study of the action of TGFβ on a variety of different cells has revealed a very important property of this molecule, its capacity to inhibit cell growth, in particular that of endothelial and epithelial cells, lymphocytes, and certain cells of the haemopoietic and neuroectodermal lineages.

Structural analysis reveals TGFβ to be processed from a precursor of 390 aa by proteolysis. A two chain species of 25 kDa (112 aa per chain) is generated and stored in a biologically latent form with a carrier protein. TGFβ is produced by many different normal cells and tumour cells and has been purified from platelets, kidney, and placenta. Seven distinct TGFβs have now been isolated; it is clear now that what were thought to be different factors [for example, cartilage inducing factor (CIF) and myeloid and thymocyte differentiation inhibitors] are in fact TGFβ.

The TGFβs are members of a superfamily of proteins known to be involved in tissue differentiation (e.g. Mullerian inhibitory substance, factors with specific action on bone development and the product of the decapentaplegic developmental gene in *Drosophila*). Related to but distinct from the TGFβs are the inhibins and activins which are heterodimeric having α and β chains. TGFβ can modulate growth regulatory genes and act via effects on production of proteins involved in the modulation of the extracellular matrix and in mediation of cell–cell interactions.

There are three distinct specific cell surface receptors (53 kDa, 73–95 kDa, and 300 kDa) to which TGFβs can be crosslinked. The TGFβ1 and TGFβ2 species differ in their affinity for the 53 kDa receptors on some cells. Massague has suggested that the 53 kDa receptor is involved in transducing the antiproliferative effects, while the 300 kDa receptor may act to present TGFβ on the cell surface (Massague *et al.* 1987). The detailed structures of these receptors are as yet unknown.

The discovery that TGFβ could have antiproliferative effects was of great significance because it emphasized that growth factors could be growth regulators capable of producing positive and negative effects on cells. Further, its ability to inhibit growth of certain tumour cells offers new hope to cancer researchers. There is currently much study on TGFβs as antproliferative factors for epithelial cells, as immunosuppressive agents for T cells, and as wound healing agents.

11.1.6 *Insulin and the insulin like growth factors*

The hormone insulin is most familiar as a regulator of glucose metabolism but can also act synergistically with a number of growth factor regulators to modulate their effects on cell proliferation. Insulin itself is a member of a family of structurally related peptides which includes the hormone relaxin, the growth regulator NGF, and the insulin like growth factors 1 and 2 (IGF-I and IGF-II). All these factors are processed from larger precursors to form 2-chain active species. The action of insulin as a synergistic growth modulator appears to be through its interaction with the IGF-I receptor to which it binds with a reduced affinity compared to IGF-I. IGF-I is a 70 aa peptide identical to the growth hormone-dependent peptide somatomedin-C and has a specificity for cells of mesenchymal origin that express specific receptors. Indeed IGF-I mediates the action of pituitary growth hormone and is the major factor controlling growth of man. It is synthesized in human fetal mesenchymal tissue predominantly by fibroblasts and may act on other cells in a paracrine manner. IGF-I (and EGF) can be considered as 'competence factors' for certain cells (for example, BALB/c 3T3 cells) which have been primed to enter the proliferative pathway by other factors such as PDGF.

IGF-II was initially purified from the conditioned medium of the Buffalo rat liver cell line (BRL 3A) as a factor called multiplication-stimulating activity (MSA). No clear physiological role for IGF-II has been identified. RNA transcripts are expressed in fetal tissues and also in certain brain cells in the adult suggesting that it could be involved in normal development. It is remarkable that the receptor for IGF-II appears to be identical to that which serves as the mannose-6-phosphate receptor. The meaning of this observation is unclear but will provoke a hunt for the underlying reason for such an overlap in receptor function.

11.1.7 *Platelet derived growth factor*

PDGF is a potent mitogen and chemoattractant for cells of mesenchymal origin including smooth muscle and glial cells. Although originally thought to be confined to platelets it is now known to be produced by several other cell types including the placenta. Platelet PDGF is thought to be involved with other growth factors such as TGFα and β in mediating repair processes at wound sites.

PDGF purified from human platelets is a highly basic protein of 28–32 kDa made up of two polypeptide chains, A and B. The elucidation of the structures of these chains was greatly helped by the finding that the oncogene v-*sis* encoded the B-chain of PDGF (see p. 318). The full structure of the A and B chain precursors has been established and sequence analysis of the factor itself has revealed the structure of the mature peptides. The A and B chains are 50 per cent homologous in sequence. A splicing variant of the A chain exists with an altered C-terminus but its role is unclear. Examination of the PDGFs elaborated by distinct cell lines and of PDGF purified from pig platelets has made it clear that the chains can be expressed together or independently and that three species of factor exist: AA and BB homodimers or the AB heterodimer. The different isoforms of PDGF have different effects on various cell types. PDGF-AB is a more potent mitogen and chemo-attractant for fibroblasts than PDGF-AA. This difference seems to be correlated with the expression of two distinct forms of the receptor, type A and type B, that may also perhaps be expressed as AB and BB dimers: skin fibroblasts express the B and A forms at a ratio of 7:1. The type A and B receptors chain are well characterized as members of a subfamily of receptors having tyrosine kinase cytoplasmic domains (see p. 313 and Chapter 9). The receptor types A and B can be separately downregulated by AA, AB, or BB forms as expected.

11.1.8 *Fibroblast growth factors*

A growth factor which acted as a mitogen for fibroblasts was first isolated from pituitary glands by Gospodarowicz and termed basic FGF (bFGF) because of its basic character. This feature distinguished it from a related 'acidic' factor later purified from neural tissue and characterized as acidic FGF (aFGF). Both factors are able to promote proliferation of a wide variety of cell types originating from meso- and neuroectoderm. Recent analysis has revealed that a number of factors purified in many different laboratories which were given descriptive names indicating their origin or specificity of action (e.g. endothelial cell, astroglial derived, brain derived, and retina derived growth factors) are in fact either a or bFGFs.

Basic and acidic FGFs share 55 per cent sequence homology and interact with receptors which have recently been characterized as tyrosine kinases. A variety of different forms of the factors are found having truncated amino terminal sequences; the biological importance of these is unknown.

bFGF appears more potent than aFGF. The FGFs are potent mitogens for capillary endothelial cells and hence are important angiogenic agents which could play a fundamental role in tumour development by stimulating the formation of blood vessels. bFGF is produced by many different cells and a closer study of aFGF, which was thought to be specifically made in neural cells, shows that it also can be produced in other tissues including the kidney.

The study of oncogenesis by mouse mammary tumour virus (MMTV) and the continuing search for oncogenes in tumour DNAs by trans-fection of DNA into NIH/3T3 test cells (see Chapters 8 and 9) has revealed further members of the FGF family. MMTV has been found to integrate close to a gene called *int-2* whose predicted sequence can encode a protein homologous to the FGFs. Two further members of the FGF family have been isolated as oncogenes, *hst* from human stomach cancer and Kaposi sarcoma DNAs, and FGF5 from a bladder tumour. In these cases, the coding region of the FGF-like genes is normal and it appears that altered expression in time or place may be involved in their action as oncogenes. The precursors for these factors do not have classical signal peptides and it has been difficult to show that they are produced as secreted products, whereas the addition of a signal sequence to the bFGF cDNA and its expression in immortalized cells will lead to transformation. Further study of the role of the five distinct FGF like genes in cancer is underway.

11.1.9 Neuromedins and neuroendocrine peptides

A variety of neuropeptides have been described which are released in brain or peripheral nervous system neurons and serve as messengers to modulate diverse secondary functions of the nervous system. An examination of these peptides using cells in culture, best exemplified by the studies of Rozengurt (1985) on Swiss 3T3 cells, have revealed their ability to modulate cell proliferation. The pituitary is a rich source of peptides which can act as mitogens. Vasopressin is an example of a pituitary endocrine hormone which is primarily known for its effects on the physiology of muscle, kidney, and liver. In addition, it can act as a mitogen synergistically with insulin for Swiss 3T3 cells by acting through one of the two types of specific receptors. Bradykinin, produced in the peripheral circulation, dilates blood vessels; it can also act with insulin to promote proliferation of some cells.

Perhaps the most relevant peptides to cancer research are those of the bombesin family which are associated principally with neurons in the gut but can be produced by some small cell lung cancer cell lines where it may act as an autocrine mitogen. This autocrine mode of action is supported by studies which have employed monoclonal antibodies or antagonists to bombesin to block growth of small cell lung cancer cell lines. Bombesin shares with PDGF the ability to act alone as a mitogen for Swiss 3T3 cells suggesting that its release under abnormal conditions could have direct effects on proliferation of cells. The recent observation that an angiotensin receptor is encoded by the *mas* oncogene and that angiotensin can also stimulate proliferation of neural cells widens the possible role of these peptides in cell regulation (see p. 317).

The importance of the growth modulating effects *in vivo* of these peptides remains to be elucidated; clearly studies of bombesin and of the angiotensin receptor suggest that their production in tumour cells could be important and hopefully amenable to attack by antagonists.

11.2 Receptors and signal transduction

The interaction between a growth factor and a target cell is mediated by high affinity receptors usually present on the surface of the cell. Receptors can be detected and quantitated using radiolabelled ligand and with specific antibodies that recognize the receptor. The numbers of receptors can vary from less than a hundred to as many as several million on certain tumour-derived cell lines. Cells generally express receptors for several different growth factors and the numbers and affinity of these receptors can be modulated by ligand binding and other mechanisms.

Following the interaction of ligand with receptor, a process called 'down regulation' can occur which involves loss of cell surface receptors through internalization via coated pits into the internal vesicle system of the cell. Receptor and ligand can be proteolysed inside the cell or the receptor, in some cases, can be recycled following dissociation of ligand at the acid pH of the internal vesicles. Experimental study of receptor recycling for some growth factors has shown that, like the low density lipoprotein (LDL) and transferrin receptors which are involved in transport functions, the receptors reappear at the cell surface after releasing the ligand into an intracellular compartment. As more receptor structures are established it has become clear that several distinct signal transduction mechanisms are employed by the growth factors (see p. 313 and Fig. 11.2).

11.2.1 *The structure of receptors*

The ability to characterize the different receptors for diverse growth factors has been revolutionized by the ability to produce large amounts

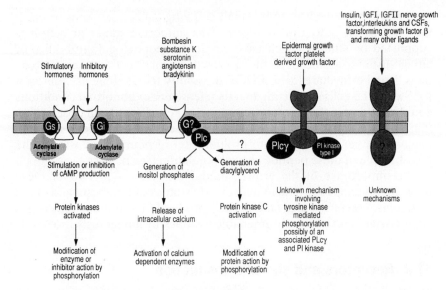

Fig. 11.2 The mechanism of signal transduction of growth regulators.

of ligand which can be used to isolate and purify receptors. It is to be expected that even those receptors present in vanishingly small amounts will be characterized to add to the rapidly growing databank of receptor structures.

Growth factor receptors are inserted in the plasma membrane of the cell surface where they transduce signals from the exterior of the cell into the membrane and cytoplasm and initiate cell growth and differentiation through the effects on gene expression and DNA synthesis. The ability to transduce transmembrane signals as a consequence of ligand binding is shared with a number of other cell surface receptors and it is not surprising that, as primary structures have been established for these receptors, a number of basic structural and functional motifs have been discovered which allow receptors to be grouped in families.

The ligand binding domain structure provides one way to classify receptors. For example, the basic structural motif of the immunoglobulin domain has been used by the PDGF family of receptors [PDGF, c-*fms* (CSF-1), and c-*kit*] and by the IL1 and IL6 receptors. The immunoglobulin building block has also appeared in diverse receptors such as the T cell antigen receptor, class I and II histocompatibility antigens, and a variety of cell surface recognition proteins. A second subset can be loosely defined by those receptors employing cysteine-rich motifs as part of their external domain. This group includes the EGF, *neu*, insulin, and IGF-I receptors. A third group with structures similar to the β-adrenergic receptor presently includes substance K, 5-HT receptors, and the *mas* oncogene product which appears to be an angiotensin receptor. Other receptors have unique external domains; this is almost certainly because many receptors remain to be discovered and also because our ability to define structural motifs is limited. Included here are the IL2, IGF-II and NGF receptors.

The process of signal transduction and amplification of growth factor triggered signals involves 'second messenger' functions which are limited in number and shared by different receptors. Thus it is to be expected that receptors for diverse growth factors will share structural features related to their use of particular second messenger systems. The largest group of growth factor receptors thus far studied has a cytoplasmic tyrosine kinase activity; these include the EGF, *neu*, insulin, PDGF, FGF, and CSF-1 receptors. This group can be subdivided on the basis of sequence insertions within or at the C terminus of the kinase domain allowing definition of subgroups related to either the EGF, PDGF, or insulin receptors. A number of other receptors contain seven trans-membrane segments (e.g. the β adrenergic, 5-HT, and substance K receptors, and the *mas* oncogene or angiotensin receptor). These receptors are linked to signal transduction systems that involve G proteins and which either generate inositol phosphate and diacylglycerol as second messengers or use the cAMP pathways. Lastly there are those for which the mechanism of signal transduction remains unknown (e.g. IL2, NGF, IL1 and IL6 receptors).

11.2.2 *The tyrosine kinase receptors*

The first growth factor receptor characterized in detail was the EGF receptor. It had been known that EGF stimulated an intrinsic tyrosine kinase similar to that first found with the Rous sarcoma virus oncogene protein pp60$^{v\text{-}src}$ (see Chapter 9) and inspection of the amino acid sequence of the EGF receptor revealed an homologous kinase domain. A large body of biochemical evidence was eventually used to generate a receptor model with an external ligand binding domain linked through a single transmembrane domain to the cytoplasmic tyrosine kinase and autophosphorylation domains. The external domain contained repeated sequence motifs which have been used to predict a more detailed structure (Fig. 11.3).

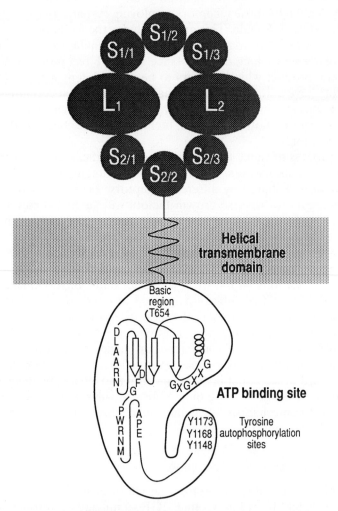

Fig. 11.3 A model for the structure of the epidermal growth factor receptor. The external ligand binding domain is thought to have two large (L) and six small (S) subdomains. The internal tyrosine kinase domain has a number of conserved residues involved in ATP binding (shown in single letter amino acid code). In addition regions of B-sheet (shown as arrows) and α-helix (shown as a coil) can be defined by analogy with nucleotide binding proteins of known structure (after McDonald *et al.* 1990).

Chemical modification and mutagenesis studies show that the kinase is essential for signal transduction and that the receptor can be auto-phosphorylated *in vivo* at a single major tyrosine residue P_1 (residue 1173) which lies in a distinct C-terminal domain (and at two further sites: P2, residue 1168 and P3, residue 1148 *in vitro*). The roles of the kinase and autophosphorylation activities in normal or abnormal growth control remain unclear. More than 15 receptor like proteins with tyrosine kinase domains are now known; these can be divided into four groups sharing structural features.

Closely related to the EGF receptor is the *neu* (also called the erb-B2 or HER2) protein for which a ligand has yet to be found. *Neu* was discovered by two routes; firstly, as an oncogene expressed in rat neuro-blastomas induced by transplacental administration of a chemical car-cinogen (see p. 321), and then as a gene closely related to the EGF receptor gene at the DNA level. The *neu* protein shares with the EGF receptor all the basic features of the ligand binding and cytoplasmic domains.

The insulin and IGF-I receptors define a group of six proteins of which four were detected as oncogenes (*met*, *trk*, *sea*, and *ros*) with no known ligand. The insulin receptor and IGF-I receptors are homodimers of 2a and 2b chains which bind either two insulin or IGF-I molecules.

The PDGF receptor group has an external domain, with five sub-domains, four of which presumably evolved by gene duplication from the gene encoding the primordial immunoglobulin folding region seen in such molecules as *Thy*-1 and of course the whole family of immuno-globulins. In addition, these receptors have inserts at a particular point in their kinases which distinguish them from the other tyrosine kinase receptors. This group includes the CSF-1 receptor, first discovered as the oncogene v-*fms*, and the related putative receptors encoded by the oncogenes *kit* and *ret* whose ligands are unknown and the FGF receptor.

The last group of tyrosine kinase receptors includes two incompletely characterized newly discovered putative receptors encoded by the *eph* (from an erythropoietin producing hepatoma) and *eck* (short for epi-thelial cell kinase) (or BR11) genes.

11.2.3 *Signal transduction by tyrosine kinase receptors*

The predicted structure of the tyrosine kinase receptors has a single transmembrane domain which separates the external and internal regions of the receptor. This short helical domain has no features which could help in signal transduction since the amino acid side chains are uncharged. This suggests that the ligand induced conformational change in the external domain generates either a push-pull or rotational signal which is transduced from the outside to the inside of the cell or that oligomerization of receptors is involved in transduction. Thus two molecules induced to dimerize by ligand binding could juxtapose their cytoplasmic domains and allow intermolecular tyrosine phosphoryl-ation. A similar mechanism could operate for the insulin receptor which is a covalent homodimer. The model has been very hard to substantiate experimentally.

The role of the kinase itself is unclear. Kinase activity is essential for the receptors to transduce mitogenic signals and may, through an auto-phosphorylation induced change, induce receptor internalization, con-sequent downregulation of bound ligand, and recycling of receptors. How the mitogenic signal is transduced is unknown. Recently a specific substrate phosphorylated on tyrosine has been identified which may mediate signal transduction. In the case of the PDGF receptor, an associated phosphatidyl inositol (3) kinase (type I), a phospholipase C (γ), and the *raf* protooncogene product can be immunoprecipitated as a complex. Tyrosine phosphorylations of these associated enzymes may generate a second messenger inositol (3, 4) or (3, 4, 5) phosphate. For EGF there also may be an associated phospholipase C.

The interaction of a growth factor such as EGF or PDGF with its receptor produces a complex cascade of morphological, physiological, and biochemical events which eventually, after a lag of eight or more hours, can result in stimulation of DNA synthesis and cell division in target cells. As yet one cannot explain how the signal cascade is directly linked to the alterations in gene expression which are clearly necessary for stimulation of cell proliferation and involve changes in action of transcription factors.

It is still controversial whether internalization plays an important signalling role. In the case of EGF, some experiments suggest that a subpopulation of receptors remains on the surface. It is important to consider that cells must be exposed to the growth factor for several hours, suggesting that a persistent signal is needed to induce prolifer-ation.

11.2.4 *Growth factor receptors with unique structures*

The structures of a number of growth factor receptors are unique perhaps because the structure of only a relatively small number of receptors is as yet known.

The IL1 and IL6 receptors, whose structures have recently been established by means of expression cloning, have external domains of 339 and 319 amino acids which include three and one immunoglobulin domains (of about 90 aa) respectively. Both show single transmembrane domains and cytoplasmic regions of 217 and 82 amino acids respectively. As yet there are no clues to their mechanism of signal transduction.

The IL2 receptor has an external domain of 20 amino acids and a short cytoplasmic domain of only 13 amino acids. It is thought that a 'converter protein' which associates with the receptor will provide a transmembrane signalling system for this receptor.

The NGF receptor has a structure with a 222 amino acid external domain that has four cysteine rich repeats, similar to the recently characterized TNF receptor, and a 147 amino acid cytoplasmic domain.

The IGF-II receptor is a remarkable case of a receptor that appears to serve two functions as it binds IGF-II and also mannose-6-phosphate. The receptor has a large external domain of 2264 amino acids with 15 repeats of a unit of 150 residues that is homologous to a second smaller type of mannose-6 phosphate receptor. The cytoplasmic domain of the IGF-II receptor is 164 amino acids. It is unclear how the scavenging function of the receptor for glycoproteins that have mannose exposed (as a result of loss of terminal glycosylation) is compatable with its role in mediating IGF-II action.

The substance K receptor, the 5-HT and *mas* oncogene or angiotensin receptor define what is expected to be a growing family of receptors having the basic structural features seen in the β-adrenergic receptor. These receptors have external and cytoplasmic domains which are linked through seven transmembrane domains. The suspicion is that the bombesin, vasopressin, and bradykinin receptors may belong to this family. Further details of this group of receptors could provide clues to the signal transduction pathways that are linked via G proteins to the cAMP or inositolphosphate pathways.

11.3 Subversion of growth control by oncogenes

The retroviruses and their viral oncogenes have already been introduced in Chapters 8 and 9. The nucleotide sequence of each oncogene has in most cases been determined and hence we know the amino acid sequence of the putative transforming proteins which these oncogenes encode. Antisera to transforming proteins allow detection of these products in virally transformed cells, tumours from experimental animals, in human tumours, tumour cell derived lines, and normal cells. Because the viral oncogenes are usually altered or aberrantly expressed counterparts of normal cellular genes, it is of great interest to determine the function of the normal genes so that their aberrant expression can be related to the causation or progression of cancer. In some cases, discrete differences in structure of the oncogene and its normal gene (proto-oncogene) are known, but in most cases their functions have until recently remained elusive. For several oncogenes, the function of the normal gene is now known and establishes a link with the pathways used by growth factors to regulate normal proliferation.

11.3.1 *Oncogenes encoding growth factors*

The discovery of a number of structurally diverse oncogenes had until 1983 provided clues to specific molecules which could subvert the function of normal cells to generate a transformed phenotype and produce tumours in experimental animals. Biochemical functions have been identified for only a few of these oncogenes. However, the first observations that linked oncogene action specifically with growth control were provided in my laboratory: the *sis* oncogene expressed a growth factor (Waterfield *et al.* 1983) and that the *erb*-B gene encoded a truncated, and presumably constitutively signalling, growth factor receptor (Downward *et al.* 1984).

During subsequent years the gradual deciphering of oncogene function has shown that these genes act on pathways which are initiated by growth factors. Here we examine the role of oncogenes which encode the factors themselves.

The sis *oncogene and PDGF.* This discovery that the v-*sis* oncogene of the acutely transforming simian sarcoma virus (SSV) encoded a transforming protein closely related in amino acid sequence to the PDGF B chain was made through the use of amino acid sequence computer databanks. The nucleotide sequence of v-*sis* which encoded a transforming protein of unknown function (deduced in the laboratories of Gallo and Aaronson) had been deposited in the sequence databank and later the amino acid sequence data from my lab for PDGF was used to search all the stored sequences for similarities. This led to the detection of an extensive identity between the predicted viral protein sequence and that of the growth factor. This result was particularly provocative because it brought together two fields of research—oncogenes and growth factors—and suggested that the reduced requirements of certain transformed and tumour derived cell lines for PDGF could be mediated by autocrine production of this factor in these cells.

Further analysis by molecular genetics and protein chemistry has shown that the sequences acquired by SSV from the normal c-*sis* B chain protooncogene on chromosome 22 can encode a functional PDGF-like molecule which is expressed as p28^{sis}—a homodimer of B chains. The hypothesis was formulated that the expression of the PDGF-like growth factor at the wrong time or in the wrong cell type could, in cells express-ing the PDGF receptor, cause transformation by autocrine action.

Many different approaches have failed to resolve any detailed mech-anism for autocrine action by the v-*sis* gene product. Simple interaction of receptor and PDGF at the cell surface does not explain the finding that exogenously added PDGF does not induce the transformation of NIH/3T3 test cells whereas these cells can be transformed by the viral oncogene. It seems likely that an internal factor–receptor complex may be activated by constitutive expression of PDGF in these cells. Also, antibodies to PDGF can inhibit proliferation in part of SSV transformed diploid fibroblasts but not in many other transformed cells.

As mentioned above PDGF can exist as a homo- or heterodimer of A and B chains, encoded by genes on chromosome 7 and 22 respectively. Both A and B chains can be expressed independently; indeed their transcription has been seen both separately and together in a number of different cell types. The A chain has however not been detected naturally as a viral oncogene, although it can transform 3T3 cells when inserted in a synthetic virus. It remains to be seen whether the effect of AA, BB, and AB chain expression and transformation of certain cells occurs because distinct receptors of the A and B type are differentially expressed in diverse cell types.

PDGF is synthesized by normal cells, the most important perhaps being the megakaryocyte. At first sight the mechanisms for synthesis, storage, and release of PDGF and platelets seem to provide a precise method for limiting the need to activate a potential 'transforming' signal. Megakaryocytes become multinucleate and incapable of proliferation during differentiation. In the lung the megakaryocyte shatters to form platelets which no longer have any capacity to synthesize proteins and PDGF stored in granules is released at wound sites in the process of blood clotting. Granule release delivers a measured dose of PDGF and also, it should be noted, of other growth factors, particularly TGFβ. Thus the cell lineage that makes the factor self destructs—an ideal situation for limiting paracrine delivery of a potent mitogen and chemoattractant to a wound site where repair is needed.

Closer examination also reveals activation of PDGF gene transcrip-tion and PDGF expression in smooth muscle cells, endothelial cells, and macrophages involved in the repair process. In smooth muscle cells, the A chain is activated and may act as an autocrine growth factor; however, in macrophages where B chains can be expressed, and endothelial cells where A or B can be elaborated, the lack of receptors means that the factor will only have a paracrine effect on other cells. Interestingly, connective tissue cells *in vivo* normally lack PDGF receptors which can be induced by a variety of treatments including growth factor action.

Clearly the control of cell proliferation and differentiation at a wound site is a complex well-regulated process. Since cancers do not commonly develop at wound sites, there must be exquisite control over the expression of such genes which can in some circumstances produce cell transformation.

Oncogenes related to FGFs. The possible role of autocrine activity of oncogenes such as that observed with studies of PDGF and the *sis* oncogene has been extended by the unravelling of the structure and action of genes which make up the FGF gene family.

Initial studies defined acidic and basic FGFs encoded by distinct genes expressed in a variety of normal tissues which could act as mitogens for cells of neuroectodermal and mesodermal origin. Later came the finding that a related FGF gene, the *int*-2 locus, could be activated by insertional mutagenesis following integration of mouse mammary tumour virus (MMTV). Transfection techniques identified two other FGF like genes: *hst*/KS3 gene in stomach tumour DNA and Kaposi's sarcoma DNA, and FGF5 isolated from bladder tumour DNA. The five genes encode members of a gene family which share a core of homologous sequences but differ in the size of the proteins they encode. Particular interest of late includes studies of normal expression of these genes during fetal development. Uncertainty remains regarding the expression mechanisms for these gene products since although *hst*/KS3 and FGF5 (and possibly *int*-2) have recognizable amino terminal leader sequences, the aFGF and bFGFs do not. This suggests that secretion and presentation of these factors to their receptors may employ novel mechanisms.

Other growth factors as potential oncogenes. Overexpression of TGFα in the presence of TGFβ produces a transformed phenotype in normal rat kidney cells. A similar study in Rat-1 cells showed that the transformed phenotype could be prevented by monoclonal antibodies to TGF supporting the concept that an autocrine loop is involved. The level of overexpression seems important and the degree of transformation varies with the cell type examined. As these studies are performed with immortalized cell lines, the results may not be relevant to primary cells. However, overexpression clearly confers a selective growth advantage and could be critical in cells already possessing other oncogenic changes.

Overexpression of haemopoietic growth factors in immortalized lines may confer a more malignant phenotype. For haemopoietic cells, however, malignant transformation appears to be more directly correlated with an altered ability to differentiate rather than with constraints conferred by growth factor requirements.

11.3.2 *Oncogenes encoding defective receptors*

The process of deciphering the molecular mechanisms of receptor signal transduction has involved the elucidation of the primary structures and analysis of the function of the receptor polypeptides. The first structure established was for the EGF receptor and it quickly became clear through a computer search for related sequences in the author's laboratory that the viral oncogene v-*erb*-B of avian erythroblastosis virus (AEV) encoded a portion of the chicken EGF receptor. AEV had acquired only the middle coding region of the receptor gene which resulted in virtual elimination of the ligand binding domain and also loss of the C-terminus of the receptor (see Fig. 11.4) suggesting that the viral oncogene could function by allowing expression of an altered receptor which generated a ligand independent signal. Support for the concept that altered receptors could be important in transformation quickly followed. Analysis of the properties of cells transformed by the v-*fms* oncogene of a feline sarcoma virus led Sherr and Stanley to deduce that this oncogene encoded a mutant receptor for CSF-1 (M-CSF) (see Sherr *et al.* 1985). The search for genes related to known receptors, the analysis of new oncogene structures, and the elucidation of an ever increasing number of diverse receptor structures has shown that the expression of altered receptors can, by several different mechanisms, generate transforming proteins.

Within the tyrosine kinase family of receptors, a number of proto-oncogenes are now known which encode proteins having all the basic structural features of receptors but for which the ligands are unknown. This includes *neu*, or the EGF receptor related protein c-*erb*-B2 (HER-2) the c-*erb*-B3 and the *trk*, *met*, and *ros* protooncogenes which are most closely related to the insulin receptor, and the *kit* and *ret* protooncogenes which are part of the PDGF/CSF-1 receptor family.

The structural modifications which can convert the protooncogene to an oncogene are diverse and incompletely analysed in all cases. For the EGF receptor, insertional mutagenesis by avian leukosis virus in chickens leads to deletion of the ligand binding domain—the consequence is erythroblastosis. Further truncation and mutation of the C-terminus in AEV *erb*-B2 is associated with the generation of sarcomas in chickens.

The conversion of the structurally closely related *neu* protooncogene to a transforming protein can be mediated by a single nucleotide change. This was first observed in the analysis of a rat neuroblastoma induced by transplacental chemical carcinogenesis using nitrosourea by Bargmann and Weinberg (1988). The conversion of a valine to a glutamic acid residue in the transmembrane domain results in the generation of the transforming protein.

(a) Receptors linked to G proteins and phospholipase C

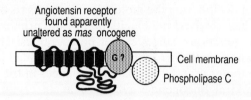

(b) Receptors with tyrosine kinase cytoplasmic domains

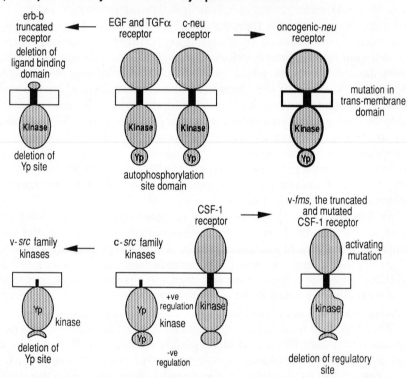

Fig. 11.4 Mechanisms for the generation of oncogenic receptors or kinases of the tyrosine kinase family. A. The *mas* oncogene. B. The tyrosine kinases.

For the CSF-1 receptor, the groups of Sherr and Rohrschneider have shown that two distinct changes are required to form a fully transforming v-*fms* protein; first, a mutation in the ligand binding domain which, although it does not eliminate ligand binding, activates the receptor, and second a truncation at the C-terminus of the kinase domain which seems to remove a negative tyrosine phosphorylation regulatory site (Roussel *et al.* 1988; Woolford *et al.* 1988). Truncation and rearrangement of the structurally related *kit* gene converts it to an oncogene but the precise nature of these changes is still ill-defined. In the case of *trk*, a DNA rearrangement ablates the ligand binding domain by replacing it with unrelated coding sequences. The truncations of *ros* and *rat* remain to be functionally characterized.

In general it seems that the alterations to the receptor may produce ligand independent signals which presumably provide constitutive second messenger signals to the cell where the altered receptor is expressed. In the case of the v-*erb*-B transforming protein, expression occurs in erythroid cells where the receptor is not thought to be normally expressed. This suggests that this cell can interpret the (as yet unknown) second messenger signal generated by this receptor.

The *mas* oncogene (angiotensin receptor) and related family members (the β-adrenergic, 5HT, and substance K receptors) encoded proteins that cross the membrane in seven helices similar to the rhodopsin photoreceptor, whose three-dimensional structure can be confidently suggested through work on the structurally related membrane protein bacteriorhodopsin. Present evidence suggests that the *mas* oncogene is structurally unaltered and that its function as a transforming protein must be a consequence of the cell type in which it is expressed (see Fig. 11.4).

The search for mutated receptors in primary human cancers has as yet not been very rewarding, perhaps mainly because it has been so limited. For the EGF receptor, it seems that only in some brain tumours are rearranged receptors detected in the primary tumour material. In these cases, gene amplification of the receptors has been detected with modification of the receptor coding regions. Their functional role remains unclear.

As yet, the type of single base change which activates the rat *neu* oncogene has not been found in human tumours. It is of course very difficult to detect small changes in receptor genes, particularly those that do not involve restriction sites or are not localized to specific bases as is the case with the rat *neu* oncogene. The availability of the polymerase chain reaction technology should revolutionize our ability to detect gene alterations and allow a more detailed followup of the experimental observations deduced through the study of oncogenes.

The number of receptors which can transduce growth regulatory signals must be much larger than the number of growth factors and regulators which are currently known and certainly must eventually include an enormous number of as yet uncharacterized receptors. It is possible that as a consequence of DNA rearrangements (or even altered expression sites) any of these receptors could be expressed as ligand independent signal generating systems which could throw the networks and systems that monitor cell proliferation and differentiation out of control and cause cancer. Such events could serve as one of a series of independent changes needed to induce the disease.

11.3.3 *Altered receptor expression in tumours*

The early studies of EGF receptor structure employed the vulval carcinoma cell A431 which was particularly useful because it displayed more than 2×10^6 receptors compared to normal cells which have approximately 1×10^5 (or less) receptors. Receptor cDNA probes showed that the receptor gene was highly amplified in this cell line. Examination of many different tumour cell lines has shown that such overexpression of EGF receptors is relatively common with or without gene amplification in as many as 25 per cent of carcinoma cell lines, although the frequency of gene amplification was lower in primary tumour tissues. Interestingly, overexpression of the receptor was consistently higher in carcinomas than in other types of tumours. Detailed studies of breast and bladder tumours suggest that overexpression may be relevant to the severity of the tumour. As yet this phenomenon is not useful as a prognostic aid. (See also p. 323.)

The *neu* gene is amplified and/or overexpressed in human breast carcinomas and this change has been correlated in some studies with poor prognosis for patients with those tumours. This correlation remains statistically controversial because of the limited numbers of tumours examined so far, but analysis of an enormous number of tissue bank specimens is in progress.

The human *neu* gene (but not the rat gene for some reason) when introduced into NIH/3T3 cells produces a transformed phenotype if expressed at high levels. In fact the degree of transformation is much higher than that produced by the overexpression of the EGF receptor which also requires EGF to exert its full effect. We do not yet know the ligand for *neu* but it appears that the *ras* transformed NIH/3T3 test cells express its ligand and hence could also mediate transformation through activation of the normal receptor as seems to be the case for the EGF receptor.

These results suggests that altered expression of a normal receptor, particularly where its ligand is available, may contribute to the oncogenic phenotype.

11.3.4 *Oncogenes acting on growth factor triggered signal pathways*

The growth factor triggered second messenger systems are rapidly being deciphered by classical biochemical approaches aided, in no small measure, by the isolation of new oncogenes or the identifcation of possible oncogene function. A brief review of the role of these oncogenes in modulating signal transduction pathways is included here.

The *ras* oncogene, which is activated by a single amino acid substitution to transform mammalian cells at low levels of expression, has been the subject of intense study and there has been speculation that it may function to modulate growth factor receptor signal transduction as a G protein. However, a direct role can be discounted with the demonstration that it is the numbers of growth factor receptors, and hence the response to various growth factors, which is changed in *ras* transformed cells. The discovery of the GTPase activating protein (GAP) product which may be the downstream effector and also the regulator of *ras*, now provides a new route to searching for a role for *ras* in modulating intracellular signal transduction initiated at as yet unknown receptors which generate the active GTP binding form of *ras*.

The search for specific genes activated in response to various growth factor signals has revealed a large family, with perhaps as many as 80–100 members, of genes which are involved at several different levels in mediating the effects on the differentiation state or division of the cell.

These genes have been found by differential screening of cDNA libraries made from treated and untreated cells. Within the family of so called early response genes which are activated within 15 min to 1 hour after growth factor treatment of cells, are the products of the *fos, jun*, and *myc* oncogenes. Recent work from several laboratories has shown that *fos* and *jun* encode proteins found in complexes that regulate gene transcription. c-*fos* encodes a nuclear protein characterized by Curran which can be extensively modified by phosphorylation and which can be induced by a diverse series of growth factors including EGF, NGF, and PDGF. This response is in part controlled by the serum response element described by Treisman which is contained within the control sequences that lie 5′ to the coding region from the *fos* protein. A specific serum response factor that also binds to this region is now known to mediate in part the induction of *fos* transcription. A number of other *fos* related proteins also exist.

The identification of the *jun* oncogene as a modified transcription factor was inferred by the observation by Vogt of a sequence which had a similarity to v-*jun* and the yeast transcription factor GCN4. The consensus binding site of GCN4 is closely related to that of the mammalian transcription factors of the AP-1 group. The AP-1 proteins have now been found to interact in complexes with the proteins of the *fos* group. These complexes, whose mechanism of action we do not yet fully understand, will provide a system that has given new insight into the processes of control and subversion of growth control by oncogenes acting as the level of gene transcription.

The additional information that proteins of the *jun* and AP-1 groups are one and the same provided the link needed to show that the *fos* and *jun* oncogenes subvert transcription factor functions. The observation that the v-*erb*-A oncogene encoded a modified cytoplasmic thyroid hormone receptor provides a related route for subversion of transcriptional control.

As yet we do not know how *fos*, AP-1 and *jun* (and presumably *myc*) are rapidly turned on by growth factor receptors but an understanding of such a mechanism could provide at least one complete cascade to explain how a growth factor signal can alter gene regulation.

11.3.5 *Links to the effects of tumour promoters*

Recent studies of the effects of phorbol esters on the physiology and biochemistry of platelet activation have provided new insight into the control of cell proliferation and the interpretation of multistage processes in causation of cancer. Nishizuka and colleagues showed that phorbol esters (such as TPA) mimic effects mediated by the ubiquitous enzyme protein kinase C, which has subsequently been shown to be the major receptor for TPA (Nishizuka 1989). Mounting evidence suggests a major role for protein kinase C as a mediator of signal transduction for various receptors, and the identification of this enzyme as the tumour promoter receptor gives a major focal point for studying the promotion stage in the generation of tumours in response to carcinogens (see Chapter 7). More recently it has been shown by Parker and Nishizuka that there are a family of protein kinase C enzymes (α, β, γ, λ, and ε) which differ in their tissue specificity for expression and in some cases perhaps their activation mechanisms.

Protein kinase C can be activated by diacylglycerol liberated as a result of phospholipase C activity generated by, for example, PDGF, bombesin, and other receptors linked to this second messenger system. In fact, the treatment of cells with a diacylglycerol analogue increases proliferation.

A great deal of effort is currently going into the characterization of phospholipase C enzymes which are responsible for generating the activator diacylglycerol. It is interesting that Hanafusa's group (Mayer *et al.* 1988) have identified an oncogene, v-*crk*, which shares structural features with phospholipases.

11.4   Abnormal growth control in cancer—a summary

Enormous progress is being made in elucidating the molecular bio-chemistry of the control of growth and differentiation of mammalian and other cells. The ability to define and manipulate the structure and function of these genes and their products through molecular biology techniques has revolutionized this area of research. Added to these advances in technical approaches are the use of oncogenes and tumour promoters as probes for specific points where subversion of function occurs during malignant transformation. It is now possible to formulate a scheme for regulation of growth and differentiation in cancer cells based on our knowledge of the biochemical pathways triggered by the growth regulators—these polypeptides which naturally are responsible for signal-ling such normal processes. Many gaps remain in the signal cascades that lead from receptors to alterations in gene expression but the way is clear for experimental approaches to fill in the detailed steps that may link these processes.

It seems evident that, in experimental cancer studies, the subversion of growth control can occur by production of a growth regulator at the wrong place, or the wrong amount or the wrong time (e.g. *sis* or the FGF related oncogenes). The need for such a growth regulator can be by-passed by alteration in receptor expression such that new or greater numbers of receptors in the presence of ligand (e.g. overexpression of TGFα) can generate an anomalous signal (e.g. EGF receptor and *erb*-B2) or alternatively the receptor structure could be changed through DNA rearrangement to deliver a constitutive signal (e.g. *erb*-B, *neu*, *fms*). Such changes can take place in cells where the receptor might not normally be expressed (e.g. *erb*-B in erythroblasts).

Alternatively the receptor could be bypassed perhaps at the level of a phospholipase or through the action of a tumour promoter (TPA and protein kinase C). The intracellular messenger system can also be sub-verted by constitutive action of *trans*acting factors (*fos*), transcription factors (*jun*), or by a *trans*regulator which is a hormone receptor (*erb*-A). How and when changes of these types are important in human cancers is the subject of great study and the next few years will reveal whether such events, if they are of primary importance, can be thera-peutically manipulated.

The availability of an ever increasing number of growth factors in large amounts and detailed knowledge of ligand–receptor interactions could make available novel therapeutics which are selective agonists and antagonists. The most immediate application is in the use of CSFs, interleukins, TNF, and interferons for direct chemotherapy, for restoration of cells damaged in chemotherapy, or for the generation of cells with antitumour specificity.

The realization that TGFβ can act as a negative regulator has opened up new horizons and reinforced the concept that surveillance mechanisms for preventing cell growth may be of central importance.

Further knowledge of interactions between dominant and recessive oncogenes (tumour suppressor genes or anti-oncogenes) provides new incentives to look for more negative regulators. It is to be expected that the actual steps which link factor, receptor, and second messenger to alterations in gene transcription and DNA synthesis will be unravelled in the near future. The observation that oncogenes provide probes for key points for subversion in cancer cells together with knowledge of their location in the cell control networks offers great hope of understanding more about the causes of cancer.

References and further reading

Bargmann, C. I. and Weinberg, R. A. (1988). Oncogene activation of the non-encoded receptor protein by point mutation and detection. *EMBO Journal* 7, 2043–52.

DeLarco, J. E. and Todaro, G. J. (1978). Growth factors from murine sarcoma virus-transformed cells. *Proceedings of the National Academy of Science, USA* **75**, 4001–5.

Dexter, T. M., Garland, J., and Testa, M. G. (ed.) (1989). *Molecular and cellular biology of hemopoietic growth factors.* M. Dekker, New York.

Downward, J. *et al.* (1984). Close similarity of epidermal growth factor receptor and v-erb-B oncogene protein sequences. *Nature* **307**, 521–7.

Green, M. R. (1989). When the products of oncogenes and anti-oncogenes meet. *Cell* **56**, 1–3.

Hanks, S. K., Quinn, A. M., and Hunter, T. (1988). The protein kinase family: concerned features and deduced phylogeny of the catalytic domains. *Science* **241**, 42–52.

Levi-Montalcini, R. (1987). The nerve growth factor 35 years later. *Science* **237**, 1154–62.

McDonald, N., Murray-Rust, J., and Blundell, T. (1990). Structure-function relationships of growth factors and their receptors. *British Medical Bulletin* **45**, 554–70.

Massague, J., Cheifetz, S., Ignotz, R. A., and Boyd, F. T. (1987). Multiple type-β transforming growth factors and their receptors. *Journal of Cell Physiology* **55**, 43–7.

Mayer, B. J., Hamaguchi, M., and Hanafusa, H. (1988). A novel viral oncogene with structural similarity to phospholipase C. *Nature* **332**, 272–5.

Metcalf, D. (1987). The molecular biology and functions of the granulocyte-macrophage colony stimulating factors. *Blood* **67**, 257–67.

Nishizuka, Y. (1989). Studies and prospectives of the protein kinase C family for cellular regulation. *Cancer* **63**, 1892–903.

Parker, P. J., Stabel, S. and Waterfield, M. D. (1984). Purification to homogeneity of protein kinase c from bovine brain—identity with the phorbol ester receptor *EMBO Journal*, **3**, 953–9.

Rosenberg, S. A. (1988). Immunotherapy of cancer using interleukin-2: current status and future prospects. *Immunology Today*, 58–62.

Roussel, M. F., Downing, J. R., Rettenmier, C. W., and Sherr, C. J. (1988). A point mutation in the extracellular domain of the human CSF-1 receptor (c-fms proto-oncogene product) activates its transforming potential. *Cell* **55**, 979–88.

Rozengurt, E. (1985). The mitogenic response of cultured 3T3 cells: integration of early signals and synergistic effects in a unified framework. In "Molecular mechanisms of transmembrane signalling" (ed. Cohen, P. and Housley, M.). Elsevier Scientific, Amsterdam.

Sach, L. (1987). The molecular control of cell development. *Science* **238**, 1374–9.

Sherr, C. J. *et al.* (1985). The c-fms proto-oncogene product is related to the receptor for mononuclear phagocyte growth factor CSF-1. *Cell* **41**, 665–76.

Sporn, M. and Roberts, A. (1988). Peptide growth factors are multifunctional. *Nature* **332**, 217–18.

Waterfield, M. D. (ed.) (1989*a*). Growth factors. *British Medical Bulletin* **45**.

Waterfield, M. D. (1989*b*). Epidermal growth factor and related molecules. *Lancet* **i**, 1243–6.

Waterfield, M. D. *et al.* (1983). Platelet-derived growth factor is structurally related to the putative transforming protein p28sis of simian sarcoma virus. *Nature* **304**, 35–9.

Woolford, J., McAuliffe, A., and Rohrschneider, L. R. (1988). Activation of the feline c-fms proto-oncogene: multiple alterations are required to generate a transforming phenotype. *Cell* **55**, 965–77.

Yarden, Y. and Ullrich, A. (1988). Molecular analysis of signal transduction by growth factors. *Biochemistry* **27**, 3113–18.

12

Biology of human leukaemia
MELVYN F. GREAVES

12.1 Introduction

In the preceding chapters the part played by many different factors in the development of tumours has been described in some detail. This chapter, concerned with the biology of human leukaemias, demonstrates how our knowledge of the basic cellular and molecular mechanisms of the control of stem cell growth and differentiation, the effects of growth factors and oncogenes, and the recognition of specific chromosome changes and differentiation markers not only increase our understanding of this group of diseases but are also invaluable in clinical diagnosis.

12.2 Cell lineages and differentiation of blood cells

12.2.1 Cellular pedigrees

Much of our current picture of the heterogeneity of leukaemias and lymphomas and the possible role played by particular molecular changes at the DNA level has been critically dependent upon our understanding of the normal cell lineage structure and differentiation programme of haemopoiesis. The different families of mature white and red cells are clearly demarcated by morphology and function but are descended, by a succession of steps, from the same population of multipotential stem cells in the bone marrow (and yolk sac or liver in embryogenesis) (Fig. 12.1).

It is important to appreciate that the blood forming tissues are arranged in lineage hierarchies with successive steps of amplification (in numbers) and genetic restriction (in differentiation options). The founder cells are considered to be stem cells by the criteria that they not only give rise to maturing progeny but can maintain their own numbers (i.e. 'self renew') almost indefinitely. Stem cells are, however, hetero-geneous in their capacity to self-renew and other, more mature cells, not normally considered as stem cells, may have very extensive self-renewal capabilities which can be revealed under certain circumstances. Mature, immunocompetent lymphocytes are particularly impressive in this respect since they may survive (as intermitotic cells) for 10 years or more and can be maintained for long periods *in vitro* provided the appropriate growth factor (interleukin 2, IL 2) is present.

Our understanding of the control of haemopoiesis, though incomplete, is still considerably in advance of what we know of the developmental physiology of other tissues. As illustrated in Fig. 12.2, interactions

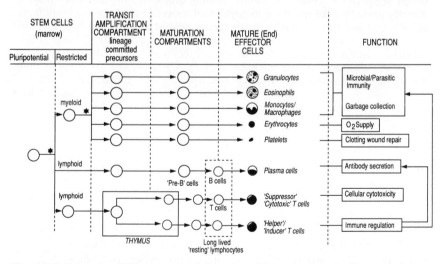

Fig. 12.1 Blood cell lineages and function. The asterisks indicate that our understanding of developmental potentialities and the sequence of pattern of commitment at these apparent 'junctions' in the lineage tree are still very incomplete.

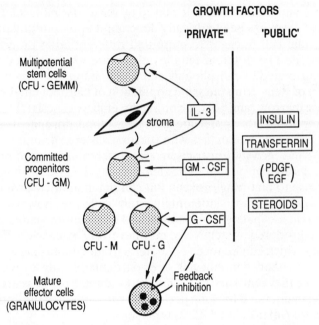

Fig. 12.2 Cellular and humoral interactions regulating growth and maturation in the haemopoietic system. 'Private' growth factors are those specific for, and acting directly upon, haemopoietic cells. 'Public' factors are circulating factors which regulate the growth of blood forming and other tissues. They may act directly upon blood forming cells or, as in the case of epidermal growth factor (EGF) and platelet derived growth factor (PDGF), influence haemopoiesis indirectly via effects on the stromal elements of blood forming tissue. CFU-GEMM, colony forming units of mixed cellularity (usually macrophage, erythroid, granulocyte and megakaryocyte); CFU-M, colony forming unit-monocyte/macrophage; CFU-G, colony forming unit composed of granulocytes only; IL3, Interleukin 3; GM-CSF, granulocyte-macrophage colony stimulating factor; G-CSF, granulocyte colony stimulating factor.

between stem cells, committed progenitors, and stromal elements (fibroblasts, epithelial cells, macrophages, and extracellular matrices) appear crucial in blood cell maturation. Several polypeptide, hormone-like growth factors regulate the growth and differentiation of haemopoietic cells (e.g. interleukin-1, 2, and 3, and granulocyte-macrophage colony stimulating factor, GM-CSF) (reviewed by Clark and Kaman 1987). These regulatory factors are frequently derived from activated T lymphocytes but may be released from a wide variety of tissues. A key, as yet unresolved, issue is whether these same proteins are mediating the control exercised by the stromal elements. This is a particularly vexing question because progenitor cells will proliferate and mature in the presence of bone marrow stroma without the addition of growth factor or the presence of any detectable soluble growth factor (Dexter 1982). A potentially important discovery in this respect is that the glycosamino-glycan (see Glossary) component of the stromal cell associated extra-cellular matrix can bind haemopoietic growth factors and 'present' them in a functional form to progenitor cells (Gordon *et al.* 1987).

It is widely anticipated that alterations in stromal interactions and/or in growth factor recognition and response will be critically involved in leukaemogenesis (see Chapter 11). Direct evidence for this is not avail-able at present although this possibility is very much encouraged by several recent findings: (i) the genes for several growth factors *and* their complementary receptors are clustered in a region on the long arm of chromosome 5 that is frequently deleted in a form of myeloid dysplasia associated with acute myeloid leukaemia; (ii) the receptor for one haemopoietic growth factor, M-CSF, has been shown to be encoded by the cellular homologue of the oncogene *fms* (see Chapter 9); (iii) trans-fection of the IL3 gene via a retrovirus vector into normal IL3 dependent haemopoietic progenitor cells (of mice) abrogates their dependence on exogenous IL3 for proliferation and, when these transfected cells are injected back into mice, a disorder strikingly similar to chronic myeloid leukaemia is observed; and (iv) primitive progenitor cells in chronic myeloid leukaemia appear to be deficient in their capacity to adhere to normal marrow stromal cells.

The precise role played by stromal cells and 'growth' factors in the differentiation process which restricts stem cell progeny to a single lineage is unknown although part of the process appears to involve random or stochastic events. This important question will only be answered when we have access to substantial numbers of essentially normal multipotential stem cells which can be maintained, cloned, and manipulated *in vitro*.

12.2.2 *Topographical compartmentalization of blood cell types*

A considerable amount of architectural organization exists in bone marrow, lymph nodes, thymus, and spleen and is probably important for cell differentiation, traffic, and function. Haemopoietic progenitor cells are found predominantly in the bone marrow with smaller numbers present in the blood and spleen, particularly during regeneration. A three-dimensional matrix and organizational heterogeneity also exists within the marrow consisting of bone, extracellular matrices, and blood cells.

A striking topographical compartmentalization exists for lymphoid cells. Thus T cells differentiate initially in the thymic cortex and are then released into the blood-lymphatic circulation with B lymphocytes derived from the bone marrow. Within lymph nodes and spleen fairly discrete areas consisting of predominantly T cells or B cells exist in association with other cells which are involved in immune responses, e.g. dendritic reticulum cells. Figure 12.3 shows a lymph node germinal centre which is a site for B cell proliferation, surrounded and infiltrated by T cells. As anticipated from this pattern of cell distribution, acute leukaemias originate from progenitor cells in the bone marrow whereas lymphomas originate predominantly from thymus or lymphoid tissue. The diversity of histopathology in lymphomas reflects in part the multiplicity of T and B cell subsets but also the topographical restriction, homing, and traffic routes of particular cell types and the variable degree of architectural disruption in neoplasia (Jaffe 1983).

12.2.3 *Markers of differentiation and proliferation*

Blood cell differentiation is characterized as in other tissues by morphological changes and functional activity. Biochemical markers of differentiation are required to identify earlier stages of maturation since immature cells are morphologically indistinguishable. Monoclonal antibodies, each of which recognizes only one specific antigen (see Chapter 15), have been especially useful in this respect (Greaves *et al.* 1981*a*; McMichael 1987). Lysosomal enzymes, enzymes involved in nucleotide metabolism (e.g. adenosine deaminase, ADA) and terminal deoxynucleotidyl transferase (TdT) have also been extensively used.

More recently, molecular markers have been developed which identify either unique, cell type products or genetic rearrangements at the immunoglobin locus or the locus for T cell receptor characteristic of particular stages of differentiation in lymphocytes. The latter 'markers' are proving of special value in the analysis of lymphoid leukaemias and lymphomas since they not only identify lineage commitment prior to its *expression* at the protein and functional level but are *clonal* in nature and so can be operationally used as tumour specific markers (Waldmann 1987). Figure 12.4 illustrates clonal rearrangements of immunoglobulin genes that can be identified in DNA from B cell leukaemias. Similar rearrangements occur in the α and β genes of the antigen receptor (T cell receptor, TCR) on T lymphocytes. These rearrangements occur within a limited region of individual chromosomes and constitute the normal mechanism of clonal diversification that accompanies lymphoid cell diferentiation. Those same loci, and others, may be involved in rearrangements (especially translocations between different chromosomes—see Chapter 10) which are intimately linked to the *leukaemic transformation* process (Showe and Croce 1987).

(a)

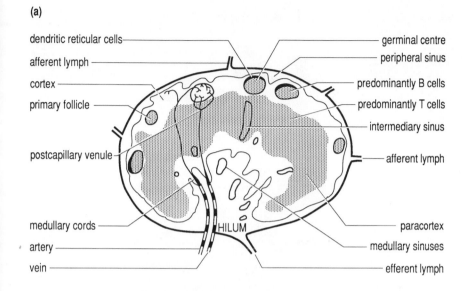

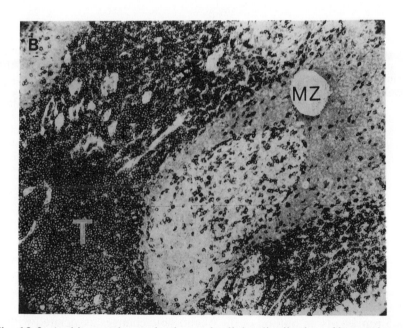

Fig. 12.3 Architectural organization and cellular distribution of human lymph nodes. (A) Topography of lymph node. (B) Immunoperoxidase staining of germinal centre area of lymph node with monoclonal anti-T cell antibody showing paracortical T cell area (T), pale germinal centre (B cell area) and its mantle zone (MZ), both showing some degree of T cell infiltration.

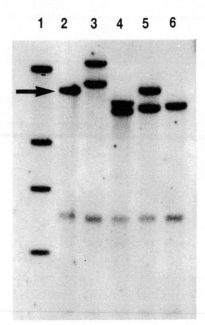

Fig. 12.4 Southern blot analysis of clonal Igμ gene rearrangements. The restriction endonuclease *Bam*HI has been used to digest DNA from human haemopoietic cells. The DNA was then electrophoresed, blotted onto nitrocellulose, and hybridized with a ^{32}P labelled Cμ probe. Track 1: molecular weight markers. Track 2: germline μ DNA (from a myeloid cell line). Tracks 3–6: DNA from different B lineage leukaemias showing various, *clonal* alterations, e.g. track 3, both alleles rearranged; track 4, both alleles rearranged; track 5, one allele rearranged, one allele remains germline; track 6, one allele deleted, one allele rearranged. Note that during normal B cell differentiation both heavy μ chain gene regions usually rearrange but only one produces a functional product (allelic exclusion) (taken from Greaves *et al.* 1986).

The cell surface contains a veritable zoo of antigens (Fig. 12.5) and multiple antigenic changes occur on the cell surface during haemopoiesis and lymphoid cell maturation. These have been extensively documented using monoclonal antibodies in conjunction with flow cytometry, tissue section staining, and immunochemical procedures. As an example, Fig. 12.5 represents an outline of the phenotypic changes associated with T cell maturation. Note that as maturation proceeds some 'markers' are lost and others gained. Some 'markers' are associated with discrete stages of differentiation, e.g. CD1 and TdT in the thymic cortex, whereas others are associated with active proliferation, e.g. the transferrin receptor on

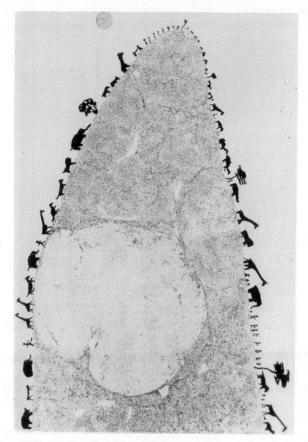

Fig. 12.5 The cell surface contains a veritable zoo of antigens.

cells in the subcapsular thymic cortex and IL 2 receptor on activated mature T cells. An appreciation of the complexity and diversity of cellular phenotypes and their modulation by differentiation and proliferation provides an important framework for the analysis of leukaemic cells, and of course the same probes (antibodies, DNA, etc.) used to analyse normal differentiation and function can be readily applied to malignant cell populations.

Many of the cell surface antigens identified by monoclonal antibodies and used as cell type specific markers have now been analysed by biochemical and molecular techniques. Most are glycoproteins and in some

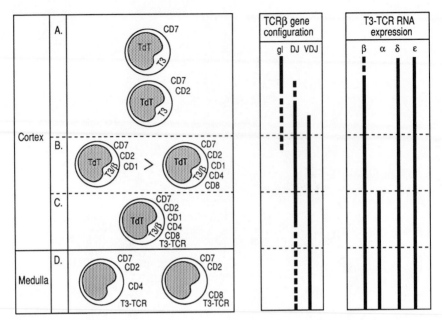

Fig. 12.6 A putative sequence of intrathymic T cell differentiation. Likely sequence of differentiation steps from anatomical compartments A through to D (arrows) is shown, see Furley *et al.* (1986) for details. Cells in the most immature compartment (subcapsular cortex A) are numerically infrequent (~1–5 per cent of total). CD numbers are the 'Cluster of Differentiation' workshop code numbering system for cell surface antigens defined by monoclonal antibodies. Note that some antigenic markers are cytoplasmic—CD3 *and* T cell receptor (TCR) β chains in immature cells, or nuclear—terminal deoxynucleotidyl transferase (TdT). Shown alongside antigenic map is the profile of T cell receptor gene rearrangement and expression in the corresponding cell compartments. Solid bar, detected; dotted bar, positive or negative result can be found; blank, not detected; gl, germ line (non-rearranged); DJ, diversity (D) and joining (J) region rearranged but not variable (V) regions; VDJ, all 3 regions rearranged; β,α, T cell receptor genes β and α. RNA expression for δ and ε is for CD3 δ and ε genes.

cases the antigenic component is carbohydrate associated with protein and/or lipid.

12.3 General features of haematological malignancy

Cancers of the blood-forming and lymphoreticular tissues can be broadly subdivided into the leukaemias, lymphomas, and myelomas. Within these major categories are subtypes identified by clinical, haematological, and histopathological features. Whilst there is broad agreement on the variety of leukaemic types and their relationship to normal cells, their classification of lymphomas is more confusing and controversial.

Leukaemias have been recognized for more than a century and are characterized, by definition, by an excess of white cells in the blood. Clinically, acute leukaemias will usually present with symptoms reflecting a disturbance of normal bone marrow function, e.g. anaemia, bruising, and infection. Myeloma will also present with manifestations of the side effects of the disease process, such as osteolytic lesions and impairment of kidney function by the immunoglobulins (myeloma proteins) secreted by the tumour cells.

Diagnosis is problematical in situations where the disease process is indolent or mild, e.g. persistent lymph node enlargement (lymphadenopathy), increase in number of blood lymphocytes (lymphocytosis), or myeloid dysplasia (bone marrow abnormalities). These conditions may be preleukaemic in nature, leading after months or years to overt malignancy. New karyotypic and molecular markers may provide a more precise means to evaluate the state of haemopoietic cells in such premalignant conditions.

12.3.1 *Incidence, geographic distribution, and aetiology*

The blood cell malignancies represent around 5 per cent of all cancers in the Western World but are major cancers in certain age groups or geographic/ethnic settings (Magrath *et al.* 1984). Thus acute leukaemia (mostly acute lymphoblastic leukaemia) is the major paediatric cancer in the West whilst Burkitt's lymphomas associated with Epstein Barr virus (EBV) are dominant childhood tumours in tropical Africa. Intestinal lymphoma has been a common cancer in certain Mediterranean countries and in the Middle East but may now be declining, at least in certain areas (e.g. Israel). There are other marked differences in the incidence and geographic distribution of blood cell cancers (Table 12.1). Thus chronic lymphocytic leukaemia (CLL) and the follicular variety of lymphoma are relatively rare in orientals whilst an aggressive form of T cell lymphoma/leukaemia, adult T cell leukaemia (ATL), is common in individuals born in southern Japan, the Caribbean, and West Africa. These latter leukaemias are associated with a C type retrovirus—HTLV-1 (human T cell lymphotropic virus type I).

Burkitt's lymphoma and ATL are at present the only human leukaemias or lymphomas whose aetiology is known to involve infectious viruses (see Chapter 8). Evidence for other environmental or genetic factors that may be 'candidate' aetiological agents or cofactors for these diseases is relatively sparse and the causative agents involved in

Table 12.1 Uneven geographic distributions of leukaemia/lymphoma

USA/Europe
  Incidence rate 2–4× that of Japan for chronic lymphocytic leukaemia (CLL), non-Hodgkin lymphoma (NHL) (follicular), Hodgkin's disease, and myeloma (except *extra-nodal* lymphoma and for ATL in Nagasaki/Kyushu region)
Japanese immigrants to USA
  B-CLL remains low; NHL (follicular) increases
Southern Japan, Caribbean, and West Africa
  HTLV associated adult T cell leukaemia (ATL) high
Tropical Africa
  EBV associated Burkitt's lymphoma (malaria associated) common
Mediterranean region
  Intestinal immunoglobulin-producing (IgA$^+$) lymphomas common
Nigeria
  Relatively high incidence of chronic lymphocytic leukaemia in young female adults
Tropical Africa
  Apparent low incidence of common ALL (cf. T-ALL and AML) in black children. Similar picture (until recently) in black children in the USA

most human leukaemias and lymphomas are unknown (Table 12.2) It is clear from the atomic bomb experience in Japan that radiation can cause or initiate acute and chronic myeloid leukaemia in humans (see Chapter 7). From the study of both experimental and naturally occurring leukaemias and lymphomas in animals we might anticipate that chemical carcinogens, viruses, and irradiation may all have a role to play in the different varieties of human disease. Genetic factors appear to be of marginal significance but there are some instructive exceptions to this, e.g. the chromosome breakage disorders such as Bloom's syndrome and Fanconi's anaemia (see Chapter 4).

It is also possible that in certain leukaemias no environmental leukaemogen or inherited susceptibility may be involved. Some of our own recent work indicates that acute lymphoblastic leukaemia (ALL) in children may have a remarkably constant, low incidence rate (approx 3.0 per 10^5/year) in 'developed' countries. We suggested that *spontaneous mutation, in utero*, might therefore be responsible for initiating this type of leukaemia (Greaves 1988). At least two spontaneous mutations may be involved and it seems likely that infections may well play a role by indirect promoting effects.

In studying the aetiology and biology of leukaemia it is important to bear in mind that, in common with most other cancers, these diseases are multifactorial and a succession of events, possibly involving different agents and cofactors, may well be necessary for the evolution of the malignant clone. Indeed, as discussed below, the multistep nature of cancer is well illustrated by (i) Burkitt's lymphoma; (ii) the transition of

Table 12.2 Causes of human leukaemia?

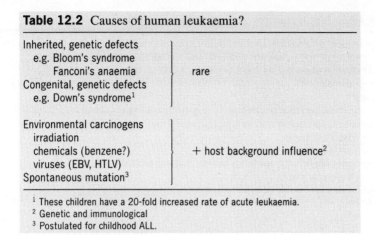

Inherited, genetic defects	
e.g. Bloom's syndrome	
Fanconi's anaemia	rare
Congenital, genetic defects	
e.g. Down's syndrome[1]	
Environmental carcinogens	
irradiation	
chemicals (benzene?)	+ host background influence[2]
viruses (EBV, HTLV)	
Spontaneous mutation[3]	

[1] These children have a 20-fold increased rate of acute leukaemia.
[2] Genetic and immunological
[3] Postulated for childhood ALL.

myeloid dysplasia to frank acute myeloblastic leukaemia (AML); and (iii) the blast crisis (acute leukaemic phase) of chronic myeloid leukaemia (CML).

Finally, in the context of aetiology, it is important to emphasize the link between immunodeficiency, immunoregulatory disorders, and lymphoma. This association has attracted increasing attention as a major problem in heart or kidney transplant recipients and in other individuals who may be immunologically compromised, e.g. patients with the acquired immunodeficiency disease syndrome (AIDS). Many of these patients develop lymphomas that are EBV associated. In these conditions, and possibly in lymphoma in general, it is considered that defective immune regulation by T cells may allow B cells to proliferate extensively and acquire, by random genetic errors, molecular changes which precipitate rapid selection of malignant clones. Some of the likely genetic events underlying such patterns of cellular evolution can now be identified.

12.4 Common biological features of leukaemic cells

Table 12.3 lists the major features of leukaemic cells that have been well documented. In many respects these characteristics of leukaemia have served as a paradigm for other human cancers. A key feature is clonality. With few exceptions (some transplant patients or AIDS patients with lymphoma), all leukaemias and lymphomas are monoclonal, i.e. they derive from a single cell. This striking observation is also a feature of most other human cancers and can be taken to indicate that an exceedingly rare genetic event(s) is normally required and produces leukaemia

Table 12.3 Major features of leukaemic cell populations

1. Non-random chromosome changes
2. Uncoupling of maturation and proliferation (maturation 'arrest')
3. Conservation of normal differentiation linked phenotypes
4. Clonal dominance:
   Markers
     X chromosome linked polymorphism e.g. hypoxanthine phosphoribosyl transferase,
     glucose-6-phosphodehydrogenase
   Chromosome changes
   Gene rearrangements:
     Immunoglobulin light and heavy chains (B lymphocytes)
     T cell receptor chains (T lymphocytes)
5. Clonal evolution

by a mechanism involving clonal selection. This interpretation does not rule out, however, an involvement of early non-clonal events contributing to leukaemogenesis.

The consistent biological features of leukaemic cells (Table 12.3) can be linked by a model of leukaemogenesis (Fig. 12.7). The key cell population consists of *clonogenic* leukaemic cells. These cells emerge and assume dominance by virtue of very rare (clonal) genetic events in the 'target cell' population (e.g. bone marrow progenitor cells). More than one genetic event may be involved and in some cases the genes involved may be already identified protooncogenes (e.g. c-*myc* in Burkitt's lymphoma, c-*abl* in CML) (see Chapters 9 and 10). In most leukaemias and lymphomas these genetic alterations are brought about by translocations between different chromosomes but other types of genetic alterations such as deletions, inversions, point mutations and gene amplifications or some combination of these mechanisms may be involved (see Chapter 10). Significantly, *different* subtypes of leukaemia and lymphoma have distinct chromosomal rearrangements and therefore, we presume, different genes are involved.

In acute leukaemia, and probably in lymphoma also, the dominant cell population deviates only minimally from the normal and retains a relatively normal phenotype in relation to its developmental and proliferative status. This fidelity of phenotype provides the basis for new classification schemes and differential diagnosis (Greaves 1982). The phenotype of the leukaemic cell is not, however, a precise or 'frozen' replica of the normal counterpart. Detailed immunophenotypic analyses have indicated that the components (i.e. different antigens) of the leukaemic cells' composite phenotype are associated in a manner that is asynchronous with respect to any normal maturation compartment (Fig. 12.7). The crucial alteration in leukaemia appears to be one which alters

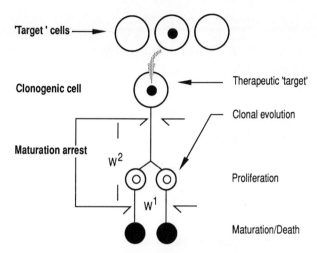

Fig. 12.7 Simplified model of leukaemogenesis. See text for discussion of this model (taken from Greaves 1982).

the probability of maturation (and thus death) versus self-renewal in favour of the latter, such that immature cells which are normally transitory accumulate in apparent maturation arrest (Greaves 1986). This partial uncoupling of maturation and proliferation can be illustrated by maturation 'windows' whose width reflects the extent of residual maturation competence or the stringency of arrest and the compartments within which cells will accumulate (W^1 versus W^2, respectively, Fig. 12.7) and thus represent the dominant leukaemic cell population. The clonogenic cells may represent a small proportion of the total leukaemic population (usually 0.1–10 per cent), but are crucial since they are responsible for the maintenance of the disease and further clonal evolution associated with malignant progression. These cells therefore represent the major therapeutic targets. Note also that since differentiation arrest is seldom absolute the clonogenic leukaemic cell can have a *different* phenotype from the bulk of leukaemic tissue as clearly illustrated by chronic myeloid leukaemia.

In the light of these concepts, leukaemic cell heterogeneity observed between patients with dissimilar or similar leukaemias and in an individual patient at diagnosis or during relapse can be ascribed to: (i) the target cell population initially involved; (ii) the stringency of maturation arrest; and (iii) the expression of abnormal cellular features associated with progression. This diversity is not only clinically important but is a crucial feature to take into account when attempting to identify underlying molecular events. Over a period of several years a sequence of different genetic events may occur progressively to uncouple the leukaemic clone

(Fig. 12.8). The developmental level of target cells is, however, of over-riding importance as it probably governs and restricts both the range of potential aetiological agents and the phenotypic and clinical features of any emerging leukaemia.

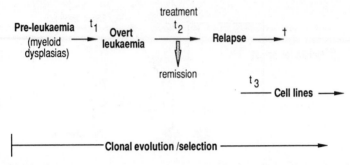

Fig. 12.8 Progression of leukaemia t_1, t_2, t_3, variable time intervals. † = death.

12.5 Phenotypic analysis of leukaemic cells and implications for mechanisms of leukaemogenesis

12.5.1 *Phenotypic diversity*

Leukaemia and lymphoma cells can be characterized and classified in relation to normal cell types using enzymatic, immunological, and molecular markers. Simple morphological correlates are also still very useful, particularly for those countries or hospitals with minimal labora-tory facilities, since they usually distinguish quite readily between most lymphoid leukaemias and the various subtypes of myeloid leukaemias, and may even identify subtypes of lymphoid leukaemias. The light microscopic appearance of stained smears of EBV associated Burkitt's lymphoma/leukaemia and HTLV-I associated adult T cell leukaemia, for example, are quite distinct.

Table 12.4 illustrates in a simplified diagrammatic form the major phenotypic differences between acute myeloblastic leukaemia and acute lymphoblastic leukaemia in humans. The latter consists of two major subtypes corresponding to progenitors of the T and B lymphocyte lineages (Greaves 1986). Within a single cell lineage, diverse cellular phenotypes can dominate and be associated with quite distinct clinical entities as illustrated by the four types of T cell leukaemia/lymphoma analysed with monoclonal antibodies and flow cytometry in Fig. 12.9.

No unequivocally leukaemia specific antigens have been identified to date despite frequent claims to the contrary and concerted attempts to

Table 12.4 Major phenotypic characteristics of blast cells in acute leukaemia

Markers	ALL		AML
	Type 1	Type 2	
	pre-B (common ALL null ALL)	pre-T (T-ALL)	
Cell surface antigens[1]			
HLA-DR	+	−	+ or −
T lineage e.g. CD7, CD2, CD3	−	+	−[2]
CALLA (CD10)	+ or −[3]	+ or −[4]	−
B lineage e.g. CD19, CD22	+	−	−
Myeloid lineage e.g. CD33	−	−	+
Enzymes[5]			
TdT	+	+	−[6]
Peroxidase	−	−	+
Focal acid phosphatase	−	+	−
Lysosomal hydrolase isoenzymes	+	−	−
Purine degradation enzymes			
5′NT (5′ nucleotidase)	H	L	L
ADA (adenosine deaminase)	I	H	L
PNP (purine nucleoside phosphorylase	I	L	(V)
Gene rearrangement[7]			
IgH	+	−	−
T cell receptor β chain	−	+	−

[1] CD—Cluster of Differentiation or Workshop number for antigen.
[2] Some AML may express the CD7 antigen.
[3] Proportion positive is related to age of patients.
[4] Approximately 15 per cent positive.
[5] H, High levels; L, low levels; I, intermediate levels; V, variable levels.
[6] Approximately 5 per cent positive.
[7] Partial or aberrant rearrangements of immunoglobulin or T cell receptor β chain genes may also be found in the inappropriate lineage, e.g. immunoglobulin in T leukaemic cells. The frequency of these 'cross-lineage' rearrangements is about 25 per cent.

find them. The most logical, though perhaps unpalatable, interpretation is that they do not exist (Greaves 1982). There may be two important exceptions to this. Immunoglobulin producing B cell tumours produce only a single cell surface immunoglobulin type which, though a normal gene product, is operationally leukaemia/lymphoma specific (see Chapter 18). The same applies to T cell receptor proteins. The other potentially leukaemia specific determinants are products of proto-oncogenes mutated or structurally altered in their coding regions. Examples of this are the p21 product of N-*ras* which is frequently

T-ALL   T-NHL

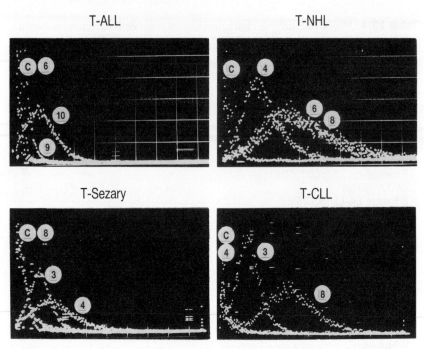

T-Sezary   T-CLL

Fig. 12.9 Fluorescence activated cell sorter analysis of cell surface antigens in various subtypes of human T cell leukaemia. Vertical axis: relative cell number. Horizontal axis: relative fluorescence intensity. T-ALL, T (or thymic) acute lymphoblastic leukaemia; T-NHL, T non-Hodgkin lymphoma (in a child and therefore of thymic subtype); T-Sezary, a cutaneous neoplasm of adults; T-CLL, T-chronic lymphocytic leukaemia, a rare neoplasm of adults; C, control antibody. 3, 4, 6, 8, 9, 10 = various monoclonal antibodies (T-series) reactive against T cell surface determinants (compare with Fig. 12.6). 3 = CD3, 4 = CD4, 6 = CD1, 8 = CD8, 9 = transferrin receptor (CD7), 10 = OKT10 (taken from Greaves *et al.* 1981*b*).

mutated by single base changes in AML and the c-*abl* encoded protein which in CML is derived from a fusion of c-*abl* with another gene, bcr (see Chapter 10).

Another clinically relevant phenotypic variable is the proliferative activity of the leukaemic clone. This can be assessed by investigation of receptors associated with proliferation, such as the transferrin receptors or, in the case of mature T cells, receptors for IL 2; an alternative and more usual procedure is to measure DNA synthesis or DNA content. The latter can be conveniently carried out at a single cell level by flow cytometry using fluorescent dyes, such as propidium iodide, which intercalate with DNA. From such analyses the proportion of cells in cycle can be estimated giving values which often differ between subtypes of a leukaemia and between leukaemic subpopulations from different sites of the same individual. In ALL and non-Hodgkin lymphoma those subtypes with the large growth fraction, or more rapid proliferation, generally respond well initially to chemotherapy (since dividing cells are more sensitive), but also more readily relapse and so have a worse prognosis. The most likely explanation of this response pattern is that rapidly proliferating cancers spawn drug resistant mutants at a relatively fast rate.

On the basis of these phenotypic studies it is possible now to align different types of leukaemia with their nearest normal counterparts in haemopoietic differentiation. Such a scheme, simplified (especially with respect to the B cell lymphomas), is illustrated for those leukaemias and lymphomas involving lymphoid cells in Fig. 12.10. A similar scheme could be drawn up for the myeloid lineages. Considerably more heterogeneity in both normal and leukaemic cell populations exists than can be easily accommodated in such a figure.

12.5.2 'Target' cells, molecular alterations, and progression in leukaemia: the lessons from CML

Phenotypic 'maps' as illustrated in Fig. 12.10 provide reasonably accurate guides to the approximate levels of maturation arrest in leukaemia in relation to the normal blood cell lineages and maturation compartments. The developmental level of a particular leukaemia may suggest, but does

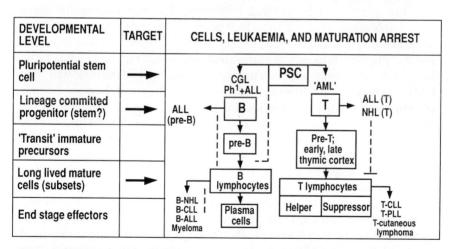

Fig. 12.10 Phenotypes of the major subtypes of lymphoid malignancy in relation to normal cells. ➡ Possible, initial 'target' cell (see text). - - - - variable extent of maturation competence (*in vivo*). T-cut. lymphoma, cutaneous T lymphoma; T-PLL, prolymphocytic leukaemia. For other abbreviations see text.

not necessarily identify, the precise target cell population involved in the genetic events responsible for initiating the disease. This point is well illustrated by a consideration of chronic myeloid leukaemia (CML).

CML occurs in association with a special chromosome marker—the Philadelphia (Ph) chromosome (see Chapter 10), which is the product of a reciprocal translocation [t(9;22)(q34;q11)]. This genetic rearrangement results in the formation of a hybrid gene *bcr* from 22q11 plus the protooncogene c-*abl* from 9q34. The resultant hybrid protein has activated protein tyrosine kinase activity (compared with normal c-*abl* protein). CML is an almost universally fatal disease. Its rate and pattern of malignant progression is extremely variable although a clear two-stage process is usually evident. A chronic phase, characterized by excessive numbers of immature granulocytes, gives rise to a second or acute phase called 'blast crisis', in which immature blast cells predominate. The latter have a diversity of phenotypes but approximately two-thirds are myeloblastic and one-third are lymphoblastic; rare blast crises involve erythroid cells or megakaryoblasts. Also, blast crises can be a mixture of different cell types or can switch from predominantly lymphoid to myeloid (or vice versa).

In 'lymphoblastic' blast crises, the blast cells are indistinguishable (except by karyotype) from common ALL cells and may synthesize immunoglobulin chains and have rearranged Ig genes (or more rarely T cell receptor genes).

Chromosome studies indicate that the changes seen in blast crisis development reflect intraclonal evolution. The explanation for these remarkable phenotypic shifts, which accompany malignant progression, lies in the original 'target cells', the selection of different clonogenic cells and the altered stringency of maturation arrest (see Fig. 12.11) and Table 12.5). Chromosome and isoenzyme studies have demonstrated that CML originates in pluripotential stem cells. In the relatively benign, chronic stage of the disease the Ph positive clone is dominant in the marrow and most, if not all, dividing progenitors of different lineages are Ph positive. However, only in the granulocytic lineage is the total number of maturing progeny elevated. Differentiation itself is relatively normal although the majority of granulocytes may not be fully mature. Blast crisis can be interpreted as a second or sequential genetic alteration involving the selection of a single clonogenic cell in which proliferation is now effectively uncoupled from maturation. If this clonogenic cell is committed to the B cell lineage, then blast crisis mimicking common or pre-B ALL results. Alternatively, if it is a granulocytic progenitor, then myeloblastic blast crisis resembling AML results. Interestingly, some patients present clinically with Ph positive acute lymphoblastic leukaemia with no evident prior chronic phase, indicating either that the

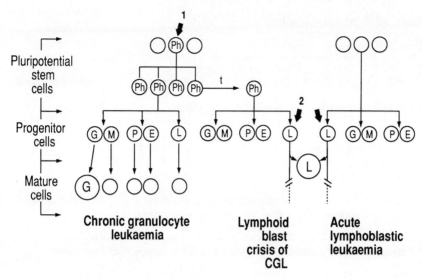

Fig. 12.11 Minimal, two step model for the evolution of blast crisis in chronic granulocytic leukaemia (compared with possible origin of acute lymphoblastic leukaemia). Ph, Philadelphia chromosome; G, M, P, E, and L, precursors for granulocytes, monocytes, platelets, erythrocytes, and lymphocytes respectively; t, variable time.

first selective event was bypassed or that the two events occurred in relatively quick succession in a preclinical phase. Recent molecular investigations indicate that both of these explanations are probably correct and that Ph positive ALL *without* a chronic phase may involve an association of *c-abl* with a different 5′ region of the *bcr* gene compared to that seen in CML.

The subsequent clinical course of CML blast crisis compared with ALL or AML is very revealing. 'Lymphoid' blast crises respond initially to therapeutic protocols appropriate for ALL, i.e. those incorporating prednisolone and vincristine; this probably reflects the steroid sensitivity of normal lymphoid progenitor cells in marrow. Remission, however, in contrast to that seen in ALL, is always short lived and relapse occurs soon thereafter with selection of new clonogenic cells from within the Ph positive stem cell pool which are not steroid sensitive. Myeloid blast crises seldom remit and, with increasing chemotherapy, severe bone marrow damage, concurrent infection, and death are likely. For these reasons, high dose chemotherapy (and irradiation) combined with stem cell replacement by marrow transplantation (see Chapters 17 and 18) is now generally considered the best therapeutic option for patients with CML.

Table 12.5 Multistep pathogenesis of CML

Step	Event/marker	Cell	Consequence	Clinical outcome
1	?	Pluripotential stem cell	Clonal advantage and normal differentiation	Preleukaemia, myeloid dysplasia
2	Ph chromosome $9q34^{c\text{-}abl} \rightarrow 22q11^{bcr}$ Activation of c-*abl* protein-tyrosine kinase activity	Multipotential stem cell	Clonal dominance + (a) Normal differentiation (b) Excessive granulocyte production	CML
3	Additional chromosome markers[1]	Multipotential stem cell or lineage restricted progeny	Clonal dominance with minimal differentiation	CML blast crisis

[1] Duplication of the Ph chromosome, or trisomy 8, or isochromosome 17 or other changes.

The dramatically different clinical course of Ph negative ALL versus Ph positive 'ALL'/CML blast crisis provides a vivid example of the clinical importance of 'target' cell and clonogenic cell identity as well as illustrating the impact that maturation competence has on the biological phenotypes of leukaemic cells which dominate during different stages of a cancer.

Stem cell origins, patterns of shifting phenotypes, and imposition of maturation arrest analogous to those observed in CML are probably common in solid tumours also. In the brain, the progression of a relatively well differentiated astrocytoma to a more anaplastic glioblastoma mimics, in cellular phenotype or cytopathology, primary glioblastoma (i.e. highly malignant glioblastoma multiforme), but this progression has often been traditionally interpreted as reflecting dedifferentiation. The more likely interpretation is that, by analogy with CML, some astrocytomas originate in normal glioblasts and that only after an additional (rare) genetic event(s) occurs which uncouples maturation from proliferation do glioblast phenotypes dominate.

Apart from its clinical relevance, the biology of CML illustrates a diversity and varied evolutionary pattern which nevertheless strikingly parallels components of normal differentiation (Table 12.6).

The sequential involvement of different genes which can confer selective (clonal) advantage is likely to be a common feature not only of CML but of most leukaemias and lymphomas. Thus in Burkitt's lymphoma, which represents the best studied 'model' at the molecular level, three distinct genetic events and other important associated factors can be identified.

Table 12.6 Lessons from CML

Cause of possible event 1 is unknown

Oncogenic events 1, 2, 3 are
  Spaced *variably* in time
  Can occur in different cell types (but within a single clone)
  Probably involve different genes

Step 1/2: the Ph chromosome involves hybrid gene formation [*bcr* (or *phl*)+c-*abl*] and
  enhanced kinase activity of a c-*abl* encoded domain

Phenotypic 'switches' reflect
  'Stem cell origin
  Normal differentiation linked phenotypes
  Maturation arrest (not dedifferentiation or aberrant differentiation)

Stem cell origin makes cure by conventional eradication therapy unlikely, although
  bone marrow transplantation is a possible alternative

1. Transformation (or immortalization) of B cells by EBV with defective T cell surveillance against EBV infected cells (as a consequence of immunodeficiency associated with malaria and malnutrition).

2. Continuous, rapid proliferation of EBV 'transformed' B cells *in vivo* predisposing to other independent genetic events including chromosome breaks and reciprocal translocations. These may occur essentially at random and are independent of the EBV genome, but alterations which juxtapose c-*myc* with immunoglobulin heavy or light chain loci have a strong selective advantage for the growth of the transformed B cells in which they occur.

3. Continued growth of the malignant clone *in vivo* (if the patient has not yet died) or *in vitro* (as established cell lines) may select for additional, genetic changes associated with further growth advantage and/or independence from exogenous growth factors.

We have seen that, in a stem cell leukaemia such as CML, successive independent genetic events can occur in distinct cell types (though intraclonally derived); in Burkitt's lymphoma and other mature lymphoid neoplasms it is more likely that the clonal progeny at approximately the same developmental level sustain the succession of molecular alterations.

12.5.3 *Age associated vulnerability of 'target' cells*

The association between malignancies of lymphoid progenitors (i.e. ALL) with childhood reciprocates with the predominant association of mature lymphoid malignancey (e.g. CLL, myeloma, most B-non-Hodgkin lymphoma) with adult life. This remarkable correlation of cancer type with age is relevant to the possible aetiology and/or pathogenic mechanisms involved. One plausible explanation for this age association invokes the concept of available 'target' cells 'at risk' of leukaemogenic events. Thus during early development, when there is considerable proliferative demand on lymphoid progenitors, these cells are relatively at risk compared with adult life where the turnover in pre-B and pre-T cells is likely to be extremely slow. Evidence for an age associated vulnerability of lymphoid progenitors to leukaemogenesis comes from observations on atomic bomb survivors. Acute lymphoblastic leukaemia was predominantly seen in individuals who were less than 15 years old at the time of exposure (see Chapter 7).

In contrast, throughout adult life, mature immunocompetent lymphoid cells are at risk due to their inherent proliferative potential and their long life span which increases opportunities for exposure to one or a succession of leukaemogenic events. In this context, breakdown of immuno-regulation (primarily by T cells) could predispose towards lymphoproliferative disorders and eventual rare 'spontaneous' genetic events could endow additional clonal growth advantage, expressed clinically as lymphoma or leukaemia. Patients with inherited or acquired immunodeficiency diseases or individuals immunosuppressed for transplantation are especially at risk of such events and it is striking that mature B cell lymphomas are the usual neoplasms in both paediatric and adult patients that have malignancy associated with immunodeficiency. Burkitt's lymphomas in children are an apparent exception to the otherwise striking association between lymphoid progenitor malignancy and childhood, but can be readily explained on the basis of the presence of special cofactors (especially malaria) which are immunosuppressive for the childhood population at risk.

In these respects lymphoid neoplasms are in principle similar to other tumours which have marked age associations. Thus the typical tumours of childhood, e.g. neuroblastoma, connective tissue sarcoma, Ewing's sarcoma, Wilm's nephroblastoma, and rhabdomyosarcoma, originate in cells which proliferate in early development but rarely if at all in adult life, whereas the more common but exclusively adult epithelial carcinomas probably originate in tissue stem cells continually proliferating and exposed over decades to potential carcinogens. In contrast to lymphoid malignancies, CML and AML do not appear to conform to this pattern. This difference could reflect the proliferative demand on multipotential and myeloid restricted progenitors both in early development and throughout adult life.

12.5.4 *Reversibility of 'maturation arrest'*

Considerable interest has focused on the possible reversibility of maturation arrest in leukaemia. If the central defect(s) is (are), as postulated, basically regulatory in nature then it might be possible to induce terminal maturation and thus exhaust the leukaemic clone as a non-toxic alternative to chemotherapy. This idea has been championed by Sachs in particular (Sachs, 1987), who has provided ample evidence for differentiation induction of rodent myeloid leukaemic cells *in vitro* and *in vivo*.

Human leukaemic cells and leukaemic cell lines can be induced by agents such as retinoic acid, vitamin D3, phorbol esters, and dimethylsulphoxide to undergo some further maturation. However, sensitivity to these inducers is extremely variable and the real clinical potential of this approach remains uncertain. This general strategy may well become more effective when we know more about the biochemical basis of the maturation-proliferation uncoupling process and the biological activity of proto-oncogene coded proteins.

12.6. Practical implications of leukaemic cell biology

Our understanding of the biology of both normal haemopoiesis and leukaemia in humans, though still incomplete, is in several respects considerably more advanced than our knowledge of epithelia and carcinomas which constitute the major human cancers. Although the relevance of leukaemia and lymphoma to the major clinical problem of metastasis (see Chapter 2) can be debated, it is important to consider to what extent a better grasp of cancer cell biology as exemplified by the leukaemias can be translated into real practical benefits for patient management.

Although still an issue of considerable debate, the benefits already derived from leukaemic cell biology are, in this author's view, considerable (Table 12.7) especially in terms of standardized, differential diagnosis and the introduction of new forms of therapy (see Chapters 15 and 18).

Monoclonal antibodies have proven particularly valuable (Table 12.7) and now have a routine place in diagnostic laboratories and expert system computer software has been developed to aid standardized interpretation of the often complicated immunophenotypic data generated (Alvey and Greaves 1987). Immunological markers are, however, only one component of the prognostically relevant biological features of leukaemic cells, and other parameters, notably growth fraction, ploidy,

Table 12.7 Clinical utility of monoclonal antibodies in leukaemia/lymphoma

Differential diagnosis
  Equivocal diagnoses improved
  Incorrect diagnoses reversed (AML/ALL)
  'Undifferentiated' leukaemias typed
  'Cryptic' erythroleukaemias revealed
  Diagnostic subgroups identified (ALL, CGL, blast crisis)
  Non-haemopoietic neoplasms distinguished (neuroblastoma, carcinomas)

Monitoring
  Extra-medullary relapse (central nervous system, testis)
  Thymic cells (T^+/TdT^+) in marrow

Therapy
  T cell removal from allogenic marrow (*ex vivo*)
  Removal of leukaemic cells from autologous marrow (*ex vivo*)
  In vivo treatment?

and specific karyotypic abnormalities, are regularly used for diagnostic classification as a basis for selection of appropriate therapy.

Where consistent molecular alterations in genes have been identified, there is an obvious potential for molecular diagnosis and for monitoring the elimination of the leukaemic clone and then screening for the possible re-emergence of the clone in relapse. Detection of rearranged genes by Southern blotting provides a particularly convenient technique for molecular diagnosis, for example in follicular lymphoma where a large proportion of cases (~ 85 per cent) have a rearranged *bcl*-2 gene (fused to IgH). CML provides a particularly vivid example since every case, including some which lack a detectable macroscopic Ph chromosome (and irrespective of ethnic group or geographic location), has an easily detectable rearrangement of the *bcr* gene (Fig. 12.12). Conventional Southern blot procedures are insufficiently sensitive for detecting minimal residual disease since they will only detect molecular markers (or rearrangements) if the cells involved constitute at least 1 per cent of the cell population being studied. A level of one per cent leukaemic blast cells in a bone marrow is equivalent to approximately 10^{10} total leukaemic cells in the body—a number which could increase to frank leukaemic relapse levels in a few days! The recently developed technique for amplifying specific gene sequences, called the polymerase chain reaction is a major innovation in this respect since altered genes (e.g. 'hybrid genes') can be detected in as few as 1 in 10^5 normal cells.

Now that most of the haemopoietic growth factors have been molecularly cloned and are available as recombinant products, some clinical

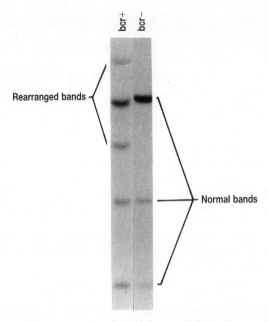

Fig. 12.12 Molecular diagnosis of chronic myeloid leukaemia: Detection of gene (bcr) rearrangement by Southern blotting. Control DNA extracted from normal cells and digested with the endonuclease Bgl II gives three restriction fragments identified with a radioactive bcr gene probe (lane bcr^{-1}). DNA from leukaemic cells has 3 bcr fragments corresponding to the unaltered bcr gene on the normal chromosome 22 plus novel bands corresponding to the rearranged gene on the 'Philadelphia' chromosome 22.

applications of these key products can be anticipated in leukaemia. It remains to be seen whether they can be safely used to promote end-cell maturation and hence exhaustion of the leukaemic clone. A more likely use already demonstrated in animal and clinical trials is the use of growth factors such as GM-CSF and G-CSF to facilitate rapid recovery of myelopoiesis following toxic chemotherapy (in leukaemia and other cancers) and hence protect the patient from infection during this very vulnerable period (Monroy *et al.* 1987).

References and further reading

Alvey, P. L. and Greaves, M. F. (1987). A computer program for interpreting immunophenotypic data as an aid to the diagnosis of leukaemia. *Leukaemia* **1**, 527–40.

Clark, S. C. and Kamen, R. (1987). The human hematopoietic colony-stimulating factors. *Science* **236**, 1229–37.

Dexter, T. M. (1982). Stromal cell associated haemopoiesis. *Journal of Cellular Physiology Supplement* **1**, 87–94.

Furley, A. J. *et al.* (1986). Developmentally regulated rearrangement and expression of genes encoding the T cell receptor-T3 complex. *Cell* **46**, 75–87.

Gordon, M. Y., Riley, G. P., Watt, S. M., and Greaves, M. F. (1987). Compartmentalization of a haematopoietic growth factor (GM-CSF) by glycosaminoglycans in the bone marrow microenvironment. *Nature* **326**, 403–5.

Greaves, M. F. (1982). 'Target' cells, cellular phenotypes and lineage fidelity in human leukaemia. *Journal of Cellular Physiology Supplement* **1**, 113–25.

Greaves, M. F. (1986). Differentiation-linked leukaemogenesis in lymphocytes. *Science* **234**, 697–704.

Greaves, M. F. (1988). Speculations on the cause of childhood acute lymphoblastic leukemia. *Leukemia* **2**, 120–5.

Greaves, M. F., Delia, D., Robinson, J., Sutherland, R., and Newman, R. (1981a). Exploitation of monoclonal antibodies: A 'Who's Who' of haemopoietic malignancy. *Blood Cells* **7**, 257–80.

Greaves, M. F., Mizutani, S., Furley, A. J. W., Sutherland, D. R., Chan, L. C., Ford, A. M., and Molgaard, H. V. (1986). Differentiation-linked gene rearrangement and expression in acute lymphoblastic leukaemia. In: *Clinics in Haematology* (ed. Hoffbrand, A. V.). W. B. Saunders Publ., Eastbourne.

Greaves, M. F., Rao, J., Hariri, G., Verbi, W., Catovsky, D., Kung, P. and Goldstein, G. (1981b). Phenotypic heterogeneity and cellular origins of T cell malignancies. *Leukemia Research* **5**, 281–99.

Jaffe, E. S. (1983). Follicular lymphomas: possibility that they are benign tumours of the lymphoid system. *Journal of the National Cancer Institute* **70**, 401–3.

Knapp, W., Dörken, B., Gilks, W. R., Rieber, E. P., Schmidt, R. E., Stein, H., and von dem Borne, A. E. G. Kr. (ed.), (1989). *Leucocyte Typing IV. White Cell Differentiation Antigens.* Oxford University Press, Oxford.

Magrath, I., O'Conor, G. T. and Ramot, B. (ed.), (1984). *Pathogenesis of leukemias and lymphomas: environmental influences.* Raven Press, New York.

Monroy, R. L. *et al.* (1987). The effect of recombinant GM-CSF on the recovery of monkeys transplanted with autologous bone marrow. *Blood* **70**, 1696–9.

Sachs, L. (1987). The Wellcome Foundation Lecture 1986: The molecular regulators of normal and leukaemic blood cells. *Proceedings of The Royal Society, London B* **231**, 289–312.

Showe, L. C. and Croce, C. M. (1987). The role of chromosomal translocations in B- and T-cell neoplasia. *Annual Review of Immunology* **5**, 253–77.

Waldmann, T. A. (1987). The arrangement of immunoglobulin and T cell receptor genes in human lymphoproliferative disorders. *Advances in Immunology* **40**, 247–321.

13

Hormones and cancer
W. I. P. MAINWARING

13.1 Introduction

Cancer is not a modern disease. Skin cancers were lucidly described in the Ebers papyrus, dating about 1770 BC, and because of its external location and ready observation, breast cancer was well known to Hippocrates and his acolytes. This was the phase of medicine based on his humoral theory of disease, and breast cancer was attributed to an 'excess of overheated black bile'. During the latter part of the 18th century the Scottish surgeon, John Hunter, established that removal of the gonads led to a dramatic shrinkage in the size of the accessory sexual glands in both sexes in animals. This glandular atrophy was particularly noticeable in the case of the prostate in males and the uterus in females. These observations provided the basis for investigations into the part played by hormones in human cancer. Indeed, Hunter subsequently devised a remarkably sound system for the classification of human breast cancers and did a vast amount of work on their pathology; many of his specimens are still in the Hunterian Museum in London. Research began in earnest in 1889 when Schinzinger proposed a relationship between ovarian function and breast cancer. In clinical terms, the years 1894–6 were the watershed between speculation and reality. In this period, Beatson described the palliation of breast cancer in some patients after ovariectomy, and Rann and White advocated castration for the arrest of prostate cancer. Research in this area then gathered real pace with the purification and characterization of the principal steroid and poly-peptide hormones in the years 1905–35. The synthesis of very active analogues and hormone antagonists has also had a profound effect on research progress, opening up genuine possibilities for the successful treatment of cancer by hormonal means. For example, in 1941 Huggins and Hodges introduced the non-steroidal oestrogen, diethylstilboestrol, for the treatment of prostate cancer, a therapy still widely used today.

Two recent reviews (Bulbrook 1986, Raynaud *et al.* 1988) bring the story up to date and consider modern concepts on the biology, mechan-isms of hormone action, and responsiveness of these tumours in some detail.

The incidence of cancers is markedly different in the two sexes and this may be largely explained by the different hormonal status of men and women. Breast cancer, for example, can occur in both sexes but is 100 times more prevalent in women. This difference is attributable to the different types of hormones secreted after birth and most strikingly after puberty. On the other hand, prostate cancer does not occur in women because during embryonic life the structural progenitors (or anlagen) for this organ disappear and persist only in the male. Similar arguments explain the absence of uterine and cervical cancer in men.

A deeper understanding of the mechanism of action of hormones in molecular terms has also provided an impetus to our knowledge of how hormones are implicated in cancer. The cornerstone of current thinking on hormone action is the target cell concept in which specific organs respond to only a restricted number of hormones. It is widely accepted that this distinctive pattern of response reflects the types of receptors present in different cells; a target cell contains the appropriate receptors whereas a non-target cell does not. Certain aspects of earlier studies on receptors have recently been challenged but the essential foundations of the target cell concept remain both persuasive and tenable. Indisputably, certain cells respond more dramatically to hormonal stimuli than others and this is particularly true for prostate, uterus, and breast. It is hardly coincidental that these are the major sites for the development of hormone sensitive tumours. While many target organs respond to the same hormone, not all of them become malignant. For example, the production of keratin by hair follicles requires a combination of hormones but tumours in these cells are rare. In the kidney the secretion of renin is under stringent hormonal control but tumours of renin secreting cells are very rare. Current evidence suggests that a target cell is at risk of malignant change only if the appropriate hormones prompt it to divide, as in the case of prostate, breast, and uterus.

It is not clear how the distinctive distribution of receptors occurs, but certainly it is laid down during differentiation in the embryo, fetus, and neonate. There is some evidence that certain hormones promote the synthesis of their own receptors; this may be true for the receptors for sex hormones in the urogenital tract just before and during puberty.

Since the malignant process is typified by dedifferentiation, carcinogenesis may result in profound changes in the concentrations of receptors. Indeed the assay of receptors has proved useful in monitoring both the appearance and successful treatment of certain forms of cancer (see p. 382).

The course of many types of cancer is profoundly influenced by hormonal imbalance. During neoplasia, several relevant intracellular changes may occur. A cell normally secreting a hormone may produce more or less as a consequence of neoplasia. On the other hand, a cell which normally cannot synthesize a hormone may acquire this ability (ectopic expression) as the result of malignant change.

Cancer endocrinology is concerned with the role of hormones in cancer induction and their effects on tumour growth. Some tumours, particularly those which arise from hormone responsive organs, may require hormones for their continued growth. Accordingly, alterations in the concentrations of these hormones can be exploited as a method of treatment.

13.2 Basic endocrinology

13.2.1 *Types of hormones*

As defined by Starling, a hormone (from Greek, to arouse) is a compound secreted into the vascular system by one organ to enhance the activity of other organs distant from the site of synthesis (endocrine). While still true in many respects, this definition needs qualification with respect to many recently described situations (see also Chapter 14). First, testosterone has a direct effect on certain organs, such as muscle and testis, but in many others, it must undergo obligatory metabolism to elicit a biological response; in cancer, its necessary conversion to 5α-dihydrotestosterone in accessory sexual organs, such as the prostate, is particularly important. Second, cholesterol is not a hormone but may be converted to the important steroid-like hormone, $1\alpha,25$-dihydroxy-cholecalciferol (the active metabolic form of vitamin D_3), by a sequence of reactions occurring in the skin, liver, and finally kidney. Third, many hormones have direct effects on their organs of synthesis, as in the case of sex hormones. Fourth, many hormones work via second messengers, as originally proposed by Sutherland with cyclic AMP as the 'middle-man'; it is now clear that cyclic granosine monophosphate, Ca^{2+} ions, phosphoinositol, and even polyamines can serve as second messengers (see Chapter 11). Last, hormones do not invariably activate the bio-chemical processes in their target cells; glucocorticoids, for example, kill T lymphocytes, granulocytes, and macrophages, and tumours derived from them.

Many hormone-like growth factors have been extracted and purified from serum in recent years; these are discussed in Chapter 11. This chapter will be concerned with those hormones secreted by the ovary, testis, adrenal cortex, and anterior pituitary. In chemical nature, these important hormones are either steroid or polypeptide hormones. In addition, the thyroid hormone, thyroxine, must be taken into account.

The secretion of all these hormones is regulated by precisely co-ordinated activity of the hypothalamus, pituitary, and the secretory (endocrine) glands themselves (Fig. 13.1). The hypothalamus is part of the central nervous system and controls the release of tropic hormones from the pituitary gland at the base of the brain. The pituitary has two parts, anterior and posterior, each of which produces different hormones. The regulatory system is a closed loop, requiring releasing hormones and stimulatory hormones or tropins; both of these classes of hormones are polypeptides. The crucial interaction is between complex nerve centres in the hypothalamus and the anterior pituitary which contains specialized cells responsible for the production of a distinctive tropin.

In response to appropriate external and internal stimuli, the hypo-thalamus secretes releasing hormones into a system of veins which drain directly into the anterior pituitary and stimulate the selective synthesis of tropins. There are different releasing hormones for thyrotropin (TRH), corticotropin (CRH), and one for the two gonadotropins (GnRH), lutropin and follitropin. The gonadotropins stimulate the production of the female sex hormones (oestrogen and progesterone) from the ovary or male sex hormones (androgens) from the testis. The tropins are secreted into the general circulation where they stimulate the synthesis of

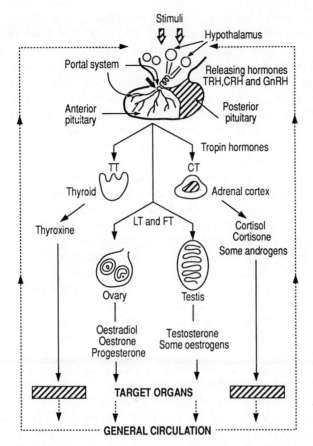

Fig. 13.1 The secretion of steroid hormones and thyroxine is under the control of the hypothalamic pituitary axis. Specific tropins are secreted by the anterior pituitary, including thyrotropin (TT), corticotropin (CT), and the gonadotropins, lutropin (LT) and follitropin (FT). The secretion of the tropins is triggered by specific releasing hormones, TRH for TT, CRH for CT, and GnRH for both LT and FT. An excess of hormones in the blood switches off the secretion of releasing hormones by the hypothalamus by closed negative feedback loops.

hormones from target cells. Thyrotropin stimulates thyroxine secretion from the thyroid. Corticotropin promotes the synthesis of the glucocorticoids, cortisol and cortisone, from the adrenal cortex, together with a small but significant amount of testosterone.

There are two mechanisms to prevent excessive hormonal stimulation. First, the steroid hormones and thyroxine are distributed strongly bound to specific transport proteins in serum; indeed, 0.5 per cent or less of free biologically active hormone may be available to the cells of the body. Second, there are powerful enzymes for steroid breakdown (catabolism) in the liver in both sexes. Should these latter control mechanisms fail, there is a dramatic rise in the concentration of free active hormone in serum with potentially dangerous consequences.

As well as the polypeptides already described, the anterior pituitary secretes two other polypeptide hormones, somatotropin and prolactin, in response to a variety of stimuli. Somatotropin promotes the growth of many cells and clearly may be important in the cancer process. Prolactin, as its name suggests, was originally identified by its powerful role in inducing lactation. Further research has shown, however, that prolactin has diverse functions and is certainly implicated in the carcinogenic process in many organs, notably breast and prostate.

As an aside, two other points should be made. The mineralocorticoid steroid hormone, aldosterone, is regulated by a novel means and although its level in serum may rise enormously in Conn's syndrome (tumour in the zona glomerulosa of the adrenal cortex), it plays no significant part in the process of carcinogenesis. The hypothalamus also secretes two polypeptide hormones, oxytocin and vasopressin, along with carrier proteins known as the neurophysins, into the posterior pituitary; again, there is no proven involvement of either hormone in neoplasia.

All classes of steroid hormone have diverse effects on a wide range of target cells, but as stated before, their most important feature as far as cancer is concerned is whether they can stimulate cell division. The oestrogens and androgens are powerful mitogens and promote growth and mitosis in many target organs, such as breast, uterus, and prostate; these are potentially dangerous hormones as can be demonstrated experimentally. By contrast, glucocorticoids and progestins are protective agents and generally inhibit cell multiplication; there are certain exceptions and these will be discussed later.

13.2.2 Mechanisms of hormone action

Steroid hormones and thyroxine. Most of the responses to steroid hormones are mediated by specific and seemingly mobile receptor proteins; a few responses related to gluconeogenesis and secretion may be promoted by other means. Steroids enter all cells from the blood by passive diffusion but there is some evidence that certain tumours may possess mechanisms for the facilitated or active transport of steroid hormones. If specific receptors are present, a high affinity hormone receptor complex is formed which, after a conformational change or 'activation', is translocated to the nucleus to occupy large numbers of so

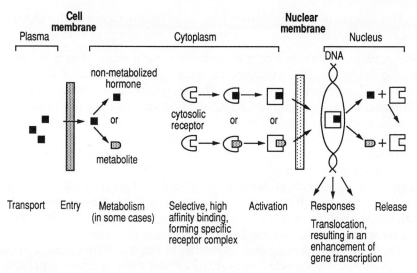

| Cell membrane | | Nuclear membrane |
| Plasma | Cytoplasm | Nucleus |

Transport   Entry   Metabolism (in some cases)   Selective, high affinity binding, forming specific receptor complex   Activation   Responses   Release

Translocation, resulting in an enhancement of gene transcription

Fig. 13.2 A scheme of the mode of action of steroid hormones. Metabolism of the secreted hormone within its target cells is important only in the case of the androgen, testosterone. (Based on Mainwaring 1980.)

called acceptor sites, composed of DNA and non-histone nuclear proteins (Fig. 13.2). Most responses are evoked by the selective acceleration of gene transcription, i.e. specific changes in gene expression. We do not really understand why different cells respond in such contrasting ways to the same hormonal stimulus or why only certain target cells can be driven into mitosis. The extreme specificity of response to steroid hormones depends largely on whether receptors are present or not, and on the binding of steroid which is a very selective process indeed. The concentration of the hormones themselves is an additional factor. Under normal circumstances, for example, oestrogen receptors bind physiological concentrations of androgens very weakly, if at all. In both sexes, however, certain tumours in the adrenal cortex produce such massive amounts of androgens (virilizing hyperplasia) that the oestrogen as well as androgen receptors are fully occupied, and elicit certain oestrogenic as well as androgenic responses, even in males.

The thyroid gland produces iodine containing hormones, thyroxine (tetraiodothyroxine) and triiodothyroxine (T_3). These have a profound effect on cell metabolism and growth of many cells, including some tumours. A deficiency in infants leads to mental and physical stunting (cretinism). Iodine deficiency causes an overgrowth of the thyroid cells leading to the formation of tumour-like nodules (goitres) but these rarely if ever become malignant.

The receptors for triiodothyroxine are nuclear rather than cytosolic (see Chapter 9 for discussion of the T_3 receptor as a protooncogene).

Polypeptide hormones and growth factors. Both these classes of compound have specific receptors as integral components of the outer surface of the plasma membrane. The receptors have a close structural and functional association with enzyme systems and ion gates capable of generating a wide range of second messengers (see Chapter 11). Occupation of the specific receptor sites enhances the synthesis of second messengers, and since the subsequent steps are operational cascades, the hormonal signal is amplified and a few molecules of hormone can evoke a very pronounced effect. Some internalization of polypeptide hormones has been detected in the Golgi apparatus and lysosomes of target cells; this may represent a mechanism for their degredation but may be a more subtle reflection of a wider influence on cellular function.

There are defences against excessive hormonal stimulation. Excess polypeptides are degraded by proteases in the walls of arteris and capillaries. A long biological life for cyclic nucleotides is prevented by the ubiquitous presence of phosphodiesterases. Finally, there is evidence that protracted exposure of specific receptors to their appropriate hormone leads to a very significant decrease in the number of receptors (down-regulation); this phenomenon is being exploited in novel approaches to the hormonal therapy of certain tumours. It follows from this general discussion that tumours may arise from hormone producing or hormone responsive cells (see section 13.5).

13.3 Hormones and carcinogenesis

The first hint that hormones were potential carcinogens came in 1932 when Lacassagne induced mammary tumours in male mice with the oestrogen, oestrone benzoate. Using a similar approach, Kirkman induced kidney tumours with oestrogens but only in hamsters and not other rodents. This unexpected finding showed that hormones may influence organs not usually thought to be hormone sensitive. The synthesis of the non-steroidal oestrogen, diethylstilboestrol, by Dodds in 1935, provided researchers with an extremely powerful research weapon. Of all the classes of steroid hormones, the sex hormones, and oestrogens in particular, are the most potent carcinogens. Work in this area is not without its controversies. There are often conflicting results when the effects of hormones are compared in different species. Of greater importance, data obtained from experimental animals often conflict with findings in the human. Taking an overall view, hormones can be carcinogenic under certain circumstances or can provide the means for the arrest of tumours. Our knowledge in this area is based on studies on animals, in man, and on cells in tissue culture.

13.3.1 *Animal studies*

The literature on this topic is now vast and will only be covered briefly. In experimental animals prolonged exposure to oestrogens and their analogues, such as stilboestrol, results in cancer formation in many organs, but most commonly in reproductive organs, kidney, liver, and anterior pituitary, but there is a considerable variation between species. It has also been reported that protracted treatment of dogs with potent androgens may lead to prostate tumours. Concomitant administration of carcinogenic hydrocarbons, such as dimethylbenzanthracene, often enhances the tumour incidence achieved by hormones alone; in addition, hydrocarbons may help to induce tumours in organs normally insensitive to hormones. So far, tumours have not been consistently induced in experimental animals even by chronic treatment with polypetide hormones including prolactin and somatotropin.

Two examples illustrate interspecies differences. Tumours of the uterine cervix can be induced in mice by extremely low doses of oestrogens; under similar dose-corrected conditions, all other species are refractory. Mammary tumours are common in some mouse strains and are often associated with a mammary tumour virus (see Chapter 8). Repeated pregnancies tend to increase the incidence of these tumours. In human breast cancer there is no evidence of a virus and pregnancy tends to protect (see Chapter 3). Prostate tumours are very rare in most animals although common in man. Such anomalies create problems when trying to assess the potential danger of hormones in humans. This problem was approached by Dunning in 1963. From a spontaneous rat prostate tumour, she developed a whole range of transplantable, androgen sensitive and androgenic insensitive tumours; these tumours are stable and can be grown in tissue culture.

While studies on experimental animals have taught us much about hormones and cancer, they have not shed any light on certain pressing human problems. For example, experimental studies have failed to explain why cancer is so prevalent in the human prostate, yet rare in adjacent accessory sexual glands which are subject to a similar if not identical hormonal milieu. Further, animal studies have failed to explain why the male dog is the only species to share a high incidence of prostatic cancer. Prostate cancer in humans is a disease of old age, and there have been some reports on a spontaneous prostate tumour arising in aged AxC strain rats. The incidence of prostate tumours was raised to 70 per cent if the animals were exposed to exogenous androgens. While of considerable interest, these studies may tell us little about the human disease, because ageing men are unlikely to be exposed to exogenous androgens and the endogenous production of testosterone tends to decline during ageing. Answers to such questions could provide invaluable insights into the aetiology of the human disease.

The use of animals for testing potentially dangerous hormones has prompted many controversies. Perhaps the most notorious centres on synthetic progestins related to 17-hydroxyprogesterone. When tested in beagle dogs, but not other animals, these compounds were found to induce mammary tumours. Despite lengthy debate, these synthetic progestins were banned from inclusion in contraceptive pills on the evidence obtained from one experimental species only.

13.3.2 *Studies in man*

Evidence that hormones are carcinogenic in man comes largely from clinical observations and epidemiological studies. Many relevant observations, still largely unexplained, have been made. For example, prostate cancer has never been recorded in eunuchs or castrati, and there is a high incidence of breast cancer in socially enclosed female communities, such as nunneries.

Hormones and cancer in women. There is a rapidly growing body of evidence that prolonged exposure to sex hormones, especially oestrogens, results in a high incidence of several forms of human cancer. Women are more usually affected since they tend to take more hormone preparations largely for obstetric and gynaecological reasons. Nonetheless, androgen containing formulations markedly increases the risk of hepatoma in men, and transvestites with oestrogen implants have a higher incidence of breast cancer than normal men. The contraceptive pill (see Chapter 3) contains a combination of oestrogen and progesterone; the critical point in terms of potential danger is the relative proportions of the two hormones. While essential for contraceptive function, the oestrogen is the potentially threatening component, whereas the progesterone is protective or 'anti-oestrogenic'. Early preparations containing a high oestrogen to progesterone ratio have all now been banned because they markedly increased the risk of endometrial (uterine) cancer. Contraceptive preparations containing a very low oestrogen to progesterone ratio are currently considered safe and indeed may well provide protection against cancer of the ovary, breast, and endometrium. There has been cautionary reports of a high incidence of breast cancer in young women who have been taking oral contraceptives since just after the menarche. One recent survey in the UK shows an increase in the risk of about 40 per cent after four to eight years of pill use, and about 70 per cent after more than eight years' use. But researchers stressed breast cancer was still uncommon below age 36— even a 70 per cent in risk would put up the chances of developing breast cancer by this age only from one in 500 to one in 300. Fortunately, such dangers to health may be obviated in the future by the wider use of contraceptive implants containing only a synthetic progestin, levonorgestrol. These subdermal implants are effective, long lasting, and have a reduced risk of cancer.

Formulations containing oestrogens alone, so called 'happy pills', became fashionable for helping certain women through the undesirable psychological and physical aspects of the menopause. Many such preparations have now been withdrawn; although achieving the proposed objectives, patients also ran a higher risk of endometrial cancer and possibly breast cancer as well. Such postmenopausal treatments are ethically acceptable only if progestins are also included in the regimen.

A most distressing illustration of the dangers of hormones is provided in reports of vaginal cancer in young women whose mothers took stilboestrol during early pregnancy to prevent threatened abortion.

Extensive epidemiological studies have established the major risk factors for breast cancer (see Chapter 3). These include early menarche, late menopause, having close relatives with the disease, infertility, obesity, geographic location, and having the first pregnancy late in life. The current strategy is to try to explain these findings in terms of hormonal imbalance, but no unequivocal conclusions can yet be drawn. Nonetheless, breast cancer is generally considered to result from over-exposure to oestrogens and underexposure to progesterone (see also Bulbrook *et al.* 1988). The risk associated with obesity can be partly explained by the ability of adipose cells to synthesize oestrogens but impaired activity of detoxification mechanisms in the liver and elsewhere may also be implicated. The beneficial effect of having a child early in life could be due to the high concentrations of progesterone-like hormones in pregnancy protecting the breast cells against oestrogens in the long term. The risks of early menarche and late menopause can be combined in that a high number of menstrual cycles may be dangerous in terms of the repeated surges of oestrogens ultimately providing the stimulus for malignancy. No convincing correlation has yet been drawn between endocrine imbalance, family history and domestic locality. Elevated concentrations of free oestradiol are present in the plasma of women with breast and endometrial cancer as compared with normal, age matched controls. Similar measurements in other cancers could be vital in the future for diagnostic and prognostic purposes.

All of these epidemiological considerations apply equally forcibly to endometrial cancer. One possible clue, as with prostate cancer in men, is provided by studies on Japanese migrants to Western cultures. Breast and endometrial cancer increase in these women, possibly a result of changing diet. Western foods tend to be richer in fat and this may cause subtle but significant changes in the hormonal milieu and even hormonal imbalance. This is considered in more detail in Chapter 3.

Hormones and cancer in men: prostate cancer. The high incidence of prostate cancer in the human male remains an enigma and, as in so many cancers, we have no real idea why the incidence rises so sharply in old age. Certainly, the precise involvement of androgens remains to be eluci-dated. The only real clue so far is that the malignant prostate has a marked ability to form and retain elevated concentrations of 5 α-dihydrosterone; this androgen is a far more powerful mitogen than testosterone itself. The disease is associated with westernized, industrialized societies and remains relatively uncommon in Mongoloid races. In several studies, it has been shown that when Chinese and Japanese emigrated to California and Hawaii, their risk of developing prostatic cancer rose significantly. The formerly low incidence of prostate cancer in Japan itself is now gradually increasing, whereas formerly common cancers, especially of stomach, are gradually decreas-ing. While not a complete explanation, there is plausible evidence suggesting that the newly acquired risk is attributable to the adoption of social customs and particularly the diet more typical of the West.

In contrast to breast cancer, epidemiological studies on prostate cancer are much less extensive. In the United States, this cancer is higher in blacks than Caucasians and lower in Jewish immigrants. Throughout Europe, no relationship has been established between the high incidence of prostate cancer and socioeconomic status, marital status, fertility, social habits, hair distribution, and physical size. There are hints of a possible connection with recurrent prostatitis, repeated infections, and venereal disease. A remarkable connection has been drawn in several reports between prostate cancer and 'sustained sexual interest' and 'sex drive'. Needless to say, these terms are difficult to qualitate, let alone quantitate. Investigators have used various parameters to measure sexual activity, and it remains possible that increased activity in some form is associated with prostate cancer.

Although a considerable literature has been amassed on the epidemi-ology of many human cancers, it is surprising that measurements of oestrogens and androgens in blood, urine, sebum, and even saliva, have so far failed to show that abnormalities in the concentrations of sex hormones are associated with the disease states. It should be stressed, however, that measurements in the past have been as total hormone concentrations. In biological terms, it is the concentration of free, unbound hormone which is important.

13.3.3 *The actions of hormones on tumour cells in culture*

Stable lines of tumour cells provide an invaluable approach for studying the direct effects of hormones. Such studies are expanding rapidly, because cells in culture, rather than in intact animals, have the advantages of easy manipulation, high biochemical activities, and ethical acceptance. Three experimental systems will be briefly reviewed.

Human breast cancer cells. Because of the understandable concern about the present scale of the breast cancer problem, a great deal of effort has been directed to the establishment of such tumour cells in culture. The driving force behind these enterprises is the ability to screen the effects of hormones and other drugs on the tumour cells directly. The MCF-7 line of breast tumour cells has been investigated in detail (Dickson and Lippmann 1986).

The growth rate of these cells is enhanced by low concentrations of oestrogens and prevented by anti-oestrogens, such as nafoxidene and tamoxifen. Such studies have helped to clarify many aspects of the hormonal management of breast cancer and certainly provided hints for the improvement of clinical treatment. First, the cells can be maintained in the complete absence of oestrogens, so that hormones exert their influence by accelerating or modulating an existing, basal rate of cell proliferation. Second, anti-oestrogens inhibit basal proliferation in the absence of oestrogens. It is now clear that anti-oestrogens, such as tamoxifen, bind extensively to the oestrogen receptor and are efficiently translocated to the nucleus, but the anti-oestrogen receptor protein complex is biologically inactive. Third, oestrogens also enhance the density to which the cells will grow. In particular, the cells can proliferate in the presence of powerful oestrogens without attaching to an artificial substrate, such as agar or collogen, which mimics the underlying stroma and mesenchyme present in the intact breast. It has been postulated that oestrogens change the social behaviour of these tumour cells in such a way that inhibitory influences on their growth, such as contact inhibition, are largely offset. Finally, the tumour cells grow best in culture medium containing serum, but this can be replaced by an artificial medium containing transferrin, insulin, cortisol, thyroxine, prolactin, and epidermal growth factor.

These observations have considerable relevance in the clinical context. It would seem that the responses are mediated by the classical receptor machinery and so treatment with anti-oestrogens can be predicted to be successful only if the tumour contains receptors (r^+) rather than lacking them (r^-). On current evidence, it would seem that oestrogen deprivation alone will only slow tumour growth to a basal rate but will not stop it completely. Nonetheless, the importance of anti-oestrogens in the palliation of breast cancer is experimentally and now clinically proven. Clearly, growth of the breast tumour cells requires a variety of polypeptide and steroid hormones. Antagonism of these hormones may therefore be useful in the management of breast cancer and such possibilities are currently the centre of active clinical trials.

As a final point, very high doses of steroids and related compounds, such as stilboestrol, inhibit the growth of MCF-7 cells. This restraint on cellular proliferation cannot be mediated by the oestrogen receptor machinery and the molecular basis for such cytocidal effects remain to be clarified.

Prostate cells in culture. Various sublines of the Dunning prostate tumour have been used to confirm the hormonal effects found in the MCF-7 cell line. The prostate cells grow best in the presence of powerful mitogens, such as 5 α-dihydrotestosterone, mibolerone, and certain 5-androstranediols, but are inhibited by anti-androgens, including the non-steroidal compound, flutamide. Autonomous sublines contain few if any androgen and prolactin receptors, and anti-androgens occupy the receptor sites of androgen sensitive sublines but the resultant complex is without biological activity. Although a number of cell lines derived from human prostate tumours are available, few respond to hormones in the same way as do cells *in vivo*.

A promising approach to endocrine studies both in the breast and prostate is the use of organ cultures in which small pieces of tissue, with epithelium and stroma in their normal relationship, can be maintained. These techniques have given useful information but are still to be exploited fully.

Rat lymphomas cells. Powerful glucocorticoids are used in transplantation surgery to suppress the immune system and there is now some evidence that these steroids not only kill some lymphoid cells but may also act as a trigger for the development of some leukaemias and lymphomas.

Studies on rat lymphoma cell lines have provided explanations for the cytocidal effects of glucocorticoids, and also led to a better understanding of the mechanism of action of steroid hormones in general. This work has been expedited by the availability of powerful analogues of cortisol. Many of these analogues, such as dexamethasone, fluoprednisolone, and triamcinolone acetonide, contain substituent fluorine atoms at positions 6 or 9 of the steroid ring system. The halogen substitutions greatly potentiate the biological activity of these synthetic glucocorticoids because they do not bind to the transport protein, corticosteroid binding globulin, in serum and also resist catabolism within glucocorticoid target cells.

The anti-inflammatory action of glucocorticoids, namely the killing of B lymphocytes and tumours derived from them, was first described by Daughaday in 1943. Subsequent research has shown that cell death is mediated by glucocorticoid receptors in the lymphocytes and lymphomas, followed by synthesis *de novo* of many species of mRNA encoding proteins responsible for the lethal process. The hormone induced cytolysis is clearly a complex process; certainly the uptake of life maintaining glucose is completely suppressed, together with the intracellular accumulation of toxic fatty acids and powerful DNAases.

Certain sublines of lymphomas have been found to be resistant to glucocorticoids and their study has been of fundamental importance. Resistance can be explained by several types of change in the receptor system for binding glucocorticoids (Fig. 13.3). The majority of resistant

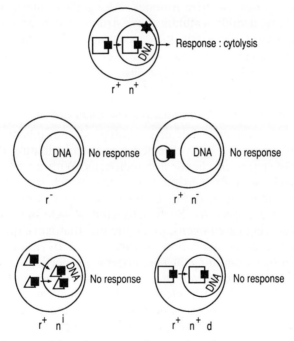

Fig. 13.3 In normal lymphocytes and many lymphomas, a receptor complex containing glucocorticoids, such as prednisolone, translocates to the nucleus and, by activating gene expression from DNA, produces proteins ($\star$) which promote cytolysis. Such cells are r^+n^+. Many lines of lymphomas resist the cytocidal effects of glucocorticoids. Some are r^-; others contain structurally different receptors, which cannot translocate to the nucleus and are r^+n^-; other modified receptors may translocate in great numbers yet not evoke a response and are r^+n^i; and finally there are resistant lines, r^+n^+d, in which only the crucial change in gene expression is impaired.

lines are termed r⁻, meaning that they have either lost the ability to synthesize the receptor protein or produce a modified and biolocially inactive receptor (unable to bind ligand). Other resistant lines contain the normal complement of receptors but translocation of the receptor complex to the nucleus is either lost (r^+n^-) or is inactive in evoking a response (r^+n^i). A final class of resistant cells has a normal receptor translocation system but complete resistance to glucocorticoids; these cells have been described as deathless (r^+n^+d).

The identification of the r^-n^i, and r^+n^+d sublines raises some very interesting questions. Clearly the occupation of surface receptor and nuclear acceptor sites is not the full explanation for hormonal responses in molecular terms. Other biochemical events after receptor occupation and translocation are clearly involved. Genes are the ultimate regulators of the biochemistry of cells in terms of their viability, sensitivity to mitogens, and response to hormones. Along these lines, it could be argued that these deathless mutants have structural defects in such genes. In the final analysis, however, these seemingly encouraging results on cultured tumour cells must be viewed with caution, tinged even with disappointment, when applied to the human equivalents of these cancers.

13.3.4 *A comparison of hormones and chemical carcinogens*

Because of their potential danger in ecological and industrial terms, a great deal of effort has been directed towards elucidating the general mechanism of action of chemical carcinogens. With a few exceptions, chemical carcinogens exist as precarcinogens which must be activated usually in the target organ, before their carcinogenicity can be maximally expressed (see Chapter 6). Such activation is necessary with many naturally occurring carcinogens, including the aflatoxins, quercitin, and cycasin. In the latter activation is carried out by intestinal micro-organisms. Before examining the carcinogenic properties of hormones in detail, it is useful to compare their mechanism of action with that of carcinogens (Table 13.1). It is clear that the mechanisms are very different except that both hormones and chemical carcinogens need dividing cells as targets for the malignant process to develop.

13.4 The mode of action of hormones in carcinogenesis

As described in Chapters 1 and 6, the original concept of Berenblum, that carcinogenesis consists of a first phase of initiation followed by various phases of promotion, is now widely accepted. With a few notable exceptions, hormones act as promoters or cocarcinogens (Moolgavkar 1986). There are only a few instances where they may be considered as

Table 13.1 A comparison of the mechanism of action of hormones and chemical carcinogens

Property	Hormones	Chemical carcinogens
Mutagenicity	Not mutagenic	Mutagenic
Site of action	mRNA and proteins	DNA
Latent period	Long	Short
Nature of tumours	Benign or locally invasive	Highly malignant and metastatic
Activation	Generally not necessary	Necessary
Species- and sex-dependency	Very marked	Barely relevant
Nature of repeated exposure	Prolonged	Single exposure may suffice
Expression of carcinogenicity	*In vivo* only	*In vivo* and *in vitro*
Regression after ceasing exposure	Regression	No regression
Need for cell division	Needed	Needed

genuine initiators, like the chemical carcinogens. Generally, hormones enhance the rate of initiation and development of tumours induced by all of the proven classes of initiators, such as chemical carcinogens, viruses, and ionizing radiations. A crucial point which argues against hormones being initiators is that they are not mutagenic, whereas almost all genuine initiators are powerful mutagens.

Perhaps one case where hormones may act as genuine initiators is the vaginal cancer in young women whose mothers took stilboestrol during pregnancy. At various stages in development, human embryos are acutely sensitive to damage by a wide range of exogenous agents. In this case, traces of stilboestrol could pass the protective barrier of the placenta and initiate carcinogenesis in oestrogen target organs, such as the vagina. The cancer may only develop after birth under the promoting influence of oestrogens after puberty.

In acting primarily as cocarcinogens, hormones could exert their influence in any of three ways, namely in classical promotion, in accelerating tumour growth, or in sensitizing target cells to initiating agents. Hormones may also have a significant role in tumour progression (see p. 376).

13.4.1 *Carcinogenesis*

Promotion. There are many examples in experimental systems where hormones act as promoters and greatly increase the incidence of tumour development by initiators. Particularly good examples, although by different mechanisms, include the promoting influence of glucocorticoids, thyroxine and somatotropin on the induction of skin tumours in mice by methylcholanthrene and croton oil, and the activation of mouse mammary tumour virus by glucocorticoids.

Another good example is that of mammary tumours induced in rats by dimethylbenzanthracene (Russo and Russo 1986). This chemical is a potent inducer of breast tumours in the rat if it is administered as precisely 50 days of age; if administered at other times, its carcinogenicity to breast epithelium is markedly decreased (Table 13.2). Seemingly, the hormonal milieu and the developmental state of the cells at this time are optimal for tumour induction. Removal of the ovaries at any time reduces the induction of tumours and this may be reversed by administration of oestrogens. In hypophysectomized rats, oestrogens are ineffective as promoters and it seems their influence is expressed via the hypothalamic pituitary axis by the accelerated secretion of prolactin. This may be an authentic example of prolactin sensitive tumour induction. Progesterone has a complex influence in this experimental system. When administered prior to the initiating carcinogen, tumour incidence decreases but, when given after the initiator, tumour incidence increases. This paradox remains to be clarified. Certainly, it would be dangerous to extrapolate these findings to breast cancer in women. As emphasized before, data from animal work are not always consistent with clinical findings, and in this particular case, the very nature of the tumours may be fundamentally different.

Table 13.2 The influence of hormones on the induction of mammary tumours by dimethylbenzanthracene

Additional treatment	Number of tumours	Day of appearance of tumours
None	Many	90–94
Ovariectomy, 30–35 days	Few	106–110
Ovariectomy, 65–70 days	Few	106–110
Progesterone, 30–35 days	Few	106–110
Progesterone, 60–70 days	Very many	79–83

Dimethylbenzanthracene was given through a stomach tube to female Sprague-Dawley rats of 50 days of age, either alone, with administration of progesterone, or to ovariectomized animals.

The involvement of oestrogens in the development and course of endometrial cancer is much clearer, although it is difficult to draw an absolute distinction between initiation and promotion. Certainly, oestrogens promote the disease, after earlier phases of hyperplasia and preneoplasia, and these are all countered by progesterone or synthetic progestins. While this is a striking example of tumour promotion in the human, it is surprising that oestrogens rarely induce endometrial cancer in experimental animals. The difference could be of some importance, because it could imply that induction of tumours by the sex hormones is not simply by evoking changes in cell proliferation and mitosis; other factors are almost certainly involved.

We know very little about the relationship between other hormones and carcinogenesis. Although it has been suggested that there is a close relationship between the androgens and tumour growth in the prostate, firm evidence that they play any role in tumour induction is lacking.

Tumour growth. There are many striking examples in the literature of the profound influence of the sex hormones on tumour growth. This was very evident in work on Shionogi mouse breast tumour cells, an unusual androgen responsive tumour, but most strikingly demonstrated in recent research on the transplantable Dunning prostate tumours in the Copenhagen rat. Growth of the androgen sensitive tumour lines is effectively suppressed by castration or a variety of anti-androgens, including cyproterone acetate and flutamide; powerful oestrogens (e.g. stilboestrol) also strongly inhibit growth of these tumours.

In considering tumour growth, two aspects are particularly important. Tumour cells may divide more slowly than many normal cells, but are distinguished by their inexorable rather than their rapid growth. Second, an increase in tumour size occurs only when cell multiplication exceeds cell death; in the prostate and breast, it has been estimated that between 25–45 per cent of new cells soon die. From these considerations, hormones could exert an influence on tumour size by maintaining exponential growth or slowing cell death. There is little sound evidence on the latter point.

Although hormones may affect tumour growth by direct action on the cells, indirect mechanisms may also play a part. The influence of hormones on the growth of common human tumours, such as prostate and breast, is complex. These tumours are composed of epithelium, stroma, and blood vessels. In the normal breast, oestrogens promote the proliferation of the endothelial cells in the capillaries and this may provide an indirect way in which these hormones modulate tumour growth by increasing the supply of nutrients. In the prostate, there has been a belief for many years that a close structural and functional interaction between the stromal and epithelial elements is mandatory for the co-ordinated normal growth of the organ. Prolonged exposure to androgens, especially 5α-dihydrotestosterone, could upset this delicate stromal–epithelial interaction. Future research in this area has been given a strong boost by the successful maintenance of separated stromal and epithelial elements of both dog and rat prostate in culture; this breakthrough should provide new insights into the mechanism controlling prostate growth and the part played by the androgens in the process.

Sensitization. Dividing cells are prime targets for carcinogens. Both classes of sex hormone are powerful mitogens and promote division, and even hyperplasia, in certain of their target organs, including prostate, breast, and cervix. These hormones, therefore, increase the potential number of cells that will be at risk on exposure to initiating agents. This process may simply be termed sensitization. Using the rat prostate as the model system, new insights have recently been gained into mechanisms by which androgens enhance DNA replication. After the rapid saturation of nuclear acceptor sites by receptor 5α-dihydrotestosterone complex and an accompanying phosphorylation of certain nuclear proteins, the subsequent responses evoked by androgens proceed in a precise biphasic manner (Fig. 13.4a). The first (Group I) responses are not implicated in DNA replication whereas the later (Group II) responses involve the synthesis de novo of the biochemical machinery for DNA unwinding proteins. After reaching a zenith just prior to and during the S phase, the synthesis of these components then ceases (Fig. 13.4b) despite continued androgenic stimulation. These changes are attributable to a temporal distinction in the expression of the structural genes representative of Group I and Group II components. Importantly, the occupation of the nuclear acceptor sites by 5α-dihydrotestosterone is also biphasic, the two phases coinciding precisely with the synthesis of Group I and Group II components (Fig. 13.4c). Similar studies need to be performed on oestrogen target cells, as in rat uterus and on androgen sensitive and insensitive Dunning tumours; but they do provide a pointer to the molecular basis of stringent control being lost and the transformed prostate cells permanently retaining the machinery necessary for continuous replication of DNA.

Hormones may influence DNA synthesis in ways other than by gene expression. They may induce the synthesis of growth promoting factors in their target cells or in other cells of the body, but there is scant evidence on these points.

13.4.2 *Tumour progression*

Progression in tumours has been unequivocally demonstrated in many experimental systems and is supported by clinical observations. In general, hormone sensitive tumours progress to an autonomous hormone insensitive state, but whether this is due to changes in the cells or to selection of resistant cells already present in the tumour is not known. A good example is seen in the BR6 mouse which develops multiple mammary cancers very frequently. The tumours appear and

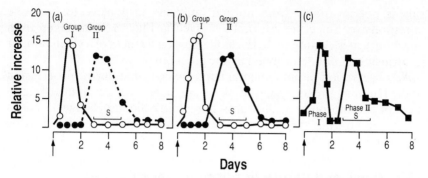

Fig. 13.4 Changes evoked in the prostate of castrated rats by testosterone are biphasic. Group I responses occur early and are not associated with DNA replication; Group II occur later, during the synthetic (S) phase of the cell cycle when DNA synthesis is in progress. The changes are presented relative to measurements in castrated animals prior to implantation with testosterone, marked by arrows. (a) Changes in the activities of aldolase (Group I, ○) and thymidilate synthetase (Group II, ●). (b) Changes in the amounts of the specific mRNAs for aldolase (○) and thymidilate synthetase (●). (c) Changes in the concentration of androgen receptors in the nucleus (■).

grow rapidly during pregnancy and may regress after birth, but grow again at precisely the same site during the next pregnancy (Fig. 13.5). The tumours each behave independently. Some regress completely after pregnancy, others show incomplete regression and others become completely autonomous, i.e. they are no longer dependent on pregnancy associated hormones for their growth. It should be stressed that progres-

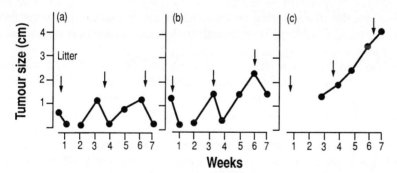

Fig. 13.5 Three mammary tumours with different properties in a single BR6 mouse. The tumours arise during pregnancy, and may disappear (responsive tumours) or persist (unresponsive tumours), after delivery of the litters, marked by arrows (data from Foulds 1969). (a) Left groin—pregnancy-responsive—complete regression. (b) Right axilla—pregnancy-responsive—incomplete regression. (c) Right neck—pregnancy-unresponsive.

sion is not invariably from normal hormone responsive to hormone unresponsive. Since the information given in Fig. 13.5 was taken from one mouse, it is clear that the hormonal responsiveness of a given tumour is intrinsic and not determined by the gross environment.

An observation of fundamental importance is that during progression of the Dunning prostate tumour, there are major changes in the complement of chromosomes, i.e. towards aneuploidy. Similar changes occur in other tumours but the precise gene changes involved have still to be discovered.

13.5 Types and nature of hormone sensitive and hormone producing tumours

From the earlier discussion it is clear that endocrine associated tumours fall into two groups, hormone sensitive from target organs, and hormone producing from endocrine glands (see also Chapter 14).

13.5.1 *Hormone sensitive tumours*

Tumours arising from target organs should be described as hormone sensitive rather than hormone dependent. This is not simply a semantic distinction. Hormone dependence implies that certain tumours could only persist during hormonal stimulation and would atrophy if the source of hormone was removed. Both clinically and experimentally this is rarely the case. All tumours are heterogeneous with respect to cell types, and it follows that certain cells may be amenable and even acutely sensitive, to hormonal manipulation whereas others certainly are not. However, by prudent handling of the hormonal environment of certain tumours, the disease may be checked in many cases, and in the most encouraging situations, even cured; such tumours are best described as hormone sensitive. When considering hormone sensitive tumours, it may well be that the future will hold some remarkable surprises. Traditional endocrinology has been set on its head over the last decade by the discovery of a bewildering array of polypeptide hormones, from coded neurotransmitters to growth regulators (see Chapters 11 and 14). Few investigators would have anticipated the identification of the naturally occurring analgesics, the endorphins, or predicted that hormones classically associated with the gastrointestinal tract, such as gastrin, glucagon, and cholecystokinin, could also be synthesized within the central nervous system. Our traditional tenets of endocrinology are probably very insecure and many growth factors and hormones of great relevance to human cancer are just being isolated and characterized.

Steroid sensitive tumours. Tumours in the accessory sexual glands are among the most common forms of cancer in men and women and these tumours are rightly described as hormone sensitive. The aetiology of these diseases is clearly complex and changing. At the turn of the century, for example, carcinoma of the uterus was the most common hormone sensitive tumour in women; since then the incidence of this tumour has decreased and has been superseded by the pronounced and increasing incidence of breast cancer. This change is baffling but could reflect hormonal changes resulting from improved hygiene, better contraception, and better medical care in general. This particular tumour highlights a major problem for modern investigators; it is clearly imperative to find a plausible explanation for such important changes, but early medical records are often wanting and experimental animal models are few. The only animal equivalent for uterine cancer is the rabbit, and even here the disease is endocrinologically and pathologically distinct from the human disease.

Breast cancer is a major threat to the health of the female population in all westernized societies. Benign tumours, fibroadenomas, are rare before puberty but represent the commonest form of breast tumour in the age span, 25–30 years. Carcinoma of the breast is age associated, and becomes more common from about 30 years of age (see Chapter 3). There is a clear endocrinological basis for the disease and, as described later, this provides the rationale for clinical manipulation and arrest of the disease in many women. Growth abnormalities in the breast can also result in papillomas, which range from well to poorly differentiated tumours. A rarer condition is Paget's disease, which is defined as a primary malignancy of the nipple but is often associated with a tumour of the breast. Tumours in other organs of the female reproductive tract, e.g. the vulva and vagina, are relatively rare, but may be induced by oestrogens under certain circumstances (see p. 366). In both cases, a hormonal imbalance, and particularly an excess of oestrogens is involved.

In Western societies, cancer of the prostate is the fourth most common male cancer. It usually appears in men of 60 years or older and increases in frequency with age. The tumours vary enormously from highly invasive poorly differentiated tumours to relatively slow growing and well differentiated forms. Cancers in other male accessory organs, such as epididymis and seminal vesicle, are rare. In all cases, there is a hormonal aetiology for the disease. Cancer of the penis and scrotum is also very rare; a hormonal basis for these tumours is possible but not proven. Circumcision certainly is accepted as the prime defence against cancer of the penis; for reasons unknown, this cancer is particularly frequent in the stallion. No other animal shares this risk.

Polypeptide sensitive tumours. It is now clear that the cells of the body are continually exposed to a wide battery of polypeptide hormones. Although many cells respond to somatotropin and virtually all cells respond to insulin by increasing uptake of glucose and amino acids, very few tumours seem to be sensitive to these polypeptide hormones. There is growing evidence for the involvement of prolactin in the induction and progression of certain tumours. The first indication of the involvement of prolactin in carcinogenesis came from studies on the growth of several lines of transplantable mammary tumours in the AxC rat. This hormone may also play a significant role in the growth of cancers in the breast and prostate; in the latter case, prolactin may enhance the uptake of androgens from the peripheral circulation. There are indirect indications but little definite information that polypeptide hormones of the pituitary may modulate the growth of certain tumours. The effects of other polypeptides are considered in Chapters 11 and 14.

13.5.2 *Hormone producing tumours*

All organs producing steroid hormones (ovary, testis, adrenal) are potential sites for malignancy. Tumours, as well as producing effects by their size and position, may cause symptoms by interfering with the normal production of hormones by the organ, by overproduction of normal hormones, or by the production of hormones not normally produced.

Ovarian tumours most commonly occur in the age range 40–60 years. Benign conditions are also quite common. Ovarian cancers are of many different types. The most common tumours arise from the surface epithelium and have a wide range of structure; the second group are of germ cell origin, and a third rare group arises from sex cord stromal cells (these cells normally surround germ cells and are responsible for the production of ovarian steroid hormones). Tumours from these latter cells are likely to be hormone producing and may produce oestrogens or androgens.

Testicular tumours, although not common, are increasing in frequency and are the most frequent form of cancer in young men. As outlined in Chapter 3 there is strong evidence that hormonal changes *in utero* may be a factor in their development. As in the ovary, tumours may arise from the germ cells or *much* less often from specialized androgen producing interstitial cells or oestrogen producing Sertoli cells. Some germ cell derived tumours, for reasons unknown, may produce a hormone, human chorionic gonadotropin (HCG), normally produced by the human placenta.

The adrenal gland has two parts, an inner medulla and an outer cortex, each having a different embryological origin. The medulla can be considered as a neuroendorcine organ and its tumours are discussed in

Chapter 14. The cortex produces steroid hormones, mainly gluco- and mineralocorticoids but it may also produce oestrogens and androgens. Tumours in the adrenal cortex have been widely investigated and have very interesting properties. Small, benign adenomas are quite common, but carcinomas are rare. The tumours show no age dependency and can occur from infancy to senility, with women tending to be more prone to the disease than men. The tumours are often well differentiated, with an almost normal glandular appearance, but in many cases they may grow to a remarkable size. The disease is usually fatal because of metastasis to vital organs, rather than a lethal deterioration in the synthesis of life supporting glucocorticoids and mineralocorticoids.

The physiological outcome of cancer in steroid producing organs can often be profound. Tumours may lead to overproduction or impairment of steroid secretion, either of hormones normally produced or of other steroids. In either event, the clinical consequences can be dramatic, often dangerous, and even socially distressing. For example, certain adrenal tumours in women result in virilization due to excessive secretion of androgens, with consequent psychological and physical disturbances such as abundant growth of bodily hair and voice changes.

Tumours in organs producing polypeptide hormones can also have dire consequences. In many pituitary tumours, the secretion of tropins is anomalous, with dramatic interference, for example, in sexual activity and fertility. Similarly, tumours of the pancreas can either curtail or over-produce the secretion of insulin and glucagon, with dangerous and even fatal consequences (see Chapter 14). Tumours of other endocrine organs, e.g. thyroid, can produce similar effects.

Human cancers can cause unexpected clinical difficulties. Some tumour cells may acquire the ability to synthesize hormones; for example, certain types of lung cancer actively secrete the normal pituitary hormone, adrenocorticotropin. This is generally referred to as ectopic hormone secretion. Similarly, in women tumours of placental origin (hydatidiform mole and choriocarcinoma) may secrete vast quantities of gonadotropins. The progression or cure of many cancers can be monitored by the cessation of abnormal hormone secretion.

13.6 Treatment of hormone sensitive human cancers

From the overview of general endocrinology presented earlier (see Fig. 13.1), there are many opportunities for curtailing the supply of hormones to tumours. Some of the approaches are far more radical than others and in recent years ingenious and less stressful approaches have been actively explored. Means have also been found to reduce the harmful effects of many drugs and hormones on normal cells.

13.6.1 *Hormonal manipulation*

Endocrine treatment for prostate cancer has been used for more than 40 years. Huggins and Hodges argued that since the normal gland required androgens for its growth, anti-androgenic treatment might inhibit the growth of prostatic tumour cells. This can be done either by removing the main source of androgen—the testes—or by antagonizing androgens with oestrogen administration, usually by giving stilboestrol orally. In many cases this proved to be effective. About 80 per cent of all cases respond at first but eventually most relapse. It was thought that this may have been due to the production of androgens by the adrenals, perhaps stimulated by pituitary hormones produced in increased amounts as a result of castration or oestrogen treatment. Removal of the adrenals or pituitary was then used but these operations are life threatening them-selves and the results were not satisfactory. Most tumours eventually become hormone independent and continue to grow. The use and dosage of stilboestrol remains a very controversial issue. Certainly, it is accepted that only low doses should be used, to reduce the risk of breast enlargement, electrolyte and cardiovascular disturbances, as well as psychological problems. Obviously better forms of treatment are required.

Great thought has been given to the targeting principle in the therapy of tumours (see Chapter 18). The ideal is to deliver the drug selectively to the tumour itself, eliminating deleterious side effects. For example, thyroid cancers can be selectively and effectively killed by low doses of radioactive ^{125}I because of the remarkable penchant of the organ for concentrating this halogen. In the prostate, the perceptive suggestion was to use phosphorylated forms of stilboestrol and oestradiol-1β. Until the phosphate groups are removed, the drugs have little if any biological activity; however, the prostate has a remarkably high activity of phos-phatases, thus ensuring that the prostatic tumour will tend to concentrate the active drug component specifically, but the clinical results are still uncertain. Another approach of great promise is the development of drugs which selectively inhibit the enzyme, 5α-reductase, responsible for forming the powerful mitogen, 5α-dihydrotestosterone, within the prostate tumour itself. Active inhibitors, such as a wide range of 4-azasteroids, are currently on clinical trial; the preliminary findings look encouraging. Prostate cancer produces extremely painful meta-stases in the bones and relief from this has recently come from a rather surprising quarter. Ketoconazole was originally developed as an anti-fungal agent and it is now the principal means for relieving bone pain. While the drug is clearly an inhibitor of cytochrome P450 systems (see

Chapter 4), the reason for its action on relief from bone pain remains unknown.

Two areas of recent and very extensive research on prostate cancer have been very disappointing. A lot of effort has been applied to the development of anti-androgens which block the function of the androgen receptors. Many such compounds have been synthesized, including the steroid, cyproterone acetate and the non-steroidal compound, flutamide. Clinically, neither have proved useful in curing prostate cancer, although they have been effective in individual cases. Indeed, the usefulness of cyproterone acetate has been limited virtually to its legal acceptance as the means for chemical castration of persistent sexual offenders throughout Western Europe. A real problem that remains is how hormone sensitive an individual prostatic tumour really is. It seems that measurement of 5α-reductase is a far better indicator of the hormone sensitivity of a given tumour rather than androgen receptors. After all, it is beyond doubt that this enzyme complex is induced and stringently regulated by androgens.

The most revolutionary proposal for the successful arrest of prostate cancer by hormonal means has come from research in Canada; suppression of androgen secretion in patients with prostatic cancer, by a drug combination of anti-androgens, such as flutamide or anandron, plus a powerful range of synthetic analogues of gonadotropin releasing hormone. The anti-androgens block the peripheral effects of any androgens while the GnRH analogues suppress the hypothalamic pituitary axis and also reduce the gonadotropin receptors in the testis by down-regulation, thus effectively suppressing the secretion of androgens. Early results on this approach seemed to be promising but later work has not supported the original enthusiasm. Nonetheless, new approaches for the hormonal manipulation of prostate cancer are needed, because radical surgical removal of the prostate is now seen as a difficult and unsatisfactory procedure; metastases of the disease cannot be treated by surgical means.

Many of the considerations on prostate cancer are relevant to breast cancer in women. Methods are being sought to replace radical surgery on the breast or on the endocrine glands. Such practices as hypophysectomy and radical mastectomy of the breast tumour are much rarer now than in the past (see Chapter 16). As in the prostate, the normal breast, and many of its tumours, respond to the relevant sex hormones, mainly oestrogens and progestins.

Two aspects of the research and management of breast cancer have been most encouraging. First, it is now accepted that measurements on the concentrations of oestrogen and progesterone receptors are valuable in the successful management of individual patients. The measurement of progesterone receptors is based on sound evidence that they are oestrogen induced proteins. If the tumour is receptor positive, r^+, then hormonal manipulation is advised; if r^-, then chemotherapy should be used. Second, there are several anti-oestrogens available now and tamoxifen in particular has been of striking benefit in the treatment of many women; about 30 per cent of all patients had an objective response and a further 20 per cent showed some clinical improvement. The results are better in postmenopausal women. For premenopausal women, ablation of the ovary, either by surgery or X-rays, remains the first choice for reduction of oestrogen levels.

Synthetic progestins, of long biological life and which may be taken orally, are finding wide and successful use in the treatment of endometrial cancer. Anti-oestrogens may also have an important role to play here in the future.

Synthetic glucocorticoids, especially prednisone, prednisolone and their fluorinated counterparts, have also been invaluable for the treatment of many types of lymphoma and leukaemia. Such approaches are only of value if the malignancy is r^+, containing receptors for glucocorticoids; this is not always the case. Even if cells are r^+, hormones alone will not eradicate the disease and additional chemotherapy is required (see Chapter 17). Novel glucocorticoids which are more specific in their action on r^+ cells are currently under study. For example, triamcinolone acetonide 21-oic acid methyl ester is a potent glucocorticoid without the side effects of any of its structural congeners; in particular, it neither suppresses the secretion of natural glucocorticoids from the adrenal glands nor promotes involution of the thymus. A major problem in treatment of hormone responsive tumours is the growth of hormone independent tumour cells.

13.6.2 Chemotherapy

We have not yet explored all the possibilities for the successful treatment of breast and prostate cancer by hormonal means. Surprises and new opportunities may be in store. For example, there is a growing and impressive body of evidence that aminoglutethimide may be very valuable in the treatment of most hormone sensitive tumours; this drug, like ketoconazole, is an inhibitor of enzyme complexes containing cytochrome P450 and thus provides a clinical means of suppressing steroid hormone synthesis. In both these tumours, the administration of pharmacological doses of steroid hormones, including glucocorticoids, has often been found to be beneficial. There is no plausible explanation of these observations in molecular terms. As mentioned earlier, the final arrest of hormone sensitive human tumours will depend on a combination of hormonal manipulation and chemotherapy.

References and further reading

Bulbrook, R. D. (ed.) (1988). *Cancer Surveys* 5.

Bulbrook, R. D. *et al.* (1988). Sex hormone-binding globulin and the natural history of breast cancer. *Annals of the New York Academy of Sciences* **538**, 248–56.

Dickson, R. B. and Lippmann, M. E. (1986). Hormonal control of human breast cancer cell lines. *Cancer Surveys* **5**, 617–24.

Foulds (1969). *Neoplastic development*. Academic Press, London.

Henderson, B. E., Ross, R. K., Pike, M. C., and Casagrande, J. T. (1982). Endogenous hormones as a major factor in human cancer. *Cancer Research* **42**, 3232–9.

Lupulesco, A. (1982). *Hormones and carcinogenesis*. Praeger Press. A book containing sound and detailed information. Highly recommended.

Mainwaring (1980). In *Cellular receptors* (ed. D. Schulster and A. Levitski), pp. 91–125. John Wiley, Chichester.

Minesita, T. and Yamaguchi, K. (1965). An androgen dependent mouse mammary tumour. *Cancer Research* **25**, 1168–75.

Moolgavkar, S. H. (1986). Hormones and multistage carcinogenesis. *Cancer Surveys* **5**, 635–48.

Raynaud, J.-P., Namer, M., and Ojasoo, T. (ed.) (1988). Medical management of prostate cancer. *American Journal of Clinical Oncology* **11**, Suppl. 2.

Russo, I. H. and Russo, J. (1986). From pathogenesis to hormone prevention of mammary carcinogenesis. *Cancer Surveys* **5**, 649–70.

14

The neuroendocrine system and its tumours

J. M. POLAK and S. R. BLOOM

The classical view of endocrinology, meaning the study of hormone producing glands and their disorders, has been challenged seriously following the discovery that many body tissues contain the so called 'diffuse neuroendocrine system' (Polak and Bloom 1983, Falkmer *et al.* 1984). The cells of this system produce 'regulatory peptides', capable of acting not only at the distance, as chemical messengers or hormones, but also locally as paracrine substances or as neurotransmitters/neuro-modulators (see Fig. 14.1). The reason for this multiplicity of actions is that regulatory peptides are produced by two different classes of tissue cells. Active peptides can be produced by endocrine cells or by neural elements. The former release the peptide either into the blood stream (circulating hormones) or locally, in a paracrine manner, whereas the latter release the peptide from the nerve terminal after neuronal depolar-ization to act as neurotransmitters/neuromodulators.

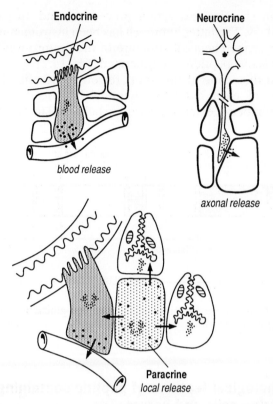

Fig. 14.1 Diagrammatic representation of the tripartite mode of action of the component cells of the diffuse neuroendocrine system.

Peptide producing endocrine cells are dispersed throughout the body, often intermingled within non-endocrine tissue. These cells were first recognized by Feyrter who called them 'clear cells' because of their weak histological staining, or cells of the 'diffuse endocrine system', in view of their dispersed localization and possible regulatory function. The special histochemical properties of these cells led Pearse to apply the term 'APUD' (amine precursor uptake and decarboxylation) to them whereas the morphological characteristics these cells share with neurons (i.e. dense core secretory granules) led Fujita to propose the term 'para-neurons'. The recognition that neural tissue, often anatomically closely associated with endocrine cells, is also capable of producing and secreting active peptides led to the expansion of the 'diffuse endocrine system' to 'diffuse neuroendocrine system'.

Regulatory peptides are being discovered at an exponential rate, due to advances in chemical extraction procedures and, in particular, to molecular biology. The latter approach has been instrumental also in the determination of the chemical structure of the pre-pro-molecules from which smaller active peptides are generated (Fig. 14.2).

It is beyond the scope of this short review to describe in detail the characteristics of each individual peptide. The reader is, however, referred to a number of extensive review articles and to Table 14.1.

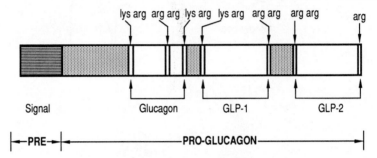

Fig. 14.2 Schematic representation of the pre-pro-glucagon molecule indicating the positions of peptides of the glucagon family.

14.1 Morphological features of peptide containing endocrine cells and nerves

Endocrine cells are found scattered throughout the body and those located in hollow organs frequently possess microvilli extruded into the lumen (Solcia *et al.* 1987). The latter can be visualized better at the electron microscope level. Endocrine cells have intracytoplasmic secretory granules, the morphology of which, including their electron density and their limiting membrane and size, permits separation into various cell types. Secretory granules tend to be located at the basal pole of cells, close to the basement membrane and to the organ blood supply. Endocrine cells frequently display a cytoplasmic elongation along the basement membrane. The arrangement of these features suggests functional properties: receptor functions are attributed to microvilli, peptide product is stored in secretory granules, and a cytoplasmic extension indicates a possible local or paracrine role for some of the peptide containing endocrine cells.

Numerous neurotransmitters have been identified in the nervous system and these include not only the classical neurotransmitters, acetylcholine and noradrenaline, but also amino acid neurotransmitters and, in particular, peptide neurotransmitters. All can be identified in cell bodies, nerve fibres, and terminals by the use of specific antibodies.

14.2 Techniques for the visualization of the 'diffuse neuroendocrine system'

14.2.1 *Light microscopy*

General markers. Secretory granules in the cytoplasm of endocrine cells have the ability to take up silver salts and precipitate them into a silver deposit with (argyrophilia) or without (argentaffinia) addition of a reducing agent. For amine containing endocrine cells, argentaffin (Masson) silver impregnation is used and, for other endocrine cells, a number of argyrophilic methods have been proposed. Among these, one of the most commonly used is the silver impregnation method of Grimelius. These techniques work well also in neuroendocrine tumour tissue.

Neuron specific enolase (NSE) is an isoenzyme of the glycolytic enzyme enolase. Antibodies to NSE stain endocrine cells and their related nerves, and are useful for the demonstration of neuroendocrine differentiation in tumours, particularly poorly granulated neuroendocrine tissue (e.g. small cell carcinoma of the lung, Merkel cell tumours of the skin).

Chromogranins are a family of peptides originally extracted from the adrenal medulla. They have been classified as chromogranin A, B, and C. The pre-pro-chromagranin molecule gives rise to three different peptides which are separately expressed in tissues. These peptides are: pancreastatin, which is included in the structure of pro-chromogranin A; GAWK, included in the structure of chromogranin B; and a peptide found at the C-terminal end of chromogranin B termed CCB (C-terminal of chromogranin B) (Fig. 14.3).

Synaptophysin is a molecule originally found in the membrane of a small secretory vesicle type containing classical neurotransmitters which is expressed differentially in neuroendocrine.

The occurrence of intermediate filaments, e.g. vimentin, keratin, and neurofilaments, is rare in neuroendocrine tumours (e.g. small cell carcinoma of the lung, Merkel cell tumour).

Peptide markers. Immunocytochemistry, using antibodies specific to regions of regulatory peptides and their pre-pro-forms, has been used extensively.

14.2.2 *Electron microscopy*

Electron microscopy has been fundamental in determining the neuroendocrine differentiation of normal and tumour tissue. By electron

Table 14.1 Neuroendocrine regulatory peptides

Peptide	Distribution	Endocrine cells (C) Nerves (N)	Main known actions
Insulin	Pancreatic islets (β cells)	C	Blood sugar
Glucagon	Pancreatic islets (α cells)	C	Blood sugar
Enteroglucagon	Intestine	C	Trophic to gut
Secretin	Small intestine	C	Pancreatic bicarbonate secretion
Gastric inhibitory polypeptide (glucose dependent insulino-trophic peptide) (GIP)	Small intestine	C	Insulin secretion, gastric acid secretion
Vasoactive intestinal polypeptide (VIP)	Central and peripheral nervous system	N	Muscle relaxation, vasodilation, secretion
Peptide with histidine	Central and peripheral nervous system	N	Muscle relaxation, secretion
Growth hormone-releasing	Hypothalamus	N	Growth hormone production
Gastrin	Pyloric stomach and small intestine	C	Gastric acid secretion
Cholecystokinin (CCK)	Small intestine	C	Gall bladder contraction, pancreatic enzyme secretion
CCK-8	Central and peripheral nervous system	N	Excitatory
Gastrin-releasing peptide (GRP) (see also Neuromedin C) (Bombesin)	Lung, central and peripheral nervous system	C, N	Trophic, release of other regulatory peptides
	Amphibian skin and gastro-intestinal tract	C	
Neuromedin B, Neuromedin C (= GRP main form)	Spinal cord (porcine)	N	
Growth hormone-releasing factor (GRF)	Hypothalamus	N	Growth hormone production
Substance P	Central and peripheral nervous system	N	Sensory: vasodilation, muscle contraction

Neurokinin A (= Substance K, Neuromedin L); Neurokinin B (Neuromedin K)	Spinal cord (porcine)	N	
Pancreatic polypeptide	Pancreatic islets	C	Pancreatic enzyme secretion, gall bladder contraction
Peptide with C- and N-terminal tyrosine (PYY)	Intestinal system	C	
Neural peptide with tyrosine (NPY)	Central and peripheral intestinal tract	N	Vasoconstriction
Motilin	Small intestine	C	Gut motility
Somatostatin	Stomach, intestine, pancreatic islets, thyroid gland, central and peripheral nervous system	C, N	Release and action of many peptides; may be antitrophic
Neurotensin	Intestine, adrenal medulla, central nervous system	C	Vasodilation, gastric acid secretion
Parathyroid hormone	Parathyroid gland	C	Blood calcium
Calcitonin	Thyroid gland	C	Blood calcium
Calcitonin gene-related peptide (CGRP)	Central and peripheral nervous system	N	Vasodilation; sensory
Adrenocorticotrphic hormone (ACTH)	Anterior and intermediate lobes of pituitary gland; central nervous system	C, N	Production of adrenal corticosteroids
Corticotrophin-like intermediate lobe peptide (CLIP)	Intermediate lobe of pituitary gland	C	
Melanocyte-stimulating hormone (MSH) (α,β,γ)	(Anterior) and intermediate lobes of pituitary gland, central nervous system	C, N	Skin pigmentation
Endorphin (α and β)	Anterior and intermediate lobes of pituitary gland, central nervous system	C, N	Opioid
β-Lipotrophin	Anterior and intermediate lobes of pituitary gland, central nervous system	C, N	

Table 14.1—*continued*

Peptide	Distribution	Endocrine cells (C) Nerves (N)	Main known actions
Dynorphin	(Anterior) and intermediate lobes of pituitary gland, central nervous system	C, N	Opioid
Enkephalin (met- and leu-)	Central and peripheral nervous system, adrenal medulla	N C	Opioid
Vasopressin	Central nervous system (neurosecretory)	N	Antidiuretic
Oxytocin	Central nervous system (neurosecretory)	N	Contraction of uterus and let-down of milk
Thyroid hormone releasing hormone (TRH)	Hypothalamus	N	Release of TSH
Luteinizing hormone releasing hormone (LHRH)	Hypothalamus	N	Release of gonadotrophins
Corticotrophin releasing factor (CRF)	Hypothalamus	N	ACTH production
Growth hormone	Anterior pituitary gland	C	Growth
Prolactin	Anterior pituitary gland	C	Milk production
Placental lactogen	Placenta	C	Lactogenic, trophic, lipolytic
Thyroid stimulating hormone (TSH)	Anterior pituitary gland	C	Thyroid hormone production
Follicle stimulating hormone (FSH)	Anterior pituitary gland	C	Ovarian follicle formation, spermatogenesis
Luteinizing hormone (LH)	Anterior pituitary gland	C	Ovulation, gonadal hormone production
Chorionic gonadotrophin	Placenta	C	Luteotrophic
Various recently discovered peptides	Heart (atrium)	?C	Sodium excretion, diuresis, blood pressure

Chromogranin family

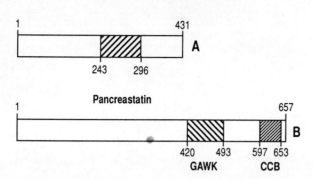

Fig. 14.3 Diagram showing the structure of chromagranins A and B. The positions of their derivative peptides pancreastatin (in chromogranin A), GAWK, and CCB (chromogranin B) are marked.

microscopy, the presence of dense core secretory granules is easily distinguishable (Fig. 14.4). Secretory granules are not destroyed by poor fixation, and thus can be distinguished in tissue fixed for routine pathology. In conventional osmium staining for electron microscopy, the limiting membrane of the granule is easily distinguishable and analysis of the size, electron density, halo, and characteristics of this membrane can thus be seen. These features are indicative, especially in normal tissue, of the production by a given cell of a particular peptide.

14.2.3 In situ *hybridization*

The use of cDNA probes, complementary to messenger RNA directing the synthesis of a given peptide in the technique of *in situ* hybridization allows morphologists to visualize the biosynthetic events prior to peptide packaging in granules (Adams *et al.* 1989). A number of different methods for *in situ* hybridization exist, using radioisotope or biotin and the end product is visualized by autoradiography or by methods permitting the visualization of biotin labels (see Chapter 10).

Table 14.2 summarizes the main applications of *in situ* hybridization. Figure 14.5 illustrates one of these applications, namely gene expression and site of synthesis. The latter application is particularly important in tumour pathology, since tumours may have an abnormally poor mechanism for peptide storage (low number of secretory granules), thus giving negative results with immunocytochemistry. Or, alternatively, the tumour may have receptors (binding sites) on the surface which are subsequently internalized providing spurious false positive results with immunocytochemistry.

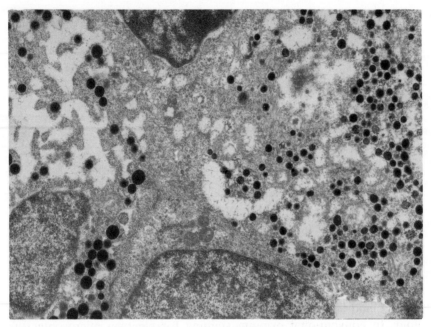

Fig. 14.4 Electron micrograph of human pancreatic glucagonoma showing two distinct secretory granule populations in neighbouring tumour cells. Glutaraldehyde fixation, uranyl acetate, and lead citrate counterstains. ($\times 8600$)

Table 14.2 Main applications of *in situ* hybridization in the study of the diffuse neuroendocrine system

---

1. Gene expression
   Sites of synthesis of hormones/neuropeptides are revealed by location of gene expression in tissues.

2. Rate of synthesis
   Physiological or pathological variations in hormone/neuropeptide synthesis are determined by monitoring changes in cells expressing specific genes.

3. Combination with other techniques
   Combination with, for example, immunocytochemistry gives more complete analysis of cellular function by allowing comparison of gene transcription with hormone/neuropeptide storage

4. Determination of co-synthesis of multiple messengers
   Coexistence of multiple regulatory peptides can be assessed by using multiple probes on a single tissue section

5. New taxonomy of normal neuroendocrine cells and tumours
   The previously unsuspected capacity of a neuroendocrine cell to synthesize a certain product provides the basis of new cellular/tumour classification

---

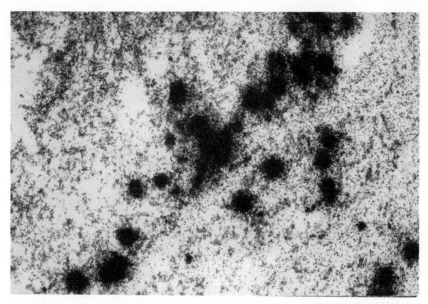

Fig. 14.5 Section of a small cell carcinoma of the lung with the sites of synthesis of human bombesin (gastrin releasing peptide) marked by black silver grains. Haematoxylin counterstain. ($\times 210$)

14.2.4 *Receptor visualization*

Peptide binding sites can be visualized by autoradiography or by immunogold staining procedures. The techniques of *in vitro* autoradiography using tissue sections and emulsion coated coverslips or sensitive film have been developed for the visualization of peptide binding sites. Densitometric image analysis of these preparations is now being used for quantitation.

14.3 Neuroendocrine tumours

14.3.1 *Nomenclature*

Various terms have been used to describe tumours arising from the diffuse endocrine system, e.g. carcinoma, APUDoma; there is no consensus on nomenclature.

Immunocytochemistry and radioimmunoassay allow further characterization of these tumours in terms of their function. Neuroendocrine tumours frequently produce more than one regulatory peptide (mixed endocrine tumours), but often one circulating peptide is responsible for the associated clinical syndrome. In such cases the tumour is frequently referred to by the name of this active secretory product, e.g. gastrinoma and insulinoma. In view of these difficulties, we propose three levels of nomenclature. First, 'neuroendocrine tumours' is to be used as a generic term to describe all tumours thought to be derived from the diffuse neuroendocrine system and showing the general histological, histochemical, and immunohistochemical features of such cells. Second, terms may refer to particular histological patterns at specific sites, e.g. islet cell tumour of pancreas or oat cell carcinoma of lung. Third, an indication of the major peptide secretions of the tumour is included. In this system each level is complementary to the others and a full description of a tumour should include mention of all three.

14.3.2 General features of endocrine tumours

Clinical features. It is difficult to assess the true incidence of endocrine tumours but well recognized, functioning tumours are clearly uncommon, with the exception of small cell carcinoma of the lung.

Clinical features of some functioning tumours may be obscured by the presence in the circulation of more than one regulatory peptide with counteracting effects. In addition to hormonal effects, non-specific manifestations, depending on site, size and invasion, may be present. Clinical suspicion is usually roused when more common diseases have been excluded. This sometimes delays diagnosis, with adverse effects on the prognosis. High levels of secretion of a regulatory peptide may provide a useful assay although circulating levels of active peptides can be elevated in diseases other than tumours. Many pre-operative, non-invasive techniques are also widely used, including a series of stimulation and suppression tests for the particular hormone. Further, a series of localization techniques are used (see Chapter 16).

Biochemistry. Highly sensitive specific radioimmunoassays for the measurement of regulatory peptides in blood and tissue are readily available and allow early diagnosis of functioning tumours. Thus, very small tumours may be diagnosed clinically and biochemically, but localization of the precise site of the tumour may still be a problem.

Histology. Peptide producing endocrine cells are frequently unevenly distributed within a tumour. It is therefore very important to sample a large number of specimens taken randomly throughout the tumour mass. The growth pattern of endocrine tumours is quite characteristic. Tumours are composed of uniform cells with few mitoses arranged in irregular masses, ribbons, or glandular structures. A glassy, amorphous degeneration of the tumour stroma (amyloid) is common and sometimes extensive. A more functional criterion of malignancy has recently been proposed. This is the production and release of alpha human chorionic gonadotropin (HCG) by many malignant endocrine tumours. This criterion is especially relevant for pancreatic endocrine tumours.

Electron microscopy. Tumours have been found to contain variable numbers of secretory granules, but often peptide producing tumour cells store less peptide than their normal counterparts. It is generally accepted that a poorly granulated tumour reflects high secretory activity, and this frequently correlates with the presence of abundant ribosomes, rough endoplasmic reticulum, and prominent Golgi apparatus. Poorly granu-lated tumours are usually associated with significantly elevated levels of circulating hormone. This suggests that one of the metabolic defects in endocrine tumour cells resides in the control of hormone secretion.

14.3.3 *Ectopic hormone production*

By definition, ectopic hormones are produced by a tumour arising in an organ which does not secrete the substance normally. For example, Cushing's syndrome, caused by the overproduction of adrenal hormones, may develop as a result of the production of adrenal cortico-trophic hormone (ACTH), a pituitary hormone which stimulates the adrenal glands, e.g. by neuroendocrine tumours of the pancreas or lung. Synthesis and sometimes secretion of a number of peptides by endocrine and non-endocrine tumours may be much more common than is realized. Symptoms of ectopic hormone secretion as tumour markers are important. Ectopic hormone production may occur in tandem with normal hormone secretion due to abnormal regulation of DNA trans-cription, mRNA processing, or even, perhaps, defective post-trans-lational modification of pre-pro-hormone.

14.3.4 *Multihormonal tumours*

Multiple hormone production may be caused by single or multiple endo-crine neoplasia (MEN). The production of more than one hormone by a single endocrine tumour was thought to be uncommon. This was due mainly to the fact that the effects of one secreted hormone were clinically predominant and thus obscured the presence of other less active peptides. It is now recognized that some patients have symptoms attribu-table to the simultaneous secretion of more than one hormone and may show a transition of effects from one to another, sometimes due to treat-ment (chemotherapy). In most instances the hormones are produced by separate cell types.

14.4 Individual tumours

14.4.1 *Tumours of the pancreas and gut*

Insulinomas. β-cells in the pancreatic islets secrete insulin, and insulinomas are one of the many causes of hypoglycaemia (low blood sugar). They constitute 70–75 per cent of all pancreatic endocrine tumours and are about equally common in both sexes. All patients with insulinomas should also be checked for other endocrine disturbances since β or non-β cell adenomas can be associated with other inherited endocrine neoplasms of the multiple endocrine neoplasia (MEN) syndrome. Clinical features (headaches, blurred vision, sweating, hunger, and palpitations) are almost always present, even with small tumours. Symptoms are usually intermittent and thus a patient can be misdiagnosed as having psychiatric, cardiac, or neurological disease. The highest incidence of insulinomas is found between ages 30–60 years. Virtually all insulinomas are localized in the pancreas. About 90 per cent of them are solitary and most (84–96 per cent) are benign.

Insulinomas, like most other endocrine tumours, produce different molecular forms of insulin, C peptide, and pro-insulin in variable proportions.

Tumours associated with the Zollinger–Ellison syndrome (gastrinomas). Gastrinomas represent 20–25 per cent of pancreatic endocrine tumours. They occur most often between 30 and 50 years of age and there is a slight male preponderance. The clinical features, as originally described by Zollinger and Ellison, are intractable gastric, duodenal, and jejunal ulceration, bleeding and very high gastric acid secretion. These features are rarely found nowadays since the tumours are usually diagnosed at an earlier stage due to increasing awareness and availability of assays. Plasma gastrin levels are usually very high, but calcium stimulation of gastrin levels, a commonly employed diagnostic test, permits the distinction between a gastrinoma and other hypergastrinaemic conditions associated with recurrent peptic ulcers. Eighty-five per cent of gastrinomas are found in the pancreas. Extra pancreatic gastrinomas are found, for instance, in the duodenum (13 per cent) and in other areas (1 per cent), including the stomach, upper jejunum and bile ducts. This distribution in the frequency of anatomical sites for gastrinomas does not fit with the distribution of gastrin-containing (G) cells in normal tissue; in particular, G cells are not found in the adult human pancreas but have been described in the fetal pancreas.

At least 60 per cent of gastrinomas are malignant. The tumours frequently metastasize, especially to the liver. Histologically these tumours have the typical appearance of neuroendocrine tumours, e.g.

immunoreactivity for neuron specific enolase, chromogranins, and various segments of gastrin and pre-pro-gastrin. Tumours producing gastrin frequently secrete other regulatory peptides, in particular pancreatic polypeptide.

Diarrhoeagenic (WDHA) tumour syndrome, VIPomas or PHMoma syndrome. Certain patients with non-β islet cell tumours and severe diarrhoea did not meet the criteria for the diagnosis of gastrinoma. The absence of gastric acid hypersecretion in such patients was also reported. Since the discovery that vasoactive intestinal polypeptide (VIP) is produced frequently and in large concentrations by such tumours, the word VIPoma was coined. VIPomas represent 3–5 per cent of all pancreatic endocrine tumours. The syndrome is characterized by watery diarrhoea. Approximately two-thirds of patients complain of abdominal colic and some have intermittent high faecal fat output. There is often a significant weight loss. At operation, a high amount of alkaline secretion from the pancreas is found. Some patients may also suffer from occasional flushing attacks. Fifty to seventy-five per cent of tumours are malignant. The tumours are often quite large (2–7 cm). VIP is known to be produced both by endocrine and neural tumours and histologically tumours show either classical features of islet cell tumours of the pancreas or features of ganglioneuroblastomas.

Glucagonomas. The typical clinical picture shows the following symptoms: a migratory skin rash localized to the lower abdomen, perineum, and legs; an abnormal glucose tolerance test; anaemia; a sore red tongue; angular stomatitis; severe weight loss; depression; tendency to develop overwhelming infection; and venous thrombosis (in about one-third of patients).

The characteristic skin rash gives as a cutaneous marker of internal malignancy. Administration of zinc induces a fast remission of the skin rash. This led to the postulate that the skin condition may be due partly to zinc deficiency. The disease occurs most often between 40–70 years of age and appears to be slightly more common in women than in men. More than 60 per cent of glucagonomas are malignant. Pancreatic poly-peptide producing cells are found frequently, but also insulin, somato-statin, and gastrin producing cells may be found. Glucagon producing pancreatic adenomas constitute part of the MEN Type I syndrome.

Tumours producing pancreatic polypeptide (PP): PPomas. Oversecretion of PP is one of the most frequent associations noted with well defined functioning neuroendocrine tumours. Tumours most frequently contain-ing PP cells include VIPomas, glucagonomas and insulinomas. There-fore, PP positive immunostaining in a metastasis of an endocrine tumour of unknown origin points to the possibility of the primary tumour being present in the pancreas.

Somatostatinomas. The association of somatostatin production by endocrine tumours, with a defined clinical syndrome, remains in dispute. Most somatostatinomas are solitary and localized in the pancreas; fewer tumours have been described in the gut. Somatostatin containing tumours are frequently malignant and their growth pattern is similar to that described for other neuroendocrine tumours.

Growth hormone releasing factor (GRF) producing tumours, or GRF-omas or acromegalic tumours. Tumours of the pancreas associated with acromegaly have frequently been described in the literature, and GRF has now been found to be produced by pancreatic endocrine neoplasms.

Carcinoids and serotonin producing tumours. As the term carcinoid has been associated at various times with a morphological appearance, a clinical syndrome, and a biochemical feature (the production of serotonin), its usage has often been confusing. In this section, it is used to describe endocrine tumours of the gut whether or not they are associated with the clinical features or serotonin production. The typical carcinoid syndrome consists of episodes of flushing, hypertension, diarrhoea, cough, wheezing, and localized capillary dilation. Fibrosis, oedema, pellagra-like lesions of the skin, peptic ulcer, and abdominal fibrosis can also be present. The symptoms described above are particularly prominent once the tumour has metastasized to the liver, allowing the tumour products to enter the circulation. Seventy-five per cent of serotonin producing tumours arise from the gastrointestinal tract. Unlike other endocrine neoplasms, the size of the tumour may be considered as a prognostic factor. When a tumour measures more than 2 cm in diameter it is considered malignant and it is likely that metastases are already present. Frequently carcinoids are multiple and sometimes associated with other malignant neoplasms. Serotonin is particularly prominent in midgut carcinoids but is usually absent from foregut and hindgut tumours. Apart from serotonin, carcinoid tumours of the bowel produce other regulatory peptides. Gastric carcinoids are a separate entity; these occur in association with mucosal atrophy (atrophic gastritis) and pernicious anaemia, or may occur in patients in which acid secretion has been blocked completely by newly discovered drugs for ulcer treatment.

Pituitary gland tumours. In man the pituitary is divided into two lobes, with the anterior lobe being known as the adenohypophysis and the posterior, or neural, lobe as the neurohypophysis. The former produces at least six hormones having major effects on metabolic processes and on other endocrine glands. Hormone production is stimulated or inhibited by hypothalamic factors and by feedback from the hormonal secretions of the target organs. The neurohypophysis is an extension of the brain containing a variety of peptides and amines, in addition to the main neurosecretory hormones, vasopressin and oxytocin.

Tumours of the pituitary gland are relatively frequent. Prolactin secreting tumours are by far the most common and are followed by growth hormone secreting tumours and then ACTH secreting tumours. Gonadotropin and thyrotropin secreting tumours are infrequent. Mixed tumours, secreting more than one hormone, are a common finding. The cell of origin of all these tumours has been disputed but the frequent occurrence of 'mixed' pituitary adenomas might be thought to favour the existence of precursor cells which give rise to all tumour cell types.

Prolactin secreting adenomas cause infertility and impairment of testicular function, milk secretion, and amenorrhoea. Tumours of this type are frequently poorly granulated. Thus, attempts to immunostain prolactin in an actively secreting, poorly granulated tumour may be un-rewarding.

Pituitary tumours associated with Cushing's syndrome react with anti-bodies to various portions of peptides of the ACTH and related hormones.

Thyroid tumours. Two classes of neuroendocrine tumour of the thyroid, medullary carcinoma, are now recognized: the sporadic and the familial types. The medullary carcinoma is one of the few malignant tumours with an inheritance of an autosomal dominant type with high penetrance (see Chapter 4). Twenty per cent of all cases of medullary carcinomas are thought to be genetically determined and these usually present with phaeochromocytomas (tumours of the adrenal medulla) and other tumours as part of the multiple endocrine neoplasia (MEN Type II) syndrome. The tumour arises from special calcitonin producing cells (C cells) in the thyroid and C cell hyperplasia precedes and accompanies inherited medullary carcinoma. The defect is probably at the level of C cell hyperplasia and the development of the malignancy is possibly a secondary defect. It is, therefore, exceedingly important to screen relatives of a patient with inherited medullary carcinoma for the possi-bility of C cell hyperplasia, by measurement of plasma calcitonin. Furthermore, it is necessary to analyse the entire thyroid of sporadic medullary carcinomas for the possible presence of C cell hyperplasia and organize subsequent screening of relatives.

Calcitonin and its flanking peptide PDN-21 or katacalcin are the peptides most frequently found in medullary carcinomas. Calcitonin gene related peptide (CGRP) is also found often. Other peptides or substances shown to be produced by medullary carcinomas are ACTH, neurotensin, bombesin, somatostatin, serotonin, prostaglandins, and substance P. The production of multiple hormones or other peptides may explain the bizarre clinical features often encountered in medullary carcinomas, including protracted diarrhoea.

Tumours of the adrenal medulla and paraganglia. Phaeochromocytoma is the most frequently found tumour of the adrenal medulla, but ganglioneuromatous differentiation can also be found. A phaeochromocytoma is clinically recognized in the following circumstances: as a familial case, hypertensive episodes, persistent hypertension with increased urine excretion of vanillyl mandelic acid, noradrenalin, or adrenalin.

The size of the tumour ranges from a microscopic lesion to over 2 kg in weight, but most are about 5–6 cm diameter and weigh an average of 9 g. Histologically, the appearance of phaeochromocytomas is quite bizarre, with very many irregular cells of variable size, structure and arrangement. Many multinucleated, gigantic cells can be found which, at the ultrastructural level, contain numerous distinctive electron dense secretory granules. Peptides of the enkephalin/dynorphin family, neurotensin and, very recently, a novel peptide known to produce marked vasoconstriction, neuropeptide Y are produced by both the normal and tumour tissue.

Phaeochromocytomas may be a component of the familial MEN II syndrome totether with medullary carcinoma of the thyroid, parathyroid hyperplasia, or adenoma. Phaeochromocytomas are usually benign tumours but malignancy can occur when tumours are found bilaterally. They occur in adults between the ages of 25–55 years.

Tumours of paraganglia are called extra-adrenal phaeochromocytomas in Europe, and paragangliomas in America. These can occur all along the sympathetic chain and in the para-aortic bodies. Paraganglia form a widely disseminated system of small sensory and perhaps local neurosecretory organs that develop in fetal life, persist in infancy, and thereafter become smaller and more difficult to find in certain parts of the body such as the fibrous tissue behind the peritoneum along the aorta. Several lie close to other major blood vessels, in or near nerves or ganglia, or are located in organs such as the lung or duodenum. The urinary bladder also contains paraganglia and phaeochromocytomas of the urinary bladder have been reported.

14.4.2 *Respiratory tract tumours*

The respiratory tract is also a common site for neuroendocrine tumours. One of the active peptides frequently produced by small cell carcinomas is bombesin/GRP which is also present in normal mucosal endocrine cells. The structure of human pro-bombesin (gastrin-releasing peptide) from a small cell carcinoma of the lung has now been established using region-specific antibodies to various segments of human pro-bombesin molecule. Antibodies directed to the C-terminal portion of human pro-bombesin are excellent, and possibly better than, antibodies to bombesin itself, for demonstrating the production of bombesin gene products by small cell carcinomas (Fig. 14.6). It is interesting that the small cell carcinoma is a poorly granulated neoplasm, suggesting that the tumour cell bypasses the secretory granule pathway.

Within neuroendocrine tumours of the lung, the carcinoid seems to represent the benign end of a continuous spectrum which ends in the small cell carcinoma. The lung is a fairly frequent site of carcinoids, in fact 12 per cent of all carcinoids arise in the lung. However, carcinoids represent only 1 per cent of all lung tumours. The highest incidence of bronchial carcinoids occurs in the 31–40 year age group, with a slight preponderance in females (62 per cent). The tumour usually arises in the main bronchi but may be located in the lung periphery. Multicentric growth has been described and carcinoids may be part of a pluri-glandular syndrome. Carcinoid tumours are generally well demarcated, grow slowly, and have a low malignant potential. Ultrastructurally, car-

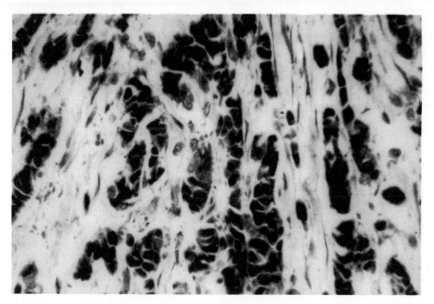

Fig. 14.6 Section of a small cell carcinoma of the lung densely immunostained (black) for the C-flanking portion of human pro-bombesin. ($\times 262$)

cinoid tumours of the lung are frequently well granulated. Immunocyto-chemistry reveals a variety of products being produced by carcinoid tumours, in particular serotonin, bombesin, PP, and occasionally ACTH, with or without overt clinical syndromes. Small cell carcinoma of the lung is the malignant counterpart of lung endocrine tumours. It is one of the most frequent neuroendocrine tumour types and comprises 20 per cent of all bronchial carcinomas. Small cell carcinomas are divided into two main types depending on cell size: (i) oat cell type, consisting of small oat shaped cells with a finely granular chromatin pattern, no nucleolus and very sparse cytoplasm; and (ii) intermediate cell type, composed of slightly larger cells with more abundant cytoplasm. There does not seem to be a correlation between the histologial subtype and patient survival. The best neuroendocrine marker for the character-ization of small cell carcinomas is the non-granular marker neuron specific enolase. Small cell carcinomas of the lung have long been associ-ated with hormonal abnormalities, particularly Cushing's syndrome and inappropriate antidiuretic hormone secretion. Extensive data indicate that bombesin released from tumour cells of the small cell carcinoma acts in an autocrine (paracrine) manner. The tumour cells release their product into the vicinity and in turn activate tumour cell receptors to enhance growth.

Small cell carcinomas of the lung are aggressive neoplasms, having the poorest prognosis of all lung tumours with an overall five year survival rate of only 2 per cent. The tumour is generally extremely sensitive to combined chemotherapy and radiotherapy, whereas other lung tumours are more amenable to surgery.

14.5 Conclusions

Studies of the 'diffuse neuroendocrine system' are a new facet of endo-crinology and expand the classical view of glandular endocrinology. Regulatory peptides are still being discovered and their role in the control of body functions is being firmly established. Advances in molecular biology make possible determination of the structure of peptide precursors and the use of region specific antibodies to the various portions of the molecule are opening up new investigative areas of peptide neuroendocrinology. The application of other modern tech-niques now allows not only investigation of the stored peptide but of all intracellular steps leading to this peptide synthesis, including visualiz-ation of the mRNA. It is predictable that more functional studies will be forthcoming, analysing the quantities of messenger in relation to peptide biosynthesis. The site of action of regulatory peptide receptors can now be analysed and therapeutic agents are being developed to interact with particular regulatory peptides specifically, either at the site of their production or release or at their receptor level, and therefore new avenues for the treatment of this very interesting group of tumours are becoming available.

References and further reading

Adams, C., Hamid, Q. A., and Polak, J. M. (1987). In situ hybridisation with radio-labelled cRNA probes, using tissue sections and smears. In *Methods in molecular biology*, Vol. 5, *Tissue culture*, (ed. J. Walker). Humana Press Inc., New Jersey.

Bloom, S. R., and Polak, J. M. (1981). *Gut hormones* (2nd edn). Churchill Livingstone, Edinburgh.

Eiden, L. E., Huttner, W. B., Mallet, J., O'Connor, D. T., Winkler, H., and Zanini, A. (1987). A nomenclature proposal for the chromogranin/secretogranin proteins. *Neuroscience* **21**, 1019–21.

Falkmer, S., Hakanson, R., and Sundler, F. (ed.) (1984). *Evolution and tumour pathology of the neuroendocrine system*. Elsevier, Amsterdam.

Hanley, M. R. (1989). Peptide regulatory factors in the nervous system. *Lancet* **i**, 1373–6.

Havemann, K., Sorensen, G., and Gropp, C. (eds.) (1985). *Recent results in cancer research: peptide hormones in lung cancer*. Springer-Verlag, Berlin.

Polak, J. M. and Marangos, P. J. (1984). Neuron-specific enolase, a marker for neuroendocrine cells. In *Evolution and tumour pathology of the neuroendocrine*, Vol. 4. (ed. S. Falkmer, R. Hakanson, and F. Sundler), pp. 433–80. Elsevier, Amsterdam.

Polak, J. M. and Bloom, S. R. (1985). *Endocrine tumours*. Churchill Livingstone, Edinburgh.

Polak, J. M. and Bloom, S. R. (1985). Pathology of peptide-producing neuroendocrine tumours. *British Journal of Hospital Medicine* **53**, 153–7.

Polak, J. M. and Bloom, S. R. (1983). Regulatory peptides: key factors in the control of bodily functions. *British Medical Journal* **286**, 1461–6.

Polak, J. M. and van Noorden, S. (1983). *Immunocytochemistry: practical applications in pathology and biology*. Wright, Bristol.

Polak, J. M. and Varndell, I. M. (1984). *Immunolabelling for electron microscopy*. Elsevier, Amsterdam.

Solcia, E. *et al.* (1987). Endocrine cells producing regulatory peptides. *Experientia* **43**, 839–50.

Varndell, I. M. *et al.* (1984). Visualization of messenger RNA directing peptide synthesis by *in situ* hybridization using a novel single-stranded cDNA probe: potential for the investigation of gene expression and endocrine cell activity. *Histochemistry* **81**, 597–601.

15

Immunology of cancer

Peter Beverley

15.1 Introduction to the immune system

Animals or men born without a properly functioning immune system suffer from multiple infections and usually die at an early age. Thus the primary function of the immune system is to defend the animal against pathogenic micro-organisms but the system will also react against other foreign substances. It was proposed, as long ago as the turn of the century by Paul Ehrlich, that the immune system might also prevent, or at least retard, the growth of many tumours. In this chapter I shall examine this hypothesis but to do this it is necessary to explain first how the immune system works.

15.2 Organization of the immune system

Many of the cells of the immune system (white cells or leucocytes) are found throughout the body because they circulate in the blood stream and also migrate into tissues, particularly at sites of inflammation. Lymphoid cells are also found in specialized lymphoid organs, the bone marrow, thymus, spleen, and lymph nodes (Fig. 15.1). The lymph nodes are widely distributed throughout the body. The tonsils and adenoids are specialized lymph node organs in the pharynx; the lymph nodes in the intestine are known as Peyer's patches. All the cells of the immune system originate from self-replacing stem cells in the bone marrow, as do red blood cells and platelets (see Fig. 12.1, Chapter 12). In this chapter we shall be mainly concerned with the function of lymphocytes because

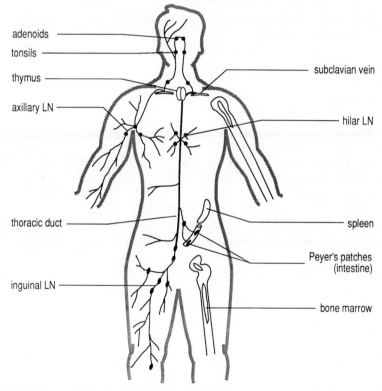

Fig. 15.1 Lymphoid organs. The organization of the lymphoid organs is shown schematically. Afferent lymphatics drain the tissues and enter regional lymph nodes (LN). Efferent lymphatics from the lymph nodes drain into the thoracic duct which enters the venous system. The spleen, thymus, and bone marrow connect to other lymphoid organs mainly via the blood circulation (not shown in the figure).

these cells are responsible for specific immunity. In contrast granulocytes mediate non-specific protection, particularly against bacterial pathogens. Granulocytes and macrophages are attracted to sites where invading micro-organisms are present and are able to ingest (phagocytose) and often kill the micro-organisms. They lack the two major properties of lymphocytes: specificity and memory. Granulocytes are also short-lived (hours) whereas lymphocytes are long lived (months or years). There are two main classes of lymphocytes. Bone marrow derived (B) lymphocytes mature in the marrow and then migrate directly to the spleen and lymph nodes. They are then ready to react to foreign substances (antigens). Thymus derived (T) lymphocytes migrate from the bone marrow to the thymus where they mature before migrating to the peripheral lymphoid organs (spleen and lymph nodes). These two types of lymphocyte although morphologically very similar, can be readily distinguished by their differing cell surface phenotypes; that is, the array of glycoproteins carried in the lipid bilayer of the cell surface membrane. T and B lymphocytes also have very different functions although they interact during most immune responses.

15.2.1 B lymphocytes and antibody

The principal function of B lymphocytes is to synthesize and secrete antibodies. When these cells are stimulated by encountering a foreign antigen, they first go through several cycles of cell division and then differentiate into specialized antibody secreting cells called plasma cells. Antibodies are globular glycoproteins (hence immunoglobulins) found in the blood plasma. The general structure of an immunoglobulin molecule is shown in Fig. 15.2. It consists of two heavy (H) and two light (L) chains. The H and L chains are each divided into two portions, the N-terminal portion (approximately 110 amino acids) being termed the variable part and the remaining portion (approximately 330 or 110 amino acids for H and L chains respectively) being termed the constant part. The variable part of a heavy and light chain form an antigen binding site while the constant part of the heavy chain determines the biological function of the molecule. Differences in the amino acid sequence of the constant part of the H chain determine the class of the immunoglobulin molecule. There are five major immunoglobulin (Ig) classes, IgA, D, E, G and M, with H chains designated by the greek letters α, δ, ε, γ, and μ (Table 15.1). The two light chain types which also differ in their constant parts are called κ and λ. Each class has different properties which are summarized in Table 15.1. IgM and IgD are found on the membrane of resting B lymphocytes and act as receptors for antigen. When B cells are stimulated they first secrete IgM. Secreted IgM is a pentamer of subunits

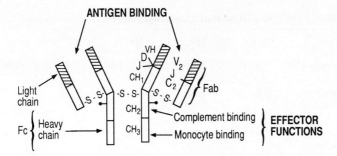

Fig. 15.2 The structure of an antibody molecule. Heavy and light chains are linked by disulphide bonds (S–S). A further disulphide bond joins the two heavy chains. There are further intrachain disulphide bonds in each variable (V) and constant (C) domain of the protein. The symbol —• denotes the site of carbohydrate attachment. The antigen binding fragment (Fab) and crystallizable fragment (Fc) of the molecule are indicated.

joined by an extra polypeptide, the J chain. Later most B cells switch to secreting IgG, A, or E. IgG is the main class present in the blood and is a monomer of the basic four chain immunoglobulin molecule. In man there are four subclasses of IgG termed IgG1, IgG2, IgG3 and IgG4 which have closely related H chains. In the mouse and rat there are corresponding IgG subclasses called IgG1, IgG2a, IgG2b and IgG3. IgA and IgE are minor components of serum and have specialized functions relating to protection of mucous membranes and allergic reactions respectively.

It is a remarkable property of the immune system that it can produce specific antibodies (able to bind with high affinity) to almost any antigen encountered. This is possible because each B lymphocyte clone synthesizes and secretes a different immunoglobulin molecule. The molecules differ mainly in the variable part and this variability is achieved by several mechanisms. These include selection from a large pool of variable (V) genes, random joining of V genes to regions coding for diversity (D) and joining (J) segments, random combination of heavy and light chains, and somatic mutation (Kocks and Rajewsky 1989). This allows for the production of millions of different antigen binding sites. Each combining site has a unique structure (idiotype) and by immunizing an animal with homogeneous Ig it is possible to raise anti-antibodies (anti-idiotypes) which react only with the immunizing Ig and not other Ig molecules (see Chapter 18).

Generally when an animal encounters an antigen, for example a virus, many B cells are stimulated. Each B cell carries on its surface membrane

Table 15.1 Properties of human immunoglobulins

Property	IgG	IgM	IgA	IgD	IgE
Molecular weight	150 000	900 000	400 000	180 000	190 000
Concentration in serum (mg/ml)	10	1	2	0.03	< 0.001
Relative amount in secretions	Low	Low	High	Very low	Low
Crosses placenta	Yes	No	No	No	No
Complement fixation	Yes	Yes	No	No	No
Transport across epithelia	No	No	Yes	No	Yes
Allergic reactions	No	No	No	No	Yes
Fc binds to cells	Yes	Yes	Yes	No	Yes

immunoglobulin molecules with different antigen binding sites. These act as the receptors for antigen and the secreted immunoglobulin of each cell has the same antigen binding site (specificity) as the membrane bound immunoglobulin. Each B cell will divide to produce a clone of daughter cells all producing immunoglobulin with the same specificity, but since many B cells bind the antigen the serum of the immunized animal will contain a mixture of the antibodies produced by many clones. A second contact with the same antigen elicits a greater and more rapid antibody response (immunological memory).

15.2.2 *Monoclonal antibodies*

Antisera made by animals in response to an antigen are the product of many different clones (polyclonal) of B lymphocytes and individual antibodies react with many different sites (epitopes) on the antigen. If the antigen is complex, as are bacterial or mammalian cells, it may be very difficult to determine exactly which molecules of the cell the antibodies are directed toward. For many years the complexity of antigens hampered efforts to produce antibodies which would distinguish between tumour and normal cells although, in principle, the exquisite specificity of antibodies made this an attractive approach to understanding the changes of malignant transformation. In 1975 Kohler and

Milstein provided a solution to this problem when they published a method for immortalizing single antibody secreting B lymphocytes by fusing them to a plasma cell tumour (myeloma) to produce a hybrid cell (hybridoma). All the progeny of such a hybridoma cell produce identical (monoclonal) antibody molecules (for futher details see Chapter 18). The ability to produce unlimited quantities of homogeneous antibody of any desired specificity has had a major impact in many areas of biology including tumour immunology.

15.2.3 T lymphocyte function

Until recently the functions of T cells were far less well understood than those of B cells. T lymphocytes do not secrete a single major protein product such as immunoglobulin but carry out their functions either by direct contact with other cells or by producing lymphokines (secreted proteins that have powerful biological effects on other cells and are active at very low concentrations). A further difficulty in understanding T cell function was that no T cell receptor for antigen was identified until 1983, but it is now clear that the T cell receptor consists of two protein chain rather similar to the light chains of immunoglobulin (Davis and Bjorkman 1988). Diversity of receptor specificity is generated by mechanisms similar to those found in B lymphocytes (described briefly in Section 15.2.1 and Chapter 11). In spite of the apparent similarities between the B and T cell receptors for antigen, the two cell types 'see' antigen in very different ways. While soluble antigen can bind to the surface immunoglobulin of B cells to stimulate a response, soluble antigen does not stimulate T cells. T cells always 'see' antigen in association with 'self': the phenomenon of genetic restriction.

The nature of this phenomenon can be made clearer by describing one of the earliest experiments to demonstrate it. If an animal (or man) is immunized with a virus, for example influenza A, immune T cells from the animal are able to kill influenza A infected cells but not uninfected target cells or cells infected with a different virus such as influenza B. The immune T cells are therefore specific for influenza A virus. The experiment, however, has a second part. The infected target cells must come from the same animal (or inbred strain) as the immune T cells or they cannot be killed. Thus the T cells do not recognize virus only; they appear to 'see' virus + self. Further genetic experiments showed that the self components required were membrane glycoproteins coded in a region of the genome called the major histocompatibility complex (MHC) (see also Chapter 4). This name derives from older experiments which showed that the same gene products are responsible for the rejection of organ grafts, a reaction also known to be initiated by T lymphocytes. Thus in both the response to foreign pathogens and to tissue grafts T lymphocytes recognize and respond to MHC antigens although in one case foreign MHC and in the other self MHC + antigen. T cells do not 'see' antigen as a whole protein molecule but rather as a small fragment consisting of as few as a dozen amino acids. These peptides lie in a cleft in the MHC molecules and the T cell receptor binds to the complex of MHC + peptide (Bjorkman *et al.* 1987). That only small fragments of antigens are recognized by T cells implies that larger molecules must be broken down (processed) before T cells can recognize them. In many cases this processing is carried out by phagocytic cells, such as macrophages, but all cells continuously turn over proteins so that in a virus-infected cell, for example, fragments of viral polypeptides are produced and can become associated with the MHC molecules of the cell. Although the way in which T cells recognize antigen is now understood, genetic restriction still creates a problem for the experimenter wishing to study T cell immunity to tumours since in order to observe a reaction to a tumour cell, the T lymphocytes and tumour cells must be MHC identical.

In contrast to B lymphocytes, T cells do not appear to secrete their surface receptor in measurable amounts and carry out their multiple functions by other means. Table 15.2 lists some of the activities of T cells. Some of these can be carried out by T cells alone. Typical of this type of reaction are graft rejection or protection against virus infection mediated by the 'cytotoxic' (cell killing) T cells described above. Many other functions involve other cell types as well as the responding T cells. Very often these responses are immunoregulatory; that is, the T cells control the responses of other cell types. Thus helper T cells stimulate B lymphocytes to secrete antibody and suppressor T cells can switch off antibody

Table 15.2 T lymphocyte functions

In vivo
  Delayed type hypersensitivity
  Graft rejection
  Graft versus host response
  Protection against viral and fungal infections

In vitro
  Help for antibody responses
  Mixed lymphocyte responses—proliferation
                            —cytotoxicity
  Proliferation in response to non-specific mitogens
  Proliferation in response to specific antigens
  Cytotoxicity against specific antigens (in association with MHC antigens)
  Suppression of antibody, proliferative or cytotoxic responses

production. This latter response is important in terminating a response to a foreign pathogen but may also prevent inappropriate immune responses to self antigens (autoimmunity).

Many immunoregulatory functions are mediated by lymphokines. They have powerful effects on the growth and differentiation of their target cells and act as local hormones within the immune system. They are similar in many respects to the growth factors discussed in Chapter 11. Table 15.3 lists some of the better characterized lymphokines as well as the monokine interleukin 1 which plays a role in the initial activation

Table 15.3 Functions and properties of cytokines

Cytokine	Main effects	Molecular weight and other properties
Cloned and characterized cytokines		
Interleukin 1 α	Activation of T cells,	17 500 ⎱ Produced by
β	pyrogenic response	17 500 ⎰ many cells
Interleukin 2	Growth of T cells	13 500, T cell product
Interleukin 3	Growth of haemopoietic stem cells	18–30 000, T cell product
Interleukin 4	Growth and differentiation of T and B cells	15–18 000, T cell product
Interleukin 5	Effects on B cells and granulocytes	20–22 000, T cell product
Interleukin 6	B cell differentiation but many other effects also	21 000, produced by many cells
Interleukin 7	Growth of B cell	25 000
Interleukin 8	Granulocyte chemotaxis	8 000, produced by many cells
Interferon α	Inhibition of viral replication	15 000, produced by many cells
Interferon γ	Increase of MHC expression Inhibition of viral replication	38 000, T cell product
Tumour necrosis factor α	Necrosis of tumours, pyrogenic	25 000 and multimers
Tumour necrosis factor β	Cytotoxic *in vitro*	25 000 and multimers
Leukaemia inhibitory factor	Inducer of myeloid differentiation	20 000, produced by many cells
Colony stimulating factors (see Chapter 11)	Promote growth and maturation of progenitor cells	Various
Less well characterized factors		
Macrophage migration inhibition factor (MIF)	Inhibition of macrophage migration *in vitro*	Heterogeneous, 12–70 000
Macrophage activation factors (MAF)	Activates macrophages to become bactericidal or tumoridal	Similar to MIF
Chemotactic factors	Attraction of macrophages and granulocytes	8–38 000

of T cells. One of the best characterized lymphokines is interleukin 2 (IL2) which is produced by activated T cells and can stimulate both the producing cell (autocrine stimulation) and other T cells to divide. Only T cells which have been first stimulated by contact with antigen express receptors for IL2. The growth promoting effects of IL2 are very powerful and it is used to grow antigen activated T cells *in vitro*. So potent a growth stimulus is IL2 that it is possible to grow a large number of cells from a single T cell. This is known as an IL2 dependent T cell clone. Such clones are being used to analyse the specificity and functions of T cells and are the T cell equivalent of monoclonal antibodies.

In contrast to IL2, which promotes T cell growth, other lymphokines induce differentiation into effector cells. Interleukins 4, 5, and 6 together regulate the proliferation of B lymphocytes after stimulation by antigen and their differentiation into high rate antibody secreting cells. In other cases the main functions of lymphokines are not yet clear. The interferons are examples of this category. Interferons have effects on many different cell types which include stimulation of cytotoxic activity, antiproliferative effects, and the induction of increased expression of MHC molecules at the cell surface. This last effect is important because it leads to more efficient presentation of antigens to T cells. Because lymphokines have such powerful regulatory effects it is possible that derangements in their production or receptors might play a role in disease processes, particularly malignancy. As yet there is little evidence for this but abnormal expression of IL2 receptors has been documented in some T cell tumours so that this is a field which requires further study.

15.3 The immune system and cancer

15.3.1 *Immune surveillance*

As the importance of the immune system in protection against infection became apparent, it was suggested that immunity to tumours might also be important. In 1909 Paul Ehrlich postulated that we might all die of tumours if the immune system did not remove 'aberrant germs' (nascent tumours). This idea led to many attempts to demonstrate immunity to tumours in which tumours were transplanted from one animal to another. The transplants were rejected and this was taken as evidence of immunity to the tumour. Only later was it recognized that these experiments demonstrated not tumour immunity but transplantation immunity directed against MHC antigens.

Only when genetically homogeneous inbred animals (mice) became available was it possible to carry out experiments on tumour immunity and it was then shown that if a growing tumour was excised from a mouse and the animal was challenged with a graft of the same tumour, the graft was rejected. The animal was thus immune to the tumour cells but not to a graft of a different tumour. At about the same time as these early experiments on tumour specific immunity, Thomas and later Burnet restated Ehrlich's hypothesis of protection against 'aberrant germs' as the theory of immune surveillance against tumours. They proposed that tumours arise frequently and that the majority were eliminated by the immune system well before becoming clinically apparent. This hypothesis stimulated a great deal of experimental work in both man and experimental animals because the theory made clear predictions which could be tested. In particular it suggested that tumours should differ antigenically from normal cells and that they would arise more frequently in circumstances when the immune response of the host is compromised. A corollary of this view was that it might be possible to use immunological means to detect tumours and perhaps to stimulate the immune system to destroy tumour cells (immunotherapy). In the remainder of this chapter I shall examine the evidence for and against immune surveillance, discuss the nature of tumours antigens and consider the possibility of using immunological methods for detection of tumours and of harnessing the immune system for tumour destruction.

15.3.2 Evidence for and against immune surveillance

Burnet summarized his view of immune surveillance as follows. (i) Most malignant cells have antigenic qualities distinct from those of the cell type from which they derive; (ii) such antigenic differences can be recognized by T cells and provoke an immune response. If this view is correct it follows that; (iii) the incidence of malignant disease should be greatest in periods of relative immunological inefficiency, particularly in the perinatal period and old age; (iv) immunosuppression whether genetic or induced by drugs, radiation, infection, or other causes should increase the incidence of cancer; (v) spontaneous regression of tumours may occur and evidence of an immune response should be apparent in these cases; and (vi) large scale histological examination of common sites of cancer should reveal a higher proportion of tumours than become clinically apparent. Burnet also suggested ways in which the theory of surveillance might be tested experimentally. Thus immunosuppressive agents should facilitate the transfer of tumours, or damage to the T cell immune response produced by surgical removal of the thymus might lead to increased tumour incidence.

At first sight a variety of clinical and experimental data do seem to be in accordance with the surveillance theory. In man some tumours show a higher incidence in the first few years of life than in early adulthood and the incidence, but of different tumour types, then rises progressively with increasing age (see Chapter 1). There is also compelling evidence in man that the incidence of tumours is greatly increased in immunosuppressed individuals. This is true both in rare patients with inherited immuno-deficiency and in the large number of kidney grafted patients who are treated with immunosuppressive drugs to prevent rejection of the grafted kidney. In these individuals the incidence of tumours may be as high as 1 per cent. However, a closer examination does not support the surveil-lance hypothesis.

The age incidence of tumours is as well explained by many other theories of cancer causation as by immunosurveillance. Tumours are caused by genetic changes in their cells of origin. These changes might be expected to occur either as errors during periods of rapid cell division (early life) or when external causes (carcinogens) have had time to take effect as in later life.

The more persuasive data derived from immunosuppressed individ-uals are similarly less convincing when re-examined. While there is indisputably a large increase in tumour incidence the overwhelming majority of these tumours are tumours of the lymphoid system, whereas in normal individuals the common tumours are of epithelial origin (see Chapter 1). The frequency of the most common tumours such as those of the lung, stomach, intestine and breast is not increased, while those of the skin, lip and cervix show some increase. These data are not in accord with a straightforward surveillance theory which would predict an overall increase in tumour frequency of the most common tumours. A very large study of over 15 000 nude mice, which lack a functional thymus and are therefore congenitally T cell immunodeficient, also contradicted the surveillance hypothesis since no tumours were seen. How then can the increased incidence of lymphoid tumours in immuno-suppressed individuals be explained, and what role, if any, does the immune system play in protection against tumours?

Another experimental study of immunosuppression, as well as a closer examination of the nature of the tumours arising in transplant patients, provide some clues. When a large group of mice were treated from birth with anti-lymphocyte serum raised in rabbits, their T cell immunity was sufficiently depressed so that skin grafts from another mouse strain were retained indefinitely. Like the nude mice, these mice did not develop spontaneous tumours but when they inadvertently became infected with polyoma virus (see Chapter 8) a number developed multiple tumours of a type characteristically caused by this virus. Similarly the lymphoid tumours seen in transplant recipients are commonly of B lymphocyte origin and have been shown to contain both DNA and proteins charac-teristic of the Epstein Barr virus (EBV). This member of the herpes virus family is implicated in the cause of Burkitt's lymphoma, a B cell tumour seen in parts of Africa, and nasopharyngeal carcinoma (see Chapter 8). The virus can also immortalize normal human B lymphocytes *in vitro*. It is very likely therefore that this virus is responsible for at least one step in the transformation of B lymphocytes into malignant lymphomas which occurs in immunosuppressed individuals. These findings suggest, therefore, that the important role of the immune system in tumour protection may be in preventing the spread of potentially oncogenic viruses. This role is in accordance with the modern view of T lymphocyte function which suggests that they recognize antigen in association with self MHC and are particularly important in combating intracellular parasites such as viruses.

This view agrees with experimental and clinical data on EBV, an ubiquitous infectious agent in human populations. More than 90 per cent of adults have evidence of past infection in the form of antibody to the virus in their serum. Infection may be symptomless but in young adults EBV causes infectious mononucleosis. Following either symptomless infection or mononucleosis the virus is carried lifelong and the individual also has lifelong immunity. Immune T lymphocytes can be demonstrated *in vitro* in assays in which they prevent the outgrowth of transformed cells from B lymphocytes which have been exposed to the virus (regression assays). Under normal circumstances there is thus a balance between virus production and the immune response while in immunosuppressed individuals the immune system may be unable to prevent virus spread. The T cells of such individuals cannot prevent outgrowth of EBV transformed B cells *in vitro*, and virus can often be isolated from body tissues and secretions such as saliva.

Some infectious agents are themselves immunosuppressive. The human immunodeficiency virus (HIV) infects mainly helper T cells and monocytes. These two cell types are extremely important in initiating immune responses so that their destruction leads to a severe acquired immunodeficiency syndrome (AIDS). AIDS patients have a high incidence of tumours, particularly of the lymphoid system. They also frequently acquire Kaposi's sarcoma, an otherwise rare tumour. Because other immunodeficient patients do not acquire Kaposi tumours, it has been suggested that HIV or another virus may cause the tumour.

If it is accepted that T cells deal mainly with infections, what relationships does the immune system have to the majority of tumours which do not have an obvious viral cause? Evidence from experimental animals and man suggests that some tumours do provoke an immune response by the host (see p. 415). Such a response would not occur until a stimulatory dose of antigen (a large enough number of tumour cells) was available. Grafts of small numbers of tumour cells may provoke a response which causes tumour rejection while large numbers overwhelm the immune response and grow progressively, but very low numbers of grafted cells may again grow progressively (a phenomenon termed 'sneaking through').

Since the grafting of small numbers of tumour cells may be similar to the early phase of growth of a spontaneous tumour from a single cell, much effort has been devoted to the elucidation of the mechanisms by which a nascent tumour manages to grow in the face of a competent T cell immune system. In experimental animals a variety of mechanisms have been demonstrated, including the masking of tumour cell antigen by 'blocking' antibody, the liberation by the tumour cells of soluble antigen, the formation of 'blocking' antigen–antibody complexes, and the development of suppressor T cells. These mechanisms are illustrated in Fig. 15.3. Nevertheless the fact that anti-tumour immune responses can be demonstrated in animals suggests that, given sufficient understanding

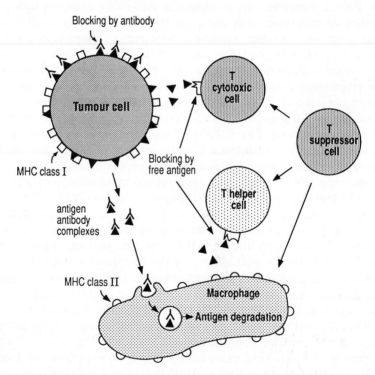

Fig. 15.3 Mechanisms which may prevent an effective immune response to a tumour. Antibody masks tumour antigens and free antigen blocks T cell receptors. Antigen–antibody complexes prevent presentation of antigen to T cells and suppressor cells switch off T cell responses.

it might be possible to manipulate the response to prevent blocking or suppression and stimulate an effective anti-tumour response. To do this effectively, it is necessary to know what target antigens the immune system 'sees' on tumour cells and which immune mechanisms are important in tumour cell destruction. While experimental models can provide useful clues, most have utilized tumours induced with viruses or carcinogens. It is important, therefore, to attempt to define the nature of the immune response to spontaneous human tumours both in terms of target antigens and immune mechanisms.

15.4 The nature of tumour antigens

15.4.1 *Animal tumour antigens*

It has been demonstrated repeatedly that laboratory animals can be immunized against tumour cells (see p. 415). Immunity is generally tumour specific, that is, directed only toward the immunizing tumour and not against other similar tumours. Immunity can be transferred to other animals by immune cells but not by serum and is mediated by T cells. Several groups have attempted to identify molecules capable of inducing such tumour specific responses. A common strategy has been to fractionate tumour cells and use fractions to immunize against tumour challenge. Fractions capable of generating protective immunity can then be used to produce antisera (in a different species) which can in turn be used to identify and purify molecules from the active subcellular fractions.

This strategy has lead to the isolation of a glycoprotein of 96 kDa (gp96) from a chemically induced murine sarcoma which is able to immunize against tumour challenge. The gene coding for gp96 has been cloned and sequenced and is not related to any other known gene. It is thought that mutations in the gene may give rise to the individually distinct antigens of at least some chemically induced sarcomas (Srivastava and Old 1988).

In another animal model in which highly immunogenic skin tumours are induced by exposure to u.v. light, the tumour antigen of one such tumour has been shown to be an altered (mutated) MHC antigen. Since MHC antigens are the target for T cell reactivity it is not surprising that this tumour is highly immunogenic. That very different molecules have been identified in different tumours as targets for tumour rejection responses is hardly surprising, given the diversity of tissues in the body and the tumours which arise from them. However, the fact that individual tumour specific antigens are the targets for protective immune responses, rather than molecules shared between many tumours, makes detection of responses to tumours in outbred species such as man very difficult. Because of genetic restriction (see p. 411) and the uniqueness of tumour antigens, detection of an anti-tumour response requires the use of tumour and lymphocytes from the same individual; this is often difficult to achieve.

15.4.2 *Human tumour antigens*

A great deal of effort has been devoted to attempts to use antibodies to detect new antigens on human tumours (see Chapter 18). A unique genetic change in a cell might lead to the expression of a new antigen unique to that tumour—a tumour specific antigen. Such an antigen may provoke a host response but the development of antibodies to it may only be useful to the tumour bearing individual because the same antigen is not found in other tumours, even of the same type. More useful from the point of view of diagnosis or immunotherapy are tumour associated antigens. These are antigens present on or in all tumour cells of a particular type but not found on normal cells. In tumours caused by viruses, proteins coded by the viral genome are effectively tumour associated because they are generally found only in very rare normal cells, e.g. some lymphomas have EBV antigens and one type of T cell leukaemia has antigens of the human T lymphotropic virus (HTLV-1) which may directly transform T lymphocytes. These are both rare tumours so we must now consider tumour associated antigens of more common tumours.

Early attempts to reveal tumour associated antigens used sera either from tumour patients or from animals deliberately immunized with human tumours. Both types of study have considerable problems. Sera from tumour patients might appear to be ideal reagents but in practice they often have in them antibodies capable of reacting with many types of cells. These include antibodies to blood group and histocompatibility antigens and autoantibodies which react with normal as well as tumour cells. To sort out a minor proportion of antibodies specific for tumour antigens is a complex task. It has usually been attempted by absorption analysis in which the serum is incubated with a succession of different normal cell types to remove antibody to them and then tested for reactivity with tumour cells. Such manoeuvres generally dilute the specific antibody and also lead to non-specific loss so that even if the resulting antiserum reacts only with tumour cells it may be so weak that it is difficult to use for identification and characterization of the antigens detected.

Antisera raised in animals, usually rabbits, present similar problems. The rabbit antiserum 'sees' many antigens on a human tumour cell but the majority of these are present on normal cells also. Such sera therefore require extensive absorption to render them specific for tumour cells and it is not surprising that reports of successful production of specific heteroantisera are few in number, nor have they in general been particularly useful in tumour diagnosis or therapy. Some exceptions are mentioned in the next section.

When the hybridoma method for producing monoclonal antibodies was developed, many investigators realized that monoclonal antibodies might provide reagents for detecting differences between cell types at a molecular level (see Chapter 18). This has indeed proved to be the case and, for example, mouse monoclonal antibodies can distinguish between human leukocytes and all other human cells, between T and B lymphocytes or between different subpopulations of T lymphocytes. The molecules identified by these monoclonal antibodies are differentiation antigens (antigens present on one cell type but not another, or related to the stage of maturation, see Chapter 12) and in some cases have been shown to be involved in functions associated with the particular cell type. These results suggested that it should be possible to produce monoclonal antibodies which could distinguish between tumour and normal cells, if differences existed. Differences have indeed been recognized but so far there is no convincing evidence of a specific tumour associated antigen.

One of the few useful polyclonal heteroantisera was raised against acute lymphoblastic leukaemia (ALL) cells. After extensive absorption this serum could distinguish ALL cells from normal lymphocytes or bone marrow. Several monoclonal antibodies have been produced which react in a similar fashion and were therefore considered initially to identify a tumour associated antigen of leukaemia cells (common acute lymphoblastic leukaemia antigen or CALLA). More careful examination of bone marrow showed that a small number of normal cells carried CALLA. Subsequent studies have shown that CALLA is a differentiation antigen transiently expressed on cells early in the B lymphocyte lineage. One category of antigens which can easily be confused with a tumour associated antigen is thus a differentiation antigen expressed only on rare normal cells.

But from the point of view of diagnosis and treatment, CALLA is a very useful antigen. CALLA positive cells are very rare in normal bone marrow (< 1 per cent) and are not found in peripheral blood so that their presence can be used to monitor disease. Furthermore, because CALLA is not present on stem cells, it is possible to remove CALLA positive cells from bone marrow without damaging the potential of the marrow cells to regenerate a functional haemopoietic system (see Chapter 18). This, therefore, is an example of a normal differentiation antigen which for practical purposes may be regarded as a tumour associated antigen.

There are a number of other examples of differentiation antigens which provide similar possibilities. Many antibodies raised against carcinomas recognize heavily glycosylated surface glycoproteins. These antigens are often restricted to particular epithelia or specialized cells within an epithelium. Tumour cells expressing the antigens can therefore often be detected when the tumour spreads outside the primary site. This has been exploited in detecting small metastases of breast cancer in the axillary lymph nodes and metastatic lung cancer cells in bone marrow.

Perhaps the most thoroughly studied tumour, from the point of view of tumour antigens, is melanoma, a tumour of pigment cells (Chapter 1). Over 40 different antigens have been defined in these tumours by mono-clonal antibodies (Herlyn and Kaprowski 1988). None of these are truly tumour associated, though some are only weakly expressed on normal cells in the adult. At least six categories of antigen have been described (Table 15.4). Several of these are likely to influence the behaviour of the tumour cells. For example increased expression of growth factor receptors may lead to uncontrolled growth, while abnormal production or abnormal distribution of adhesion molecules will influence the ability of the cells to metastasize (Chapter 2). Antibodies to several of these molecules may be potential targets for therapy because of their high expression on the tumour and low expression on normal pigment or other cells.

Table 15.4 Melanoma antigens

Antigen	Comment
MHC antigens	Melanoma frequently expresses MHC class II as well as class I
Pigment antigens	Intracellular, sometimes of diagnostic value
Growth factor receptors	Receptors for EGF, TGFα, TGFβ, NGF, and FGF
Adhesion molecules	Chondroitin sulphate proteoglycan, melanoma associated adhesion molecule, gangliosides
Transport proteins	Melanotransferrin, S-100 calcium binding protein
Extracellular matrix proteins	Laminin and collagen

Another category of antigens which can appear tumour associated are those which are expressed in dividing cells only. Some monoclonal anti-bodies raised against tumour cell lines appeared initially to distinguish between tumour and normal cells. More extensive studies showed that these antibodies also detected an antigen present on normal cells when these were actively dividing. This antigen has now been shown to be the cell surface receptor for transferrin. Transferrin is a serum protein which transports iron required for cell division. The expression of the trans-ferrin receptor is therefore correlated with the cell cycle, not with malignancy. The proportion of cells in a population which express transferrin receptors may therefore indicate how many cells are actively dividing. By no means all cells in a tumour divide, but in general the greater the percentage of dividing cells the more rapid will be the growth of the tumour. In a study of lymphomas it was indeed found that the more rapidly progressive tumours did have a higher proportion of transferrin receptor positive cells. The transferrin receptor is therefore useful in indicating the prognosis in tumour patients.

While the transferrin receptor is widely distributed on actively dividing normal cells, other differentiation antigens which appear on dividing cells are restricted to certain cell types. One example of this is the receptor for interleukin 2. This is expressed on T lymphocytes (and some B lymphocytes) after these have been activated by contact with antigen. Some T cell tumours express this receptor and, as in the case of CALLA, for practical purposes it may be regarded as a tumour associated antigen because normal T cells have been replaced by the malignant population. This has encouraged attempts to treat some patients with antibodies to the receptor.

Well before the advent of monoclonal antibodies, conventional heteroantisera were used to define a rather different category of tumour associated antigens to which monoclonal antibodies have now been raised: these are differentiation antigens with a restricted tissue distribution and expressed during fetal life but not found in the adult (oncofetal antigens). Certain tumours, however, re-express these antigens. Two examples have been studied particularly extensively: carcinoembryonic antigen (CEA) and α-fetoprotein (AFP). The former is found in fetal intestine and in colonic tumours, whereas the latter is found in fetal liver and adult liver tumours. Both antigens can be detected in the serum of tumour patients and sensitive immunoassays for their detection have been developed, but neither has proved satisfactory for diagnosis. In the case of CEA, this is because it is difficult to establish normal and abnormal levels since, in contrast to earlier data, more recent results have shown that many different tumours produce CEA and that levels may also be raised in a variety of non-malignant conditions. In the case of AFP, raised levels are associated with liver cancer but also with other liver diseases and false negative results are sometimes obtained. Both CEA and AFP are probably most useful in monitoring the effects of treatment on CEA or AFP producing tumours. A fall in the serum level occurs following successful treatment and a rise may be detectable before clinical recurrence of tumour. Antibodies to both CEA and AFP have been used also in studies of tumour localization and immunotherapy (see Chapter 18).

It is notable that there are no solely tumour associated antigens, but studies of cellular oncogenes suggest that tumour cells do express altered proteins (Chapters 9 and 11) so that tumour associated antigens should exist. Why then are they not detected? Several explanations may be offered. First, if the genetic changes in a tumour are unique to that tumour the antigen would be tumour specific. While this might provoke a host response, the specific antibody in a polyclonal antiserum may be difficult to detect especially as it would only react with that particular tumour. Second, cell bound surface antigens may provoke a T cell rather than antibody response (see p. 411). Third, attempts to detect tumour antigens with heteroantisera or monoclonal antibodies, are likely to be difficult because of the large number of foreign proteins seen by the immunized rabbit or mouse. In spite of this there have been some reports of detection of human tumour specific antigens with monoclonal antibodies. These are of course difficult to confirm because the antibody reacts only with the immunizing tumour. Such antibodies are of use for diagnosis or therapy only in the original patient and may be most useful in studying genetic changes in tumour cells. In the foreseeable future the most useful antibodies are likely to be those against differentiation antigens with a restricted distribution which are also expressed on tumour cells. How these can be exploited is discussed in Chapter 18.

15.5 Cell mediated immune responses to tumours

15.5.1 *Cell transfer experiments*

The discovery that rejection of foreign tissue grafts was mediated by T lymphocytes stimulated attempts to identify the immune mechanisms responsible for protection against tumours in animals. The early experiments were performed *in vivo* by passive transfer; lymphocytes or serum from a donor animal immune to a tumour were transferred into a genetically identical but non-immune host which was then challenged with the same tumour. In such experiments, both cells and serum could sometimes transfer immunity. Antibody is most effective against leukaemia cells while lymphocytes could protect against solid tumours (usually carcinogen induced fibrosarcomas). The reason for this may be that leukaemia cells generally remain within the blood or lymph so that the cells are easily reached by antibody. In contrast carcinoma or sarcoma cells are extravascular and antibody may not always easily penetrate the tumour (see Chapter 18) while lymphocytes are able to leave blood vessels and enter the tumour tissue.

More detailed analysis of the mechanisms responsible for tumour protection by cells depended on technical advances. When separation of T and B lymphocytes became possible, these populations could be passively transferred before challenge with tumour and, in later experiments, a similar approach was used to study the roles of the T helper and T suppressor/cytotoxic subsets in tumour immunity. These methods are cumbersome and time consuming and it is difficult to assess the contribution of host non-immune cells to the anti-tumour response. More importantly, they cannot be applied to studies of tumour immunity in man so that further technical advance was required before progress could be made.

15.5.2 In vitro *cytotoxicity*

The first method which allowed the study of the effect of lymphocytes on tumour cells *in vitro* was the colony inhibition technique. In the original method, lymphocytes were mixed with tumour cells and seeded in culture dishes. After several days the lymphocytes can be washed out of the dish leaving adherent tumour cells behind. By this time each tumour cell originally seeded will have divided several times forming a colony. After appropriate staining these colonies can be counted with the naked eye. If lymphocytes from immune animals are used, there is a reduction in the number of colonies compared to controls. The reduction may reflect either death of some tumour cells or inhibition of growth. This test was the forerunner of others in which the lymphocytes and tumours were incubated for a shorter time (48 h) and surviving single tumour cells counted microscopically. A further development was to label the tumour cells with a radioisotope and measure cell death by the release of the radiolabel into the medium. These cytotoxicity assays could of course be applied to human cells provided that appropriate target tumour cells could be obtained. Tumour material is often available from cancer patients at operation and it is possible to disperse this into a suspension of single cells. Generally tumour cells obtained in this way will survive long enough in tissue culture for cytotoxicity assays to be performed. In addition, a proportion of tumours explanted *in vitro* can be grown into permanent cell lines. These lines may retain phenotypic characteristics of their cell of origin for long periods of culture.

The early data obtained from colony inhibition assays using tumour cells and lymphocytes from cancer patients showed that human tumour cells could be killed by autologous (self) lymphocytes. Surprisingly the apparent specificity of the cytotoxicity differed from that found in animals. In man lymphocytes from a patient with a lung tumour could kill all lung tumour cells but not colon, breast, or other target cells from other individuals. In animals in contrast the cytotoxicity was specific for the original tumour only. Subsequently much of the human data was invalidated by a study in which the lymphocytes from a large number of tumour patients and normal individuals were tested on a panel of target cells of varying origin. The results showed that all lymphocytes could kill most *in vitro* grown target cells. This type of cytotoxicity came to be called natural killer (NK) activity because prior immunization was not required (Takasugi *et al.* 1973).

15.5.3 *Natural killer cells*

Much effort has been expended in the identification of the cells responsible for NK activity, their relationship to other lymphocytes, their biological role, and the identification of the antigens recognized by them on target cells. Only the first question can be answered with any confidence. NK cells have a characteristic phenotype. They are larger lymphocytes than most T and B cells and have characteristic cytoplasmic granules (large granular lymphocytes). They share surface antigens with T lymphocytes and also with monocytes. There are also some surface molecules unique to NK cells. Just as their phenotype makes it difficult to determine their origin so is it difficult to assign these cells a definite function though various possibilities have been suggested; for example, that they regulate haemopoiesis or are an early non-specific response system for combating viral infections. Because they can kill many tumour cell lines, NK cells have also been suggested to play a role in surveillance against tumours.

Irrespective of their exact function, NK cells have made it very difficult to study specific immune responses to human tumours using cytotoxicity assays. A further problem not at first appreciated is posed by the necessity for antigens to be seen by T cells in association with self MHC antigen (genetic restriction). Even if tumour specific immune T lymphocytes are present in a patient, they would be expected to kill only the patient's own tumour or another tumour carrying the same tumour antigen and the correct MHC antigen. NK cells in contrast show no genetic restriction. It is therefore very difficult to perform adequately controlled experiments to reveal specific T cell immunity, especially as this is likely to be a weak effect since in any patient with a growing tumour the immune response must have been overwhelmed or suppressed. To overcome these technical problems and amplify a weak specific response, tumour immunologists have attempted to use the in-vitro boosting method developed for studying T lymphocyte responses to non-tumour antigens.

15.5.4 In vitro *boosting of T cell immunity*

In both experimental animals and man it has proved possible to restimulate immune T lymphocytes *in vitro* with foreign MHC antigens (in a mixed lymphocyte culture) or antigens such as tetanus toxoid or influenza virus. The stimulated T cells proliferate and express receptors for the growth stimulatory factor IL2. Long term growth of the cells is possible if the T cells are alternately exposed to their specific antigen and more IL2. In principle very large numbers of antigen specific T cells can be grown in this way from a single cell. Cloned T cell populations have been used in studies in T lymphocyte function, to examine how T cells interact with monocytes and B lymphocytes, the nature and number of lymphokines produced by different types of T cells, the role of T cell surface molecules in T cell function, and the specificity of T cell responses to antigen. Therefore T cell clones which respond to tumours can be used to determine the nature of antigens in a tumour which can be recognized by the host T cells. In practice the technical problems are formidable. T lymphocytes from all donors do not seem to grow equally well and the amount of antigen for stimulation and assay of the T cells may be limited, unless the tumour cells can be grown as a permanent line. Although it is possible to stimulate a patient's T cells and then grow these as a polyclonal population, it has often proved difficult to obtain clones. Nevertheless some information on the specificity of tumour reactive T cells has been obtained.

In the earliest studies in man intriguing results were obtained. Two sorts of T lymphocyte lines were grown. Cytotoxic lines were able to kill autologous tumour cells but not others of the same type while T cells which proliferated in response to tumour (presumed to be helper/inducer cells) could respond to autologous and other tumours of the same type. These results suggested that the proliferating T cells might be recognizing tumour antigens processed and displayed in association with MHC antigen by accessory cells (monocytes). Cytotoxic T cells in contrast see only autologous tumour because only this has the correct tumour and MHC antigens (Fig. 15.4). In these early experiments the T cells were not cloned so that the dominant specificity in the population would be observed. Results with T cell clones have revealed further complexity.

When a set of clones which could proliferate a response to a lung tumour extract was examined, it was found that individual clones reacted in a variety of ways. They might react to autologous tumour extract only, all lung tumour extracts, extracts of any dividing cell, all autologous extracts, or extracts of tumour but not normal tissue. While the antigens detected by these clones have not been identified in molecular terms, the results suggest that part of the T cell response to a tumour may be against tumour specific or tumour associated antigens but much of the response is to normal tissue components. The tumour response may thus be partly an autoimmune response, perhaps induced by the breakdown of tissue caused by tumour growth (Moss *et al.* 1984).

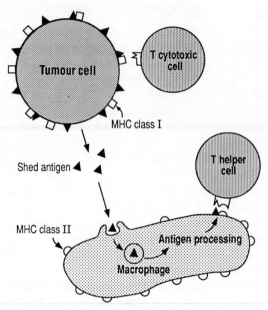

Fig. 15.4 Presentation of tumour antigen to T lymphocytes. Tumour antigen may be displayed in association with MHC Class I antigen (□) on the cell surface so that it is recognized directly by cytotoxic T cells or processed by macrophages and displayed with MHC Class II antigen (○) to stimulate T helper cells.

A rather different result was obtained when cytotoxic clones reacting to melanoma cells were analysed. Some of these seemed to kill all melanoma targets but not most other cells. The target structure for these T cells has been partially characterized as a highly glycosylated molecule on the surface of melanoma cells. Whether T cells recognize this target using the same type of receptor that normally sees antigens in association with self MHC antigens is not clear, but it is intriguing that the specificity of these killer T cells is similar to that observed in the earliest experiments in humans using the colony inhibition assay.

It is clear that the T cell response to tumours is complex in terms of the variety of molecules recognized on the tumour cells and the different types of T cells activated during the response. As techniques improve it is likely that many more tumour reactive clones will be obtained and that these will increasingly be used to identify molecules in tumours to which the host can respond. Some of these molecules may be sufficiently restricted in their tissue distribution to be useful in diagnosis or therapy.

15.6 Immunodiagnosis and immunotherapy

15.6.1 Immunodiagnosis

Ideal reagents for immunodiagnosis or immunotherapy would discriminate absolutely between tumour and normal cells. In addition they would distinguish between benign and malignant tumours. However, as discussed earlier, most if not all antibodies raised against tumours identify differentiation antigens. Nevertheless, antibodies can be useful in cancer diagnosis just because they identify differentiation antigens and thus the origin of a cell. Panels of monoclonal antibodies are finding a role in pathology and haematology laboratories where they are used to improve the identification and classification of tumours. For example in the diagnosis of acute lymphoblastic leukaemia (see Chapter 12) antibodies have allowed clear distinctions to be made between T and B cell forms of the disease which have very different prognoses with conventional chemotherapy. Identification of bad prognosis patients is important because it is sensible to try new forms of therapy in individuals with little chance of survival with current treatment.

Monoclonal antibodies are also useful in identifying the origin of tumour cells when this is difficult by conventional histological methods; antibodies to cytoskeletal proteins (cytokeratins), epithelial membrane antigens, and leukocyte antigens can usually identify the origin of metastatic tumour cells even if these are cytologically undifferentiated. This has important implications for the management of the patient because secondary carcinoma is often chemotherapy resistant while lymphoid tumours are often sensitive.

Monoclonal antibodies are also being used in attempts to localize tumours *in vitro* (see Chapter 18).

15.6.2 Immunotherapy

Immunotherapy is treatment by immunological means. In active immunotherapy the tumour bearer's own immune system is stimulated to respond to the tumour while, in passive immunotherapy, immune cells or their products are given. Table 15.5 summarizes the possibilities. The aim of treatment in cancer is to eliminate the tumour without harming the host. Because immune responses are highly specific, immunologists have long hoped that immune cells or antibodies might be used in this way. Unfortunately as discussed earlier, few cells or antibodies are truly tumour specific so that some side effects on normal cells must be expected. Nevertheless many experiments have been carried out. Since the use of monoclonal antibodies is discussed in detail in Chapter 18 only other means will be considered here.

Table 15.5 Immunotherapy

Approach	Method	Agent
Non-specific		
Local, active	Intra-tumour injection	BCG[1], viruses
Systemic, active	Immunostimulants	BCG, MER[1], C. parvum[1], levamisole
Systemic, passive	Mediators or lymphokines	Thymic factors, interleukins, interferons
Specific		
Systemic, active	Immunization	Tumour cells or extract + adjuvant
Systemic, passive	Specific factors	Transfer factor
	Serotherapy	Polyclonal or monoclonal antibodies, coupled to drugs radioisotopes, or toxins
	Cells	In-vitro grown specific T cells, LAK cells
Ex vivo	Bone marrow purging	Antibodies + complement or antibodies coupled to magnetic beads or toxins

[1] BCG, Bacille Calmette-Guerin, a non-pathogenic strain of tubercle bacillus; MER, the methanol extraction residue of BCG; C. parvum, Corynebacterium parvum. BCG, MER, and C. parvum all have adjuvant activity. LAK, Lymphokine Activated Killer.

Attempts to treat human tumours by active immunization (Table 15.5) with tumour, or the administration of immunostimulating agents, have met with little success. It could be argued that this is because the presence of a growing tumour blocks immune effector mechanisms (see Fig. 15.3). This may be untrue because even when immunization has been tried after conventional treatment to reduce tumour load, it has still not been successful. A more likely explanation for failure of active immunization is that there are few host lymphocytes capable of responding to tumour antigens. Because the results of active immunization have been disappointing and yet some tumour responding T cells can be detected *in vitro*, experimenters have tried to expand these populations for therapeutic use. What then are the possibilities and difficulties of this approach?

First, the technical difficulties of isolating and growing large numbers of tumour specific T cells are formidable. *In vitro* responses to tumour cells or extracts are weak, presumably reflecting a low frequency of responding T cells, and while a few groups have succeeded in isolating tumour reactive clones, it has proved extremely difficult in most cases to grow large numbers of cells. Unfortunately the factors governing the extent of growth *in vitro* of T cell clones are not yet understood but may include the genetic makeup of the donor as well as the type and state of differentiation of the responding T cells. Other problems remain, not the least being the possibility that the *in vitro* grown T cells may not migrate normally *in vivo* and thus the majority may fail to reach a tumour site. Data concerning the *in vivo* function of T cell clones in experimental animals are conflicting but at least in some cases protection against tumour challenge has been observed. Much more experimental data on the most effective type of T cell, the most effective way to use them, and how best to induce and grow tumour immune cells needs to be gathered before this type of therapy becomes a practical proposition. It is also likely to be extremely expensive!

When lymphocytes from tumour patients (or normal individuals) are cultured with high doses of the lymphokine IL2, effector cells are produced which are able to kill many different tumour targets *in vitro*. In recent experiments these lymphokine activated killer (LAK) cells have been reinfused into tumour patients who are also given IL2. Although some tumours appear to regress, the efficacy of this form of treatment remains to be established (Rosenberg and Lotze 1986).

Although the use of immune cells for immunotherapy has not progressed very far, their secreted products (lymphokines) are now becoming available. The techniques of molecular biology have made it possible to produce sufficient quantities for *in vivo* studies of these substances which are normally secreted in minute quantities during immune responses. So far the most extensively studied lymphokines are the interferons, a family of glycoproteins which were first identified because they inhibit the replication of viruses. Interferons also inhibit cell division and stimulate NK activity. All of these properties suggested that they might have therapeutic effects on tumours. A number of clinical trials on different types of tumour have been carried out and it is clear that interferons are not strikingly effective anti-tumour agents for treatment of most tumours. Some effects have been documented for certain rare tumours (e.g. hairy cell leukaemia), and in experimental models; however, it may be that further research will better define when interferons can be useful and how they should be combined with other treatments. Less data on the anti-tumour effects of other lymphokines are available but IL2, tumour necrosis factor, and lymphotoxin are being tested in clinical trials. As in the case of interferon, it seems likely that these agents may be effective against particular tumour types rather than all tumours. All of these agents suffer from the problem of non-specificity, that is, they are equally likely to have effects on normal as well as malignant cells.

15.7 Conclusions

The advent of the hybridoma technique for production of monoclonal antibodies has provided reagents to identify and purify molecules present in lymphocytes, other normal cells, or tumour cells. The techniques of molecular biology make it possible to isolate the genes coding for these molecules, to sequence them, and, if required, to produce the molecule *in vitro*. These are powerful techniques for investigating the function of the immune system and for attempting to define how tumour cells differ from normal cells. The immune system also provides a model for studying the growth and differentiation of cells. Lymphocytes can be readily obtained even from humans, and may be cultured *in vitro*. In the long term it is likely that immunologists may contribute more to cancer research by providing research tools and insights into cell function than by experiments on immune responses to tumours, which may well turn out to be illusory.

In the more immediate future the use of monoclonal antibodies in immunodiagnosis and immunotherapy is likely to expand. For therapy, human monoclonal antibodies would be advantageous since rodent antibodies provoke an antibody response to the foreign protein which limits the duration of treatment. So far it has proved difficult to produce human antibodies of a desired specificity with any regularity and in large quantities. Genetically engineered hybrid antibodies, part mouse part human, may be an alternative solution.

The next few years are likely to see many studies of the effect of purified lymphokines. While it seems unlikely that most of these will provide 'magic bullets' which will be tumour specific, these agents do have powerful biological effects and an understanding of these will contribute to elucidating the mechanism of regulation of growth and differentiation in cells.

I have left to the end an area in which immunology may well contribute to cancer treatment, or rather prevention. It is clear that a number of viruses play a role in the induction of tumours. These include hepatitis B virus in liver cancer, EBV in Burkitt's lymphoma and nasopharyngeal cancer, papilloma viruses in genital tumours, and HTLV-1 and 2 in some lymphoid tumours (see Chapter 8). Prophylactic immunization against these and perhaps other as yet undiscovered agents is likely to prevent or reduce the incidence of these tumours. Already hepatitis B vaccine is available and under trial; much effort is also being expended on a vaccine to EBV. As is the case with infectious disease, prophylactic immunization rather than treatment may be the immunologists most direct contribution to the reduction of cancer mortality.

References and further reading

Bjorkman, P. J. *et al.* (1987). Structure of the human histocompatibility antigen, HLA-A2. *Nature* **329**, 506–12.

Davis, M. M. and Bjorkman, P. J. (1988). T-cell antigen receptor genes and T-cell recognition. *Nature* **334**, 395–402.

Ghosh, B. C. and Ghosh, L. (1987). *Tumour markers and tumour associated antigens.* McGraw-Hill, New York.

Herlyn, M. and Koprowski, H. (1988). Melanoma antigens: immunological and biological characterisation and clinical significance. *Annual Reviews of Immunology* **6**, 283–308.

Hood, L., Weissman, I., Wood, W., and Wilson, J. H. (1984). *Immunology* (2nd edn). Benjamin/Cummings, Menlo Park, California.

Kawano, M. *et al.* (1988). Autocrine generation and requirements of BSF-2/IL-6 for human multiple myelomas. *Nature* **332**, 83–5.

Kocks, C. and Rajewsky, K. (1989). Stable expression and somatic hypermutation of antibody V regions in B cell differentiation. *Annual Review of Immunology* **7**, 537–60.

McMichael, A. J., and Fabre, J. (1982). *Monoclonal antibodies in clinical medicine.* Academic Press, London.

Moss, F. M., Lamb, J., and Souhami, R. L. (1984). Heterogeneity of specificity of oligoclonal human T lymphocyte lines induced with autologous pulmonary tumour. *British Journal of Cancer* **49**, 659–61.

Rosenberg, S. A. and Lotze, M. T. (1986). Cancer immunotherapy using interleukin-2 and interleukin-2-activated lymphocytes. *Annual Review of Immunology* **4**, 681–710.

Sell, S. and Reisfield, R. (1985). *Monoclonal antibodies in cancer.* Humana Press, Clifton, New Jersey.

Sell, S. (1987). *Immunology, immunopathology and immunity.* (4th edn). Elsevier, New York.

Sell, S. (1988). *Basic Immunology: Immune mechanisms in health and disease.* Elsevier, New York.

Srivastava, P. K. and Old, L. J. (1988). Individually distinct transplantation antigens of chemically induced mouse tumours. *Immunology Today* **9**, 74–83.

Takasugi, M., Mickey, M. R., and Terasaki, P. I. (1973). Reactivity of lymphocytes from normal persons on cultured tumour cells. *Cancer Research* **33**, 2898–902.

Waterfield, M. D. (ed.) (1989). Growth factors. *British Medical Bulletin.* **45**, 541–53.

16

The local treatment of cancer

I. S. Fentiman

16.1 How tumours present

Almost all patients with cancer seek treatment because of abnormalities that arise as a direct or indirect result of the malignant process, the most common symptoms being a consequence of the infiltrating growth pattern of the tumours. Invasion of surrounding connective tissue by cancer cells elicits a stromal reaction resulting in fibrous tissue formation. This dense scar tissue is responsible for the hard irregular character of cancer in organs such as the thyroid and breast. Contraction of fibrous tissue in tubular organs like the oesophagus or colon produces strictures which impede or obstruct the flow of normal contents. Direct infiltration of blood vessels by malignant cells may give rise to local haemorrhage so that the patient with lung cancer may expectorate blood or in the case of a bladder carcinoma may pass bloodstained urine. It must be stressed that these symptoms frequently arise from non-malignant causes. Nevertheless, they act as a signal to the clinician that further investigation is necessary. Indeed, rather non-specific symptoms may sometimes lead to the early diagnosis of malignancy.

Our knowledge of cancer biology suggests that by the time a cancer becomes detectable (at say a volume of 1 ml) it contains about 1×10^9 cells, and there is a high probability that tumour metastasis has occurred. To increase the chance of cure, it is necessary to diagnose tumours earlier or, better still, to diagnose tumours in which spread through the basement membrane has not occurred e.g. *in situ* carcinomas (see Chapter 1).

16.2 Screening

For some tumours, screening seems to be of value since it aims to detect *in situ* and asymptomatic tumours, particularly in the uterine cervix and breast. Screening for carcinoma of the cervix by examination of cytological smears has been in use for several decades and decreases in mortality in countries with extensive screen programmes seem to be well established (Knox and Woodman 1988). Several studies in various countries have demonstrated that breast screening saves lives in women aged over 50. Approximately one-third of the cancers diagnosed by mammographic screening are ductal carcinomas *in situ* (DCIS), a malignancy for which there is a high probability of cure. Available data have been reviewed by the Forrest Committee (1986) who have recommended the establishment in Britain of a national screening service for women aged between 50 and 65.

For other common cancers the evidence that screening works is less compelling but positive results have been reported in other countries. Attempts to screen populations for colonic carcinoma by testing for traces of blood from the tumour in the faeces (faecal occult blood) have not been successful; national changes in mortality rates for colonic carcinoma have not yet been demonstrated. At present, the technique may be of value when applied to patients with familial colonic disease, such as polyposis coli (familial adenomatous polyposis, see Chapter 4), or first degree relatives of known cases and in the follow-up of patients with treated colonic carcinoma.

Other screening methods depend on the detection of tumour markers in the blood or other body fluids. Unfortunately no tumour specific substances have yet been isolated (see Chapters 12, 15, and 18) but the presence of abnormal tumour associated proteins or normal tissue specific proteins not usually present in the blood may be detected. Abnormal amounts of hormones in the blood may also suggest the presence of tumours of the endocrine system (see Chapters 13 and 14). None of these markers is sensitive enough or sufficiently specific to be used for the detection of cancers in patients without symptoms of disease but in some cases, particularly in prostate, ovarian, and testicular cancers, they are useful in predicting prognosis and monitoring the response to treatment.

16.3 Diagnosis of cancer

Superficially located cancers of the tongue, skin, breast, and thyroid may be detected by inspection alone or by palpation. Other internal cancers such as those arising from pelvic organs (e.g. prostate, rectum, uterus, and ovary) may be found by internal digital examination. Other internal tumours may be detected by using special instruments (endoscopes) which allow direct optical examination and removal of pieces of tumour or doubtful tissue for microscopic examination. More deeply located tumours may require a variety of imaging techniques before they can be demonstrated. Some of these are described below but see Britton (1987) for a more detailed review.

16.3.1 Radiography

Straightforward X-rays may be very important in suggesting a diagnosis of cancer in bone or lung because of the differences in X-ray densities between normal and malignant tissue in these sites. To assess the extent of the disease accurately it is necessary to obtain additional information from a computerized tomography (CT) scan (see p. 437). Plain X-rays of the breast (mammography) are of use in screening in women over 50. Younger women have such dense breast tissue that signs of malignancy, such as an irregular mass possibly associated with microcalcification, may be undetectable. Many cancers of the upper and lower gastro-intestinal tract may be delineated by radio-opaque barium meals or enemas, tumours being seen as strictures, ulcers, or filling defects (regions from which the radio-opaque material is excluded).

Demonstration of certain cancers requires the administration of contrast medium through a needle (cannula) either intravenously or intra-arterially. A bladder carcinoma may be seen after an intravenous urogram in which a special radio-opaque dye is injected into a vein. The dye is concentrated in the kidneys and excreted in the urine; sequential radiographs of kidney and bladder are taken. For direct visualization of a renal carcinoma, injection of the material directly into an artery (arteriogram) may be necessary. These techniques involve not only radiation exposure but also the risk of allergic reaction to contrast media and haemorrhage after arterial cannulation. For these reasons a variety of other imaging techniques have been developed.

16.3.2 Computerized tomography (CT) scanning

This imaging technique has profoundly altered the accuracy of monitoring patients with cancer. It is still expensive and involves a relatively high radiation exposure. Nevertheless, using CT scans, serial 'cuts' can be taken through the area of interest. This is particularly useful in patients with apparently isolated lung tumours which CT scanning may often reveal to be multifocal and more extensive than suspected on routine chest radiography. In this way, inappropriate surgery may be avoided. Also, for patients with dubious lesions in bone, particularly the vertebrae, CT scanning enables an accurate delineation of the lesion and the exclusion or confirmation of maliganancy. In patients treated by a combination of surgery and radiation sometimes it can be impossible to decide clinically whether symptoms are due to the late results of therapy or to recurrent disease. CT scanning may demonstrate recurrent carcinoma which would be otherwise undetectable. CT scanning can be used not only for assessment by also to monitor the response to treatment.

16.3.3 Ultrasound scanning

Ultrasonic scanning is a very useful technique which does not involve any ionizing radiation and therefore can be repeated frequently and can detect lesions as small as 5 mm in diameter. It is of particular value in imaging the liver where it can demonstrate small metastases or the presence of stones within the bile ducts. Another use is in the screening for and evaluation of gynaecological malignancies, particularly tumours of the ovary. Ultrasound enables solid tumours to be distinguished from cystic lesions and has been used for the investigation of breast masses in young women in whom X-ray mammography is of little value.

16.3.4 Magnetic resonance imaging (MRI)

This new imaging technique is still under clinical investigation. The patient is placed in a high magnetic field and a short radiofrequency pulse is applied. Subsequently a signal is emitted which can be transformed into images (Pykett *et al.* 1982). The signals from tumours differ from those of normal tissues and can be characterized by two constants, the longitudinal relaxation and transverse relaxation times. So far it has been found to be of particular value in the imaging of midline structures such as the brain, spine, pelvis, and the contents of the neck. As further development occurs, this may prove to be a valuable monitoring technique because of lack of radiation exposure.

16.3.5 *Cellular diagnosis*

It must be stressed that although cancer may be strongly suspected on clinical and radiological grounds, the diagnosis cannot be confirmed without cytological or histological evidence. Unless pathological proof of malignancy has been obtained it is not possible to plan rational treatments, nor to compare various therapeutic measures, nor to give a prognosis of any value. Even the most experienced clinicians may sometimes be misled by inflammatory lesions that mimic malignancy. Only under the most exceptional circumstances should a patient have treatment for cancer without pathological confirmation of the diagnosis.

16.3.6 *Cytology*

For the interpretation of cytological specimens, an experienced cytologist is required; there is at present a nationwide shortage in the UK. Cells from the suspected or potentially malignant site are removed, smeared on a slide, and then examined by the cytologist. Malignant cells may be recognized by alteration in cell size and shape and changes in nuclear morphology.

Exfoliative cytology. This is the examination of shed cells and is particularly applied to the uterine cervix. Cells are removed by a special spatula and then smeared, fixed, and examined. Other uses are the monitoring of urine from patients at high risk of urinary tract malignancy such as aniline dye workers, and of sputum in smokers. In addition, serous fluid from patients with undiagnosed pleural effusions or ascites may be examined after centrifugation and staining. The identification of malignant cells is an indication for further investigation.

Fine needle aspiration (FNA) cytology. In this technique, cells are aspirated from the tumour site through a fine needle attached to a syringe. This as been particularly employed for the investigation of lumps in breast and thyroid. In the former case, when malignancy is found, it is also possible to determine the hormone sensitivity of the tumour cells by an immunoperoxidase-linked monoclonal antibody stain (Magdelenat *et al.* 1987). FNA is becoming increasingly used because of its simplicity although the evaluation sometimes may be very difficult.

Occasionally, false positive cases are encountered. In impalpable breast lesions detected by mammography, FNA may enable a diagnosis of malignancy to be made after radiographic direction of the needle. A drawback to FNA is that it is not possible to distinguish between in-situ and invasive carcinoma cells because the topography of the aspirated cells cannot be established always.

16.3.7 *Histopathology*

Histology remains the cornerstone for the diagnosis of malignancy. Given the right material a pathologist can distinguish between in-situ and invasive carcinoma and provide information on the differentiation of the cancer (tumour grade) and also on the extent of spread of the disease (tumour stage). Whether the tumour extends to the margins of excision may be particularly important where less extensive surgery is being used together with irradiation. Tissue is obtained either as a sample of the tumour (incision biopsy) or by complete removal of the tumour (excision biopsy). The method of processing the tissue will depend upon the urgency with which a diagnosis is required As part of a planned major procedure an incision biopsy may be taken and the tissue immediately frozen in solid carbon dioxide to allow rapid sectioning by microtome after which the sections are stained and examined (frozen section). Using this frozen section technique a definite diagnosis of malignancy can be obtained within 10 min of biopsy and the proposed operation can be performed secure in the knowledge that the patient does have confirmed cancer. However, the use of frozen section does carry certain disadvantages. The rapid processing of the specimen may provide a suboptimal tissue section and subtle changes that distinguish malignancy from non-malignancy may be missed resulting in a mistaken diagnosis of carcinoma. Fortunately this is rare.

In the laboratory the tissue is fixed, usually with formalin. This procedure takes approximately 24 h after which the tissue can be dehydrated, embedded in paraffin wax, and then sliced with a microtome. The wax is removed and the slide usually stained with haematoxylin and eosin (H&E stain). The entire procedure takes approximately 48 h but does provide much better material for pathological examination. Furthermore, the examination of multiple specimens allows determination of the extent of the tumour and assessment of lymphatic and blood vessel invasion. The delay involved in waiting for the results of a paraffin section has no effect on the eventual outcome of treatment. Most important from the patient's point of view is the uncertainty of the outcome. Thus a woman with a breast lump will not be sure whether she will recover from the anaesthetic with two breasts or one. It is becoming more evident to surgeons that time is required for patients to come to terms with the diagnosis of malignancy and the potential treatment options. Pressurizing a patient into a decision on treatment may prejudice their psychological response to the diagnosis and subsequent recovery from the procedure. Thus, when possible, it is better for a full pathological investigation of the biopsy specimen to be carried out. This also gives an opportunity for patients with suspected malignancy to discuss and come to terms with their possible treatment options.

New approaches. Until recently the pathological evaluation of biopsy specimens has been based largely on morphology and topography, but functionally important changes may be present without apparent morphological alteration. The use of markers to characterize cell products has become more frequent, largely as a result of the development and availability of a panel of monoclonal antibodies (see Chapters 15 and 18). It is now possible to separate morphologically similar cell types into discrete groups with possible implications for both therapy and prognosis. Under certain circumstances, where the cells are poorly differentiated, the use of monoclonal markers may distinguish between cells of similar appearance and enable the pathologist to determine whether the tumour is a carcinoma, a melanoma, a sarcoma, or a lymphoma. This may be particularly important in assessment of lymph nodes excised from axilla or groin where the primary site is unknown.

Another recent area of collaboration between the pathologist and the basic scientist is the use of flow cytometry. Cytofluorimetry enables the determination of both ploidy and S-phase fraction of the tumour cells which again may distinguish between morphologically similar tumours. Early reports suggest that tumours with a normal 'diploid' quantity of DNA carry a better prognosis than those which are aneuploid. Such investigations can be carried out not only on freshly obtained specimens but also on archival material enabling the retrospective evaluation of the influence of ploidy on prognosis (Merkel *et al.* 1987).

16.4 Staging

Staging attempts to determine the extent of spread of a particular tumour, both to select the most appropriate therapy and also for prognostic purposes. A rather crude system of staging forms part of the clinical examination of the patient with suspected or proven malignancy. More accurate evaluation is obtained with laboratory or imaging investigations although these techniques still have insufficient sensitivity to detect most micrometastases.

16.4.1 *Clinical staging*

The accuracy of clinical staging will depend upon a variety of factors including the anatomical location of the primary lesion, the physical characteristics of the individual patient, and the clinical experience of the examiner. Even under the most favourable circumstances the disease will often be either under- or overstaged. As an example, the major lymph nodes draining the breast are located in the axilla. Palpation of the axilla yields false negative results in 30 per cent of patients and false positive results in another 30 per cent. Testicular tumours drain to the para-aortic lymph nodes which are impalpable unless grossly enlarged so that clinical examination frequently results in understaging.

Despite these drawbacks, clinical staging remains the most widely used system since it does enable international comparisons of cases and treatments under circumstances where sophisticated staging techniques may be unavailable. Several staging systems have been devised for different tumour types but the main international classification is the TNM system (Table 16.1). This comprises an assessment of the tumour (T), nodes (N), and metastases (M) (Table 16.1) and can be very convoluted but nevertheless provides a framework of international comparisons ranging from the very simple to the exceedingly complex.

Table 16.1 TNM staging

Stage	Clinical manifestations
I	Small localized tumour
II	Spread to local lymph nodes
III	Large local tumour and/or spread to further lymph nodes
IV	Presence of distant metastases

As a broad outline, Stage 1 and 2 tumours are deemed operable and are usually treated with surgery and/or radiotherapy. Stage 3 cancers are technically inoperable and the major treatment is radiotherapy, often combined with subsequent surgery together with systemic treatment. Patients with stage 4 disease are incurable by local techniques and some form of systemic therapy is usually employed as first line treatment although either surgery or radiotherapy may be used to deal with specific local or metastatic complications.

16.4.2 *Other staging investigations*

The selection of the most appropriate staging investigation for a particular tumour will depend upon the primary site together with a knowledge of the most likely pattern of spread of metastases which is due in part to the topographical anatomy of the organ involved. Gastro-intestinal cancers metastasize early to local lymph nodes followed by venous embolism into the hepatic portal system resulting in the development of liver metastases. In contrast, breast cancer rarely metastasizes to the liver as an early event and is more likely, because of blood-borne spread, to produce metastases in bone or lung. Primary lung cancers may invade the pulmonary vein and thereby frequently metastasize to the brain. Approximately 50 per cent of brain tumours are metastatic, of which the commonest primary site is lung.

The simplest staging investigation is a blood test to determine the haemoglobin concentration and the white cell count. This is carried out to exclude the presence of bone marrow metastases which may manifest as a leukoerythroblastic anaemia. Metastatic tumour cells may also be detected in bone marrow aspirates and for tumours such as lymphomas this is an essential part of the staging. Another relatively simple test is the measurement of blood electrolytes together with bone and liver derived enzymes (biochemical screen) which may be elevated in response to metastatic disease. Finally, a chest X-ray is usually taken to exclude gross disease of the lung, pleura, and ribs. In patients with cancers of the lung, thyroid, prostate, breast, and kidney, which have a predilection to metastasize to bone, it was common practice to carry out a radiographic skeletal survey, in particular focusing on pelvis, vertebrae, and femora. However, this involved a fairly high radiation exposure with a limited pick-up rate and has now been largely superceded by radio-isotopic scans, which are often more sensitive.

16.4.3 *Radioisotopic scans (scintiscans)*

These scans are images derived from gamma ray emission of radioactive isotopes tagged to compounds which are taken up selectively by particular organs. Using a gamma camera, a scintiscan can be obtained of the organ of interest. The presence of either filling defects or areas of increased uptake may suggest the presence of metastatic malignancy. Bone scans of patients with metastases usually show areas of increased uptake (hot spots) although this is a non-specific finding whenever there is bone turnover and thus X-rays of the particular abnormalities may be necessary to determine whether the hot spots are a result of old trauma, degenerative disease or truly do represent metastases. Isotopic scans are particularly useful in the investigation of patients with suspected thyroid tumours. Radionuclide scans can be carried out on the lung, liver, and brain but lung and brain lesions are usually better demonstrated by CT scanning; ultrasound is probably the most effective method of excluding metastatic hepatic malignancy.

16.4.4 *Lymphangiography*

To visualize the lymph nodes draining certain tumours, a lymphatic vessel is cannulated and an iodine-containing contrast medium injected. After this technically difficult procedure, tumour deposits may be recognized as filling defects. This is particularly suitable for patients with tumour of the testis, prostate, or bladder in whom cannulation is made in the foot.

Lymphangiography has also been used to demonstrate internal mammary lymph nodes in patients with breast carcinoma and used too for the accurate planning of radiotherapy fields. However, for the common solid tumours, present techniques do not provide good visualization of tumour deposits in lymph nodes, and research continues to find monoclonal antibodies to assist in lymphoscintigraphy. In patients with carcinoma of the breast this could be particularly useful because those with verified absence of axillary lymph node metastases might be spared an axillary clearance which is at present the only accurate method of determining whether the nodes contain metastases.

16.4.5 *Operative staging*

Although some tumours can be accurately staged preoperatively, for the majority of patients with cancer this information is only available when the pathologist has examined the surgical specimen. It can then be accurately determined whether the cancer has penetrated the wall of a tubular organ such as the oesophagus, stomach, colon, or rectum. Furthermore, because the operative procedure will have cleared surrounding tissue, the presence or absence of lymph node metastases can be determined together with information on lymphatic or vascular invasion. This then enables a more accurate prognosis to be given. An example is carcinoma of the rectum where the Dukes staging system (Table 16.2) provides the most accurate measure of prognosis at present. Patients with Dukes stage A disease have cancer which does not infiltrate through the rectal wall, and 80 per cent of these will be alive at 5 years.

Table 16.2 Dukes staging system for cancer of the rectum

Stage	Description	Proportion of patients (%)	Crude 5-year survival (%)
A	Cancer not infiltrating through rectal wall	15	80
B	Tumour spread through rectal wall	35	60
C	Spread to local lymph	50	25

Of those patients with Dukes stage B, with tumour spread through the rectal wall, only 60 per cent will be alive at 5 years. Those with stage C, with tumour spread to local lymph nodes, have a poor survival of only 25 per cent at 5 years (Table 16.2).

16.5 Treatment

In discussion of the local treatment of cancer it is important to stress that many patients are cured by local therapy. It is most important that this potential for cure is not diminished by inappropriate attempts to carry out less than adequate surgery or radiotherapy in the hope that the situation may be salvaged with the use of systemic therapy. Nevertheless, since it is not always possible to exclude distant metastases there is an increasing shift towards combined treatment by surgery and radiotherapy, frequently accompanied by systemic therapy using either cytotoxic drugs or endocrine agents.

16.5.1 *Which cancers are curable?*

There are at present two major groups of cancers which are almost invariably curable. The first comprises patients with *in situ* cancers. As one example, patients with familial polyposis who invariably develop colonic carcinomas are curable by total colectomy. Another example is ductal carcinoma *in situ* of the breast which is detected as microcalcification on mammography and will be increasingly diagnosed as screening becomes more readily available. Until recently, total mastectomy was regarded as the treatment of choice for such cancers. With proof of the safety of breast conservation for infiltrating carcinomas there is increasing pressure for similar techniques to be used for mammographically detected in-situ cancers and this is the subject of continuing clinical trials (Fentiman 1988).

Other potentially curable cancers are those which are locally invasive and yet do not metastasize. The commonest example of this is the basal cell carcinoma, or rodent ulcer, of the skin, which typically involves the facial area and is curable by either excisional surgery or radiotherapy. Provided that rodent ulcers are widely excised or adequately irradiated then cure is possible. Another example of a tumour with locally aggressive behaviour but minimal metastatic potential is the fibroma which can involve any organ. Wide excision with confirmation of clear boundaries is necessary to cure this tumour. Failure to achieve local clearance or destruction of the tumour results in an inexorable ulcerative process producing surrounding tissue destruction. As many of these lesions are located on the face, a very unpleasant disfigurement occurs after suboptimal treatment.

Sacral chordoma, which is a rare tumour, is another example of a locally invasive cancer which does not usually metastasize. Unfortunately because of its pelvic location and propensity to invade surrounding nerves and viscera, the tumour may be only curable by very extensive surgery, the Procrustean measure of hemicorporectomy. Although this might result in a clinical cure the end result will daunt the majority of patients. This extreme case does illustrate one of the most difficult aspects of cancer surgery. What is the patient prepared to accept in order to have a high probability of cancer cure? Thus, Dukes Stage A rectal cancers may be cured by excision of the rectum and anus leaving the patient with a permanent colostomy, but for some patients this is unacceptable and so a non-curative but anal sphincter conserving procedure may be performed. Although there are patients who can be predictively cured, in the majority of solid tumours our present pathological and biochemical methods are insufficiently precise to determine those that will be cured by local treatment.

For the majority of solid tumours, those that are histologically better differentiated are more likely to be cured by local therapy but nevertheless our present techniques of assessment of prognosis are insufficiently accurate for a good individual prognosis to be given with absolute certainty.

16.5.2 En bloc resection

This remains the fundamental principle of much cancer surgery. The aim of en bloc resection is to remove the entire tumour with its draining lymphatics and lymph nodes. In this way it is hoped that the entire tumour bearing field can be removed. The underlying assumption is of a progression from local infiltration into local lymphatics with subsequent retention of tumour cells in lymph nodes and eventual distant metastasis and blood-borne spread. While this does sometimes occur, unfortunately vascular invasion may be an early event so that the presence of metastases in lymph nodes may be a marker of metastatic disease elsewhere rather than just the local extent of tumour spread.

To achieve an en bloc resection for certain gastrointestinal cancers requires technically difficult reconstructive surgery with formation of anastomoses which can be associated with the risk of leakage, haemorrhage, and death. In experienced hands, the mortality rate after colonic surgery can be reduced to very low levels. In contrast, carcinomas of the head of the pancreas in patients who may present with obstructive jaundice is rarely cured by extensive surgery. The very high complication rate and associated mortality are such that for the majority of patients a curative procedure is not possible and better palliation may be obtained with a bypass procedure to relieve the jaundice.

16.5.3 *Radiotherapy*

A great change has occurred in both medical and patient attitudes to radiotherapy. Originally regarded as a palliative treatment for inoperable or incurable cancers, it is now seen as a partner and partial replacement for surgery in the curative treatment of certain cancers such as those in skin, larynx, and cervix. Advances in both technology and technique have meant that radiotherapy can be accurately applied to the region of the cancer with minimal damage to surrounding tissue, especially the skin. How radiation kills cancer cells selectively is not fully understood but all dividing cells are particularly sensitive to radiation damage and consequently rapidly proliferating tumour cells are especially vulnerable. One of the drawbacks of radiation treatment is that therapeutic doses may also kill dividing cells in normal tissues.

Two main techniques are used for the delivery of radiation which is given either as an external beam or as short range radiation from an implanted radioactive source. External beam radiation usually involves megavoltage produced by linear accelerator as photons or electrons or from cobalt sources in the form of relatively low energy X-rays or gamma rays. The latter are often used to treat relatively superficial lesions such as basal cell carcinoma or recurrences within the skin. High energy radiation can be used to treat deeply located lesions such as prostatic carcinomas without delivering an excessive dose to adjacent normal tissue. The total dose of radiation is usually delivered in several fractions to maximize the tumour killing while sparing normal tissues, the exact method of fractionation differing from institution to institution.

Interstitial (implant) irradiation gives a high local dose to the tumour and usually employs sources such as radium, iridium, or caesium used in the form of needles or wires implanted in the tumour. This technique is widely used in the treatment of head and neck cancers to deliver a high tumour dose without irradiation to sensitive organs such as the lens of the eye or the spinal cord. A combination of interstitial and external radiation may be used to treat the potentially malignant peritumoural field as well as the site of the primary carcinoma. Insertion of the radioactive sources may be carried out under general anaesthesia. As an alternative, tubes to contain the radioactive sources may be inserted under anaesthesia and, after measurements of the implant characteristics, loading of appropriate doses of isotopes can be carried out subsequently (afterloading). This technique of afterloading has been used with caesium sources to treat cancers of the body of the uterus.

16.5.4 *Combined approaches*

To treat cancer effectively but with less disfigurement, combined approaches using surgery, radiotherapy, and sometimes chemotherapy are being used increasingly. It is because of this multidisciplinary approach that more patients with malignancy are being treated in specialized centres. An example of this approach is the management of Hodgkin's disease, a lymphoma which was previously incurable and is now a curable disease in many patients. The diagnosis is usually made by biopsy of an abnormal lymph node and determination of the exact type of Hodgkin's disease requires an experienced pathologist. Part of the staging investigations may include a laparotomy. In this procedure the spleen is removed to determine whether it is involved with disease as no imaging test is reliable. Biopsies are taken from the liver together with representative intra-abdominal lymph nodes. In females, the ovaries are fixed with clips behind the uterus to diminish the risk of their subsequent damage by irradiation. Following this a plan of treatment can be made. Those with localized disease are effectively treated with radiation alone, while those with widespread disease require chemotherapy, sometimes with radiotherapy to bulky masses of tumour.

Another example of the combined approach is the treatment of follicular carcinoma of the thyroid. Once this relatively rare disease has been confirmed histologically, a total thyroidectomy is carried out but with attempted preservation of the calcium controlling parathyroid glands. This is followed by a thyroid scan using radioactive iodine, I^{131}, which is selectively taken up by both normal and neoplastic thyroid cells. If persistent thyroid tissue is identified this is destroyed by subsequent therapeutic doses of I^{131}. After this, replacement thyroxine is given to supply physiological requirements and also to inhibit any further thyroid activity (see Chapters 13 and 14).

One of the most important uses of a combined approach has been in the treatment of breast carcinoma. Until recently, most surgeons treated breast cancer by some form of mastectomy, often followed by radio-therapy. This was believed to be the best method of achieving local control of disease and offered the best chance of cure. However, this was effected at the cost of mutilation that was unacceptable to many patients and may in part have been responsible for the delay in women consulting doctors about breast lumps. One of the techniques which has been developed and tested in Europe is a combined procedure for patients with tumours up to 4 cm in diameter. The tumour is excised by without removal of surrounding normal tissue and the axillary lymph nodes are removed usually through a separate small incision. The radiotherapist then implants the tumour site with plastic or metal tubes which are afterloaded with iridium^{192} which delivers a high local radiation dose. Subsequently, the entire breast is irradiated by external beam. Adjuvant systemic therapy can be given to patients with histologically confirmed axillary node involvement. The early results of this treatment are very encouraging with good or excellent cosmetic results in 80 per cent of patients. A controlled trial has compared this technique with modified radical mastectomy and shown that there were no differences in terms of local control or overall survival in the two groups of patients (Fentiman 1988). Research continues to determine whether the entire dosage of radiation can be given by an iridium implant, thus abolishing the need for external irradiation.

16.5.5 *Palliative treatment*

Although the aim of treatment is to achieve cure as often as possible, there will be circumstances in which patients present with advanced cancer or develop recurrences and cannot be cured by local therapy. Under these circumstances local treatment with surgery or irradiation may play an important role in the palliation of distressing symptoms. Thus surgery can relieve bowel obstruction, jaundice, haemorrhage, and ulceration. Chronic haemorrhage and pain can be treated by external radiotherapy. The latter may arise from symptomatic bone metastases in which local irradiation may achieve rapid pain relief as well as reduce the risk of pathological fractures at the site of the tumour deposit. Sometimes when a pathological fracture occurs in a weight-bearing bone an orthopaedic procedure such as an internal fixation followed by radiotherapy will provide pain relief and make the patient mobile again.

Another common site of metastasis is the pleural space, particularly in patients with primary tumours of lung, breast, and ovary. This gives rise to an accumulation of serous fluid (pleural effusion) which, by compression of the underlying lung, gives rise to shortness of breath. Immediate relief can be obtained by simple drainage of the fluid, but nevertheless in the majority of cases the fluid reaccumulates. Good long-term control can be achieved by implanting irritant compounds into the pleural space. One of the most effective techniques is the use of sterile talc which stimulates a fibrous reaction and obliterates the pleural space, preventing any reaccumulation of the effusion, thereby abolishing the shortness of breath.

Another distressing symptom is the intra-abdominal accumulation of serous fluid (ascites) which may follow recurrence of carcinomas of the ovary and large bowel. This fluid produces abdominal swelling, discomfort, and vomiting and can be palliated using a peritoneo-venous shunt. This device collects the ascitic fluid from the abdominal cavity and directs it through a cannula into the superior vena cava.

Certain tumours, in particular those arising from the breast and prostate glands, are sensitive to steroid hormones. Significant remissions may be obtained in patients with advanced disease by removing the source of such hormones. In the case of breast cancer, metastases may regress after ovarian ablation (oophorectomy) and for patients with prostatic carcinoma, testicular excision (orchidectomy) may relieve metastatic symptoms. Further remissions may occur after surgical removal of either the adrenal or pituitary glands. However, such procedures are now used rarely because similar effects can be achieved using a variety of hormonal or antihormonal drugs (see Chapter 13).

16.6 The problem areas

In Britain, the standard of surgical training is exceedingly high so that throughout the country patients have access to good surgical advice and treatment. Unfortunately not all surgeons have close working relationships with radiotherapists and take part in joint clinics. In part this is a reason why cancer centres have been set up in major cities to develop combined approaches to treatment. Such techniques may not necessarily be applicable to patients in district general hospitals and a pressing need is for methods of, for example, breast conservation which can be carried out throughout the country. It is also necessary that the attitude of some clinicians be changed so that appropriate patients are treated effectively by less mutilating surgery.

As part of the process of change, healthy individuals need to be educated in the recognition of warning signs of cancer so that they consult their general practitioners at the earliest possible stage. In addition a heightened health awareness (without induction of undue anxiety) will lead to more individuals taking part in screening programmes. Because of financial implications it is going to be necessary to limit screening. Thus only women in the age group 50–65 will be routinely screened for breast cancer in Britain. The omission of older women, because of the potential lack of compliance, is regrettable since one-third of cases will be in patients aged over 70. Identification of a smaller high risk group would be of great value in enabling fewer women to be more effectively screened.

It was hoped that the development of monclonal antibodies would revolutionize the search for tumour markers but the early results have not been encouraging (but see Chapter 18). The ideal tumour marker, likely to be a tumour-associated product, would be present in detectable levels in the tumour and blood of patients, disappear on excision of the tumour, and only reappear if the cancer recurred. One example is the hormone chorionic gonadotrophin which is produced by the tumour, choriocarcinoma. Development of this marker has enabled great progress to be made in both the treatment and monitoring of this rare tumour. Unfortunately similar markers for the more common tumours have not yet been found.

Recent developments in molecular biology have enabled gene expression to be examined within primary human tumours and encouraging early results have been obtained. A variety of oncogenes have been found to be expressed in solid tumours (see Chapter 9). There may be a relationship between the presence of the oncogene and the prognosis of the tumour but it has yet to be determined whether this is a better prognostic indicator than that obtained by histopathological grading of the tumour.

At present, some improvements in health could be obtained by wider dissemination of available techniques among both the medical profession and the public. Nevertheless, for major improvements in mortality from cancer, a combination of patient education, screening, development of markers for staging, and follow-up, together with more interdisciplinary techniques will be required.

References and further reading

Britton, K. E. (ed.) (1987). New approaches to tumour identification. *Cancer Surveys* **6**(2).

Fentiman, I. S. (1988). Surgery in the management of early breast cancer: a review. *European Journal of Cancer and Clinical Oncology* **24**, 73–6.

Fentiman, I. S., Cuzick, J., Millis, R. R., and Hayward, J. L. (1984). Which patients are cured of breast cancer? *British Medical Journal* **289**, 1108–11.

Forrest, P. (ed.) (1986). Report to the Health Ministers of England, Wales, Scotland and Northern Ireland. Breast Cancer Screening. HMSO.

Hayward, J. L. *et al.* (1984). A new combined approach to the conservative treatment of early breast cancer. *Surgery* **95**, 270–4.

Knox, G. and Woodman, C. (ed.) (1988). Prospects for primary and secondary prevention of cervix cancer. *Cancer Surveys* **7**(3).

McKenna, R. J. and Murphy, G. P. (1986). *Fundamentals of surgical oncology.* Collier Macmillan.

Magdelenat, H., Laine-Bidron, C., Mate, S., and Zajdela, A. (1987). Estrogen and progestin receptor assay in fine needle aspirates of breast cancer: methodological aspects. *European Journal of Cancer and Clinical Oncology* **23**, 425–31.

Merkel, D. E., Dresser, L. G., and McGuire, W. L. (1987). Flow cytometry, cellular DNA content and prognosis in human malignancy. *Journal of Clinical Oncology* **5**, 1690–703.

Pykett, I. L. *et al.* (1982). Principles of magnetic resonance imaging. *Radiology* **143**, 157–68.

17

Chemotherapy
J. S. MALPAS

17.1 Introduction

Chemotherapy is a relatively new method of treating cancer. The tradition of treating cancer with surgery had been in existence for over a century, and radiotherapy had been used for at least a quarter of a century before chemotherapy made its appearance in the middle of the Second World War. Malignant disease, such as leukaemia, which was generalized when it first presented, or tumours which had become disseminated, were incurable up to that time. Chemotherapy gave a first promise of cure.

The development of analogues of mustard gas and their profound effects on the lymphoid system suggested that they might be effective in treating lymphomas or tumours of the lymphatic system. The use of these agents, which were called radiomimetic drugs, as they were very similar to radiotherapy in their effect, was already quite advanced when the second success was reported in 1948 with the use of anti-metabolic agent amethopterin in the treatment of childhood lymphoblastic leukaemia. In the next decade, a 'golden age' for the discovery of and introduction to clinical use of alkylating agents and anti-metabolites, a wide range of common and rare malignancies were treated with these agents.

It was soon apparent that for a drug to be successfully introduced for use in the clinic, it was necessary to progress through a series of tests. Initially, its toxicity and the effects on various organs were established by testing in small animals. Toxicity in the liver or kidneys, for example, could be assessed, and these features looked for carefully when it was first administered to patients. The sequence of events in clinical trials will be discussed in more detail later, but essentially, Phase I studies of drugs establish the dose, route of administration, excretion pattern, and salient toxic features. Once this information is available, Phase II studies, in which patients with a variety of malignancies receive the drug, are carried out, and the spectrum of tumours that are sensitive to the agent can be defined. More advanced studies, sometimes called Phase III studies, allow for in depth evaluation of patients, usually being treated for one form of malignancy, in which the effect of the new drug plus the best established previous method of treatment is compared to a control group of patients in a randomized manner, the control group having the previously most successful form of treatment. The difficulties, errors, and constraints on the introduction of new drugs will be discussed in more detail at the end of this chapter.

17.2 Classification of chemotherapeutic drugs

Using this approach, a whole range of clinically effective drugs was made available over the next two decades. Some of these act directly on tumour cells whereas others must be activated by metabolic processes, either in the tumour cells or in organs such as the liver. A list of some of the most important drugs is given in Table 17.1. Increasing knowledge of the mechanism of cell division was being gained *pari passu* with the introduction of these agents. In the hope that greater knowledge of how these agents worked would enable them to be used more effectively clinically, questions as to whether they were active on the dividing cell and, if this was so, whether they were active at a particular stage of cell division, became of great interest.

It is possible to divide drugs into those that are active only on dividing cells, those that are active on dividing cells and affect a very particular phase of cell division (phase specific drugs), and those that affect all or most of the phases of the cell cycle (Table 17.2).

The particular phases where some of these drugs act are shown in Fig. 17.1. Whatever the mode of action of the chemotherapeutic agent, a very important finding is that it destroys malignant cells according to first

Table 17.1 Anti-cancer drugs

Classification	Drug
Alkylating agents	Mechlorethamine
	Busulphan
	Chlorambucil
	Cyclophosphamide
	Melphalan
	Thiotepa
Antimetabolites	Methotrexate
	6-Mercaptopurine
	Thioguanine
	5-Fluorouracil
	Cytosine arabinoside
	5-Azacytidine
Plant alkaloids	Vinblastine
	Vincristine
	VP-16
Antibiotics	Actinomycin D
	Doxorubicin
	Bleomycin
	Daunorubicin
	Mithramycin
	Mitomycin C
Nitrosoureas	Carmustine
	Lomustine
	Semustine
	Streptozotocin
Enzymes	L-Asparaginase
Random synthetics	*Cis*-Platinum diammine dichloride
	Dacarbazine
	Dibromomannitol
	Hexamethylmelamine
	Hydroxyurea
	Mitotane
	Procarbazine

order kinetics; in other words the same proportion of cells is killed for each dose of the agent.

It is very important to recognize that chemotherapeutic drugs damage normal tissues as well; indeed, if they were non-toxic to normal tissues, cancer would have been cured long ago. The only reason that chemotherapy is feasible is that normal tissues may recover, sometimes more rapidly than the tumour cells. It is on this narrow difference in behaviour that practical clinical chemotherapy rests.

From the point of view of the practising clinician treating patients, the

Table 17.2 Phase and cycle specific drugs

Phase specificity	Acts on	Drug
S phase specific	DNA synthesis	Methotrexate Cytosine arabinoside Hydroxyurea
Relatively S phase specific	DNA, RNA, and protein synthesis	5-Fluorouracil 6-Mercaptopurine
Cycle specific	DNA at all phases of cycle	Nitrogen Mustard Nitrosourea Cyclophosphamide

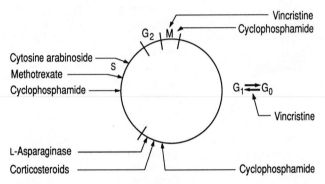

Fig. 17.1 Cell cycle showing the site of action of some phase specific drugs.

contribution that the knowledge of the kinetics of cancer cells and the effect of the drug on the cell cycle has made has been disappointing, probably because human tumours are very different in their behaviour from the rapidly growing experimental animal tumours in which the proportion of dividing malignant cells in the tumour is very high.

A more helpful classification, based on the mode of action of particular groups of agents, is probably of more immediate help (Table 17.3). The remarkable feature about this table is the wide variety of sources for chemotherapeutic agents. Some were produced deliberately on the basis of interfering with the metabolic pathway, but many were discovered by pure chance—vincristine, for example, in the search for a new anti-diabetic agent, procarbazine in the development of a tranquillizing drug, and the anti-tumour antibiotics were often discovered in the search for new anti-bacterial agents. Futhermore, the source of these drugs gives very little indication of the likelihood of activity against a particular tumour.

Table 17.3 Groups of chemotherapeutic agents

Class of compound	Examples	Tumours against which drug is most active
Anti-folic compounds	Methotrexate	Acute leukaemia Choriocarcinoma
Anti-purines and pyrimidines	6-Mercaptopurine 5-Fluorouracil	Acute leukaemia Breast cancer
Alkylating agents	Melphalan Cyclophosphamide Busulphan	Myeloma Lymphoma Chronic myeloid leukaemia
Sex hormones	Oestrogens Androgens	Prostate cancer Breast cancer
Steroids	Corticosteroids	Leukaemia Lymphoma
Anti-tumour antibiotics	Actinomycin Anthracyclines Daunorubicin Doxorubicin Mitomycin	Wilms' tumour Leukaemia Lymphoma
Plant extracts	Vinblastine Vincristine Vindesine	Hodgkin's disease Leukaemia

17.3 Drug toxicity

Information on toxicity is most useful to the clinician when considering the choice of chemotherapeutic agents for cancer therapy. Toxicity can be divided into that occurring shortly after administration, that which is somewhat delayed, and the very long term toxic effects. Before a drug or combination of drugs is administered to a patient, the gain in terms of clinical benefit must be balanced against these toxicities, examples of which are shown in Table 17.4. The table has been arranged to show the common non-specific toxicity that may be encountered with most therepeutic agents. Specific toxicities are shown on the right hand side of the table. The clinician must know in detail the general and specific toxicities of all the chemotherapeutic agents that he uses.

17.4 Principles of chemotherapy

It is important to be quite clear about the reason for giving a cytotoxic drug or combination of drugs. From experience we now know that it is possible for chemotherapy to cure various malignancies. Children with

Table 17.4 Drug toxicity

Common	Special
Immediate (within hours)	
Nausea and vomiting	Haemorrhagic cystitis (cyclophosphamide)
Phlebitis	Radiation recall (actinomycin)
Hyperuricaemia	Fever (bleomycin)
Renal failure	
Early (days to weeks)	
Reduction in white blood cells (leucopenia)	Paralytic illness (vinca alkaloids)
Reduction in blood platelets (thrombocytopenia)	Pancreatitis (asparaginase)
Hair loss (alopecia)	Cerebellar disorders (high dose cytosine arabinoside)
Diarrhoea	Ear toxicity (platinum)
Delayed (weeks or months)	
Anaemia	Peripheral nerve damage (vinca alkaloids)
Aspermia	Inappropriate ADH (cyclophosphamide)
Liver cell damage	Jaundice (6-mercaptopurine)
Fibrosis of lung	Adrenal deficiency (busulphan)
Late (months to years)	
Sterility	Liver fibrosis (methotrexate)
Testicular or ovarian atrophy	Brain damage (methotrexate)
Second malignancies	Bladder cancer (cyclophosphamide)

acute leukaemia. Wilms' tumour, soft tissue sarcomas, lymphomas, and other tumours are curable. Adults with acute myelogenous leukaemia, lymphoblastic leukaemia, non-Hodgkin's lymphoma, Hodgkin's disease, and teratoma of the testis are potentially curable. Drugs, therefore, have to be given to the limits imposed by toxicity, on a definite schedule, which may have to be tailored around other methods of treatment such as surgery or radiotherapy.

In some tumours, cure cannot be realistically contemplated, and in these patients effective drugs or drug combinations may be able to extend useful life considerably. Good examples are breast carcinoma, ovarian carcinoma, small cell carcinoma of the bronchus, and myeloma.

There are some forms of malignant disease where it is known that there has been no improvement in the duration of survival since chemotherapy was introduced, but the quality of life has been immeasureably improved. Such a disease is chronic myelocytic leukaemia, where the median duration of survival is still only about 34 months some 30 years after busulphan was introduced for the condition, but patients lead normal lives during most of the course of the disease, and the benefit of what is palliative therapy is immense.

Finally, there are occasions when chemotherapy may be used as an effective palliative agent in a tumour which has been previously treated with radiotherapy or surgery. The bone pain associated with disseminated breast cancer may be rapidly relieved, the fearful symptoms of strangulation occasioned by a large carcinoma in the region of the trachea may be dispelled in a few hours by the injection of nitrogen mustard. These beneficial effects should not be underrated.

17.5 Administration of chemotherapeutic drugs

17.5.1 Combination chemotherapy

From time to time in the clinic it is very evident that a tumour will not respond to a chemotherapeutic drug which is usually effective in the condition, or possibly, after a period of time in which the tumour has responded, it again becomes unresponsive. This phenomenon of resistance will be discussed in more detail later. An immediate practical solution has been to add a second drug, which ideally would not have the same toxicity. Such an example was the so called 'induction' therapy for acute lymphoblastic leukaemia in children. Vincristine and prednisolone were both found to be effective in producing remission, and when given singly, achieved remission in about half the patients; that is, they could return the child to normal health with no clinical evidence of leukaemia, a normal blood count, and a bone marrow test in which the leukaemic blast cells had gone. When the two drugs were combined, this percentage rose to over 90 per cent. Futhermore, as the drugs had very different toxicities, this was achieved without great hazard. In this instance the anti-tumour effect doubled, but the toxicity did not.

Addition of even more drugs was found in some tumours to be more effective. An example of response rates to one, two, and then four drugs in Hodgkin's disease is shown in Fig. 17.2. In both acute lymphoblastic leukaemia in children and in Hodgkin's disease, this considerable increase in remission rate has been accompanied by a notable increase in survival (Fig. 17.3).

The latest adult tumour to show sensitivity to this approach is teratoma of the testis, the commonest tumour of young men between the ages of 20 and 35. This may be cured, even when it is disseminated, by a combination of vinblastine, bleomycin, and *cis*-platinum. It is probable that, in the case of Hodgkin's disease and teratoma of the testis, the multi-drug combination prevents the emergence of any drug resistant strains of tumour cells (Goldstein *et al.* 1989), a feature which is not seen in many other tumours.

The response to the application of multiple drug regimes in the case

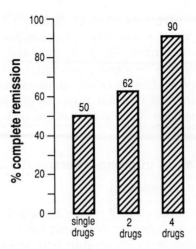

Fig. 17.2 Improvement in remission rate in Hodgkin's disease as the number of drugs used in combination increases.

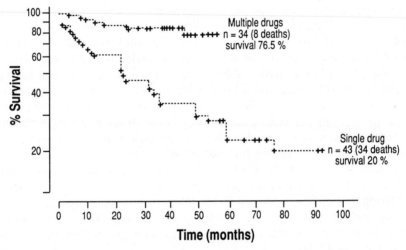

Fig. 17.3 Improvement in survival using multidrug combinations compared with the use of single agents in Hodgkin's disease.

of bladder carcinoma and bowel carcinoma has not been so successful, and more effective agents are awaited.

17.5.2 Adjuvant chemotherapy

This form of chemotherapy was first introduced into the treatment of Wilms' tumour, a tumour of the kidney in young children, on the assumption that metastases were already widely disseminated when the patient was first seen. This hypothesis considered that these metastases are composed of a few cells, which have an increased mitotic rate and a good blood supply, so that although they are not detectable they are nevertheless very susceptible to the action of chemotherapeutic agents. More recently, this degree of sensitivity has been queried, but nevertheless, when this hypothesis was tested in Wilms' tumour, it was shown that survival has increased considerably. With surgery and radiotherapy, survival was seen in only about 40 per cent of patients. Giving 'adjuvant' therapy, that is, treatment directed at the micrometastases from the time of presentation of the tumour, led to an increase in survival, which is now regularly seen in most clinical series at between 85 and 90 per cent. This principle has been applied, with considerable success, to other childhood tumours. It has been much less successful in adult solid tumours, but this is only to be expected, since the response to chemotherapy of most carcinomas affecting adults is much less satisfactory.

17.5.3 *Sanctuary sites*

A major problem in the administration of anticancer agents is that they may not always be distributed throughout the body. Thus, in the induction of remission in children with acute lymphoblastic leukaemia, the drugs vincristine and prednisolone do not enter the meningeal spaces surrounding the brain and spinal cord, certainly not in sufficient concentration to be tumoricidal. This was responsible for the high frequency of relapse seen with the meningeal leukaemia in these children, until it was demonstrated in the early 1970s that the administration of methotrexate directly into the meningeal space, combined with cranial radiotherapy, could eliminate these cells lodged in so-called 'sanctuary sites'. There is considerable argument as to whether there are other sanctuary sites, such as the testes of boys who have lymphoblastic leukaemia, or localized peritoneal areas in women with ovarian cancer.

17.5.4 *Principles of high dose therapy*

Studies on experimental tumour systems have shown that for some drugs, alkylating agents for example, the number of tumour cells that are killed is directly related to the dose of the drug that is given. There is also evidence that the emergence of drug resistance is also related to the intensity of the initial drug therapy, and that resistance is far less likely to develop if effective drugs are given in high dosage early in the course of the disease. Drug resistance most often develop if repeated small doses of drug are given over a long period of time. The range of drugs that can be used in the clinic for high dose therapy is limited. There are two kinds:

one in which there is a specific antidote to the drug (for example, methotrexate, where the patient may be 'rescued' by the use of a specific agent, citrovorum factor or folinic acid), and the other, in which the dose limiting toxicity is such that the patient can be supported through the period of the acute damage to normal tissues produced by the drug. An example of the latter is the alkylating agent melphalan, which in normal circumstances is very toxic to bone marrow, but has no other deleterious effects, even when quite high doses are given. If bone marrow is removed, and stored before the drug is given, it can be replaced as soon as the drug has been excreted. Repopulation of the bone marrow follows, and the patient's blood count recovers rapidly. Drugs which may be useful in high dose therapy include cyclophosphamide, cytosine arabinoside, and new drugs such as etoposide. Use in the clinic is still at a relatively early stage, but it appears that increased response rates and possibly increased length of survival may be seen in some children treated in this way, and in adults who, for example, are being treated for the plasma cell tumour myeloma. Further studies will be necessary before the usefulness of high dose chemotherapy can be properly assessed.

If high dose chemotherapy or a combination of this procedure and whole body irradiation prove to be effective, it becomes even more important to improve methods for removing any tumour cells from the bone marrow which had been removed and stored. Intensive research is now going on in an attempt to do this. Physical methods for removing tumour cells have not been very successful, but new methods using monoclonal antibodies are now being tried. In one technique, tumour associated antibodies are attached to magnetic beads. Tumour cells adhere to the antibodies and the beads are then removed by passing the cell suspension through a magnetic field induced by powerful electro-magnets. In another technique a toxic material, ricin, is coupled to the monoclonal antibodies so that the drug is delivered only to the tumour cells. The use of this technique in chemotherapy is discussed in detail in Chapter 18. Another approach is to use a drug, 4-hydroxyperoxycyclo-phosphamide. This is related to cyclophosphamide but, unlike its parent compound, does not require activation by passage through the liver *in vivo*, but can be safely incubated with human bone marrow *in vitro*.

There is now great interest in the extension of chemotherapy using these manoeuvres, but it is not yet possible to say whether the approaches will be useful.

17.6 Drug resistance

The occurrence of drug resistance is one of the greatest obstacles to curing cancer by chemotherapy, and consequently is the focus of much attention. The common cancers in man (squamous cell cancer of the lung, breast, and colon cancer) may show little response to drugs. They are *inherently* resistant. Other cancers, for example small cell lung cancer, show a high rate of response with a visible decrease in the size of the tumour, but often after a few months the tumour is no longer responsive and has *acquired* resistance.

As with many other illnesses, a patient may not respond to a drug when it is given, or possibly, having responded, after a short time ceases to do so. In the clinic, failure to respond may be simply that the patient has not taken the drug (failure of compliance), and although this seems unlikely to occur in the management of cancer, surprisingly, studies have shown that some patients do not take the drugs, and even children are not given the oral medication for their leukaemia by their parents. If the drug has been taken, it may not be absorbed. There is a large variation in the rate of absorption of melphalan, for example, between different individuals. Even if the drug is absorbed, it may not reach the appropriate site, as seen already in the case of meningeal leukaemia. Although all these phenomena could be included in a discussion of resistance, what the chemotherapist and biochemist is most interested in is the reason why a cell becomes resistant. This topic is discussed in detail by Stark and Calvert (1986).

Many types of drug resistance are genetic in origin. The evidence can be summarized as follows: the characteristics of the drug resistant cells are stable from one generation to the next; resistant cells are generated at a rate which is consistent with the mutation rate for the cell; drug resistance can be induced by serial exposure to the drugs *in vitro*, and altered gene products have been identified; finally, and very convincingly, DNA which is overexpressed in resistant cell lines has been successfully transfected into drug-sensitive cells, making them resistant. Furthermore, they are resistant to the same spectrum of chemotherapeutic drugs as the original cell.

These facts have important clinical implications for the physician. They indicate that the problem is largely a biochemical one and may therefore be overcome by a knowledge of the cell's biochemistry and its manipulation. Another fact that must be remembered is that the larger the tumour the greater the likelihood that drug resistant cells will be present as a consequence of mutation or selection, perhaps induced by the drugs used. Theoretically, therefore, tumours should be treated as soon as possible, while they are small, and exposed to a series of effective drugs used in an alternating sequence.

Among the biochemical changes that enhance resistance, the following mechanisms have been identified. Firstly, an increase in the repair enzymes for drug damaged DNA strands; an increase in the efflux of cytotoxic drugs through the cell membrane; this latter is a particular feature in multidrug resistance (mdr) and will be discussed later; increase in levels of target enzyme due to related gene amplification; decreasing drug activation and increasing drug degradation. Examples of the last two mechanisms are the demonstration that decreased phophoribosyl transferase impairs the activation of 6-mercaptopurine and 6-thioguanine, and uridine phosphorylase and kinase impair the activity of fluorouracil. On the other hand, generation of more enzyme may be induced in the resistant cell, of which an example is the increase in deaminases which break down cytosine arabinoside. If gram doses (rather than milligrams) of cytosine arabinoside are given to the patient, the leukaemia may again become sensitive and go into remission.

17.6.1 *Multiple drug resistance*

Cells which show resistance to one group of drugs very often are resistant to others. It is of interest that the groups of drugs concerned are usually those which are natural products (for example, the anthracyclines, doxorubicin, daunomycin, the derivatives of podophyllin, etoposide VP16 and VM26, and the vinca alkaloids vincristine and vinblastine). It is likely that these cells have conserved important mechanisms for detoxification to enable them to survive exposure to ingested toxic natural products. Whatever the reason, this ability is linked to the presence of a 170 kDa glycoprotein (gp170) (or P-glycoprotein) in the cell membrane (Ling 1989). Multidrug resistant human, mouse, and hamster tumour cell lines have been found that amplify a small group of related genes called *mdr*. The mRNA encoded by these amplified genes is over-expressed in multidrug resistant cells. The higher the concentration of P-glycoprotein, the greater the degree of resistance to transport of a wide variety of unrelated chemical substances across the membrane. When very resistant cells have been produced in the laboratory expressing a high level of P-glycoprotein, this high degree of resistance can be conferred on drug sensitive cells by DNA mediated gene transfer (see Chapter 9 for methods of gene transfer). More recently, drugs which block the uptake of calcium into the cell (such as verapamil) have been shown to oppose the action of the glycoprotein in allowing rapid egress of drugs from the cell. Verapamil can therefore restore sensitivity of a cell to a chemotherapeutic drug. An example is melphalan, whose effect is restored in multiple myeloma by verapamil infusion.

More than one mechanism may be involved in drug resistance. Methotrexate acts principally during the S phase of the cell cycle by inhibiting the enzyme dihydrofolate reductase (DHFR) which is necessary for the production of an essential metabolite, tetrahydrofolate, necessary for

DNA synthesis. Different rates of methotrexate transmembrane passage in resistant and non-resistant cells have been shown, and the relationship between sensitivity and rate of uptake of methotrexate has been demonstrated in murine leukaemia cells. Two mechanisms, one active and one passive, have been demonstrated, and the presence of this latter mechanism forms the basis for the use of therapy with high dose methotrexate.

Another fundamental finding necessary to the understanding of resistance to methotrexate is the ability of even a small amount of DHFR to stimulate enough thymidylate synthesis for cell activity. Thus, unless there is a continuous presence of free drug, synthesis will start again. Consequently, even minor changes in binding or the ability to produce more DHFR will profoundly affect the action of the drug, and render it ineffective. While drug binding to reductase has been found to be related to resistance in mouse leukaemias, the evidence for this in man is weak. What has been shown more recently is that with the steady increase in exposure to methotrexate, amplification of the gene controlling formation of the enzyme occurs, and homogeneously staining regions (see Chapter 10) on a specific marker chromosome can be seen to be increased in the resistant cell. Amplification is usually stepwise until many hundreds of the copy of the gene are present. This has been observed in human tumours, and is thus probably relevant to the clinical situation. Other anti-metabolites may exploit similar changes in their target cells, but these have not yet been identified.

This resistance which comes about as the result of gene amplification may of course be occurring in both normal cells and tumour cells, and only appears under selective pressures. It is possible that mechanisms for resistance which develop to other anti-metabolic drugs are of this nature.

The spontaneous occurrence of drug resistance in malignant tumour cells is of such fundamental importance that it is not too great an exaggeration to say that the future for chemotherapy depends on its understanding and exploitation. Normal cells never become resistant to chemotherapy, a fact of equal importance to our understanding of the differences between normal and malignant cells.

17.7 New approaches

The development of new agents depends either on luck or on a more precise knowledge of the growth requirements of the tumour cells and of the mechanism of resistance. Recombinant DNA techniques have been used to product two biological response modifiers, tumour necrosis factor (TNF) and lymphotoxin in sufficient quantity for clinical trial (see Chapter 11). Preliminary results are not striking but further work on the effects of these agents together with other factors such as interferon and interleukin 2 suggest that this may be a profitable area for further study (Balkwill and Fiers 1989 for a full review).

Isolation of the genes and gene products associated with resistance is also a promising area since their identification may allow the design of agents that interfere with their biological activity.

Another approach to the control of gene expression is by interfering with the enzyme systems involved in DNA function or repair. A good example of this approach is the work described on DNA topoisomerases 1 and 2. These two enzymes are concerned with the relaxation and recoiling of DNA, when for example the double helix separates into single strands during transcription by RNA polymerases. They do this by creating breaks in one strand (topoisomerase 1) or two (topoisomerase 2) and then catalyse the resealing. We now know that some cytotoxic drugs such as doxorubicin that intercalate in DNA seem to act by binding to topoisomerase 2 preventing resealing of double strand breaks induced by the enzyme. Since the topoisomerases are present in low concentration in resting cells but increase in amount in rapidly growing cells the sensitivity of tumour cells to the drugs may depend on the enzyme concentration. This work has opened up many new approaches to our understanding of the problems of tumour chemotherapy (Potmesil and Ross 1987).

Yet another approach is being tried in the treatment of the rare group of neuroendocrine tumours (see Chapter 14) which may produce a variety of peptides and hormones. Long acting analogues of agents such as somatostatin are being used to reduce the release of active substances and to block their peripheral activity. Hormones are also being used as carriers. Cytotoxins are being coupled to hormones which then localize at their sites of action and in theory should destroy those tumour cells which still retain some of their normal functions.

Although few of these techniques are yet of general clinical value they illustrate the logical approach of designing drugs to interfere with specific aspects of tumour growth and function.

17.8 Clinical trials

The vast array of synthetic and non-synthetic compounds which might be effective in cancer chemotherapy has led to the definition of some common sense principles of assessment, so that time, expense and other resources are not used wastefully.

The introduction of a new agent, whether it has been developed in the pharmaceutical industry or university laboratory, is preceded by a time-consuming and expensive series of toxicity trials in animals. A Phase I trial on a human subject is designed to determine the best route of administration, the features of the pharmacokinetics such as the routes of excretion and rapidity of excretion, details of toxicity such as dose limiting toxic side effects, and the range of toxic effects that may occur. Many patients may have had previously all the conventional treatment available, so that the response may be minimal. Even so, a response in one patient in a consecutive series of, say, 14 or 15 patients who have been previously treated, would encourage a Phase II study.

Phase II trials concentrate on the clinical response, and these are defined in Table 17.5. In Phase II trials, the drug is given to patients who have a variety of different tumours. Some will have already been previously treated. The aim of the trial is to determine the spectrum of activity of the agent. It very often happens that the tumour chosen is one that responds relatively infrequently to a chemotherapeutic agent of any kind, and in this case, so called 'randomized Phase II trials' are justified, in which the overall response to the new agent is compared on a random basis with the ones used previously. Phase II studies give more information about toxicity and the possible therapeutic benefit of the drug. Their chief danger is that they underestimate the efficacy of the drug, and may deter its investigation in the more difficult Phase III studies.

Table 17.5 Criteria for response in solid tumours

Response	Criteria
Complete response	Complete disappearance of all demonstrable disease
Partial response	More than 50 per cent reduction in the sum of the products of the longest perpendicular diameters of tumour with no disease progression elsewhere
No response	No change or less than 50 per cent reduction
Progression	Increase in size of tumour at any site

A great deal has been written about the advantages and disadvantages of randomized trials. Penicillin, for example, would not have needed a randomized study to show its efficacy. Unfortunately, because of the less dramatic effects of most anti-cancer agents, benefit is not easily detected, and physicians have been misled by small non-randomized studies. What is also quite apparent is that carefully conducted randomized studies can occasionally produce contradictory answers. This is most likely to be due to the fact that the two populations, the patients treated with the new drug and the controls who have been treated with the standard regimens, differ in some subtle and perhaps unrecognized way that biases response. The only way in which this bias can be eliminated in a randomized Phase III study is by recruiting large numbers of patients. The question is then whether one is ethically justified in doing this; if the difference is small, perhaps the answer doesn't matter, and a large number of patients may be locked into a trial which may contribute relatively little to knowledge about the disease and its management.

It is very tempting in this case to turn to comparing results with historical controls. It is very difficult to conclude that the historical controls are ever matched adequately. If nothing else has changed, the physician will have improved his ability to manage the condition under investigation (or at least one would hope so). Criticisms about variation in the general supportive care, ancillary services such as blood transfusion, treatment of bacterial infections, etc. are valid, but are less likely to be true in the case of a Unit which admits a lot of patients suffering from one particular disease, and where the supportive care is relatively unchanged over a number of years. Any marked improvement in response or survival is then probably a true finding, and has the advantage that any new modification can be built into the therapeutic programme quite rapidly. New methods of analysing data are now being studied. This activity, and the debate on how one should measure what is done, will certainly be welcomed, for the critical attitude that it encourages is our only real hope for continued progress (see also Calvert 1989).

17.9 Conclusion

It will be seen that the agents currently employed in the treatment of cancer are imperfect, achieve their effects by methods which are poorly understood, and may produce lasting damage to normal tissues. Despite this, the number of patients who are being cured of their cancers as a result of chemotherapy is increasing gradually every year. By 1990 it is estimated that one in every 2000 young adults reaching the age of 21 will have been cured of cancer. This is a remarkable achievement when it is remembered that it has occurred within the space of one professional lifetime.

References and further reading

Balkwill, F. R. and Fiers, W. (eds.) (1989). Biological response modifiers. *Cancer Surveys* **8**, No. 4.

Calman, K. C., Smyth, J. F., and Tattersall, M. H. N. (1980). *Basic principles of cancer chemotherapy*. Macmillan Press, London.

Calvert, H. (ed.) (1989). A critical assessment of cancer chemotherapy. *Cancer Surveys* **8** (3).

Chabner, B. (ed.) (1982). *Pharmacologic principles of cancer treatment* W. B. Saunders, Philadelphia.

Goldstein, L. J. *et al.* (1989). Expression of a multidrug resistance gene in human cancers. *Journal of the National Cancer Institute* **81**, 116–23.

Ling, V. (1989). Does P-glycoprotein predict response to chemotherapy? *Journal of the National Cancer Institute* **81**, 84–85.

Monfardini, S., *et al.* (ed.) *Manual of cancer chemotherapy* (3rd edn), UICC Technical Report Series, Vol. **56**, UICC, Geneva.

Potmesil, M. and Ross, W. E. (ed.) (1987). *Topoisomerases in cancer treatment.* NCI Monograph No. 4. National Cancer Institute, Bethesda.

Stark, G. R. and Calvert, H. (ed.) (1986). Drug resistance. *Cancer Surveys* **5** (2).

Tannock, I. F. and Hill, R. P. (ed.) (1987). *The basic science of oncology.* Pergamon Press, New York.

van der Bliek, A. M., Borst, P., (1989). Multidrug Resistance. *Advances in Cancer Research.* **52**, 165–204.

For detailed information on specific topics, consult the following specialized review series.

Seminars in Oncology. Grune and Stratton, 111 5th Avenue, New York, NY 1003, USA.

Journal of the National Cancer Institute. Public Health Service, National Institute of Health, Bethesda, Maryland 20892, USA.

18

Monoclonal antibodies and therapy
Edward J. Wawrzynczak and Philip E. Thorpe

18.1 Introduction

Antibodies are proteins made by higher animals when infected with bacteria, viruses, or other foreign substances. They are secreted into the blood by plasma cells as part of a complex immune response against the invading 'organism' (see Chapter 15).

Animals in the laboratory similarly can produce antibodies against a wide range of foreign molecules that need differ only slightly in structure from the animal's own molecules to be immunogenic. Thus, homologous proteins from closely related species may stimulate antibody production. Tumour cells often bear surface molecules which are immunogenic in the natural host or in other animal species. By immunizing animals with preparations of tumour cells, antisera can be raised that contain antibodies recognizing the various antigens found on the tumour cell surface. Antibodies that bind to normal cells can then be removed from the antiserum by suitable absorption procedures leaving behind those antibodies that react strongly with tumour cells.

Attempts to use these antibody preparations for cancer therapy have been disappointing. This is partly because the antiserum contains many antibodies with different specificities, which increases the likelihood of cross reaction with normal cells, and partly because the anti-tumour antibody is diluted with a large excess of antibodies with irrelevant specificity. In addition, the anti-tumour antibodies are often difficult to purify in adequate quantities and preparations from different animals are not equally effective. These problems were largely overcome by the technique devised by Kohler and Milstein (1975) for immortalizing and cloning individual antibody forming cells. Each clone of cells produces a single type of antibody called a 'monoclonal antibody' having a single type of antigen binding site. These cells can be grown up in large numbers allowing the reproducible manufacture of the monoclonal antibody in essentially unlimited amount and in a highly pure state. Such antibodies have a number of important uses in the diagnosis and therapy of cancer (see Chapters 12, 15, 16, and 17).

One application of monoclonal anti-tumour antibodies is to identify the type of malignant cell by analysing the tumour cells with a panel of monoclonal antibodies against different antigens. This helps the clinician to decide the most appropriate course of treatment. Second, antibodies can be used to measure the levels of soluble antigens which some tumour cells shed into the blood (see Chapters 14, 15, and 17). Third, anti-tumour antibodies injected into the patient can selectively concentrate within the tumour tissue. If the antibodies are radiolabelled, then it is possible to identify the sites of the primary tumour and large metastatic deposits within the body from the radiation that they emit. Moreover, anti-tumour antibodies injected into a patient can exert specific thera-peutic effects by stimulating the various defence systems of the body to attack the tumour. Similarly, powerful cell poisons can be attached to the antibody which then carries the poison to the tumour cells and kills them. Lastly, monoclonal antibodies can be used to remove malignant cells or T lymphocytes from bone marrow *ex vivo* as part of the treatment of leukaemia and other malignant diseases. Here, we review the present and possible future applications of monoclonal antibodies in the therapy of cancer.

18.1.1 *Antibody structure and function*

As discussed in Chapter 15, the antibody molecule, in its simplest form, consists of four polypeptide chains linked by disulphide bonds: two identical 'heavy' chains and two identical 'light' chains (Fig. 18.1). Each heavy chain consists of a variable domain and three or four constant domains, whereas the light chain consists of a variable domain and a single constant domain. The variable and constant domains of one light chain fold over the variable and first constant domain of one heavy chain to give an 'Fab' arm which contains a single antigen binding site and hence, each antibody molecule has two reactive sites (bivalent). The other constant domains of the heavy chain constitute the 'Fc' portion which mediates the effector functions of the antibody according to the immunoglobulin (Ig) class to which it belongs.

The variable region domains contain 'framework' regions and 'hyper-variable' regions. Framework regions of the primary sequence are very similar in antibodies of the same class and are believed to include amino acid residues that are concerned with maintaining the three-dimensional structure of the domain. Within hypervariable regions there is much greater diversity of the amino acid sequence between antibodies of the same class. It is principally the spatial distribution of these amino acid

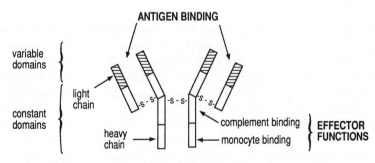

Fig. 18.1 The structure of IgG.

residues that determines the individual antigen binding specificity of an antibody molecule. The antigen binding sites of antibody molecules constitute unique structures that can be distinguished immunologically and are called 'idiotypes'.

Immunoglobulin classes are defined according to the kind of heavy chain the molecule contains. There are five major types of heavy chain, γ, μ, δ, α, and ε, which differ in their constant domains, and two types of light chain, κ and λ. The most abundant immunoglobulin in the bloodstream is IgG. It exists as a monomer with two γ chains and either two κ or two λ chains, which have a combined molecular weight of 150 000 Da. In humans, there are four IgG subclasses, IgGl, IgG2, IgG3, and IgG4, which have closely related constant domains encoded by different genes. In the mouse and rat, the subclasses are IgGl, IgG2a, IgG2b, and IgG3. IgM (containing μ chains) exists in the bloodstream as a pentamer of subunits linked by a joining or J chain and having a combined molecular weight of 900 000 Da. Each of the subunits resembles an IgG monomer except for the presence of an extra constant domain on the heavy chain to which the J chain attaches. IgD (together with IgM) is present on the surface of B lymphocytes and acts as a receptor for the antigen. IgD, IgA, and IgE are minor components of serum.

Antibodies of different classes and subclasses activate different effector mechanisms in the animal (see p. 484) and may differ in their activity as anti-tumour agents depending on which effector mechanism is the most active against any particular tumour. Since it is the Fc portion that activates the effector arm of the immune response, the intact antibody molecule is needed for immunotherapy. For radioimaging or targeting of cytotoxic agents, however, only the antigen binding portion of the antibody is needed and so it is possible to use antibody fragments, i.e. Fab or $F(ab')_2$ generated by the proteolytic action of papain or pepsin respectively (Fig. 18.2).

18.1.2 *Monoclonal antibodies*

When an antigen binds to a B lymphocyte bearing surface immunoglobulin that recognizes the antigen (in the presence of macrophages, helper T cells, and their soluble products), it triggers proliferation and gives rise to a clone of cells all of which, when mature, secrete the same antibody. A single antigen usually stimulates many different B cells. Thus, an immunized animal produces many different antigen specific antibodies. Since normal B lymphocytes soon die out in culture, there was no way to obtain an individual clone of B cells secreting a single kind of antibody until Kohler and Milstein ingeniously preserved the specific antibody-producing characteristics of individual B lymphocytes by

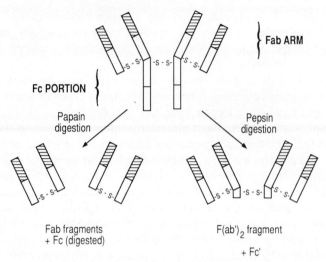

Fig. 18.2 Proteolytic fragments of IgG.

converting them into continuously growing tumour cell lines. Their procedure has three stages (see Fig. 18.3).

In the first stage, a suspension of lymphocytes is prepared from the spleen of a suitably immunized animal, usually a mouse or rat. These lymphocytes are mixed with 'myeloma' cells in the presence of an agent such as polyethylene glycol, which causes the cells to fuse together. A myeloma is a malignant plasma cell tumour whose cells secrete immunoglobulins. The myeloma cell lines used for fusion are variants that have been selected for a number of useful characteristics including resistance to 8-azaguanine, inability to synthesize or secrete immunoglobulin encoded by the myeloma genes, high fusion rates and stable growth in culture or in rodents. The fusion of a splenic B lymphocyte with a myeloma cell gives rise to a hybrid cell called a 'hybridoma'. This hybrid inherits both the capacity of the myeloma partner for continuous proliferation and the capacity of the B lymphocyte partner to synthesize and secrete the specific antibody.

In the second stage of the protocol, hybridoma cells are isolated from unfused myeloma cells in the fusion mixture by drug selection. This step is necessary because myeloma cells often divide more rapidly than hybridoma cells and would soon outgrow them in culture. Normally, myeloma cells cannot grow in the presence of 8-azaguanine. However, the variant myeloma cells used for fusion are resistant to this drug because they lack hypoxanthine phosphoribosyl transferase (HPRT), an enzyme which catalyses a step in the synthetic pathway from hypoxanthine to purines (see Chapter 6). These resistant myeloma cells use

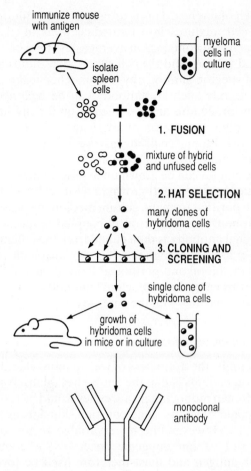

Fig. 18.3 Production of monoclonal antibody.

alternative synthetic pathways for DNA synthesis and replication. Accordingly, the cell mixture is grown in medium containing hypoxanthine, aminopterin and thymidine (HAT). Aminopterin blocks the main pathway of purine and pyrimidine synthesis so that the myeloma cells (lacking HPRT) are unable to grow. The hybridoma cells divide normally since they have HPRT activity contributed by the parental splenic B lymphocyte and can use hypoxanthine as a substrate for DNA synthesis. Unfused splenic B lymphocytes do not grow in culture and only the hybridoma cells persist when cultured continuously in the HAT selection medium.

Finally, the mixture of purified hybridoma cells is screened to identify and isolate individual hybridoma clones which secrete an antibody that binds the antigen of interest. This is usually done by diluting suspensions of hybridoma cell mixtures into individual wells of plastic microtitre plates so that each well contains, on average, less than one cell. Clones of cells are allowed to grow and the cell supernatant fluids are assayed for the presence of specific antibody by immunochemical techniques. Cells of clones that secrete such an antibody can be recloned by the same limiting dilution procedure and rescreened until truly monoclonal cell suspensions are obtained. An alternative way to isolate individual clones is to disperse hybridoma cells in agarose mixture and pick out the individual clones that grow.

Hybridoma cell lines can be grown up in large numbers either in tissue culture or as tumours in mice. If grown in tissue culture, the monoclonal antibody can be purified from the culture medium by passing it through a column of anti-mouse Ig or anti-rat Ig coupled to an inert matrix. The bound monoclonal antibody is then eluted from the column, usually with buffers of high or low pH or high salt. If grown in animals, the antibody is extracted from the blood and peritoneal fluid of the animal by standard methods but in this case, the monoclonal antibody is contaminated with normal Ig which copurifies with the antibody.

18.1.3 *Human monoclonal antibodies*

One factor that limits the usefulness of rodent monoclonal antibodies for cancer therapy in patients is that the antibodies themselves are immunogenic in man. Human monoclonal antibodies would be less immunogenic because they would have the same constant domains as normal human immunoglobulins. However, the hypervariable region of human antibodies would still be immunogenic because it is complementary in structure to the antigen and must therefore itself be foreign. The production of human monoclonal antibodies is more difficult than that of murine monoclonal antibodies for a number of reasons. First, mouse myeloma cell lines are not ideal fusion partners for human B lymphocytes because mouse:human hybrid cells tend to lose the human chromosomes encoding the antibody molecule during prolonged growth in culture. Second, there are at present few human myeloma cell lines which fuse with human B lymphocytes to produce hybridomas that secrete monoclonal antibodies in useful amounts. Third, immune B lymphocytes are difficult to obtain from man although such cells may be extracted from within and around tumour sites in patients. This problem may be solved by the techniques now being developed for immunizing human B lymphocytes with the antigen *in vitro*. Finally, not all human:human hybrid cells grow as tumours even in immunologically deficient mice. Other ways of immortalizing human antibody forming cells are also being sought; for example, B lymphocytes can be trans-formed into continuously growing cell lines by infecting them with viruses such as the Epstein Barr virus.

18.1.4 *Genetically engineered antibodies*

Recently, genetic engineering techniques have been used to construct vectors with altered immunoglobulin genes. When a suitable non-producing mouse myeloma cell is transfected with one of these vectors, it expresses the manipulated genes and is able to secrete the modified antibody. Thus, it is possible to create antibodies in which the variable region domains of both heavy and light chains are encoded by genes of a mouse antibody-secreting myeloma and the constant region domains are encoded by human genes. These 'chimaeric' antibodies have the antigen binding specificity of the antibody produced by the parental mouse myeloma and effector functions determined by the Fc portion of the human molecule. In a development of this approach, the six hyper-variable regions from the heavy and light chain variable domains of a rat antibody recognizing a widely distributed leukocyte surface antigen (CDw52) on human lymphocytes were successfully 'grafted' onto the

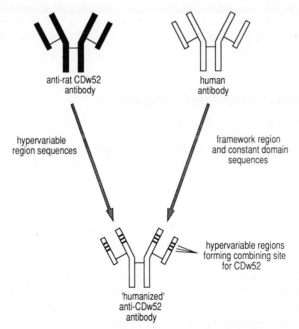

Fig. 18.4 Genesis of a 'humanized' recombinant anti-CDw52 antibody.

framework of a human immunoglobulin molecule (Fig. 18.4). The re-fashioned human antibody proved more effective than the parent rat antibody at inducing the lysis of target cells by recruitment of human effector mechanisms (see p. 484).

Further manipulation of immunoglobulin structure by site directed mutagenesis will allow the substitution of amino acid residues both in the antigen binding site to alter the affinity or specificity of binding, and in the constant domains to enhance or eliminate the various physiologically relevant functions of the antibody. More extensive reconstructions are also possible. Neuberger and his colleagues (1986) replaced the constant region domains in the Fc portion of a mouse antibody with a nuclease from *Staphylococcus aureus*. The resulting bifunctional chimaeric molecule retained both antigen binding ability and nuclease activity. This example demonstrates the promise of these techniques for combining the antigen recognition properties of antibodies with novel effector functions. For cancer therapy, it would be useful to produce chimaeric molecules made with proteins that possess cytotoxic activity. However, not all the factors which determine the correct folding and assembly of antibody molecules, or indeed of chimaeric antibodies, are understood. The *S. aureus* nuclease is a secreted monomeric protein which contains no disulphide bonds and is easily refolded after denaturation. Many of the best candidates for fusion do not have these useful properties and it is not clear whether present methodology will allow more complex poly-peptides to be incorporated into bifunctional antibodies in a biologically active form.

18.2 Tumour associated antigens

18.2.1 *The tumour cell surface*

The plasma membrane of normal cells is a mosaic consisting of different glycoproteins and glycolipids which are partially embedded in a fluid lipid bilayer. Some membrane molecules serve a transport function, others are involved in cell to cell contact, and many act as receptors for growth and differentiation factors. Some of these molecules are shared by many cell types while others have a more restricted distribution, occasionally being limited to a single cell type or even being only transiently expressed at a particular stage of cell maturation.

When a normal cell becomes malignant, it continues to express many of the cell surface molecules that are characteristic of its normal counter-part. However, these 'normal' molecules can be expressed in abnormal amounts. The malignant cell can also express cell surface molecules that are not produced by the parent cell, for example, viral antigens or antigens normally expressed only during embryological development. Although these antigens may not be tumour specific, they can often be highly selective for particular types of tumour cells (see Chapter 15).

When a malignant tumour cell is injected into an animal of another species, the animal produces a mixture of antibodies against those cell surface components which it recognizes as foreign. The immunogenic determinants (epitopes) of glycoproteins can be structures formed by either the amino acid residues or by the carbohydrate side chains. Only the carbohydrate component of glycolipids tends to be antigenic, however, since the lipid component is sequestered within the membrane bilayer.

18.2.2 *Target antigens*

It is the rule rather than the exception that a monoclonal antibody against a tumour associated antigen binds to some normal cells in addition to the target cell. The normal tissue may express either the same antigen as the tumour cell or a molecule structurally related to that recognized by the antibody. This is not a serious problem for tumour imaging provided that the normal cells have an anatomical distribution that does not obscure the image of the tumour cells. For therapy, however, binding to normal cells reduces the amount of antibody that reaches the tumour and also increases the risk of damage to normal tissues.

Antibodies that react with normal tissues may still be used for therapy in a number of situations. First, this is possible if the normal tissue is not life sustaining; for example, tumours of lymphoid origin can be attacked using antibodies against antigens on T or B lymphocytes because the healthy lymphoid cells that are killed by the treatment are soon replaced by new cells which emerge from the bone marrow. Second, the normal tissue may express the target antigen at a much lower density than the malignant cells and so escape damage. For example, leukaemic cells induced by human T cell leukaemia virus, HTLV-1, have 10- to 100-fold more surface receptors for interleukin 2 than do normal T lymphocytes. Similarly, malignant cells often proliferate more rapidly than normal cells and so express a relatively high level of the transferrin receptor and other molecules needed for cell division. Lastly, anatomical barriers can sometimes prevent the antibody from gaining access to the normal tissue bearing the target antigen. An example of this is the carcinoembryonic antigen (CEA) which is often expressed by colonic carcinoma cells. CEA is also found on the luminal epithelial cells of the gut but anti-CEA antibody injected intravenously rarely traverses the gut wall to reach the normal cells.

Tumour associated antigens can arise in a number of different ways

(Fig. 18.5). CEA is one example of an 'oncofetal' antigen that is normally associated with fetal cells of the same cellular lineage as the tumour cell and reflects the tendency of malignant cells to be or become more primitive. A second type of tumour associated antigenic determinant is found on molecules which are structural variants of normal cellular products. The p97 antigen associated with melanoma is structurally related to serotransferrin and lactotransferrin. Mutant proteins can result from random mutations in the DNA of the cell sustained after exposure to chemical carcinogens or ionizing radiation. Thus, each individual rodent fibrosarcoma induced by exposure to 3-methylcholan-threne has a different unique antigen. Cells transformed by some onco-genic viruses synthesize mutant proteins that are apparently related to the normal products of certain cellular genes (protooncogenes). Of these, a number are expressed at the cell surface, for example, the truncated avian epidermal growth factor (EGF) receptor on cells trans-formed by avian erythroblastosis virus (see Chapters 9 and 11).

A large number of the monoclonal antibodies which detect tumour associated antigens recognize unusual carbohydrate portions in glyco-proteins and glycolipids. These antigens result from derangements in the pathways of the tumour cell which synthesize oligosaccharide side chains. Human tumours which display characteristic glycolipid struc-tures include melanoma (GD3), Burkitt's lymphoma (GB3), Hodgkin's disease (asialo GM2 and Gg3) and cancers of neuroectodermal origin. A glycolipid related to the Le^a blood group antigen is present on human fetal intestinal cells as well as on gastrointestinal adenocarcinomas.

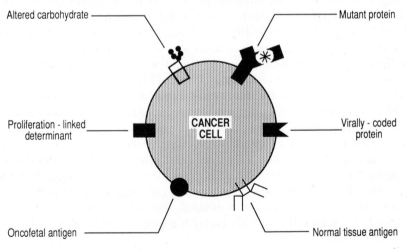

Fig. 18.5 Target antigens.

Changes in carbohydrate expression on the cell surface may also reveal cryptic (masked) antigens.

18.2.3 *Problems of targeting to tumour associated antigens*

Only a proportion of the cells in a tumour have the capacity for self renewal and are truly malignant. In some diseases, such as chronic myelocytic leukaemia, these clonogenic malignant cells represent only a small proportion of the tumour mass. Although the bulk of cells which carry the target antigen can be successfully imaged using antibodies, in therapy it is imperative that the antibodies reach and kill the clonogenic subpopulation. However, it has proved difficult to identify these rare cells and demonstrate that they also bear the target antigen.

In the heterogeneous tumour cell population, some cells express only low amounts of the target antigen and may escape being killed because less antibody binds to them. If these cells are clonogenic, the tumour will regrow. Moreover, as tumour cells continue to proliferate, mutant malignant cells that lack the target antigen frequently emerge. Thus, there are foci of cells within the primary tumour and its metastatic progeny which cannot be recognized by the antibody. This presents a serious obstacle to successful imaging or therapy that can sometimes be countered by using 'cocktails' of monoclonal antibodies which recognize several different tumour associated antigens.

The cells of a tumour which express high levels of the target antigen may also evade being coated by antibody in a number of ways. First, some antigens, which are loosely associated with the cell membrane, are shed from the tumour cell surface. The antigen in the circulation can then form complexes with the antibody and this greatly reduces the quantity of antibody that actually reaches the tumour cells. Second, the host's immune system may raise its own antibodies against the tumour associated antigen which then coat the malignant cells and 'mask' the antigen from the murine monoclonal antibody. Lastly, antibody may be physically prevented from gaining access to the majority of cells within the tumour. Thus, the most accessible tumour cells are those which surround the blood capillaries or, like leukaemia cells, are in blood vessels.

18.3 Tumour imaging

18.3.1 *Principles*

Tumour imaging is a procedure which allows the clinician to identify both primary and secondary sites of tumour growth in a patient and to estimate the overall tumour burden without recourse to surgery. An antitumour antibody labelled with a radioactive isotope, usually ^{131}Iodine

(^{131}I), is injected into the patient intravenously and, after allowing the antibody time to localize within the tumour, the patient is scanned using a radiation detector called a gamma camera. A computer compiles an image of the radioactivity detected in the patient's body and colour codes it according to the intensity of the radiation. Zones of high radioactivity in the regions of the body that are not expected to accumulate the antibody or its metabolites indicate the possible presence of a tumour. This is illustrated in Fig. 18.6. In combination with other clinical evidence, these results can help determine the most appropriate form of surgical, radiological, or chemotherapeutic treatment.

The principle of radioimmunolocalization was first demonstrated by experiments in which tumour-bearing animals were injected with radio-labelled polyclonal antibodies directed against malignant cells. The concentration of anti-tumour antibodies in the tumour typically reached levels up to four times higher than those in miscellaneous normal tissues. The absolute amount of radioactivity associated with tumour increased in the first few hours after injection reflecting the time taken by the antibody to get out of the bloodstream and to diffuse to the site of the tumour. Maximal specific uptake by tumour is generally reached between one and four days after injection. Most of the antibody that is not cleared remains in the blood and extravascular compartment. Consequently, there is a background of radiation, particularly in blood rich organs such as the liver and spleen. Tumours in these sites are not easy to

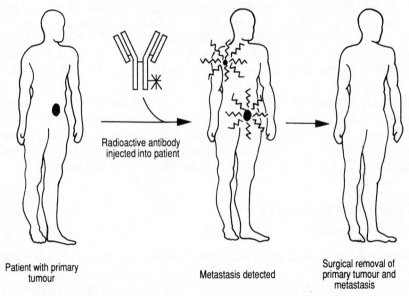

Radioactive antibody
injected into patient

Patient with primary
tumour

Metastasis detected

Surgical removal of
primary tumour and
metastasis

Fig. 18.6 The principle of radioimmunolocalization.

detect unless the image is corrected for background radiation (see p. 483).

18.3.2 *Clinical studies*

The first clinical study using radioimaging to detect tumour tissue was reported by Goldenberg *et al.* in 1978. Patients were injected with ^{131}I-labelled anti-CEA antibodies and were then scanned after an interval of 24–72 h. To correct the image for the non-specific background of radiation, a subtraction technique was used. Before injection of ^{131}I-antibody and scanning, patients were injected with ^{99m}Technetium (^{99m}Tc) either bound to human serum albumin or free as pertechnetate, ^{99m}TcO$_4^-$. After computer subtraction of the ^{99m}Tc-image from the ^{131}I-image, areas of specific antibody uptake were highlighted.

A number of groups have now used anti-CEA antibodies to image a total of several hundred human carcinomas. Similarly trophoblastic tumours and germ cell tumours expressing human chorionic gonadotrophin (HCG) or α-fetoprotein (AFP) have been imaged using labelled anti-HCG antibodies and anti-AFP antibodies respectively. There have been only a few reported uses of radiolabelled murine monoclonal antibodies for imaging of tumours: colorectal cancer using anti-CEA antibodies, ovarian and breast cancer with antibodies against human milk fat globule (HMFG), and melanoma using anti-p97 antibody and its Fab fragments. Figure 18.7 shows an example of tumour imaging in a patient. These clinical trials have identified a number of problems that limit the usefulness of the technique.

The first limitation of present radioimaging techniques is that not all tumours are detected. The amount of radioisotope which reaches tumour sites is small, as a rule 0.001–0.1 per cent of the injected dose per gram of tumour and so small tumour masses are poorly resolved. In practice, the tumours which are routinely diagnosed have diameters of at least 1 cm but metastatic microfoci, which may be more relevant to the long-term survival of the patient, are unlikely to be detected. Even the larger tumours escape detection if they express a low level of the target antigen or if they are poorly vascularized. For these tumours, computerized tomography (CT) scans (see Chapter 16) give more reliable results. A further problem is that intravenously injected antibody cannot normally penetrate sanctuary sites in the body such as the brain.

A second limitation of radioimaging is that areas of high radioactivity which are free of tumour are detected. These false positives occur when immune complexes formed in the blood are trapped by the liver and when free ^{131}I released from the antibody is eliminated from the bloodstream predominantly into the kidneys, urinary bladder, stomach, and

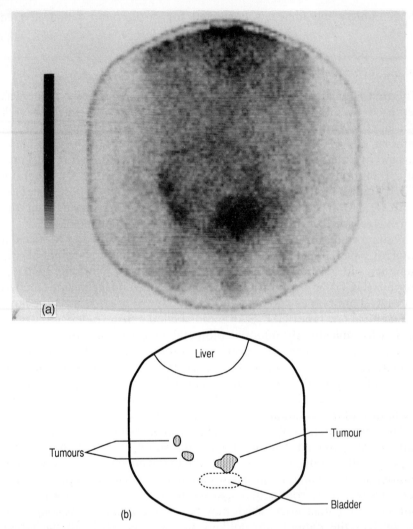

Fig. 18.7 Tumour imaging in a patient. Ovarian tumours in a patient localized by a monoclonal antibody H17E2 labelled with ^{123}I. This antibody recognizes a placental-type alkaline phosphatase associated with ovarian tumours of epithelial origin.

intestine. Accumulation of iodine by the thyroid gland is usually blocked by co-administering Lugol's iodine or potassium iodide. Subtraction of the background radiation does not always eliminate false positives. The radiation emitted by ^{99m}Tc has a lower energy than that of ^{131}I and penetrates tissue more poorly resulting in the appearance of 'halo' effects in the image of the ^{131}I radioactivity after subtraction.

18.3.3 *Strategies to improve tumour targeting*

The sensitivity of tumour detection is improved by using radionuclides with better imaging characteristics than ^{131}I and by better methods for removing background radiation. [131]I has too high an energy for optimal imaging with current gamma cameras; radiation with an energy of 0.2–0.4 MeV is most suitable. Its half-life is rather long (8 days) so that the patient's normal tissues are excessively exposed to radioactivity and the decay process also produces potentially harmful β-particles. [131]I is being superceded by other isotopes of iodine (^{123}I) and by radionuclides that have shorter half-lives, better energy characteristics, and emit no β particles (Table 18.1). The radioactive metal ions of indium (^{111}In) and gallium (^{67}Ga) can now be stably coupled to antibodies by means of chelating molecules.

Table 18.1 Isotopes for imaging

Isotope	Energy of principle γ emission (MeV)	Half-life (hours)	Method of coupling to antibodies
[131]I	0.36	192 ⎫	Direct halogenation
[123]I	0.16	13 ⎭	
[111]In	0.17–0.25	67 ⎫	
[67]Ga	0.09	78 ⎬	Chelation
[99m]Tc	0.14	6 ⎭	

Background radiation due to the residue of circulating radiolabelled antibody can be cleared in a number of ways. First, synthetic lipid vesicles called 'liposomes' coated with anti-mouse immunoglobulin bind mouse monoclonal antibodies in peripheral blood and accelerate their clearance probably because liposome particles are rapidly removed by cells of the reticuloendothelial system. Second, clearance can be enhanced by the injection of a second antibody directed against radiolabelled antibody. Lastly, the use of F(ab')$_2$ or Fab fragments may improve tumour discrimination since antibody fragments are cleared from the circulation much more rapidly than intact Ig. Moreover, since these antibody fragments lack the Fc portion, there is less non-specific accumulation of radioactivity by the reticuloendothelial system.

A new approach to background subtraction depends on the slower rate of loss of radioactivity from tumour tissue as opposed to normal tissue. A series of scans is taken at intervals after injection of the radiolabelled antibody. Zones in the body where the antibody persists show up as 'hot spots' of radiation. This technique avoids the imaging problems associated with the use of a second radionuclide.

Finally, the use of single photon emission computerized tomography, analogous to CT scanning using X irradiation, may allow a better discrimination between valid and false positive tumour images. The computer constructs images of transverse sections of the patient's body and allows examination of putative tumour sites from several angles.

New approaches to tumour identification have been reviewed recently (Britton 1987).

18.4 Immunotherapy

18.4.1 Mechanisms of tumour cell killing

The goal of antibody mediated immunotherapy is to treat cancer patients with anti-tumour antibodies that will bind selectively to tumours and bring about the destruction of malignant cells either directly or by stimulating the natural defence mechanisms of the recipient.

An antibody can be directly cytotoxic (or more accurately, cytostatic) if it binds to and inactivates a cell surface molecule which is necessary for survival or proliferation. For example, an anti-transferrin receptor antibody can inhibit the growth of human tumour cells both *in vitro* and in athymic mice. Other antibodies which might directly suppress tumour growth are those against the receptors for interleukin 2, EGF, and platelet derived growth factor (see Chapter 11) which are particularly abundant on certain types of tumour cell. Such examples are rare; antibodies themselves do not usually kill cells and need some accessory mechanism to do so.

There are two main ways by which antibody molecules bound to the surface of a cell can activate host defence mechanisms (Fig. 18.8).

Complement mediated cytotoxicity. Complement is the name given to a series of proteins (C1–C9) present in serum in inactive form. The C1q complement component binds to the Fc part of cell bound antibody and triggers a cascade reaction in which several of the remaining complement components are activated by proteolytic cleavage. The other complement proteins (C5b–9) then form a 'membrane attack complex' which inserts into the cell membrane and induces lysis.

Antibody dependent cell mediated cytotoxicity. Several distinct types of defence cells kill antibody coated cells *in vitro* and it is presumed that they also do this *in vivo*. The effector cells include polymorphonuclear cells (e.g. neutrophils), macrophages, and a type of mononuclear cell called the NK (or natural killer) cell which is present in lymphoid tissues but which lacks mature T and B lymphocyte markers (see Chapter 15). All these cells adhere to antibody coated cells by means of receptors for

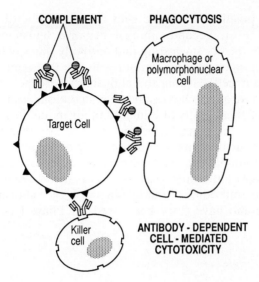

Fig. 18.8 Antibody mediated mechanisms of cell killing.

the Fc portion of the immunoglobulin molecule. Polymorphonuclear cells and macrophages also have receptors for the C3 component of complement. If the antibody is of a type that binds complement, this gives dual recognition which strengthens the adherence of the effector cell to the target cell. The various effector cells then kill the target cell by phagocytosis or direct lysis.

The heavy chain class or subclass of the monoclonal antibody determines which effector systems are activated. With monoclonal antibodies raised in the mouse, IgGs of all major subclasses can elicit cell mediated cytotoxicity, whereas IgA and IgM cannot. By contrast, IgM is the most powerful activator of complement whereas IgG2a, IgG2b, and IgG3 fix complement weakly, and IgG1 not at all. The therapeutic effect of an antibody therefore depends on which host defence system is most effective at killing tumour cells.

18.4.2 *Studies in experimental animals*

The feasibility of immunotherapy has been demonstrated by several studies in experimental animals. For example, Bernstein and his colleagues (1980) showed that up to 3×10^5 leukaemia cells transplanted into mice could be eliminated by intravenous injection of antibody against the Thy1.1 surface antigen expressed by the leukaemia cells. When mice carrying 3×10^6 tumour cells subcutaneously were treated in the same fashion, the mice were not cured apparently for two reasons.

First, antigen positive leukaemic cells coated with antibody were not killed by the host. Second, antigen negative leukaemic cells were detected in the spleen, suggesting that antibody therapy had prevented the spread of antigen positive tumour cells but that a mutant subpopulation of tumour cells lacking the antigen had escaped. Other studies have tested mouse antibodies against human tumour associated antigens in mice carrying xenografts of human tumour tissue. In general, significant regression of the transplanted tumour was only observed in animals with small tumour burdens, a situation very different from the clinical situation in human patients with spontaneous tumours and large tumour cell burdens. IgM antibodies were generally ineffective in these *in vivo* studies. Mouse antibodies of the IgG2a and IgG2b subclasses were the most effective and have since been used for Phase I clinical trials in humans.

18.4.3 *Clinical trials*

Trials with mouse monoclonal antibodies have been performed in patients suffering from advanced malignant melanoma and gastrointestinal tumours. However, only tumours of the haematopoietic system, including B cell and T cell leukaemias and lymphomas, have responded well to this treatment and some patients remain disease free after several years.

Miller *et al.* (1982) raised monoclonal antibodies against the idiotypic determinants of the neoplastic cells of a patient with a type of B cell lymphoma called 'follicular lymphoma', In this form of cancer, a single clone of B cells proliferates and so all the cells of the tumour bear surface Ig with the same idiotype. Since each clone of normal B lymphocytes in the body expresses a molecule with a unique antigen binding site, the idiotype of the surface Ig of the lymphoma cells provides a unique tumour specific target antigen. Tumour regression occurred after repeated intravenous infusions of the anti-idiotype monoclonal antibody and continued when treatment was stopped, leading to a complete remission (see Fig. 18.9). The mechanism of this remarkable therapeutic effect is obscure. It is possible that the malignant cells were not killed by the patient's defence systems but that the antibody directly suppressed their proliferation in a manner analagous with the way in which anti-idiotypic antibodies normally control the proliferation of cells that produce antibody in response to an antigenic stimulus. In support of this notion is one recent observation that, of 12 patients treated in a similar fashion with anti-idiotypic antibodies, those with the largest number of T cells infiltrating the tumour showed the best response.

More typically, in therapeutic trials against lymphomas and leukaemias, the infusion of anti-tumour monoclonal antibodies directed against

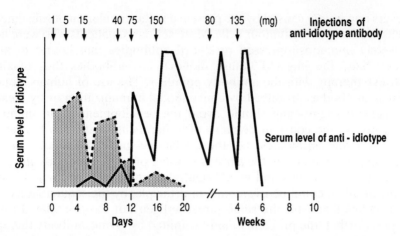

Fig. 18.9 Antibody mediated immunotherapy of B cell lymphoma. The effect of anti-idiotypic antibody on the level of serum idiotype in a patient with B cell lymphoma (see text). The onset of tumour regression coincided with the antibody induced clearance of serum idiotype.

different lymphocyte-specific antigens induced a rapid decrease in the numbers of circulating tumour cells. These effects generally persisted provided that the level of antibody was maintained by repeated infusions. However, once the antibody was allowed to clear completely, the leukaemic cell counts returned to pretreatment levels or higher. The temporary improvements observed may reflect an inherent limit to the capacity of host effector mechanisms to kill large numbers of tumour cells. It is known, for example, that the capacity of the hepatic reticulo-endothelial system to clear antibody coated tumour cells from the blood can become saturated. The tumour deposits in the lymph nodes may have been more resistant than tumour cells free in the circulation because they were less accessible to the antibody. Alternatively, the lymph nodes may have contained too few effector cells; immunotherapy using monoclonal antibodies relies on the patient being able to destroy the antibody coated tumour cells and it is likely that the effector systems of a patient with advanced malignant disease will have been compromised either by prior therapy or by the disease itself.

18.4.4 *Problems encountered during clinical trials*

Several problems undermining the therapeutic value of the anti-tumour antibodies arose during the clinical trials. These included hypersensitivity reactions to the murine protein in some patients, and transient fevers apparently caused by the presence in the circulation of substances released from dying tumour cells. More seriously, patients who were not severely immunosuppressed produced antibodies that bound to and neutralized the injected mouse monoclonal antibodies thus making further therapy with the antibody pointless. The use of human monoclonal antibodies produced by conventional immunization or by means of genetic engineering should avoid the development of an immune response to the constant domains of a therapeutic antibody. However, anti-idiotypic antibodies, i.e. those that recognize the unique determinants associated with the antigen-binding site of the antibody, are also generated in patients and would neutralize the effects of the antibody. This problem could be overcome, in principle, by the sequential use of antibodies that bind the same target antigen but have unrelated idiotypes. Further, the problem could be limited by giving antibody therapy in combination with immunosuppressive drugs. On the other hand, if host defence systems are necessary for a therapeutic effect, immunosuppression may prevent tumour destruction. The effectiveness of the antibody treatment was also reduced by the formation of complexes between the infused antibody and target antigens shed from the tumour cell surface. In some patients, this occurred even with target antigens that are not normally shed at high levels because the tumour cell destruction which followed attachment of antibodies and subsequent attack by host effector systems itself tended to raise the systemic level of free antigen. There is no good solution to the problem of antigen shedding.

Another serious obstacle encountered in the clinical trials was the phenomenon of antigenic modulation. Some cell surface components, when cross linked by antibodies, are actively internalized by the cell in a matter of minutes or are shed from the cell surface. This reduces the number of molecules of the target antigen available for binding by the antibody and may allow the cell to escape killing since the modulated antigen is re-expressed only when antibody is no longer present some days later. There are a number of ways to bypass the effect of antigenic modulation. First, not all surface components are subject to modulation and so a non-modulating antigen would be the preferred target. Second, treatment with antibody can be tailored to maximise the chances of killing the tumour cell. For instance, repeated infusions of antibody can be scheduled to coincide with the reappearance of the antigen that occurs once the antibody has cleared. Lastly, since the cross linking of target antigens that leads to modulation is a consequence of the bivalency of the antibody molecule, the way to avoid this problem may be to use monovalent anti-tumour antibodies.

Monavalent antibodies have been produced either by protein chemical methods or by exploiting the biology of hybridoma fusion. Limited proteolysis of bivalent rabbit antibody yields an Fab/c fragment which contains a single antigen binding site (Fig. 18.10a). However, this procedure is not generally applicable to antibodies from other animal species. Instead, one can construct chimaeric antibodies that are operationally monovalent by attaching a purified Fab′ fragment with the appropriate target specificity to normal human immunoglobulin (Fig. 18.10b). Alternatively, hybrid cells made by fusing B lymphocytes with a myeloma secreting Ig light chains, manufacture a proportion of antibodies with only one functional antigen binding site (Fig. 18.11). The heavy chain has an equal chance of combining with the irrelevant myeloma light chain or the splenic light chain so that about half of the antibody molecules produced by the hybrid will contain one Fab arm that can bind the target antigen and one which cannot. These monovalent antibodies have normal Fc regions and so are able to activate host effector systems. Since twice as many monovalent antibodies can, in theory, bind to the cell surface, their therapeutic effect might be greater than that of bivalent antibodies. However, the strength or 'avidity' of binding to the cell surface is greater for a bivalent antibody molecule than for a monovalent one, since both arms of the bivalent antibody contribute to the energy of binding.

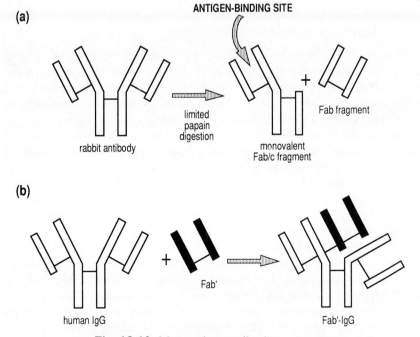

Fig. 18.10 Monovalent antibody constructs.

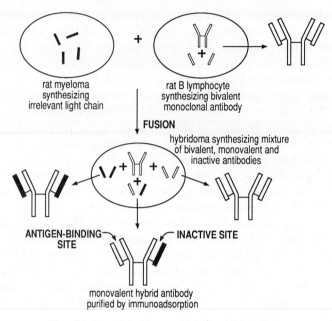

rat myeloma
synthesizing
irrelevant light chain

rat B lymphocyte
synthesizing bivalent
monoclonal antibody

FUSION

hybridoma synthesizing mixture
of bivalent, monovalent and
inactive antibodies

ANTIGEN-BINDING
SITE

INACTIVE SITE

monovalent hybrid antibody
purified by immunoadsorption

Fig. 18.11 Hybrid rat monoclonal antibodies.

18.4.5 *Novel approaches to immunotherapy*

Cytotoxic T lymphocytes are normally activated by antigen only in association with surface antigens of the major histocompatibility complex (MHC) (see Chapter 15). However, T cells can also be activated independently of MHC antigen involvement by monoclonal antibodies that bind to the T cell receptor/CD3 complex (see Chapter 15). Bi-specific antibody constructs which combine the activating antibody and an anti-tumour monoclonal antibody have been created by chemical methods and by hybridoma technology. These hybrid antibodies bind the T cell to the tumour cell (Fig. 18.12) and activate the T cell to lyse the tumour cell *in vitro*. In this way it might be possible to recruit T cells *in vivo* as a mechanism for the antibody-directed killing of tumour cells in addition to complement lysis and antibody-dependent cell-mediated cytolysis (see p. 484). Alternatively, a patient's T cells could be pre-activated with the bispecific antibody *ex vivo* before being returned to the patient. Detailed studies in animal model systems are necessary to establish the practicality of this novel approach to immunotherapy.

The immune response of patients to administered antibody has been considered a major problem of this form of therapy (see p. 488). However, the long-term regression of solid tumours in some patients was

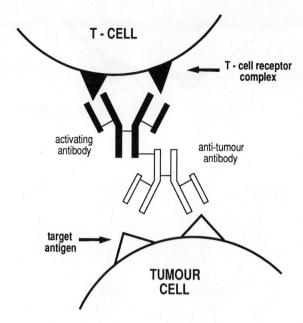

T - CELL

T - cell receptor
complex

activating
antibody

anti-tumour
antibody

target
antigen

TUMOUR
CELL

Fig. 18.12 Tumour cell killing by recruitment of cytotoxic T cells.

accompanied by the development of an antibody response against the idiotype of the therapeutic antibody. In experimental studies, an anti-tumour antibody can be used to raise a monoclonal anti-idiotypic antibody whose antigen combining site mimics the structure of the antigen on the tumour cells recognized by the anti-tumour antibody (Fig. 18.13). Animals receiving the anti-idiotypic antibody become immune to subsequent challenge with the tumour. This is because the animal produces antibodies against the anti-idiotypic antibody which recognize the site that mimics the target antigen and, therefore, also recognize the antigen on the tumour cell itself. It remains to be demonstrated that the administration to patients of anti-idiotypic antibodies raised against antibodies recognizing human tumour associated antigens can induce therapeutically useful responses.

18.5 Targeting of cytotoxic agents

18.5.1 *Principles*

In general, the anti-cancer drugs now in clinical use do not discriminate between malignant and normal cells (see Chapter 17). Most exert their effects on dividing cells so that tumour growth is inhibited but the dose of

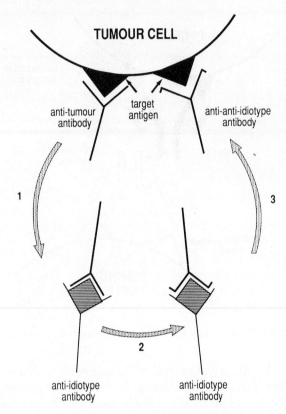

Fig. 18.13 Recognition of tumour cell antigen by anti-anti-idiotype antibodies. 1. Anti-idiotype antibody raised against anti-tumour antibody. 2. Anti-idiotype antibody used to raise anti-anti-idiotype antibody. 3. Anti-anti-idiotype antibody binds to target antigen on tumour cells.

drug which can be used is limited by its toxicity to the normal tissues of the body which need to divide frequently, e.g. gastrointestinal epithelium and the haematopoietic cells of the bone marrow.

The aim of drug targeting is to deliver the cytotoxic drug to the tumour and only expose the rest of the body to a low level of the drug so preventing harm to normal tissues. Several systems for drug delivery have been explored. One approach is to package the drug in liposomes which are then coated with the anti-tumour antibody. The liposomes attach specifically to tumour cells *in vitro* and kill them, probably by fusing with the cell membrane and discharging their drug content into the cell. Unfortunately, *in vivo*, the large size of the liposomes causes them to be engulfed by phagocytic cells in the bloodstream. The anti-tumour effects of drug loaded liposomes in experimental animals have thus been unremarkable.

Attempts to endow cytotoxic agents with target specificity by simply linking them directly to the antibody have been more successful. Drugs of low molecular weight, radionuclides, toxins of bacterial and plant origin, and cell surface active agents have all been coupled to specific antibodies and the conjugates found to exert selective cytotoxic effects.

18.5.2 *Chemotherapeutic drugs with intracellular sites of action*

In early studies, non-covalent complexes made by physically adsorbing chlorambucil to anti-tumour antibodies were found to kill appropriate tumour cell targets in tissue culture and in mice more efficiently than the free drug or antibody alone. The anti-tumour effects were later shown to be due not to drug targeting but to synergism between free drug and antibody after dissociation of the complex. The synergy probably arises because the drug impairs the cell's capacity to repair plasma membrane damage caused by the antibody and complement.

To prevent dissociation, covalent bonds have been used to link several anti-cancer agents (e.g. adriamycin, daunomycin, methotrexate, and vindesine) directly to anti-tumour antibodies. In most cases, the cyto-toxicity of the conjugate to target cells *in vitro* exceeded that to cells lacking the target antigen, but the cytotoxic potency of the conjugate was invariably low. These drugs have intracellular sites of action and enter cells by simple diffusion or active transport. Once linked to the antibody, however, they cannot enter the cell by their usual route. Antibody-drug conjugates bind to target antigens on the cell surface, enter the cell in endocytic vesicles, and are degraded by enzymes when the vesicles coalesce with lysosomes (see Fig. 18.14). Thus, the low potency of these conjugates may have been because too few drug molecules had originally localized at the cell surface to enter by this route or because drug release by lysosomal enzymes was inefficient.

Much progress has been made in devising ways to accelerate the rate of drug release once internalization has taken place by increasing the susceptibility of the covalent linkage between drug and carrier to hydrolysis. Daunomycin directly linked to albumin is not released by lysosomal enzymes but if a proteinase sensitive peptide spacer is inter-spersed between the drug and the carrier, drug release occurs rapidly. Similarly, acid labile linkages between drug and antibody can improve potency, presumably because the drug dissociates from the conjugate when exposed to the acidic milieu of the lysosome.

Only about 10 drug molecules can be attached covalently to an anti-body molecule without diminishing its antigen binding capacity. To raise the potency of antibody drug conjugates, a carrier molecule such as human serum albumin, poly-L-glutamic acid, or dextran can be used.

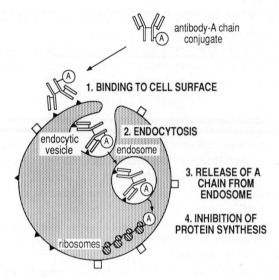

Fig. 18.14 Entry of antibody drug conjugates into the cell.

Such conjugates are more effective anti-tumour agents *in vitro* and *in vivo*. For example, Tsukada and his co-workers (1987) induced long-term remission in 60 per cent of rats with hepatoma by intravenously administering daunomycin dextran coupled to anti-AFP antibodies. The effect was specific since daunomycin alone, daunomycin-dextran plus antibody in unconjugated form, or control conjugates of irrelevant specificity were all less effective.

A few attempts to treat human malignancies with cytotoxic agents linked to polyclonal antibodies have been reported. Patients with disseminated malignant melanoma received multiple injections of goat anti-melanoma antibodies to which chlorambucil had been adsorbed non-covalently. A few patients showed regression of cutaneous and nodal metastases, but treatment had to be discontinued when they developed anaphylactic reactions to the globulin. Tumour regression (complete or partial) was also observed in children treated with a covalent conjugate of chlorambucil or daunorubicin and an antibody against neuroblastoma. In contrast, vindesine conjugated anti-CEA antibodies had no therapeutic effect in patients with colorectal or ovarian carcinoma.

18.5.3 *Radionuclides*

The radionuclide most commonly used for therapy is ^{131}I, which in addition to emitting γ radiation that is useful for radioimaging, emits high energy β particles which can penetrate tissues to a distance spanning many cell diameters. This is important for two reasons. First, the radioisotope does not need to be internalized to kill the cell. Second, the emissions can kill cells which surround the targeted cell, allowing the destruction of neighbouring malignant cells which do not bear the target antigen and of cells in poorly vascularized tumours. Other β-emitters which could be useful for radiotherapy include ^{32}Phosphorus and ^{90}Yttrium.

Pilot studies, in which patients were given antibodies radioiodinated to high specific activity, produced evidence of tumour regression with few toxic side effects. For example, Order and his colleagues (1987) treated patients with hepatocellular carcinoma with a succession of ^{131}I-labelled polyclonal anti-ferritin antibodies raised in different animal species to avoid the problem of neutralization by the patient's immune system. Anti-ferritin antibodies localize well in hepatoma because the tumour synthesizes high levels of apoferritin, a 400 kDa protein that becomes trapped in the tumour neovasculature. In 105 patients, the response rate was 48 per cent with 7 per cent complete responders. The longest complete remission continued for more than four years after therapy. Interestingly, mouse anti-ferritin antibodies were found to have a shorter tumour half-life than some other species and were clinically ineffective. Monoclonal antibodies have also been used in trials of radio-immunotherapy. Carrasquillo and his colleagues (1984) treated patients with advanced metastatic melanoma with ^{131}I-Fab fragments of two anti-bodies against melanoma associated antigens, p97 and 'high molecular weight antigen'. In some patients, a stabilization or transient reduction of tumour size was observed but most developed an antibody reaction against the mouse globulin that precluded further treatment. The other problems encountered were complexing of antibody with circulating p97, dissociation of the conjugate, and poor localization of the radio-labelled antibody within the tumour.

Radionuclides also irradiate normal tissues as they circulate through the body in conjugated form and as metabolites. For effective use, the antibody radionuclide conjugate should localize well within tumour and not be retained within normal tissues. The recent improvements in methods for tumour imaging (see p. 483) offer ways in which non-specific irradiation during therapy can be minimized. A general approach to increase the dose of radiolabelled antibody that localizes in tumour and to reduce the exposure of uninvolved tissues to radiation is to administer the antibody to the patient by routes allowing either more direct or more complete access to the tumour than the intravenous route. First, regional therapy is possible for neoplasms which are localized in discrete body regions such as the cerebrospinal space (neuroblastoma) or in the peritoneum (ovarian carcinoma). In patients with ovarian carcinoma treated with radiolabelled monoclonal antibody to HMFG antigen the intraperitoneal route of administration gave consistently higher uptake of radionuclide by ascitic tumour cells than the intravenous route. However, the intravenous route gave better uptake by solid tumour deposits in the peritoneum because the antibody was unable to reach the deposits except through the vasculature supplying the tumour. Second, extensive studies have shown that antibody localization to lymph nodes is enhanced by intra-lymphatic administration. Finally, infusion via the carotid artery may be exploited for the radiotherapy of intracranial malignancies such as gliomas which are not readily reached by antibodies in the general circulation. It should be noted, however, that there is no ready alternative to systemic therapy in the case of widespread metastatic disease.

Alpha particles may be more effective cytotoxic agents since they dissipate more energy than β particles in a path length of only one or two cell diameters and therefore have an exceedingly powerful cytotoxic action. Possible candidates for targeting include ^{211}Astatine and ^{212}Bismuth. An alternative is to target atoms of a non-radioactive isotope of Boron (^{10}B) to cells. When this nuclide is subsequently irradiated with low energy thermal neutrons, it undergoes nuclear fission liberating a high-energy α particle *in situ*. Unfortunately, thermal neutrons penetrate tissue so poorly that this approach to targeting seems to have limited application.

18.5.4 *Agents mediating membrane lysis*

Cell membrane-disrupting toxins have long been known to mediate the direct cytolysis of human erythrocytes and nucleated cells but there have been few experimental studies to assess the potential of antibody–cytolysin conjugates, or 'immunolysins', for cancer therapy. The phospholipase C from *Clostridium perfringens*, which is lytic to both human and sheep erythrocytes, became selectively lytic to sheep cells after attachment to antibody against the Forssman antigen on sheep red blood cells. A different approach is to target cobra venom factor (CVF) which mimics the action of complement component C3b and induces formation of the cell-lysing membrane-attack complex (see p. 484). CVF linked to a monoclonal antibody against human melanoma cells induced immune cytolysis of the target cells *in vitro* in the presence of added complement and in a mouse tumour model system.

18.5.5 *Toxins and related proteins*

Bacterial toxins (e.g. diphtheria toxin) and the plant toxins, ricin and abrin, are extremely potent; a single molecule appears to be sufficient to kill a cell if it enters the cytoplasm. These toxins have a similar molecular architecture, each consisting of two polypeptide chains, A and B, joined by a single disulphide bond and they bind to virtually all cells of higher animals by means of the B chain. The membrane-bound toxin is taken into the cell by endocytosis and the A chain is then somehow translocated across the membrane of the endocytic vesicle into the cytoplasm where it inactivates the cell's machinery for protein synthesis.

When targeting a toxin to a cell, the objective is to confer upon it the specificity of the antibody without attenuating the capacity of its A chain to enter cells and inhibit protein synthesis. Attempts have been made to achieve this aim by targeting intact toxins, toxin A chains only, and members of a group of ribosome inactivating proteins which act in a manner seemingly identical to the A chains of abrin and ricin. Conjugates between antibodies and toxins or their fragments are known collectively as 'immunotoxins'.

Many plants have been found to contain proteins that inhibit protein synthesis on eukaryotic ribosomes. These single chain proteins are similar in size and action to the toxin A chains and are probably evolutionarily related but they are virtually non-toxic to intact cells because they lack the equivalent of the B chain for binding to the cell surface. Gelonin, the ribosome inactivating protein from *Gelonium multiflorum*, acquires potent and specific cytotoxic activity after covalent linkage to a monoclonal antibody. Pokeweed antiviral protein, from *Phytolacca americana*, and saporin, from *Saponaria officinalis*, are likewise rendered cytotoxic when conjugated with antibody. A great advantage of using the single chain inhibitors over the toxin A chains is that they are safer to handle in quantity and need not undergo the same rigorous purification to exclude traces of intact toxin or contaminating B chain.

Research into the targeting of cytotoxic proteins has made rapid progress in recent years and is considered in detail in the next section.

18.6 Immunotoxins

18.6.1 *Design of immunotoxins for in-vivo use*

Immunotoxins made from intact toxins are outstandingly powerful cytotoxic agents for cells with appropriate antigens, often matching or even surpassing the potency of the native toxin *in vitro*. Since the B chain of the toxin can still bind to non-target cells, these immunotoxins are highly poisonous and show little evidence of selective cytotoxicity. The non-specific and undesirable binding properties of conjugates made with intact ricin or abrin (which bind to galactose-containing oligosaccharides on the cell surface) can be blocked *in vitro* with high concentrations of galactose or lactose. In animals, however, lactose is excreted rapidly and cannot protect the animal against non-specific toxicity. For therapy in animals, a permanent or semipermanent blockade is needed. Intact ricin conjugates in which the galactose binding sites of the toxin are apparently sterically (spatially) hindered by the antibody moiety show improved anti-tumour effects *in vivo* but, for reasons that are not understood, retain high toxicity to animals.

The most widely used way to eliminate non-specific binding is to attach the A chain of the toxin directly to the antibody. The intact toxin is first reduced and the A chain is then purified free of contaminating B chain. The antibody is modified with a chemical reagent introducing a reactive group that forms a disulphide bond with the single free sulphydryl group of the A chain (Fig. 18.15). The disulphide bond is necessary to allow release of the A chain from the antibody, presumably after reduction of the bond by endogenous thiols, in order to inhibit ribosomal protein synthesis. This step takes place once the immunotoxin has bound to the cell and been internalized and may be required to allow escape of the A chain from the endosome during the process of intoxication (Fig. 18.16). A chain immunotoxins are almost wholly selective in their toxic effect upon target cells in tissue culture and, in several studies, have produced impressive anti-tumour effects *in vivo*. One of the most dramatic therapeutic effects was obtained by Vitetta and her colleagues (1987) who induced prolonged remissions in mice with advanced lymphocytic leukaemia by administering a ricin A chain immunotoxin in conjunction with total lymphoid irradiation and splenectomy.

The main problem with A chain immunotoxins is that their toxicity to target cells varies greatly depending on the antigenic determinant recognized, the type of target cell, and its stage in the cell cycle. The reasons underlying these differences are not understood but are presumed to reflect the pathway by which the A chain immunotoxin enters the cell and the influence of the metabolic status of the cell upon that process. This variable cytotoxicity contrasts with the consistent and generally superior toxicity of conjugates made with the intact toxins. The cell-killing effect of weakly cytotoxic ricin A chain immunotoxins can be markedly enhanced *in vitro* by free ricin B chain, suggesting that the B chain also plays a role in facilitating the entry of the A chain portion of the conjugate into the cytoplasm of the cell. A recent finding with therapeutic potential is that toxicity is also potentiated by a 'piggyback' treatment with a B chain immunotoxin directed against the A chain immunotoxin. In addition, the weak cytotoxicity of some A chain immunotoxins is enhanced *in vitro* by the use of agents which elevate

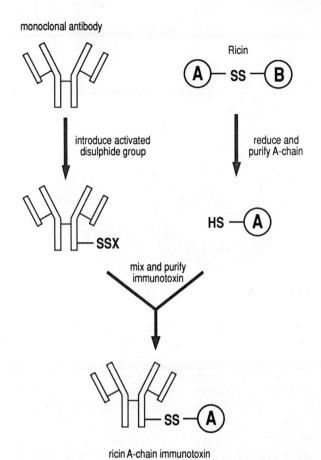

Fig. 18.15 Construction of a ricin A chain immunotoxin.

endosomal and lysosomal pH such as ammonium chloride, chloroquine, or monensin.

18.6.2 *Clinical trials*

Some clinical trials of immunotoxins in man have been reported. In a pilot study, two patients with leukaemia were treated with a ricin A chain immunotoxin against the CD5 antigen expressed by the tumour cells. As previously observed with monoclonal antibody therapy, infusion of immunotoxin led to a rapid but transient decrease in the level of circulating tumour cells. Importantly, no adverse toxicity was observed. Spitler and her colleagues (1987) conducted a phase I trial of a mouse monoclonal anti-melanoma antibody coupled to ricin A chain in 22

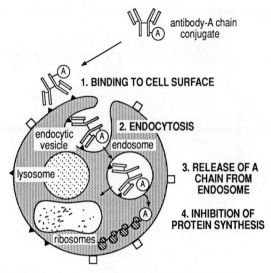

Fig. 18.16 Mechanism of action of ricin A chain immunotoxins.

patients with advanced metastatic melanoma. Half of the patients showed some response to the treatment and one patient had a complete remission. Of 46 patients treated with the same immunotoxin in a subsequent phase II trial, one patient had a complete remission and three patients had partial responses. The observed tumour regression took place slowly over a period of several months suggesting that mechanisms other than direct cell killing by the immunotoxin played an important part in the response to therapy. In several patients, hypoalbuminaemia, oedema, and allergic reactions were seen but these side effects were generally transient and reversible. Finally, an anti-CD5 ricin A chain immunotoxin has been used in patients to treat graft-versus-host disease after bone marrow transplantation (see p. 503). Immunotoxins developed against B cell leukaemia and lymphoma, breast carcinoma, and colorectal carcinoma are also being evaluated in patients.

18.6.3 *Problems of* in vivo *therapy*

The therapeutic effectiveness of the ricin A chain immunotoxins used in clinical trials may have been diminished by two problems first encountered in animals and now substantially resolved. First, the conventional type of disulphide bond that links the A chain to antibody is labile in the animal because the bond is split prematurely by reduction. This results in a decrease in the level of active immunotoxin and the release of free antibody. Unfortunately, the released antibody has a longer serum half-life than intact immunotoxin and competes with the conjugate for binding to target cells. Immunotoxins made with new crosslinking reagents that generate hindered disulphide linkages retain full cytotoxic activity yet split less rapidly in the bloodstream of the animal. Second, the terminal mannose and fucose residues present in the oligosaccharide side chains of ricin A chain are recognized by liver cells which then rapidly remove ricin A chain immunotoxins from the circulation. This clearance can be avoided by destroying the carbohydrate residues chemically or by removing them with enzymes. Recently, nonglycosylated ricin A chain, which retains full ribosome-inactivating activity, has been produced in *Escherichia coli* by genetic engineering.

Another problem with immunotoxins is that the A chain evokes an immune response in animals and in patients. Fortunately, since the single chain inhibitor proteins appear to be immunologically distinct from the toxin A chains and from one another, it should be possible to avoid immunological neutralization during therapy by using a series of several immunotoxins made from different inhibitors. Alternatively, it may prove possible to manipulate toxin genes to produce functional proteins which have altered antigenic determinants and are not neutralized when given to patients who have been immunized against the toxin.

Subcutaneous tumour cells have proven more difficult to attack than tumour cells that are free in the bloodstream or peritoneal cavity, suggesting that the major strength of immunotoxins may be in their ability to locate and destroy metastatic tumour microfoci rather than to treat sizeable tumour masses. A possible way to improve penetration into solid tumour would be to make immunotoxins with univalent Fab fragments of antibody. Such Fab-ricin A chain conjugates would be smaller than immunotoxins made with intact antibodies, allowing more rapid diffusion between cells. Furthermore, these immunotoxins would bind the target antigen less avidly because they are univalent. This might allow the Fab conjugate to penetrate more deeply into the tumour than an intact antibody conjugate with higher avidity which becomes irreversibly bound to the tumour cells most accessible to the bloodstream. However, Fab fragments and their A chain conjugates are cleared from the bloodstream of animals more rapidly than intact antibody immunotoxins.

Finally, treatment of tumour cells with immunotoxin *in vitro* and *in vivo* has led to the outgrowth of immunotoxin-resistant tumour cells. In some cases, resistance resulted from the selection of a tumour cell population with a diminished antigen density. Cocktails of immunotoxins directed against two or more tumour antigens could prevent the outgrowth of such cells. However, mutants were found that had normal levels of the antigen to which the immunotoxin bound but were unable to transport the A chain to the cytosol in intact form.

18.6.4 *Recent developments*

Monoclonal antibodies raised against tumour cells are generally screened by procedures that favour the identification of antibodies against antigens present at high density or to which the antibody binds with high affinity. However, cell types which express a similar density of a given target antigen often differ in their susceptibility to immunotoxin treatment. Thus, anti-tumour antibodies selected for strong binding to target cells are not always the best candidates for the synthesis of potent immunotoxins. There is now a simple, reliable method to screen large numbers of monoclonal antibodies for their suitability as cytotoxic carriers of A chains. Cells are pretreated with purified antibody or with hybridoma ascites or tissue culture supernatant and are then cultured *in vitro* with a conjugate of ricin A chain linked to the Fab fragment of anti-murine immunoglobulin antibody to coat the primary antibody layer (Fig. 18.17). The univalent Fab fragment is used to prevent crosslinking of antigen–antibody complexes on the cell surface because this can cause a change in the route by which the antigen is internalized leading to decreased toxicity. Antibodies which mediate cell killing by the non-covalently bound Fab-ricin A chain conjugate form covalently linked ricin A chain immunotoxins with similar cytotoxic potency.

So far, intact toxin immunotoxins have been too toxic to be considered for safe use in patients. Recently, recombinant DNA techniques have

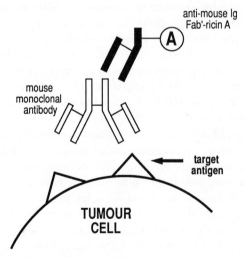

Fig. 18.17 Screening monoclonal antibodies for their ability to make potent immunotoxins.

been applied to the cloned diphtheria toxin gene to create a novel toxin molecule with reduced non-specific toxicity. The ability of the toxin to bind and intoxicate human cells was abolished by single site mutations but these changes did not impair the ability of the refashioned toxin to kill cells when targeted by covalently linked antibody. It will also be possible to eliminate the non-specific galactose binding capacity of ricin by replacing binding site amino acid residues of the B chain. Fusion proteins have been generated at the gene level between bacterial toxins and peptide hormones or interleukin 2 and found to be selectively cyto-toxic to receptor positive cells in tissue culture. By similar genetic manoeuvres, immunotoxins comprising manipulated toxin structures and immunoglobulin variable domains could be made directly.

18.7 Bone marrow transplantation

18.7.1 *Principles*

A number of malignant diseases, including leukaemia and lymphoma respond to high dose chemotherapy and whole body irradiation. This treatment is not more widely adopted because it destroys the haema-topoietic stem cells of the patient's bone marrow. The limited treatment with drugs and radiation that can be given to the patient may not eradicate all tumour cells and frequently the patient relapses. If higher and potentially curative doses of radiochemotherapy are to be used, the bone marrow of the patient must be reconstituted after treatment (see Chapter 16).

There are two ways to do this. In autologous bone marrow trans-plantation, a sample of the patient's own marrow is removed before therapy and later reimplanted on completion of the treatment. With diseases such as leukaemia or lymphoma, the marrow is infiltrated with malignant cells and so it is imperative to clear the marrow of these cells before it is returned to the patient. The alternative method, allogeneic bone marrow transplantation, is to give the patient normal bone marrow from another individual. In this case, the donor and recipient must be matched for histocompatibility antigens in order to reduce the risk of provoking graft versus host disease (GVHD), a life threatening complication that results from the attack of host cells by T lymphocytes in the allograft. GVHD is a common occurrence even with closely matched donors and recipients. The solution to this problem is to remove or kill the T lymphocytes in the marrow of the donor before it is infused into the patient.

18.7.2 *Methods for purging bone marrow*

The aim of bone marrow purging is to eliminate malignant cells from autologous marrow grafts or T lymphocytes from allogeneic marrow grafts and leave intact the haematopoietic stem cells. This has been made possible by using antibodies recognizing antigens that are present on malignant cells or T lymphocytes but are absent from the stem cells. Three approaches have been used clinically.

The first method is complement mediated cytolysis of antibody coated cells. In one study of patients receiving matched allogeneic grafts, Prentice and his co-workers (1984) pretreated the donor marrow with two mouse anti T cell monoclonal antibodies of the IgM class and followed this by two incubations with rabbit complement. This eliminated more than 99 per cent of the T cells and prevented acute GVHD in all 13 evaluable patients. In a study by Bast and his colleagues (1983), the bone marrow of 16 patients with common acute leukaemia was purged with a monoclonal antibody against the CD10 antigen in combination with rabbit complement. After therapy and reimplantation of the treated autologous bone marrow, six of the patients have remained in complete clinical remission for over one year. The action of complement is rapid but one serious drawback is the inherent variability between different batches of rabbit complement. This means that each preparation has to be screened for its ability to lyse cells in the presence of antibody and for lack of toxicity in the absence of antibody.

The second method for purging bone marrow uses monoclonal antibody toxin conjugates. Filipovich and her colleagues (1984) treated two patients undergoing marrow transplantation for acute lymphoblastic leukaemia in third remission. Bone marrow cells from a histocompatible sibling were treated with a cocktail of anti T cell antibody ricin conjugates in the presence of lactose (to prevent non-specific binding to normal cells via the ricin B chain), then washed to remove the immunotoxin and finally infused into the patient. The treated marrow was successfully grafted and no symptoms of toxicity or GVHD were evident. This study was expanded to a further 17 patients. Four of the patients had grade II GVHD primarily involving the skin but no instances of the more severe grade III and grade IV disease were seen. However, the donor marrow failed to engraft in another four of the patients. Using a different approach, 15 patients who had developed steroid-resistant GVHD after allogeneic bone marrow transplantation were given an intravenous infusion of an anti-CD5 ricin A chain immunotoxin. Twelve of the patients showed significant reduction in their disease in at least one organ only 7 days after treatment and by 28 to 40 days, 3 of 10 evaluable patients showed complete remission.

A third technique has been developed by Kemshead and his group (1986) to remove neuroblastoma cells from bone marrow. Tumour cells in the isolated marrow are first coated with a cocktail of several mouse monoclonal anti-neuroblastoma antibodies and then with a second layer of sheep anti-mouse Ig antibodies bound to polystyrene microspheres containing magnetite. The tumour cells can be removed by passing the marrow between the poles of a series of magnets (Fig. 18.18). The separation procedure removed 99.9 per cent of tumour cells from the bone marrow of patients. The cells which passed through the separation apparatus were free from polystyrene microspheres and were grafted successfully in four patients. The advantage of this approach is that the same antibody magnetite beads can be used to withdraw tumour cells coated with any mouse monoclonal antibody or with cocktails of anti-bodies.

Two main problems have arisen with T cell purging. First, the extensive depletion of T cells from allogeneic marrow results in a higher incidence of graft failure or rejection after transplantation. The reasons for this are unclear. Some T cells may need to be present in the marrow to suppress residual host immunity or to elaborate lymphokines needed for successful grafting. Second, there is evidence that depleting marrow grafts of T cells increases the risk of leukaemic recurrence. It is not clear whether specific T cell subsets are responsible for these problems. By choosing antigens restricted to various T cell subsets, the appropriate

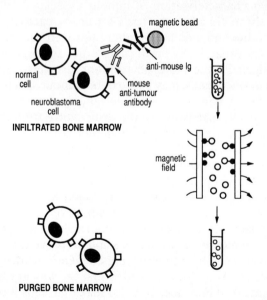

Fig. 18.18 Bone marrow purging using magnetic beads.

combinations of immunotoxins could selectively eliminate the subsets that promote GVHD without impairing the beneficial properties of other T cells in the marrow.

It is unclear whether the procedure will be of value in the treatment of other malignant diseases in which there is bone marrow involvement. To be successful, intense radiochemotherapy must eradicate all the clonogenic tumour cells in the patient's body or reduce them to a number which can be contained by the patient's immune and other defence mechanisms when they recover from the effects of the treatment. Other diseases that are responsive to radiochemotherapy and in which benefit might be expected from more intense treatment are lymphoma, oat cell carcinoma of the lung, carcinoma of the breast, testis, and ovary, and paediatric malignancies such as Wilms' tumour.

18.8 Prospects

Conventional chemotherapeutic or radiotherapeutic regimens often fail to cure patients because they do not kill all malignant cells. One reason is that the dose needed to eradicate the tumour is usually so large that the patient's own tissues would be severely damaged. The usefulness of monoclonal antibodies or antibody conjugates for therapy is also limited by the extent to which they cross-react with, and harm, normal tissues. The molecular characterization of tumour associated antigens may, in the future, help to elucidate the basis for cross reactivity of some anti-tumour antibodies and indicate whether improvements in selectivity can be obtained, for example, by raising antibodies against other epitopes of the target antigen. Furthermore, new tumour associated antigens, such as the products of oncogenes, may prove to be tumour specific.

A second factor contributing to the failure of conventional chemotherapy is that cells insensitive to the drug evolve from the tumour cell population during treatment (see Chapter 16). Antibody mediated therapy has to contend with the same problem since tumours normally contain a proportion of cells that lack the target antigen or express it at a low level. Similarly, a proportion of tumour cells expressing a high level of the target antigen may escape being killed by antibody conjugates either because they are resistant to the action of the cytotoxic agent or because they lack the mechanism to internalize the conjugate and to release the toxic moiety. It may be possible to prevent the outgrowth of resistant tumour cells by administering conventional drugs in conjunction with antibody conjugates provided that the way in which resistance to the conjugate develops does not also make the cells more resistant to the drug.

Tumour cells within solid tumours are less accessible to antibody because antibody molecules may only permeate the tumour to a depth of a few cell layers from the vasculature. Unless antibodies or antibody conjugates can gain access to the core of solid tumours and other sanctuary sites, they may be effective only against smaller deposits of tumour cells such as metastatic microfoci. It may be possible to kill cells in the solid tumour core with antibody radionuclide conjugates. Alternatively, the permeability of these tumours might be increased by agents that cause inflammatory damage or that disrupt cell to cell contacts and the extracellular matrix.

18.9 Conclusion

Monoclonal antibodies have revived earlier optimism for the prospect of treating cancer by immunotherapy. Anti-tumour monoclonal antibodies and their conjugates with cytotoxic agents have been evaluated in patients whose tumours are resistant to conventional forms of therapy. The main value of such trials has been to ascertain the side effects of the therapy and to identify the inherent limitations to these novel pharmaco-logical agents. The true therapeutic value may only emerge when trials are performed with antibody constructs made to improved design in patients with disease at a less advanced stage.

The principle of using antibody molecules as carriers of cytotoxic agents with no inherent tumour cell specificity is likely to be extended to other biologically active molecules such as interleukins, interferons, and inducers of cell differentiation. Further developments in hybridoma technology and genetic engineering will allow the manipulation of the structures of therapeutically important immunoglobulins, polypeptide hormones, toxins, and enzymes, and the creation of novel antibody molecules with characteristics desirable for cancer therapy in man. Improvements in the effectiveness of targeting are likely to stem from a better understanding of the behaviour of antibody based agents in animals and the physiological barriers to their movement and action. Furthermore, advances in our understanding of the fundamental pro-cesses that determine tumour growth and metastasis at the molecular level should provide new and possibly more suitable targets for antibody mediated therapy.

References and further reading

Baldwin, R. W. and Byers, V. S. (ed.) (1985). *Monoclonal antibodies for cancer detection and therapy.* Academic Press, London.

Bast, R. C., Jr *et al.* (1983). Elimination of leukemic cells from human bone marrow using monoclonal antibody and complement. *Cancer Research* **43**, 1389–94.

Bernstein, I. D., Tam, M. R., and Nowinski, R. C. (1980). Mouse leukaemia: therapy with monoclonal antibodies against a thymus differentiation antigen. *Science* **207**, 68–71.

Boss, B. D., Longman, R., Trowbridge, I., and Dulbecco, R. (ed.) (1983). *Monoclonal antibodies and cancer.* Academic Press, London.

Britton, K. E. (ed.) (1987). New approaches to tumour identification. *Cancer Surveys* **6**(2).

Carrasquillo, J. A. *et al.* (1984). Diagnosis of and therapy for solid tumours with radiolabelled antibodies and immune fragments. *Cancer Treatment Reports* **68**, 317–28.

Filipovich, A. H. *et al.* (1984). Ex vivo treatment of donor bone marrow with anti-T cell immunotoxins for prevention of graft-versus-host disease. *Lancet* **i**, 469–72.

Goldenberg, D. M. *et al.* (1978). Use of radiolabelled antibodies to carcino-embryonic antigen for the detection and localisation of diverse cancers by external photoscanning. *New England Journal of Medicine* **298**, 1384–8.

Kemshead, J. T. *et al.* (1986). Magnetic microspheres and monoclonal anti-bodies for the depletion of neuroblastoma cells from bone marrow: experiences, improvements and observations. *British Journal of Cancer* **54**, 771–8.

Kohler, G. and Milstein, C. (1975). Continuous cultures of fused cells producing antibodies of predefined specificity. *Nature* **256**, 495–7.

Law, L. W. (ed.) (1985). Tumour antigens in experimental and human systems. *Cancer Surveys* **4**(1).

Lennox, E. S. (ed.) (1984). Clinical applications of monoclonal antibodies. *British Medical Bulletin* **40**(3).

McMichael, A. J. and Fabre, J. W. (ed.) (1982). *Monoclonal antibodies in clinical medicine.* Academic Press, London.

Miller, R. A., Maloney, D. G., Warnke, R., and Levy, R. (1982). Treatment of B-cell lymphoma with monoclonal anti-idiotype antibody. *New England Journal of Medicine* **306**, 517–22.

Neuberger, M. S. and Williams, G. T. (1986). Construction of novel antibodies by use of DNA transfection: design of plasmid vectors. *Philosophical Transactions Royal Society of London* **317**, 425–32.

Order, S. E., Klein, J. L., and Leichner, P. K. (1987). Hepatoma: model for radiolabelled antibody in cancer treatment. *National Cancer Institute Monographs* **3**, 37–41.

Prentice, H. G. *et al.* (1984). Depletion of T lymphocytes in donor marrow prevents significant graft-versus-host disease in matched allogeneic leukaemia marrow transplant recipients. *Lancet* **i**, 472–6.

Spitler, L. E. *et al.* (1987). Therapy of patients with malignant melanoma using a monoclonal antimelanoma antibody-ricin A chain immunotoxin. *Cancer Research* **47**, 1717–23.

Tsukada, Y., Ohkawa, K., and Hibi, N. (1987). Therapeutic effect of treatment with polyclonal or monoclonal antibodies to α-fetoprotein that have been conjugated to daunomycin via a dextran bridge: studies with an α-fetoprotein-producing rat hepatoma tumour model. *Cancer Research* **47**, 4293–5.

Vitetta, E. S., Fulton, R. J., May, R. D., Till, M., and Uhr, J. W. (1987). Redesigning nature's poisons to create anti-tumour reagents. *Science* **238**, 1098–1104.

Vogel, C.-W. (ed.) (1987). *Immunoconjugates: antibody conjugates in radio-imaging and therapy of cancer.* Oxford University Press, New York.

19

Some conclusions and prospects
Robin A. Weiss

19.1 Understanding carcinogenesis

It is evident from the foregoing chapters that we are gaining major insight into the nature of cancer. The excitement in this field derives from the realization that a finite, analysable set of genes is involved in carcinogenesis. These are the genes called oncogenes and tumour suppressor genes (see Chapter 9). Their identification has brought together seemingly distinct areas of cancer research such as chemical and radiation carcinogenesis (see Chapters 6 and 7), oncogenic viruses (see Chapter 8), chromosome anomalies (see Chapter 10), and growth factors (see Chapter 11). While it is possible that oncogenes may also underlie some of the specific inherited predispositions to cancer (see Chapter 4), they have been chiefly identified in association with the mutation and rearrangement of genes during carcinogenesis; in other words cancer is usually the result of genetic changes in somatic cells.

The somatic mutation hypothesis of cancer has a venerable pedigree. The notion that carcinogenesis involves genetic change dates from the turn of the century when Boveri identified chromosomes as the repository of genetic material and speculated on chromosome imbalance in cancer. The link between mutation and cancer was first drawnn by Muller in 1927, who demonstrated the mutagenicity of ionizing radiation in fruit flies (radium and X–rays were already known to cause cancer). Later Auerbach showed that a carcinogenic chemical, mustard gas, also induced mutations, and it now appears that the great majority of chemical carcinogens or their active metabolites are mutagens (see Chapter 6).

Thus the concept that certain stages in carcinogenesis, in particular initiating events, were mutational has been firmly embedded in our thinking about cancer for many decades. But this gave no indication as to which genes became altered, or indeed whether the same genes were regularly affected. It was thought that the accumulation of mutations might allow otherwise recessive traits to appear, by removing the action of some of the very large number of genes involved in cellular control processes. The recent studies of oncogenes has focused attention on just 20 out of the 50 000 or so genes in the human genome.

The repeated identification of the same set of genes—whether captured by retroviruses, or revealed by DNA transfection, DNA amplification, or chromosomes rearrangement—indicates that these genes are of major importance in understanding cancer. Moreover, oncogenes are active genes, and tend to be expressed in a dominant manner. Mutation may cause overexpression of a normal protein or normal levels of expression of an altered protein, but in each case the gene products contribute positively to the neoplastic phenotype. Oncogenes challenge the earlier assumption that deletion of active genes led to cancer, that normality is dominant and cancer recessive.

Carcinogenesis is, of course, a much more complicated, multistep process than the activation or mutation of a single oncogene (see Chapters 1 and 6). Mutated oncogenes are not clearly dominant over their normal alleles, and expression of several different oncogenes may act synergistically in producing the malignant phenotype (see Chapter 9). Moreover, some somatic cell hybrids between normal and malignant cells have a normal phenotype. Nevertheless, the evidence that identifiable proteins encoded by oncogenes contribute to neoplastic transformation is a major step forward in our understanding of cancer, and suggest new means of treatment by blocking the function of such proteins.

The oncogene concept contrasts with theories that gene deletion leads to cancer. However, this does appear to be the case for tumour suppressor genes, sometimes called anti-oncogenes. These genes have mainly been identified through rare, familial, inherited predispositions to cancer, among which the gene for familial retinoblastoma *Rbl*, has recently been cloned. Tumour suppressor genes encode proteins which are necessary to prevent uncontrolled cell proliferation, whereas the proteins of oncogenes actively promote such growth. Thus specific genes may be involved in carcinogenesis either by stimulating cell growth or by failing to restrain cell growth.

In general, oncogenes may be thought of as dominant, because a single gene will give rise to a protein acting to stimulate growth. In contrast, tumour suppressor genes tend to be recessive, because one must lose the function of both copies in a diploid cell to release the cell from its control. In heritable predisposition to cancers, only one good copy of a tumour suppressor gene is inherited in the first place, so that a mutation in the remaining one may lead to cancer (see Chapter 4). Recent investigations of non-heritable tumours suggest that both copies may be lost in the malignant clone. For example, many sporadic colon carcinomas have chromosome deletions around the same locus that determines familial polyposis coli.

New insight into viral carcinogenesis (Chapter 8) also points to the interaction of tumour suppressor genes. For example, the protein encoded by one of the transforming genes of human papilloma virus actually binds to and inactivates the protein encoded by the cell's retinoblastoma gene. The net result is the same as losing the gene, because its protein is no longer present in a biologically active form sufficient to restrain cell proliferation.

The somatic mutation hypothesis of cancer presupposes the clonal origin of the cells comprising the malignant population. Clearly, a mutation or genetic change occurring in one cell can only be passed on to that cell's linear descendants, i.e. its derivative clone. By and large, where cancers have been amenable to analysis, clonality has been upheld. For example, there is a polymorphism for an enzyme coded by an X linked gene which happens to be relatively common in black people. Following inactivation of one or other X chromosomes early in female development, each cell will express only one form of the enzyme. Thus normal tissues display a fine mosaic of cells expressing one or other enzymic form (see Chapter 4). Fialkow's studies show that the malignant cells in a tumour usually express one form of the enzyme only, demonstrating the clonal derivation of the tumour cells after X chromosome inactivation has taken place.

Other evidence also points to a clonal evolution of tumour cells at early stages of carcinogenesis. In chronic granulocytic leukaemia, a chromosome translocation (the 'Philadelphia' chromosome) involving the *abl* oncogene (see Chapters 9, 10, and 12) is present in the majority of apparently normal and premalignant cells of the blood, indicating that a clonal stem cell carrying the translocated chromosomes has populated almost all the bone marrow before malignancy appears. Experimental chemical carcinogenesis in mouse skin indicates that a *ras* gene is mutated in codon 12 in premalignant stages of papillomatosis (see Chapters 6 and 9), again indicating clonality.

It is not clear, however, that all premalignant changes are clonal in origin. Pathologists frequently recognize 'field changes' in tissues which look abnormal and from which a malignant tumour may emerge (see Chapter 1). For example, cancer of the stomach frequently arises in patches of stomach epithelium that have undergone intestinal metaplasia, in which the tissue resembles the epithelium of the intestine rather than the stomach and this observation has been used in early screening for stomach cancer in Japan where this tumour is common. We do not know exactly what triggers metaplasia and whether such altered patches of epithelium are clonal in origin.

In concluding a discussion on the genetic basis of carcinogenesis one should mention that the clonal nature of the cancer in which mutations in a single cell lead to a malignant cell population is not universally accepted. An epigenetic view of cancer has long been held by the radiotherapist, Sir David Smithers, and more recently by Dr Harry Rubin. 'Cancer is no more a disease of cells than a traffic jam is a disease of cars', wrote Smithers in 1962, continuing, 'A lifetime study of the internal combustion engine would not help anyone to understand our traffic problems. A traffic jam is due to a failure of the normal relationship between driven cars and their environment and can occur whether they themselves are running normally or not'.

This is an attractive argument, but a deceptive one. The body's environment can be of crucial importance in allowing tumours to develop. Many remain dependent on hormones (see Chapters 13 and 14) and growth factors (see Chapter 11), and possibly the host immune system (see Chapter 15). Local interactions between cells also remain important to all but the most anaplastic tumour cells. Thus the cancer 'seed' requires fertile 'soil' in which to proliferate, as is shown by the relative inefficiency of the process of metastasis (see Chapter 2). Although thousands of tumour cells may be released from a primary tumour only a very small number will develop into secondary tumours. Nevertheless, the interaction of tumour cells with the host environment is an elective or permissive process of stimulation and response in which an essential lesion always exists in the tumour cell itself. This after all is the basis of surgery, radiotherapy and chemotherapy (see Chapters 16 and 17) directed to the elimination of the tumour. Drug resistance following chemotherapy is another example of clonal evolution and progression of tumours in which genetic changes within the malignant cell (for example gene amplification) determine the response to and escape from therapy.

Smithers based his views on observations that human tumours, particularly embyonal tumours occuring early in childhood, may occasionally regress spontaneously, and he concluded that the malignant phenotype was reversible. Certainly tumour cells differ from normal cells in subtle ways, such as arrested differentiation (see Chapter 12) in which less malignant behaviour may develop if the tumour cells can be made to follow a normal cell maturation pathway. Indeed there has recently been increased emphasis in introducing so called biological response modifiers into cancer therapy. But the natural history of tumour progression (see Chapter 6) unfortunately promotes the selection of subclones with greater malignant potential.

Is there any experimental evidence for the reversibility of malignancy, as one might expect if carcinogenesis were a non-genetic phenomenon? There are some suggestive findings that merit more detailed investigation. For example, in leopard frogs, isolated nuclei from Lucke carcinoma cells (a kidney tumour induced by a herpesvirus) have been introduced into activated, enucleated frog eggs. Following nuclear transplantation, a small proportion of the nuclei allowed the development of tadpoles with normal, differentiated tissues, but it has not been unequivocally demonstrated that these nuclei came from tumour cells rather than from the stroma. Mouse teratocarcinomas containing pluripotential embryonal carcinoma cells have been the subject of extensive study as these cancer cells can differentiate into many types of apparently normal cells. The introduction of such cells into the embryonic blastocyst allows incorporation into the inner cell mass, and hence into the embryo. Using genetic markers, it has been shown that the embryonal carcinoma cells contribute to most of the normal tissues of the mouse, although such mice have a higher frequency of teratocarcinoma. It has also been found that the descendants of embryonal cell lines can form a normal germ line, but this remains to be confirmed with fully malignant teratocarcinoma cells. Similarly, crown gall tumours in tobacco plants, which are induced by transfer of a bacterial plasmid, can give rise to normal tissues and to whole plants. Thus heritable normal behaviour can be restored to certain pluripotent cancer cells by transplantation into, or cultivation in, a suitable environment.

These experiments suggest that the cancer phenotype is not inexorably irreversible but do not refute the somatic mutation hypothesis. The chromosome changes in many tumour cells, the mutagenic properties of most carcinogens, and the alteration of specific genes in oncogenesis, strongly suggest that carcinogenesis involves genetic changes in the cells that become malignant. In addition, activated oncogenes introduced into the germ line of mice cause a genetic predisposition to cancer.

19.2 Prospects for prevention and screening

One of the biggest problems in attempting to reduce the incidence of cancer is that of changing the life style of the people at risk. In the 30 years since it became firmly established that cigarette smoking is causally linked to lung cancer, little progress has been made in reducing cigarette smoking or in stopping new cohorts of youngsters aquiring this addictive habit. Dietary factors (see Chapter 3) may be equally important, though less clearly defined, in human carcinogenesis, particularly in the digestive tract itself, and changing patterns of cancer are evident, such as the decrease of stomach cancer and rise of colorectal cancer. It is difficult to change dietary habits, though the availability of fresh or frozen vegetables in all seasons and the increasing recognition of the importance of dietary fibre is leading to better health.

With occupational cancers due in part to industrial exposure, it is much easier to prevent access to the carcinogens once they have been identified. The recognition of β-naphthylamine as a bladder carcinogen (see Chapter 6) soon led to protection of workers who were exposed to this chemical in the rubber and dye industries. Similar measures have been undertaken to reduce exposure to asbestos and related carcinogenic fibres.

One of the brightest prospects for prevention arises from the recognition of viruses as important environmental carcinogens for human cancer (see Chapter 8). As shown in Chapter 3 (Table 3.10), some 10 per cent of cancers in the USA are listed as being possibly attributable to infection, chiefly by viruses. Across the world, this percentage is likely to be much higher, as the prevalence of oncogenic viruses has been relatively low in the Western world.

Carcinoma of the uterine cervix is becoming the most common cancer of women, especially in South America. Two recently identified strains of human papillomavirus (HPV-16 and HPV-18) appear to be causally linked to flat cervical warts and neoplasia. It is interesting to note that these agents have not been isolated *in vitro* as infectious viruses; rather they have been identified through molecular genetic techniques so that the viral genes have been cloned before virus isolation. Cervical cancer is a disease easily curable by minor surgery if recognized at an early stage before invasion of the subepithelial tissue of the cervix has taken place. Furthermore the cancer, as well as preneoplastic dysplasia can be recognized by direct inspection (colposcopy) and by the examination of cervical smears. The advantages of regular screening for cervical cancer are proven in reducing cancer mortality and morbidity. Yet the cost of preventative screening as a public health measure are considerable and to date the awareness of its value is largely restricted to those women least at risk of developing the disease. With the identification of the incriminating virus, the possibility of developing immunizing vaccines is being actively pursued, though for human papillomaviruses this approach is still in its infancy.

Preventive immunization does appear to be effective in the transmission of hepatitis B virus (HBV). The virus is very prevalent in Africa and Asia. It is associated with primary liver cancer, a rare tumour in the West, but ranking sixth in cancer mortality worldwide. Dietary aflatoxins (see Chapter 7) may also play a role in human liver cancer, but there is strong evidence that HBV is the major causative factor. HBV is typically transmitted perinatally. Protection from infection can be achieved with preparations of pooled antibody to HBV surface antigen administered postnatally, following some weeks later with vaccination with surface antigen itself, manufactured by recombinant DNA methods. Trials in Taiwan and Japan suggest that the cycle of maternal transmission to infants can be broken by these measures, and hopefully this promises a dramatic reduction in the incidence of liver cancer in 40–60 years' time.

There is intensive research on vaccines to other viruses, such as Epstein Barr virus (EBV) and human T cell leukaemia virus type 1 (HTLV-1), but much development is still required before vaccines can be given preliminary trials. Infection by EBV is ubiquitous, and of the two malignancies associated with EBV, a major risk factor for one, Burkitt's lymphoma, may be malaria. Given the high morbidity and mortality of malaria itself, this is the disease that must be tackled. With EBV, however, it is undifferentiated nasopharyngeal carcinoma (NPC) that causes at least 50-fold more premature deaths than those from Burkitt's lymphoma.

NPC is prevalent among Chinese living in southern China and throughout South East Asia. Like most cancers, NPC is multifactorial in causation, with evidence of a slight heritable predisposition (increased risk with certain HLA haplotypes) and evidence of dietary nitrosamines or nickel salts early in life too. But the presence of EBV genomes in the tumour cells and of raised antibody levels to viral antigens in NPC patients implicates the virus in having the major causative role, and gives hopes for future prevention.

Like cervical carcinoma, NPC is readily treatable if recognized at an early stage in its growth. Early diagnosis is being introduced in China by mass screening for serum IgA specific to an EBV antigen. The IgA can be recognized in a drop of blood from a pin pricked ear lobe, blotted onto paper. Since EBV IgA is elevated at an early stage of NPC growth, healthy subjects with elevated IgA are then investigated for carcinoma *in situ* and are treated curatively at this stage by simple surgery or radiotherapy. The mass screening for EBV IgA in the communes of southern China promises to be the most effective scheme of preventing cancer deaths.

19.3 Prospects for new treatment

Modern methods of cancer treatment have been reviewed in Chapter 16, 17, and 18. Surgical removal and radiotherapy remain the major treatments for solid tumours, although chemotherapy plays an increasingly important role. The newer physical methods of imaging, employing computerized tomography (CT) and magnetic resonance imaging (MRI) are proving of great importance not only for diagnosis, but for conformational radiotherapy planning too. Immunological methods of imaging with radiolabelled antibodies (see Chapter 18) are still in a very primitive state compared to CT and NMR but may one day prove to be of greater use for detecting small tumours, especially metastases where the nature of the primary tumour is already known.

There have been very considerable advances in chemotherapeutic treatment (see Chapter 16) especially in leukaemia and in tumours of children and young adults. The major cancers of older patients, such as carcinoma of the breast, stomach, and colon, have been disappointingly refractory to curative chemotherapy.

Much discussion of alternative treatment modalities is devoted to immunological mechanisms (see Chapter 15) and immunological vehicles of targeting cytotoxic agents or toxins to the tumour (see Chapter 18). Monoclonal antibodies that bind to the surface of tumour cells but not to haemopoietic stem cells are already proving to be clinically useful in 'cleansing' bone marrow in autologous transplantation. It remains to be seen whether immunological targeting will be successful in tackling solid tumours *in vivo*, but it seems well worth the current investment in this field.

As with more conventional chemotherapy, a problem with immunologically targeted therapy is the likelihood of resistant clones emerging, especially as tumour cells are known to modulate the expression of cell surface antigens. Perhaps the development of new radionuclides emitting alpha particles will be of most promise with antibody treatment; such nuclides may kill all cells within a short range of the targeted antibodies and the need to translocate antibody or toxin across the cell membrane is obviated. However, as related in Chapter 18, there is a need for more research into appropriate radionuclides for conjugation with monoclonal antibodies.

Where expression of surface antigen is a necessary part of the malignant phenotype, there is more hope for immunotherapy, as cells that express the antigen can be killed, and cells that cease to express it should lose their malignancy. The overamplified expression of receptors for epidermal growth factor in many squamous cell carcinomas (see Chapter 11) may also be appropriate targets for immunotherapy. In the next few years, much effort will be made to develop drugs or other reagents that directly and selectively block the function of oncogene products.

An area of basic research which is rapidly being applied to cancer therapy is that of locally acting growth and differentiation factors, called cytokins (Chapter 11). These proteins are produced by one cell type and usually act as 'paracrine' factors stimulating the proliferation, differentiation, or degeneration of other, neighbouring cells that express specific receptors. The best known cytokines are the interleukins (IL1 to -6) affecting lymphoid cells, and the specific growth factors for myeloid cells. Interferon can also be classified as a cytokine.

A few years ago, great publicity was given to the potential of interferons, as anti-tumour therapy. Interferons were originally discovered as endogenous anti-viral agents, but they can promote terminal differentiation or arrest the growth of certain tumour cells too. Now that abundant interferon has become available through gene cloning, it is apparent that most cancers are unresponsive and further that high doses of interferon are extremely unpleasant and toxic for the patient. Therefore a disillusionment has set in that has been perhaps exacerbated by the overenthusiastic claims for interferon before it could be tested in appropriate clinical trials. Yet amid this disappointment, one rare kind of malignancy, hairy cell leukaemia, is responding well to interferon in giving a complete response (regression) in the majority of patients, and interferon has become the preferred means of treatment of these patients. We do not understand why this leukaemia is so sensitive to interferon.

Another cytokine, tumour necrosis factor (TNF), took the place of interferon in the media as the magic cure-all for cancer. Like interferon, the gene encoding TNF has been cloned so that large amounts of this factor can be manufactured in bacteria. TNF is naturally produced in small amounts by macrophages in response to bacterial endotoxins. It is probably identical to a protein known as cachectin, which is thought to play a role in the wasting of normal tissues (cachexia) seen in many advanced cancer patients. Certain animal tumours are cured by TNF administration but preliminary clinical trials on human malignancy have been disappointing. If it promotes cachexia TNF may do more harm than good, but like interferon it is hoped that some human tumours will be eliminated by this biological product.

Currently, immunological control is enjoying a renaissance in considering tumour treatment, through the agency of another cytokine, interleukin 2 (IL2). Preliminary reports suggest that interleukin 2 may also be of value in stimulating the host immune T cell response to some tumours. This locally acting factor promotes the growth of T lymphocytes. By extracting tumour infiltrated lymphocytes and amplifying the numbers in culture by IL2, re-inoculation may help to arrest the growth of the malignant tissue.

Lastly, we should not forget the progress that has been made in recent years in improving palliative treatment, designed not to cure cancer, but to improve the quality of life remaining to the affected individual. Most chronic diseases, such as cardiovascular disease, arthritis, diabetes, etc. are not curable, and we have, perhaps, expected too much of oncologists in hoping for the magic bullet. In the meantime, great advances have been made in symptomatic treatment, pain control, and psychological management. With one quarter of deaths in the Western world attributable to malignant disease, improvements in the care of incurable cancer patients deserves considerable devotion and resources alongside the search for preventative measures and curative treatment.

19.4 Conclusion

Cancer is an exciting field of research, for it presents challenging questions in our basic understanding of cell physiology and development, coupled with the prospect of helping to alleviate a major source of human suffering and death. The study of cancer is multidisciplinary, involving scientists and clinicians in many different specialties. To ensure further progress in combating cancer we must practice what we preach, and be open to the new leads that the cross-fertilization of multidisciplinary research affords.

References and further reading

Heller, H., Davey, B., and Bailey, L. (ed.) (1989). *Reducing the risk of cancers.* Hodder & Stoughton, London.

Vessey, M. P. and Gray, M. (ed.) (1987). *Cancer risks and prevention.* Oxford University Press, Oxford.

Glossary

α-particle The nucleus of the helium atom (charge of $+2$) and mass 4.

abl An oncogene originally discovered in the Abelson strain of murine leukaemia virus. This virus causes pre-B cell tumours, whereas a related feline virus isolate causes sarcomas. The cellular *abl* gene is important in chronic granulocytic leukaemia of humans. The gene product displays tyrosine protein kinase (q.v.) activity.

acentric chromosome A chromosome lacking a centromere and, hence, incapable of attaching to the spindle at mitosis; consequently these chromosomes are randomly distributed between daughter cells and mitosis.

actin A cytoskeletal protein family found in all cell types. In muscle cells it is involved in the contractile function of the tissue.

active immunization Induction of a state of immunity by administering antigen.

acute leukaemia A tumour of the reticuloendothelial system that usually runs a rapid course from disease onset to death.

adduct The addition product between two molecules, e.g. a carcinogen (or its metabolite) and a nucleic acid in DNA.

adenocarcinoma A malignant tumour of glandular epithelium.

adenoma A solid, benign glandular tumour.

adjuvant A substance which increases the immune response to an antigen given at the same time.

affinity Used herein for the degree of binding or interaction, e.g. between a ligand and its receptor, or between an antigen and an antibody specific for that antigen. One can speak of high or low affinity receptors or antibodies.

aflatoxin A toxic food contaminant produced by fungi of the genus *Aspergillus*.

alkylating The process of attaching alkyl (usually methyl or ethyl) groups to a chemical structure such as a DNA base.

allele Short for allelomorph, meaning alternative form of the same gene.

alloantigen Differences which distinguish one individual of a species from another.

allogenic Genetically dissimilar.

allogenic graft (allograft) Graft between genetically dissimilar individuals of the same species.

amino acid One of twenty molecules used as the "building blocks" of a protein.

anaplasia (dedifferentiation) Loss of differentiated characters.

anastomosis The surgical re-uniting of the severed ends of a tubular structure such as gut or a major blood vessel.

anchorage independent growth Growth (cell division) in suspension in a liquid or semisolid medium, without attachment to a solid substrate. This form of growth is a property of many tumours.

androgens Steroid hormones secreted by the testis; maintain male characteristics.

aneuploidy The presence of extra chromosomes or the absence of chromosomes, so that the karyotype is neither haploid (q.v.) nor an exact multiple thereof.

angiogenesis New growth of blood vessels.

antibody A globular protein produced by animals in response to an antigen (q.v.) and which binds specifically to the antigen.

antigen A molecule which is capable of stimulating the formation of antibodies (q.v.)

aromatic amines Polycyclic compounds consisting of hydrocarbon ring structures with at least one nitrogen containing amine group.

ascites Accumulation of serous fluid in the peritoneal cavity.

attenuated virus A virus whose pathogenicity is reduced by passage outside its natural host.

autocrine Self stimulation of a cell through production of both a factor and its specific receptor.

autophosphorylation A process by which a protein has the ability to add phosphate groups to amino acids within the protein (*see also* protein kinase).

auxotroph An organism (e.g. a bacterium) capable of synthesizing from simple nutrients all the complex biochemical compounds needed for its own growth.

axilla The armpit which contains the majority of the lymph nodes draining the breast and arm.

axon Process of a nerve cell which carries impulses from the nerve body to a nerve or other target cell.

base change (in nucleotides) *See* mutation.

base The purine or pyrimidine component of nucleotides (q.v.).

B cell B lymphocyte: the immune cells of the lineage that makes antibodies.

benign tumour Tumour which does not invade or metastasize; an absolute distinction from malignant tumours is not possible (*see* Chapter 1).

bracken The fern *Pteridium aquilinum.*

Burkitt's lymphoma A tumour in humans of mature immunoglobulin producing B lymphocytes.

burst forming unit-erythroid (BFU-E) A colony of immature and mature red blood cells growing in a semi-solid medium.

bypass The re-routing of the normal contents of a tubular structure, such as the bile duct or bowel where an inoperable blockage is present.

cDNA A complementary (copy) DNA transcribed (q.v.) from an RNA template by the enzyme RNA dependant DNA polymerase (reverse transcriptase).

c-*onc* The cellular gene corresponding to a viral oncogene and sharing close DNA sequence homology. It is thought that the virus 'picked up' the cellular gene by chance.

cannula A hollow needle.

carcinogen A chemical substance, or physical agent such as X-ray irradiation or ultraviolet irradiation, that causes cancer.

carcinogenesis Process of tumour induction and development.

carcinoma A malignant tumour of epithelium.

carrier A host with an asymptomatic infection who serves as a source of infection to others.

centromere Constricted portion of chromosome at which the chromatids (q.v.) are joined. It may be central (metacentric), off centre (submetacentric) or near one end (acrocentric).

chemical repair Conversion of a free radical to a stable molecule (usually by hydrogen atom transfer).

chemotaxis Response of cells or organisms to chemical stimuli: attraction towards is positive and repulsion negative chemotaxis.

chorioallantois Embryonic membrane formed by the fusion of the wall of the allantoic sac to the chorion. It underlies the porous egg shell in birds.

choriocarcinoma Cancer of uterus derived from placental tissue.

chromatid One of the two spiral filaments joined at the centromere (q.v) which make up a chromosome and which separate in cell division, each going to a different pole of the dividing cell and each becoming a chromosome of one of the two daughter cells.

chromatin A complex of fibres in the cell nucleus, made up of DNA histones and other proteins. During cell division the chromatin condenses into coils to form chromosomes (q.v.)

chromosome A structure in the nucleus made up of a linear strand of DNA associated with histones, RNA, and other proteins. In cell division the material (chromatin) becomes compactly coiled and clearly visible either after staining or by special microscopic techniques. Each organism of a species has the same number (diploid number) of chromosomes in its somatic cells, e.g. 46 in man.

cirrhosis Fibrous repair in response to damage in the liver.

class (of antibody) Antibody class is determined by differences in the structure of the constant part of the molecule. Different classes have different biological functions.

clastogenic Giving rise to or inducing breakages, e.g. in chromosomes.

clonal chromosomal aberrations Structural chromosomes defects observed in at least two cells within a tumour, or numerical defects observed in at least three cells within a tumour.

clone (*noun*) A collection of cells, organisms, nucleic acid sequences etc. that are all derived from the same ancestor and are thus more or less identical; a population of cells derived from a single cell by division, and so usually presumed to contain the same genetic information subject to the occurrence of mutations.

clone (*verb*) To make a clone of nucleic acid sequence by genetic manipulation, replicating the desired sequence in a micro-organism, or to isolate single cells.

codon A sequence of three nucleic acids or bases in DNA that together specify the encoded amino acid. Thus a sequence of DNA codons specifies the amino acid sequence of the encoded protein.

collagenase Enzyme which digests collagen.

colony stimulating factor (CSF) A group of proteins that stimulate the proliferation and differentiation of haemopoietic cells growing as colonies in a semi-solid medium; e.g. M-CSF (q.v.), G-CSF (for granulocytes), GM-CSF (q.v.), multi-CSF, etc.

colostomy Artificial external opening of large bowel.

complement A group of serum proteins which can be activated by cell bound antibody (q.v.) to lyse antibody coated cells.

complete carcinogen An agent, usually a chemical compound, which is able to induce cancer alone, i.e. without the need for subsequent treatment by a tumour promoting agent.

connective tissue (mesenchyme) Supporting tissues, made up of mesenchymal cells, collagen fibres, and interstitial substances. Bone, cartilage, fatty tissue, and blood vessels are specialized forms of connective tissue.

contrast medium X-ray dense material for delineating internal structure.

cystic lesion A cyst is a fluid containing sac, usually with a thin wall.

cytochrome (P450) Iron containing, electron accepting molecule involved in oxidation reduction reactions requiring energy, in this case the oxidative metabolism of carcinogens. P450 defines the characteristic absorption wavelength of a particular cytochrome on which these re-actions are dependent.

cytokine Generally a low molecular weight protein involved in cell regulation. Cytokines are secreted transiently, and generally locally, during embryogenesis, tissue remodelling and repair, immunity, and inflammation. Cytokines interact in a complex network where induction of one cytokine results in the production of other members of the network; binding of one cytokine to its cell surface receptor can alter expression of other cytokine receptors, and can up-regulate or down-regulate each others' action.

cytolytic Lysing (killing) cells.

cytopathic Damaging cells, often leading to their death.

cytoskeleton The network of insoluble, multifunctional filaments (e.g. actin, vimentin, cytokeratin) remaining after extraction of cells in detergent.

cytotoxicity Cell killing.

deletion The removal of a segment of a chromosome or gene.

dielectric relaxation time The time required for the molecules in a medium to readjust following passage of radiation.

diethylstilboestrol A powerful, synthetic oestrogen.

differentiation Process of development of new characters in cells or tissues.

differentiation antigen Molecule detected by an antibody on or in one cell type which is associated with state of maturation or position in a developmental lineage.

diploid Having twice the haploid number of chromosomes. Normal human somatic cells are diploid, having 46 chromosomes (23 pairs).

dominant An allele that manifests its effects in heterozygotes (q.v.), i.e. when present in a single dose, as well as in homozygotes. A characteristic determined by a dominant allele.

double minutes Pairs of small acentric chromosomes believed to carry amplified genes.

down-regulation The decrease in the number of available (exposed) cell surface receptors following interaction with the specific factor which binds to the particular receptor.

downstream By convention a nucleotide sequence is read from the 5' end to the 3' end, or from the beginning to the end in terms of coding potential, then one designates those sequences more 3' or more 'end-ward' as being downstream.

downstream promotion An insertional mutagenic event (q.v.) involving specifically a sequence that promotes gene transcription (hence 'promoter'). For this effect, the promoter must be located before the 5' end (beginning) of the gene; the promoter is thus 'upstream' and it promotes the transcription of the gene elements that are 'downstream' (or toward the 3' end).

ectoderm Outer of the three primary germ layers in animal embryos, which develops mainly into epidermis, nervous tissue.

electrophilic Literally, 'electron loving'. Applied to electron deficient chemical compounds which carry a net positive charge and are attracted to chemicals with an excess of electrons with which they bond covalently.

embolus Blood clot, tumour cells or other particles in blood stream.

endocrine Stimulation by factor produced at one site e.g. specific cells in a gland, and acting at a distant site.

endocytosis The process by which molecules bound to specific receptors on the cell surface are internalized by the cell.

endoderm Innermost of the three primary germ layers in the animal embryo which gives rise to the lining of the gut from pharynx to rectum and derivatives from it such as liver, pancreas, etc. and the respiratory epithelium.

endoscope Optical instrument, passed through an orifice, e.g. mouth, anus, for visualizing internal organs.

endothelium Cells lining blood vessels.

enhancer insertion An insertional mutation (q.v.) involving specifically a sequence that enhances (increases) the level of gene transcription (hence 'enhancer').

epidermal growth factor (EGF) A factor originally described for its mitogenic activity for epidermal cells. It is growth promoting for other cells as well.

epigenetic mechanisms Regulating expression of gene activity but not involving alterations in gene structure.

epithelium Tissue specific surface or glandular cells.

epitope A structural region of an antigen (q.v.) that is recognized by an antibody (q.v.) molecule.

epoxide A derivative of an aromatic (ring structured) molecule in which an oxygen molecule forms a bridge across an opened double bond between two carbon atoms.

epoxide hydrase An enzyme which utilizes an epoxide substrate, breaks open the oxygen bridge and adds one molecule of water to form a diol derivative.

erbA A gene sequence originally found in an avian erythroleukaemia virus as part of a fusion to *erbB* (q.v.). It may not be an oncogene *per se*.

erbB An oncogene originally found in two isolates of avian erythroleukaemia virus. it represents part of the gene for the receptor for epidermal growth factor (EGF, q.v.).

erythroblastosis A leukaemia of red blood cell precursors.

erythrocyte Most mature red cell from which the nucleus has been extruded (in mammals).

erythropoietin A growth factor that stimulates the proliferation and differentiation of erythrocyte precursor cells.

ets An oncogene originally found in an avian myeloblastosis virus (strain E26) fused to the *myb* oncogene (q.v.).

eukaryote Higher organisms whose cells are complex, containing nuclei and other organelles.

exocrine Secreting externally; exocrine glands deliver their secretions to an epithelial surface through ducts.

exon The parts of a gene sequence that are found in the messenger RNA (transcript) molecule which will be used for producing (translating) the protein product of the gene. Parts of the gene not found in the transcripts are known as introns (q.v.).

faecal occult blood The presence of small amounts of blood in faeces not detectable with the naked eye, which may be an indicator of colonic cancer.

familial polyposis coli Inherited condition where multiple polyps (small benign tumours) of the colon and rectum develop and which predisposes to colon cancer.

fes An oncogene originally found in a feline sarcoma virus (Gardner–Arnstein strain). The gene product displays tyrosine protein kinase (q.v.) activity.

fgr An oncogene originally found in a feline sarcoma virus (Gardner–Rasheed strain). Its gene product has tyrosine protein kinase (q.v.) activity.

fibrosis Formation of scar tissue.

filling defect A region of an X-ray or scan with absence of medium or isotope often due to the presence of tumour.

fms An oncogene originally found in a feline sarcoma virus (McDonough strain). Its product displays tyrosine kinase (q.v.) activity. This gene may be related or identical to the gene encoding the receptor for macrophage colony stimulating factor (q.v.).

fos An oncogene originally found in two murine osteosarcoma viruses. Its product is found in the nucleus and is more abundant in cells responding to growth factor stimulation, and in amnion and mature monocytes.

fps an oncogene originally found in two avian sarcoma viruses (strains Fujinami and PRCII). Its product displays tyrosine protein kinase (q.v.) activity.

frameshift mutation A mutation in which, by addition or deletion (other than in multiples of three) of nucleic acids, the reading frame of the triplet code is disrupted. All downstream codons are thereby altered and as a rule no functional protein is produced.

free radical An unstable molecular fragment caused by breakage of a chemical bond in a molecule.

gene activation Generally meant to imply that a dormant gene has been 'turned on' so that messenger RNA is made (transcribed).

gene amplification A mutation event whereby a gene is found in greater than the normal number of copies. Amplification generally involves very long stretches of chromosomes (i.e. many genes) and may occur so that the genes are amplified in number many times. (*See also* homogeneously staining regions and double minutes.)

gene cloning Making many copies of a single isolated gene by inserting it into bacteria or other cells.

gene library A collection of fragments of DNA cloned into bacteria or viruses. DNA that is cloned directly from the genome of an organism is known as a genomic library whereas DNA that is transcribed from messenger RNA from a particular organism or tissue is known as a cDNA library.

gene therapy Replacing a defective gene with a normal gene in somatic cells or in germ cells.

genetic marker An allele of a gene which follows standard Mendelian segregation and so can be used to study linkage in families or associations with a characteristic within populations.

genome The total genetic material of an organism.

genotoxic An agent that is toxic through its ability to damage DNA. At high levels of damage, this is lethal for the cell.

germ (cell) line Cells which give rise to the sperm or egg (ovum) which transmit genes from one generation to another and are thus potentially immortal. All other cells are somatic and die with the individual. Only mutations in the germ cell line can be passed on to future generations. Also used to designate unaltered gene configurations.

glucocorticoids Steroid hormones secreted by the adrenal cortex for the control of carbohydrate metabolism; they also have anti-inflammatory properties and kill certain types of lymphocytes.

glycolipid A lipid molecule having covalently attached oligosaccharide (q.v.).

glycoprotein A protein molecule having covalently attached oligosaccharide (q.v.).

glycosaminoglycan A polysaccharide component of a mucoprotein (q.v.).

gonadotrophins Hormones secreted by the anterior lobe of the pituitary (q.v.).

granulocyte White blood cells (various maturation stages include the neutrophils and polymorphonuclear cells) involved in the defence system against foreign matter.

granulocyte/macrophage colony stimulating factor (GM–CSF) A factor that stimulates the proliferation and differentiation of precursor cells of the granulocyte and macrophage lineages. Note that these two lineages share a common precursor cell, one that is already more differentiated than the haemopoietic stem cell (known as the CFUs or pluripotential stem cell) that gives rise to all the blood cells.

HLA The major histocompatibility system of man. It is a complex genetic region coding for two major classes of cell surface determinants involved in the control of immune function. There are also other genes in the region, in particular controlling certain of the complement components.

Ha-*ras* An oncogene originally found in two mouse sarcoma viruses (strains Harvey and BALB). Its product is an enzyme catalysing guanosine triphosphate (GTP). It is a member of a multigene family (Ki-*ras*, N-*ras*, q.v.), members of which are often found in mutated form in tumours.

haematuria Blood in the urine.

haemoccult test A clinical biochemical test which detects the presence of blood in the faeces.

haemopoietic Blood cell forming.

half-life The time taken for half a given quantity of a chemical (e.g. radio-isotopes, proteins, nucleic acids) to degrade or decay. Short half-lives reflect an unstable chemical.

haploid Having one copy of each chromosome as in the gametes. The human haploid chromosome number is 23.

hemicorporectomy Surgical removal of lower half of the body.

hemizygous Having only one copy of a given genetic locus (or of a number of loci in the case of deletions or loss of whole chromosomes).

heteroantiserum Antiserum raised in one animal species against cells or molecules of another.

heterozygous Having different alleles at a given locus on homologous chromosomes.

histocompatibility genes Genes that determine susceptibility or resistance to tissue or tumour transplants.

histones Highly basic proteins usually associated with DNA in chromatin (q.v.)

histopathology The study and diagnosis of disease by microscopy

homogeneously staining region (HSR) Region of a chromosome staining with an intermediate intensity throughout is length and without the normal pattern of light and dark bands.

homologous chromosomes Chromosomes that carry genes governing the same characteristics and that pair during the cell division which produces eggs and sperm. Individuals receive one member of a homologous pair of chromosomes from their father and the other member from their mother.

homozygous Having the same allele at a given locus on homologous chromosomes.

hormone A specific chemical substance produced by cells and carried to other cells or organs, often in the blood, to produce a specific effect at a distance.

hybridization, nucleic acid *See* Northern and Southern blotting/ hybridization.

hybridoma A hybrid tumour cell created by the fusion of two or more cells of different type. The term is commonly used to refer to monoclonal antibody producing cells formed by fusion of a sensitized B lymphocyte and a myeloma (q.v.) cell.

hydrophilic Literally, 'water loving'. Describes property of a substance preferentially found in water extracts as opposed to lipids.

hydrophobic Literally, 'water fearing'. Describes property of a substance preferentially found in lipid extracts (such as membranes) as opposed to water.

hyperdiploid Having more than the diploid number of chromosomes.

hyperplasia Increase in number of cells in response to stimulus—a reversible process.

hypertrophy Increase in size of cell or tissue.

hypodiploid Having fewer than the diploid number of chromosomes.

IL1 Interleukin 1. A factor produced by monocytes, involved in immune response reactions of antigen presentation by monocytes to T lymphocytes.

IL2 Interleukin 2. T lymphocyte growth factor (previously called TCGF).

IL3 Interleukin 3. A factor that stimulates growth of early multilineage haemopoietic cells.

immunogen A substance which stimulates an animal to produce antibodies (q.v.).

immunoglobulin An antibody (q.v.) molecule.

immunotherapy A form of therapy in which the host's immune system is activated or augmented, e.g. by the use of anti-tumour antibodies (q.v.).

immunotoxin A hybrid protein molecule, formed by the conjugation of an antibody and a toxic protein, capable of binding to cells bearing surface antigens recognized by the antibody and made with the intention of killing the cells selectively.

infectious mononucleosis A disease characterized by an imbalance of the immune cells in the body and a massive proliferation of T cells (q.v.).

inflammation Tissue response to infection or damage.

initiation The primary step in tumour induction caused by a carcinogen.

insertional mutagenesis A mutation caused by the insertion of new genetic material into a normal gene or sequences surrounding it. The term is generally used in relation to integrations of retroviruses (q.v.) into chromosomal DNA, although it applies to other inserted material, including DNA sequences moved from one chromosomal site to another ('transposition').

insulin A hormone produced by B cells of the pancreas. It is concerned with the regulation of carbohydrate, fat, and protein metabolism. A deficiency of insulin causes diabetes.

integration Insertion of viral DNA within a host chromosome. First applied to integration of phage DNA in the bacterial chromosome.

interferon A group of proteins produced by cells in response to viral infection and other stimuli. Their effects are complex and include antiviral actions exerted at various stages.

interleukin Substance produced by a cell of the immune system which acts on other cells (*see also* IL1, IL2, IL3).

internal mammary lymph nodes Deeply located lymph nodes which drain the breast and arm found at the sides of the sternum (breastbone).

intravenous urogram X-ray of urinary tract after injection of radio-opaque material which is concentrated in the kidney.

intron The parts of a gene sequence that are not found in the messenger RNA (transcript) molecule which will be used to produce the protein product of the gene. Introns are removed from the initial RNA copy of the gene by a process known as splicing (q.v.).

inversion A chromosomal aberration that arises when two breaks occur in the same chromosome and the region between the breaks is re-inserted after a 180° rotation resulting in a reversed gene order.

ionization potential Energy required to liberate a non-nuclear electron from an atom or molecule.

isochromosome A chromosomal aberration in which one of the arms of a chromosome is deleted and the other arm is duplicated. The two arms of an isochromosome are therefore of equal length and contain the same genes.

isogenic genetically identical.

isogenic graft (isograft) Graft between genetically identical individuals, e.g. between identical twins.

isotope One or two or more atoms having the same atomic number but different mass, e.g. ^{123}I and ^{131}I are radioactive isotopes of iodine.

Kaposi's sarcoma A tumour, possibly arising from endothelial cells, often found as skin lesions, purple or blue in colour.

karyotype The chromosome complement analysed and set out according to size, shape, and banding patterns of each chromosome.

keratinocyte Major epithelial cell type of the skin or epidermis, which contains large quantities of cytoskeletal proteins of the cytokeratin family.

kilobase (kb) 1000 bases (nucleotides) of single-stranded DNA or RNA; kilobase pairs (kbp) for double-stranded DNA; conversion factor to daltons depends on the exact base composition but, in general, average values of 345 daltons and 330 daltons per nucleotide of RNA and DNA, respectively, give reasonable estimates. Thus, 10 kb = 3.45×10^6 daltons (RNA) or 3.30×10^6 daltons (single-stranded DNA). To estimate coding potential from an RNA molecule, assuming an average of 110 daltons per amino acid, use 1 kb (RNA) = 37 kilodaltons (protein).

kilodalton (kDa) 1000 daltons, where a dalton is the standard unit of atomic mass very nearly equal to that of a hydrogen atom. The molecular weight of a protein is based on the atomic weights of the individual elements.

Ki-*ras* An oncogene originally identified in a mouse sarcoma virus (Kirsten strain). It is a member of a multigene family (Ha-*ras*, N-*ras*, q.v.), members of which are often found in mutated form in tumours. Its product is an enzyme catalysing guanosine triphosphate (GTP).

kit An oncogene originally found in a feline sarcoma virus (strain HZ4).

L-*myc* An oncogene identified by gene amplification (q.v.) in a lung carcinoma. Member of the *myc* multigene family.

laparotomy Surgical incision in flank or abdomen, usually for operations on abdominal contents.

LET (Linear Energy Transfer) Rate of energy loss to the surrounding medium, in a radiation track (unit: keV/micron).

leukaemia A malignant disease of the blood forming organs leading to the overproduction of neoplastic white blood cells or their precursors, in the blood and bone marrow.

leukoerythroblastic anaemia An anaemia in which immature red cells and increased numbers of white cells are present in the blood.

leukosis A proliferative disease of leukocytes.

lifespan study An ongoing epidemiological study e.g., of excess cancer incidence in atomic bomb survivors from the Japanese cities of Hiroshima and Nagasaki.

ligand The substance with which a receptor acts specifically to form a complex at least transiently.

linkage The presence of two or more genes on the same chromosome causing a tendency of alleles of these genes to be inherited together. Linkage occurs only when the genes are sufficiently close to one another on the same chromosome.

lipid bilayer Description of the membranes of cells, including the plasma membrane, of two apposed layers each containing lipid molecules.

liposome A synthetic lipid vesicle.

locus (plural, **loci**) Position of a gene on a chromosome.

LTR A structure found at both ends of the provirus DNA of a retrovirus (q.v.), hence 'long terminal repeat'.

lymphangiogram X-ray of lymphatic vessels.

lymphocyte White blood cell of the T or B lineage.

lymphokine Substance produced by lymphocytes which acts on other cells (*see also* interleukin).

lymphoma A solid tumour of T or B lymphocytes, e.g. in the lymph nodes, thymus, or spleen.

lysosome The cellular organelle which degrades molecules that have been taken into a cell by endocytosis (q.v.).

macrophage White cell found in all tissues which can ingest and break down antigens.

macrophage/monocyte colony stimulating factor (M-CSF) A factor that stimulates the growth and differentiation of cells of the macrophage/monocyte (q.v.) lineage.

malignant tumour Tumour which is capable of invading surrounding tissues and of metastasizing.

mammography X-ray of breast.

marker chromosome Abnormal or rearranged chromosome which can be recognized as a characteristic feature.

mastectomy Surgical removal of breast.

megakaryocyte White blood cell that undergoes nuclear division without cell division, leading to a multinucleated cell. It shatters to produce the subcellular entities known as platelets (q.v.) that participate in wound healing.

meiosis Two successive cell divisions from a diploid cell, in which the chromosomes are duplicated only once so that each of the new daughter cells have only half the diploid chromosome number, e.g. in the formation of gametes (sperm and ova).

mesenchyme *See* connective tissue.

mesoderm The germ layer lying between ectoderm (q.v.) and endoderm (q.v.). It gives rise to an epithelial component (epithelium of genital and most of urinary system), striated muscle including heart, and mesenchyme (connective tissue, cartilage, bone, smooth muscle, and blood cells).

met An oncogene originally identified by transfection (q.v.) of DNA from a human osteosarcoma (q.v.) cell line treated with a chemical carcinogen (q.v.).

metaphase The phase of mitosis or meiosis in which the condensed chromosomes attached to the spindle fibres can line up on an equatorial plate between the two poles of the cell.

metaplasia Abnormal alteration in the structure of cells.

metastasis A secondary tumour arising from cells carried to a distant site from a primary tumour.

microsome (-al) subcellular membranous particles with membrane bound enzyme activities, derived from the endoplasmic reticulum, in cell fractionation procedures.

microtome A cutting device which produces thin sections of tissue for microscopic examination.

mineralocorticoids Steroid hormones secreted by the adrenal cortex for the maintenance of ion balance, especially Na^+ ions.

mitogen A compound capable of causing cells to divide.

mitosis Process of cell division.

modal chromosome number The number of chromosomes per cell which is found most frequently in a mixed cell population.

modified radical mastectomy Surgical treatment for breast cancer involving removal of breast, overlying skin, small pectoral muscle and the axillary lymph nodes.

molecular excitation A change in the electronic configuration of an atom or molecule (accompanied by absorption of energy).

molecular mass (M_r) The relative molecular weight of a protein based on its in-vitro properties as calculated, for example, by its migration through a gel separation system where charge, tertiary structure and post-translational modification may influence the pattern of migration. This value is distinct from the true molecular weight based on the individual atomic weights of component elements.

monoclonal Derived from a single cell.

monocyte Similar cell to the macrophage but found in the blood. Both cells assist in immune responses.

monosomy The presence of only one member of a chromosome pair.

mos An oncogene originally found in a mouse sarcoma virus (Moloney strain). Its gene product has serine protein kinase (q.v.) activity.

mucoprotein Protein–polysaccharide complexes which are major components of mucins.

mutagen A chemical or physical agent that causes changes in the structure of DNA (mutations).

mutagenesis Effecting a heritable change in a nucleic acid.

mutagenicity The capacity of an agent (e.g. a chemical) to cause DNA mutation.

mutation A heritable change in the genetic material. Mutations in the broadest sense include any change, from a single base pair change in the DNA to substantial deletions or rearrangements of the DNA even involving major parts or the whole of chromosomes, and including chromosome translocations.

mutation frequency The ratio of mutated cells to normal cells following, for example, radiation (expressed either as per cell irradiated or per cell surviving).

myb An oncogene originally found in two avian myeloblastosis (myeloid leukaemia) viruses. Its product is found in the nucleus. The gene has been activated by adjacent insertion of retroviruses (q.v.) in several rodent myeloid leukaemias.

myc An oncogene originally found in four avian 'myelocytoma' viruses. Its product is found in the nucleus and appears to be a DNA binding protein. It is a member of a multigene family (including L-*myc* and N-*myc*, q.v.).

myeloid Literally, 'of the bone marrow'. However, in speaking of myeloid leukaemia, one is denoting cells of the granulocytic or monocytic lineages.

myeloma A plasma cell tumour.

myxoid Mucinous.

N-*myc* An oncogene identified by gene amplification (q.v.) in a neuroblastoma. Member of the *myc* (q.v.) multigene family.

N-*ras* An oncogene originally identified as a gene amplified in a human neuroblastoma (q.v.) and subsequently in other tumour types. It is a member of a multigene family (Ha-*ras*, Ki-*ras*, q.v.), members of which are often found in mutated form in tumours.

nasopharyngeal carcinoma Carcinoma of the squamous epithelium of the postnasal space.

natural killer cells Cells that are believed to kill non-specifically tumour and virus infected cells.

necrosis Tissue death.

neonatal thymectomy Removal of the thymus at birth, leaving the animal without functioning T cells.

neoplasia Tumour growth.

neoplasm Tumour.

neu An oncogene originally identified by transfection (q.v.) of DNA from a rodent neuroblastoma.

neuroblastoma A malignant tumour of neural crest origin. It may arise in any site containing sympathetic nervous tissue but is most commonly found in and around the adrenals.

neuroendrocine system A system of cells widely distributed throughout the body, which produces regulatory peptides. One group acts as hormones on distant organs, or on cells locally. A second group is produced at nerve endings and acts as neurotransmitters.

neurotransmitter A substance released from the axon terminal (nerve ending) of a nerve cell on excitation which stimulates or inhibits a target cell.

Northern blotting/hybridization A technique wherein RNA is fractionated according to size by electrophoresis in agarose-formaldehyde gels and then the RNA is transferred ('blotted') onto nitrocellulose paper. The RNA is then fixed to the paper by baking and then can be identified by using specific nucleic acid sequences ('probes', these may be RNA or DNA) of a particular gene or part of a gene ('hybridization'). This technique allows identification of genes being expressed (i.e. transcribed into RNA), quantitation of levels expressed, etc.

nucleotide The unit component of a nucleic acid comprising a base (q.v.), a sugar and a phosphate group.

oestrogens Steroid hormones secreted by the ovary; maintain female characteristics.

oligosaccharide A structure composed of three or more sugar residues covalently attached to protein in glycoproteins (q.v.) and glycolipids (q.v.).

oncogene Literally, a gene causing cancer. Oncogenes were identified by their presence in cancer causing viruses, where their function is critical for viral transforming properties.

osteolytic Bone destroying.

osteosarcoma A malignant tumour of bone composed of proliferating spindle cells which directly form tumour osteoid.

P450 A class of enzymes involved in oxygenating various compounds, the mono-oxygenases or mixed function oxidases.

PAP test Named after Papanicolaou who devised a staining method for cells shed (or scraped) from the surface of epithelial tissues; particularly used for the routine screening of uterine cervix for cancer cells.

palliation Treatment to ease symptoms.

papilloma A benign tumour usually on a surface, with a frond like structure.

paraaortic Alongside the aorta.

paracentesis Puncture or tapping of a fluid filled space with a hollow needle to draw off the contained liquid.

paracrine Stimulation of a cell through interaction with a substance produced by a neighbouring cell.

parasitism A relationship in which one member (the parasite) benefits at the cost of the other (the host). (*See also* symbiosis.)

partial hepatectomy Surgical removal of part of the liver.

particle accelerators Machines which emit pulses or continuous beams of high energy particulate radiation (cyclotrons, Van de Graff accelerators, Betatrons, etc.).

passive immunization Immunity acquired by receiving immune cells or antibody.

penetration Passage of a virus from outside to inside the host cell.

phagocytosis The process by which large particles or cells are engulfed by cells of the reticuloendothelial system (q.v.), such as macrophages.

phenotype The combination of characteristics expressed by a particular cell or organism.

phorbol ester A variety of compounds originally isolated from plants that are active as tumour promoting (q.v.) agents in multistep carcinogenesis assays.

phosphorylation The process of addition of phosphate groups to a protein (*see also* protein kinase).

pituitary gland (hypophysis) A small gland at the base of the brain with two main lobes, anterior (adenohypophysis) and posterior (neurohypophysis). The anterior lobe produces hormones which control secretion in sex glands, thyroid, adrenal etc. The posterior lobe secretes other hormones which are formed in nerve cells in the hypothalamus. These hormones influence blood pressure, water balance etc.

plasma cell A fully differentiated B lymphocyte synthesizing and secreting antibody (q.v.).

plasma The fluid component of blood.

platelet A subcellular component of the blood formed from megakaryocytes (q.v.) concerned with blood clotting; release mitogens at the site of the wound.

platelet derived growth factor (PDGF) A factor in platelets (q.v.) that is mitogenic for cells at the site of a wound (e.g. endothelial cells, q.v.). Part of the gene for PDGF has been captured by a simian sarcoma virus in which it is known as *sis* (q.v.).

pleural effusion Fluid in the pleural space (surrounding the lungs) in abnormal amounts.

polyclonal Derived from more than one cell.

polycyclic hydrocarbons A class of organic chemicals consisting of carbon and hydrogen molecules arranged in a series of linked ring structures.

polymerase chain reaction (PCR) A relatively new molecular biological technique whereby one or two small oligonucleotides (ca. 15–25 bases long) of a known sequence are used as a 'primer' to drive a reaction containing a DNA molecule with the primer sequence, excess single nucleotides of A, T, C, and G, and polymerase. The polymerase copies the sequence adjacent to the primer and creates double-stranded DNA. The double-stranded DNA is then heated to separate the strands, and then the polymerase reaction is repeated. This procedure is followed for many cycles (up to 50–60 times), giving many thousands of copies of the sequence adjacent to the primer from the DNA used as a 'template'. In the case of using two primers, one is using oligonucleotides known to be found at the ends of the DNA of interest such that the intervening sequence is the one that is amplified. One can also use polyT oligonucleotides to amplify up the cDNA (complementary DNA) formed from a small amount of messenger (polyA-containing) RNA.

polymorphism The occurrence of two or more alleles for a given locus in a population where at least two alleles appear with frequencies of more than one per cent.

polyp A tumour projecting from a surface, usually epithelial.

pre-clinical disease Disease detected in individuals without symptoms.

prednisolone A powerful, synthetic glucocorticoid.

probe Generally a single strand of labelled DNA (usually with a radioactive marker). The probe binds to single-stranded RNA or DNA which has a complementary strand. The hybrids can then be isolated from mixtures, or detected by radioautography or other methods. Nowadays, probes can also be made from RNA.

progestins Steroid hormones secreted by the ovary during the menstrual cycle and pregnancy.

promoter insertion An insertional mutagenic (q.v.) event involving specifically a sequence that promotes gene transcription (hence 'promoter'). *See also* downstream promotion.

promotion Stages following initiation; may be caused by specific non-carcinogenic promoting agents.

prostaglandins A group of compounds of similar structure found in the prostate (hence the name) but now known to be widely distributed in the body and affecting blood vessels, nervous system, uterus, etc.

protease Protein digesting enzyme.

protein kinase An enzyme that adds phosphate groups to amino acids (generally serine, threonine and tyrosine); the process is known as phosphorylation (q.v.).

protein kinase C A serine and threonine protein kinase (q.v.) important in the cell's response to growth factor stimulation.

protooncogene The normal cellular counterpart of a gene identified as causing a tumour (c-*onc* and v-*onc*, q.v.).

provirus The intracellular double stranded DNA form of the retrovirus genome, either free or integrated.

proximate metabolite/carcinogen The carcinogen or its metabolite which undergoes metabolic or spontaneous conversion to a further derivative, the ultimate metabolite/carcinogen which binds to the target macromolecule, usually DNA.

pseudodiploid Diploid (q.v.) chromosome number but with numerical or structural chromosome abnormalities.

pseudogene A gene thought to be permanently inactive. Many lack introns (q.v.) compared to related genes that are active. May have arisen by gene duplication events.

RBE (Relative Biological Efficiency or Effectiveness) The ratio of absorbed doses of two different types of radiation required to give the same level of effect.

RNA dependent DNA polymerase *See* reverse transcriptase.

RNA splicing A maturation process that removes internal sequences from polyribonucleotide chains (splicing, q.v.).

radiation fractionation Exposure to radiation over two or more periods separated by intervals of time.

radioimaging A technique that uses radioactive substances to visualize sites within the patient's body, e.g. the sites of tumour.

radionuclide An atom which undergoes spontaneous decay with the release of radiation. Radionuclides can be used for the radioimaging (q.v.) or therapy of cancer.

radioprotector A chemical substance that reduces the biological effects of radiation.

radon A chemically inert radioactive gas: a product of radium disintegration.

raf An oncogene originally identified in a murine sarcoma virus (strain 3611). Its product has protein kinase (q.v.) activity.

receptor A structural molecule that binds to a specific factor (e.g. hormone, vitamin, antigen, etc.). These may be found at the cell surface (cell surface receptors, membrane receptor) or within the cytoplasm (cytosolic receptor).

receptor translocation Movement of a receptor from one part of the cell (e.g. the cell surface) to another site in the cell. This may be called internalization if the complex is taken inside the cell.

recessive An allele that is expressed only when present in the homozygous or hemizygous state (compare with dominant).

recessive mutation A mutation which is not expressed in the phenotype if it is balanced by a corresponding normal gene on the homologous chromosome, i.e. it is expressed only when both homologous genes are affected or if one is deleted.

recombinant DNA (gene splicing) DNA (genes) from different sources, e.g. bacterial yeast or mammalian, can be cut into fragments and ligated to form hybrid DNA. This can be inserted into another organism, e.g. bacterial DNA carrying a human gene can be inserted into bacteria which then produce proteins coded for by the human gene.

recombination Process of exchange of genetic information between two homologous chromosomes, presumed to occur through breakage of both chromosomes at homologous sites followed by reunion after exchange. Also can occur between other RNA or DNA molecules (as with viruses).

rel An oncogene originally found in an avian reticuloendotheliosis (q.v.) virus.

restriction endonucleases Enzymes produced by bacteria that cut double-stranded DNA by recognizing a specific sequence of nucleotides (q.v.).

restriction enzymes (restriction endonucleases) These enzymes isolated chiefly from prokaryotes recognize a specific sequence usually about four to six base pairs long within double-stranded DNA and cut the DNA specifically at a precise distance from that site only. Different enzymes recognize different sequences as a rule; there are several examples of enzymes isolated from different sources that cleave within the same target sequences. They are known as isoschizomers.

restriction fragment length polymorphisms (RFLPs) Restriction enzymes (q.v.) cut samples of DNA consistently into fragments of different lengths. If individuals differ in the distribution of sequences recognized by the enzymes, the DNA is cut into different sized fragments. These consistent minor variations in length can be used to map genes to specific chromosomes and help to trace genetic diseases in families.

reticuloendothelial system 'Defence cells'. A diffuse system of phagocytic cells involved with the clearance of foreign proteins, immune complexes, large particles, and cells from the bloodstream.

reticuloendotheliosis A proliferative disease of reticuloendothelial cells.

retroperitoneal Lying behind the peritoneum.

retrovirus A member of a family of viruses sharing several physicochemical properties including the type of organization of the RNA genome and an enzyme (known as reverse transcriptase) which makes a DNA copy of the genome (provirus, q.v.) to become incorporated into the chromosomal DNA of the infected host cell.

reverse transcriptase An enzyme, RNA dependent DNA polymerase, encoded by retroviruses. In the viral life cycle, the enzyme carried in the virus particle facilitates the synthesis of a DNA copy of the viral RNA genome. It then makes a second DNA strand complementary to the first and also degrades the viral RNA (template). Used in laboratory procedures to prepare complementary DNA copies (cDNA) of other RNAs.

reversion (reverted) The process by which a muation is corrected, by 'back mutation' to the original (wild type) DNA sequence. The reverted cell (or virus, etc.) now has a normal phenotype.

ribosomes Cellular particles at which proteins are synthesized.

risk factors Environmental or inherited influences that increase the chance of cells progressing from a normal to neoplastic phenotype.

ros An oncogene originally found in an avian sarcoma virus. Its product has tyrosine protein kinase (q.v.) activity.

salvage pathway A biochemical pathway in the synthesis of DNA which reutilises preexisting nucleic acids and their precursors rather than synthesizing them *de novo* from single carbon units.

sarcoma A malignant tumour of mesenchyme.

screening The testing of a normal population to detect clinical or pre-clinical disease.

sequencing A technique for establishing the structure of DNA by recognizing the order of the nucleotides (q.v.) or of protein by the order of its amino acids.

serum The fluid left after plasma has clotted due to substances released from platelets; some plasma proteins have been removed.

signal transduction Used in reference to the process by which the interactions between a growth factor and its specific receptor at the cell surface generate a signal that is transmitted to the cell nucleus as a stimulus to proliferate.

sis An oncogene originally found in a simian sarcoma virus. The gene is part of the cellular gene encoding platelet derived growth factor (PDGF, q.v.).

sister chromatid exchange Exchange of genetic material between individual sister chromatids (q.v.).

ski An oncogene originally found in an avian carcinoma virus. Its product is found in the nucleus of cells.

somatic Refers to cells that are not part of the germ line and so their genetic complement is not passed from parent to offspring.

Southern blotting/hybridization A technique wherein DNA is fractionated according to size by electrophoresis in agarose gels (the DNA may be precut using restriction endonucleases, q.v.), and then the

DNA is transferred ('blotted') onto nitrocellulose paper. The DNA is then fixed to the paper by baking and may then be identified by using specific nucleic acid sequences ('probes'; these may be RNA or DNA) of a particular gene or part of a gene ('hybridization'). The double-stranded DNA in the gel is treated with alkali before blotting; this provides single stranded nucleic acid which can then bind covalently ('anneal') to the probes. The technique allows identification of the size of a gene, a map of restriction enzyme sites, gene amplification, gene deletion, gene rearrangement, etc.

specificity (of antibody or cells) The ability to distinguish between different antigens.

splicing An event that removes specifically the sequences of the non-coding parts of a gene (introns, q.v.) from the initial RNA molecule complementary in sequence to the gene. This is part of the 'processing' of the RNA molecule to its active form (the messenger RNA transcript) used to produce the gene product.

squamous Scaly or flattened, usually applied to epithelium.

src An oncogene originally found in an avian sarcoma virus. Its product has tyrosine protein kinase (q.v.) activity.

stem cell Precursor cell capable of growing and differentiating in response to a stimulus.

stilboestrol A powerful, synthetic oestrogen.

symbiosis 'Living together': an association of two organisms that confers mutual benefit.

synergism (synergy) Two or more compounds whose effects are additive or more than additive.

syngeneic graft Graft between individuals that have been inbred until their genetic similarity allows acceptance of grafts between members of the strain e.g. between mice of an inbred strain.

targeting The directing of substances such as drugs or toxins to target cells by attaching them to carriers, such as antibodies (q.v.), which bind to the target cell specifically.

T cell T lymphocytes: a group of cells responsible for cell mediated immunity.

telomere Each of the extremities of a chromosome.

testosterone The main male sex hormone, secreted by the testis.

tetraploid Having four times the haploid number of chromosomes.

transcription A base pairing process that copies one nucleic acid into a complementary polynucleotide chain.

transduction The process by which a virus genome recombines with other nucleic acids leading to the capture of new sequences or genes. The new material is 'transduced'.

transfect To introduce foreign DNA into a cell by experimental means so that the DNA integrates into the genome and is expressed in its new host cell. Transfection, the process.

transformation A word used in many ways. In this book, it means a change of morphological appearance of a cell ('morphological transformation') generally by a virus, chemical carcinogen, physical reagent, or even some unidentified ('spontaneous') event.

transforming growth factor (TGF) A factor that can cause a reversible change in normal cell phenotype such that cells grow into colonies in semi-solid media (anchorage independent growth). TGF-β requires EGF to produce this phenotype whereas TGF-α can work alone.

transgenic Transferring a gene across species or strain barriers. In modern usage to distinguish animals carrying a gene introduced artificially during the early embryonic stage and now incorporated into the genome.

translation A term describing the process whereby the triplet code sequence of a messenger RNA molecule is used to produce a specific order (chain) of amino acids which form a protein.

translocation A chromosomal aberration in which part of one chromosome joins onto another chromosome. A reciprocal translocation refers to the transfer of material between two chromosomes.

transmembrane Something spanning the membrane. Generally in regard to plasma membrane; e.g. a protein found external to, within, and internal to the plasma membrane.

triplet code The code by which a sequence of three nucleic acids (bases) specifies one amino acid. Each triplet thus forms one codon.

triploid Having three times the haploid number of chromosomes.

trisomy The presence of three chromosomes of one type.

truncation Literally, a shortening. In the context of this book, it means that a gene has been shortened by some event and thereby the protein product of the gene is smaller.

tumour promoter An agent capable of providing the second step in two step carcinogenesis systems whereby an initiated (preneoplastic) cell is converted to a neoplastic one (*see* initiation and promotion). Generally a specific tumour promoter is required for a particular initiating agent.

ultimate metabolite/carcinogen *See* proximate metabolite/carcinogen.

upstream By convention a sequence from the 5' end to the 3' end, or from the beginning to the end in terms of coding potential, then one designates those sequences more 5' or more 'towards the beginning' as being upstream.

urothelium The epithelial cells lining the urinary tract.

v-*onc* An oncogene identified in a virus (*see* oncogene).

Western blot A technique similar to Northern or Southern blotting (q.v.) but used to separate and identify proteins.

xenograft Graft between individuals of different species.

yes An oncogene originally found in two avian sarcoma viruses. Its product has tyrosine protein kinase (q.v.) activity.

Index

abrin 496-7
acetylaminofluorene 160, 167-8
acidic fibroblast growth factor 40, 301
    (table)
acquired immunodeficiency syndrome
    (AIDS) 215, 341, 417
acromegalic tumours (growth hormone
    releasing-factor producing tumours)
    400
adenocarcinoma 20
adenoma of bowel 172
adenomatous polyposis coli (familial
    adenomatous polyposis) 111-13
adenosine triphosphate 131
adenovirus 219 (table)
adrenal gland 380-1
  adenoma 381
  carcinoma 381
  medullary tumours 402
  virilization, tumour-induced 381
adrenocorticotrophic hormone (ACTH)
    391 (table), 397, 402
adrenocorticotropin 381
adriamycin 493
aflatoxin 372, 515
  B_1 93, 150 (table), 158, 160-1
aflatoxin-2,3-epoxide 160
age 10
air pollution 65
alcohol 87
aldosterone 362
aliphatic N-nitrosamines 161-2
alkylating agents, simple 142
allograft 26
Ames test 102, 149, 169-70
amino acids 136
4-amino-biphenyl 92, 150 (table)
aminoglutethimide 384
anandron 383
anaplasia (dedifferentiation) 12, 21
anchorage independent growth 163
angiogenic factors 40
angiogenin 40
angioma 21
angiosarcoma 21
ankylosing spondylitis 117, 185-6

anti-androgens 383, 384
antibodies 408, 469
  against tumour antigens 43
  anti-idiotypic 488, 491
  anti-tumour 491
  chimaeric 475
  function 470-1
  genetically engineered 475-6
  monovalent 488-9
  structure 470-1
antigenic modulation 488
antigens
  class I 42
  target 477-9
  tumour associated 476-9
anti-oestrogens 384
anti-oncogenes (tumour suppressor
    genes) 257-60
  p53 259-60
  Rb-1 258-9
anti-oxidants 66-8
antisera 420-1
aplysiatoxin 156
APUDoma 387
Arg-Gly-Asp sequence 38
aromatic amines 92, 149
arsenic compounds 150 (table)
aryl hydrocarbon hydroxylase 63
asbestos 64-5, 92, 149, 150 (table)
ascites 448
ascitic fluid 39
ascorbic acid 66-8
Askin's tumour 289 (table)
[211]Astatine 496
astrocytoma 350
ataxia telangiectasia 116, 162, 172, 194
  T cell neoplasia 194
atomic bomb 177
  Life Span Study 182-3
auramine manufacture 150 (table)
autocrine motility factor 35-6
autocrine system 287
autophosphorylastion 246
avian erythroblastosis virus 321
avian leukosis viruses 227
avian reticuloendotheliosis viruses 222-3